# FIREMON

Fire Effects Monitoring and Inventory System

## Integration of Standardized Field Data Collection Techniques and Sampling Design With Remote Sensing to Assess Fire Effects

This project was funded by the USDA and USDI Joint Fire Sciences Plan.
Agreement number 98-IA-189
Joint Fire Science Program, National Interagency Fire Center,
3833 S. Development Ave., Boise, ID 83705

## Contents

**FIREMON Introduction**
    The Authors
    Acknowledgments
  Executive Summary
  Introduction
  General Description
    What FIREMON Is...
    What FIREMON Is NOT...
    The Four FIREMON Components
    Integrated Sampling Strategy
    Field Assessment
    Landscape Assessment
  Firemon Documentation Structure

**Integrated Sampling Strategy (ISS) Guide**
  Summary
  Introduction
    Advanced Alternative to the FIREMON Integrated Sampling Strategy
    Resampling Existing FIREMON Plots
    Terminology
    Sampling Intensities
    Sampling Approaches
    Sampling Design Keys
    Step-by-Step Procedures
  Preliminary Sample Design Activities
    Stating Monitoring Goals and Objectives
      S.M.A.R.T. Objectives
    Determining the Sample Area and Spatial Stratification
    Determining Sampling Resources
    Determining Sampling Design
      Calculating FIREMON sample statistics
      Calculating the number of polygons to be sampled
      Determining the sampling approach
      Selecting the level of sampling intensity
      Choosing sampling methods
    Considering Tradeoffs
    Monitoring Prescribed Burn Projects
    Comparing Objectives and Sampling Design
  Designing a Fire Monitoring Project
    Determining Polygon Locations and Building a Summary Table
    Implementing the Relevé Approach
      Background
      Polygon selection
      Relevé establishment
      Using the Simple sampling intensity (Level I)
      Using the Alternative sampling intensity (Level II)
      Using the Detailed sampling intensity (Level III)
    Implementing the Statistical Approach
      Sample size determination
      Coefficient of variation for assessing variability
      Determining the attribute variability used to calculate sample size

    Calculating sample size
    Using the Simple sampling intensity (Level I)
    Using the Alternative sampling intensity (Level II)
    Using the Detailed sampling intensity (Level III)
    Plot distribution when using the statistical approach
  Establishing Control Plots
  Firemon Guides and Keys
    Sampling Strategy Checklist
      Suggested step-by-sep procedure for designing a firemon monitoring sampling effort
      Suggested step-by-step procedure for implementing a FIREMON monitoring sampling effort
    Field Assessment
      Sampling protocols
    Sample Approach Classification Key
    Sampling Intensity Key
      Simple sampling intensity (Level I)
      Alternative sampling intensity (Level II)
      Detailed sampling intensity (Level III)
    Methods Classification Key
    Vegetation Sampling Overview
    Analysis Tools

## Plot Description (PD) Sampling Method
  Summary
  Introduction
  Sampling Procedure
    Required PD Fields—Database Key
    Organization Code Fields
    Plot Information Fields
      Plot size
      Sampling information
      Linking fields
      Georeferenced plot positions
    Biophysical Setting Fields
      Topography
      Geology and soils fields
    Vegetation Fields
      Vegetation—trees
      Vegetation—shrubs
      Vegetation—herbaceous
      Vegetation—composition
      Potential vegetation
    Ground Cover Fields
      Ground cover
      General fuel characteristics
    Fire Behavior and Effects Fields
      Fire behavior
      Fire effects
    Common Fields
    Comments Fields
      Local codes
      Comments
    Precision Standards

        Sampling Design Customization
            Recommended PD Sampling Design
            Streamlined PD Sampling Design
            Comprehensive PD Sampling Design
            User-Specific PD Sampling Design
            Sampling Hints and Techniques
        Plot Description (PD) Field Descriptions
        Plot Description (PD) Equipment List
        Plot Description (PD) Cheat Sheet
        Plot Description (PD) Form

**Tree Data (TD) Sampling Method**
        Summary
        Introduction
        Sampling Procedure
            Preliminary Sampling Design
                Plot size selection
                Macroplot area
                Subplot area
                Snag plot area
                Breakpoint diameter
            Preliminary Sampling Tasks
            Sampling Tasks
                Define macroplot boundary
                Define initial sample position
                Sampling seedlings
                Saplings and mature trees
                Measuring DBH
                Sampling saplings
                Sampling mature trees
                Sampling snags
            Precision Standards
        Sampling Design Customization
            Recommended TD Sampling Design
            Streamlined TD Sampling Design
            Comprehensive TD Sampling Design
            User-Specific TD Sampling Design
            Sampling Hints and Techniques
        Tree Data (TD) Field Descriptions
        Tree Data (TD) Cheat Sheet
        Tree Data (TD) Form

**Fuel Load (FI) Sampling Method**
        Summary
        Introduction
        Sampling Procedure
            Preliminary Sampling Tasks
            Determining Piece Size
            Modifying FL Sampling
            Laying Out the Measuring Tape
            Determining the Slope of the Measuring Tape
            What Are "Woody," "Dead," and "Down" Debris?
                CWD
                FWD

            DWD Sampling Distances
            Sampling FWD
            Sampling CWD
            What Are, "Duff," "Litter," and the "Duff/Litter Profile"?
            Sampling Duff and Litter
            What Is "Woody" and "Nonwoody" Vegetation?
            Sampling Vegetation Cover and Height
            Finishing Tasks
            Successive Sampling Planes
            Determining the Number of Sampling Planes
            Resampling FL Plots
            Precision Standards
        Sampling Design Customization
            Alternative FL Sampling Design
            Streamlined FL Sampling Design
            Comprehensive FL Sampling Design
            Optional Fields
        Fuel Load (FL) Field Descriptions
        Fuel Load (FL) Equipment List
        Fuel Load (FL) CHEAT SHEET
        Fuel Load (FL) Form

## Species Composition (SC) Sampling Method
    Summary
    Introduction
    Sampling Procedure
        Preliminary Sampling Tasks
        Designing the SC Sampling Method
            Plot ID construction
            Macroplot size
            Plant species ID level
        Conducting the SC Sampling Tasks
            Initial plot survey
            Sampling cover
            Size class
            Estimating cover
            Measuring average height
            Using the optional fields
        Precision Standards
    Sampling Design Customization
        Recommended SC Sampling Design
        Streamlined SC Sampling Design
        Comprehensive SC Sampling Design
        User-Specific SC Sampling Design
        Sampling Hints and Techniques
    Species Composition (SC) Field Descriptions
    Species Composition (SC) Equipment List
    Species Composition (SC) Cheat Sheet
    Species Composition (SC) Form

## Cover/Frequency (CF) Sampling Method
    Summary
    Introduction

- Estimating Cover and Height
- Estimating Frequency
- Sampling Procedure
  - Preliminary Sampling Tasks
  - Designing the CF Sampling Method
    - Plot ID construction
    - Determining the sample size
    - Determining the quadrat size
    - Recording the subplot size ratio and NRF numbers
  - Conducting CF Sampling Tasks
    - Establishing the baseline for transects
    - Locating the transects
    - Locating the quadrats
  - Quadrat Sampling
    - Cover
    - Nested rooted frequency
  - Estimating Height
  - Precision Standards
- Sampling Design Customization
  - Recommended CF Sampling Design
  - Streamlined CF Sampling Design
  - Comprehensive CF Sampling Design
  - User-Specific CF Sampling Design
  - Sampling Hints and Techniques
- Cover/Frequency (CF) Field Descriptions
- Cover/Frequency (CF) Equipment List
- Cover/Frequency (CF) Cheat Sheet
- Cover/Frequency (CF) Form

## Point Intercept (PO) Sampling Method
- Summary
- Introduction
  - Point Sampling Techniques
    - Single points
    - Point frame
    - Grid quadrat frame
- Sampling Procedure
  - Preliminary Sampling Tasks
  - Designing the PO Sampling Method
    - Plot ID construction
    - Determining sample size
    - Conducting PO Sampling Tasks
    - Establish the baseline for transects
    - Locating the transects
    - Sampling points
  - Point Intercept Sampling
    - Recording hits
    - Recording plant status
    - Estimating average height
  - Precision Standards
- Sampling Design Customization
  - Recommended PO Sampling Design
  - Streamlined PO Sampling Design

            Comprehensive PO Sampling Design
            User-Specific PO Sampling Design
            Sampling Hints and Techniques
        Point Intercept (PO) Field Descriptions
        Point Intercept (PO) Equipment List
        Point Intercept (PO) Cheat Sheet
        Point Intercept (PO) Form

**Line Intercept (LI) Sampling Method**
    Summary
    Introduction
    Sampling Procedure
        Preliminary Sampling Tasks
        Designing the LI Sampling Method
            Plot ID construction
            Determining the sample size
        Conducting LI Sampling Tasks
            Locating the baseline for transects
            Locating the transects
        Line Intercept Sampling
            Size class
            Estimating cover (intercept)
            Estimating average height
        Precision Standards
        Sampling Design Customization
            Recommended LI Sampling Design
            Streamlined LI Sampling Design
            Comprehensive LI Sampling Design
            User-Specific LI Sampling Design
            Sampling Hints and Techniques
        Line Intercept (LI) Field Descriptions
        Line Intercept (LI) Equipment List
        Line Intercept (LI) Cheat Sheet
        Line Intercept (LI) Form

**Density (DE) Sampling Method**
    Summary
    Introduction
    Sampling Procedure
        Preliminary Sampling Tasks
        Designing the DE Sampling Design
            Plot ID construction
            Determining sample size
            Determining the belt transect size and quadrat size
        Conducting DE Sampling Tasks
            Establish the baseline for transects
            Locating the transects
            Locating the quadrats
        Density Sampling
            Size class
            Density
            Estimating average height
        Precision Standards

Sampling Design Customization
      Recommended DE Sampling Design
      Streamlined DE Sampling Design
      Comprehensive DE Sampling Design
      User-Specific DE Sampling Design
      Sampling Hints and Techniques
   Density (DE) Field Descriptions
   Density (DE) Equipment List
   Density (DE) Cheat Sheet
   Density (DE) Field Form

**Rare Species (RS) Sampling Method**
   Summary
   Introduction
   Sampling Procedure
      Preliminary Sampling Tasks
      Designing the RS Sampling Design
         Plot ID construction
         Determining the sample size
      Conducting RS Sampling Tasks
         Establish the baseline for locating individual plants
         Locating individual plants
         Tagging the individual plants
      Rare Species Sampling
         Plant identity
         Status
         Stage
         Size
         Reproduction
      Precision Standards
   Sampling Design Customization
      Recommended RS Sampling Design
      Streamlined RS Sampling Design
      Comprehensive RS Sampling Design
      User-Specific RS Sampling Design
      Sampling Hints and Techniques
   Rare Species (RS) Field Descriptions
   Rare Species (RS) Equipment List
   Rare Species (RS) Cheat Sheet
   Rare Species (RS) Field Form

**Landscape Assessment (LA) Sampling And Analysis Methods**
   Summary
   Introduction
   Definition of Burn Severity for Landscape Assessment
      Discussion of Ecological Burn Severity
   Ground Measure Of Severity: The Composite Burn Index
      What Do CBI Values Mean?
      Sampling for the Composite Burn Index
      Sample Design and Site Selection
      Plot Layout
      Plot Sampling, Using the Field Form
      Plot Photos

    Field Documentation
        Strata Definitions
            Strata hierarchical structure
        Initial Summary of Area Burned
        Rating Factor Definitions
            Substrate rating factors
            Herbs, low shrubs, and trees less than 3 ft (1m) rating factors
            Tall shrub and trees 3 to 16 ft (1 to 5 m) rating factors
            Intermediate and big tree rating factors (combined)
        Community Notes and Comments
        Data Sheets
    Remote Sensing Measure Of Severity: The Normalized Burn Ratio
        If Unfamiliar with the Science of Remote Sensing
        Introduction to the Normalized Burn Ratio (NBR)
        Timing of Landsat Acquisitions
        Two Strategies, Initial vs. Extended Assessment
            Initial assessment
            Extended assessment
        Prefire Considerations
            General software requirements
            General hardware requirements
            Digital geographic data needs
        Ordering the Data
        Steps to Process NBR and ΔNBR
        NBR Responses
        Interpreting Results of ΔNBR
            Continuous data
            Burn perimeter
            Discrete ordinal data
            Severity thresholds
    Landscape Assessment How To
        How To Record a GPS Location
            Acceptable accuracy
            Geodetic datum
            When digitizing a single point
        How To Take Plot Photos
        How To Name Files for Burn Remote Sensing
        How To Handle Reflectance
            Reflectance incorporating topography
            Atmospheric normalization
    Landscape Assessment Glossary
    Recent Landsat Satellites
        Landsat 5
        Landsat 7
    Other Remote Sensing Data Sources
    Landscape Assessment (LA) BI Cheat Sheet
    Landscape Assessment (LA) BR Cheat Sheet
    Burn Severity—Composite Burn Index (BI) Field Form

**Metadata (MD) Description**
    Summary
    Introduction

    Sampling Procedure

**Fire Behavior (FB) Sampling Method**
    Summary
    Introduction
    Sampling Procedure
        Fire Information
        Recording Ambient Weather Conditions
        Recording Fuel Moistures
        Recording Fire Behavior
        Local Codes and Comments
    Fire Behavior (FB) Equipment List
    Fire Behavior (Fb) Field Form

**Analysis Tools (AT) User Manual**
    Summary
    Introduction
    Firemon Analysis
        Getting Help
        Getting Started
        The Settings Dialog Box
        How to Run—A Brief Overview
            Example report
            Example graph
        Grouping Your Plots
        Forest Vegetation Simulator Option
        Some Hints

**FIREMON Database (DB) User Manual**
    Summary
    Introduction
    Firemon Database Installation And Configuration
        Installing the FIREMON Software
        Configuring the FIREMON Database Application
        Populating the Plant Species Codes Lookup Table
        Adding Other Items to the Plant Species Lookup Table
    Firemon Data Entry Forms
        Important Instructions Common to all Data Entry Forms
        Plot Description (PD) Data Entry Form
        Tree Data (TD) Data Entry Form
        Fuel Loading (FL) Data Entry Form
        Species Composition (SC) Data Entry Form
        Cover/Frequency (CF) Data Entry Form
        Line Intercept (LI) Data Entry Form
        Point Intercept (PO) Data Entry Forms
        Density (DE) Data Entry Forms
        Rare Species (RS) Data Entry Form
        Composite Burn Index (CBI) Data Entry Form
        Fire Behavior (FB) Data Entry Form
        Metadata (MD) Data Entry Form
    FIREMON Data Summary Reports
        Tree Data (TD) Summary Reports
        Fuel Loading (FL) Summary Reports
        Species Composition (SC) Summary Report

   Cover/Frequency (CF) Summary Report
   Line Intercept (LI) Summary Report
   Point Cover (PO) Summary Report
   Density (DE) Summary Report
   Composite Burn Index (CBI) Summary Report
  Firemon Data Analysis Program
  Simple Query Builder
  Gis Point File
  Random Transect Locator Program
  Customizing the Firemon Database
   Customizing FIREMON Codes
   Developing Custom Queries in FIREMON

## How to Guide
  How to Construct a Unique Plot Identifier
  How to Locate a Firemon Plot
   Identifying the Appropriate Place for a FIREMON Plot
  FIREMON Plot Center
   Written Directions to the FIREMON Plot
   Monumenting the FIREMON Plot
  How to Permanently Establish a Firemon Plot
  How to Define the Boundaries of a Macroplot
  How to Establish Plots With Multiple Methods
   Plot Layout
   Sampling Order
  How to Determine Sample Size
   Plotting Graphs Of Mean Values for Varying Sample Sizes
  How to Estimate Cover
   General Cover Estimation Techniques
   Additional Hints For Estimating Cover When Using Quadrts
  How to Use A Compass—Sighting And Setting Declination
  How to Establish a Baseline for Transects
  How to Locate Transects and Quadrats
  How to Offset aTransect
  How to Construct a Quadrat Frame
   Cover/Frequency Frames
   Density Frames and Belt Transects
  How to Construct Point Frames and Grid Frames
  How to Count Boundary Plants
  How to Dot Tally
  How to Measure Plant Height
   Plants Greater Than 20 ft (6 m)
   Plants Between 10 and 20 ft (3 to 6 m) Tall
   Plants Less Than 10 ft (3 m) Tall
   Other Height Measuring Tips
  How to Measure DBH
  How to Measure Diameter With a Ruler
  How to Age aTree
  How to Measure Slope
  How to Adjust For Slope
  How to Document Plot Location and Fire Effects With Photos
  How to Record a GPS Location

Acceptable Accuracy
Geodetic Datum
When Digitizing a Single Point

**FIREMON References**

**FIREMON Glossary**

**FIREMON Appendices**
Appendix A: NRIS Damage Categories, Agents, Severity Ratings, and Tree Parts
Appendix B. NRIS Lithology Codes
Appendix C: NRIS Landform Codes
Appendix D: Rick Miller Method for Sampling Shrub Dominated Systems
  Plot layout
  Tree measurements
  Shrub measurements
  Herbaceous and ground measurements
Appendix E: FIREMON Table Names, Field Names, and Field Descriptions
Appendix F: FIREMON Database Relationships

# FIREMON Introduction

## THE AUTHORS

**Duncan C. Lutes** is a Research Forester with Systems for Environmental Management, stationed at the USDA Forest Service, Rocky Mountain Research Station, Fire Sciences Lab, P.O. Box 8089, Missoula, MT 59807; phone: 406.329.4761; FAX: 406.329.4877; e-mail: dlutes@fs.fed.us. He has a B.S. and M.S. degree in forestry from the University of Montana, Missoula. His background is in fuels, principally coarse woody debris, primarily studying spatial and temporal distributions. He has been involved in the development of the First Order Fire Effects Model and several Fire and Fuels Extension variants to the Forest Vegetation Simulator. He is the Technical Editor and contributing author for this publication.

**Robert E. Keane** is a Research Ecologist with the USDA Forest Service, Rocky Mountain Research Station at the Fire Sciences Laboratory, P.O. Box 8089, Missoula, MT 59807; phone: 406.329.4846; FAX: 406.329.4877; e-mail: rkeane@fs.fed.us. Since 1985, Bob has conducted ecological research into fuel dynamics, ecosystem simulation, ecosystem restoration, and spatial modeling for the Fire Effects project. His most recent research includes 1) developing ecological computer models for exploring landscape, fire, and climate dynamics, 2) the mapping of fuel characteristics for spatially explicit fire growth and fire effects evaluation, 3) synthesis of a First Order Fire Effects Model to predict the direct consequences of a fire, 4) exploring the ecology and restoration of whitebark pine in the Northern Rocky Mountains, and 5) classification and simulation of vegetation communities on the landscape using GIS and satellite imagery. He received his B.S. degree in forest engineering from the University of Maine, Orono; his M.S. degree in forest ecology from the University of Montana, Missoula; and his Ph.D. degree in forest ecology from the University of Idaho, Moscow.

**John F. Caratti** is a Systems Ecologist with Systems for Environmental Management, P.O. Box 8868, Missoula, MT 59807; phone: 406.549.7478. Since 1988, John has developed computer software for ecological data analysis and wildlife population modeling. His most recent work includes developing database applications for georeferenced field data, vegetation classification and mapping, and fire effects monitoring. He received his B.A. degree in ecology from the University of California, San Diego, in 1988, and his M.S. degree in wildlife biology from the University of Montana in 1993. John has worked as an ecologist with the USDA Forest Service, Northern Region, a quantitative ecologist with The Nature Conservancy, and as an independent contractor.

**Carl H. Key** is a Geographer with the USDI U.S. Geological Survey at the Northern Rocky Mountain Science Center, Glacier Field Station, West Glacier, MT 59936; phone: 406.888.7991; FAX: 406.888.7990; e-mail: carl_key@usgs.gov. Carl developed the Geographic Information System currently used by Glacier National Park, and continues to support its application and advancement. Since 1976 he has coordinated and conducted several studies in fire ecology, land type classification, exotic plants, biological monitoring, and satellite telemetry. He also contributed to various other wildlife studies, has advanced software development, and technology transfer, and has consulted on management issues for the National Park Service and other agencies. He was instrumental in establishing the global change research program at Glacier in the early 1990s. Since 1996, he has been researching the characteristics of fire severity using remote sensing, and helping to implement burn assessment on a national level.

**Nathan C. Benson** is a Fire Ecologist with the USDI National Park Service at the Natural Resource Program Center, 1201 Oak Ridge Drive, Suite 200, Fort Collins, CO 80525-5596; phone: 970.267.2121; FAX: 970.225.3585; e-mail: nate_benson@nps.gov. Nate has worked for the National Park Service for more than 15 years in a variety of positions. He started his NPS fire career as a fire effects monitor at Glacier National Park, and then moved to Yellowstone and Great Smoky Mountains National Parks as a Fire Use Module Leader. Most recently he was the Prescribed Fire Specialist at Everglades National Park and is currently working for the NPS Fire Management Program Center as a Fire Ecologist. Since 1996, he has worked with Carl Key on researching characteristics of fire severity using remote sensing and has been the lead for the National Park Service in developing the National Burn Severity Mapping Project, a cooperative project between the National Park Service and U.S. Geological Survey. Nate has an M.S. degree in land resources from the University of Wisconsin-Madison's Institute for Environmental Studies.

**Steve Sutherland** is a Research Ecologist with the Fire Effects Project at the USDA Forest Service, Rocky Mountain Research Station, Fire Sciences Laboratory, P.O. Box 8089, Missoula, MT 59807; phone: 406.329.4813; FAX: 406.329.4877; e-mail: ssutherland@fs.fed.us. Since joining the lab, he has been involved with the Southern Utah Fuels Management Demonstration Project, expanding weed summaries for the Fire Effects Information System, and researching Postfire Weed Response in western Montana. Before joining the Fire Lab, Steve was the State Ecologist for the Ohio Chapter of The Nature Conservancy where he oversaw the plant community and rare species monitoring program. Prior to that, he was an Assistant Research Professor in the Department of Biology at the University of Utah. He received B.S. and M.S. degrees in forestry from Utah State University, a Ph.D. degree in evolutionary ecology from the University of Arizona, and a postdoctoral fellowship in population genetics from the University of Utah.

**Larry H. Gangi** is a Programmer with Systems for Environmental Management, P.O. 8868, Missoula, MT 59807; phone: 406.549.7478. Larry has developed software for a variety of natural resource applications including fire modeling, ecological survey data analysis and conversions, timber stand, and water resources. He received his B.S. and M.S. degrees in computer science from the University of Montana, Missoula.

## ACKNOWLEDGMENTS

Funding for FIREMON was provided by the Joint Fire Science Program. Additional support was provided by the USDA Forest Service, Rocky Mountain Research Station; USDI United States Geological Survey, and Systems for Environmental Management. USDA Forest Service Statisticians Rudy King and Dave Turner provided critical input regarding the Integrated Sampling Strategy (ISS) and the analysis tools software. Courtney Couch worked diligently to take the authors' stick-figure pictures and create many of the illustrations for FIREMON. Further valuable support came from Tim Sexton, Dick Bahr, Brian Sorbel, and Tom Zimmerman. Many at the USGS EROS Data Center, in particular, Don Ohlen, Zhi Liang Zhu and Stephen Howard, have been instrumental in the national implementation of the burn severity mapping for the NPS. Axiom IT solutions, especially Jeffrey Heng, Douglas Wissenbach, and Marc Dousset, used their expertise to build a robust data structure, data entry system, and enhanced documentation. Virginia Arensberg searched numerous publications to build a thorough glossary of terms. Finally, we thank the fire monitors, fire ecologists, fire GIS specialists, and others, from many agencies, who have enthusiastically supported development of FIREMON.

## EXECUTIVE SUMMARY

Monitoring and inventory to assess the effects of wildland fire is critical for 1) documenting fire effects, 2) assessing ecosystem damage and benefit, 3) evaluating the success or failure of a burn, and 4) appraising the potential for future treatments. However, monitoring fire effects is often difficult because data collection requires abundant funds, resources, and sampling experience. Often, the reason fire monitoring projects are not implemented is because fire management agencies do not have scientifically based, standardized protocols for inventorying pre- and postfire conditions that satisfy their monitoring and management objectives. We have developed a comprehensive system, called the ***Fire Effects Monitoring and Inventory System (FIREMON)***, which is designed to satisfy fire management agencies' monitoring and inventory requirements for most ecosystems, fuel types, and geographic areas in the United States. FIREMON consists of standardized sampling methods and manuals, field forms, database, analysis program, and an image analysis guide so that fire managers can 1) design a fire effects monitoring project, 2) collect and store the sampled data, 3) statistically analyze and summarize the data, 4) link the data with satellite imagery, and 5) map the sampled data across the landscape using image processing. FIREMON allows flexible but comprehensive sampling of fire effects so data can be evaluated for significant impacts, shared across agencies, and used to update and refine fire management plans and prescriptions.

The key to successful implementation of FIREMON requires the fire manager to succinctly state the objectives of the proposed fire monitoring project and accurately determine the available monitoring or inventory project resources. Using this information, the manager uses a series of FIREMON keys to decide the sampling strategy, methods, and intensity needed to accomplish the objectives with the resources on hand. Next, the necessary sampling equipment is gathered and dispersed to sampling crews. Field crews then collect FIREMON data using the detailed methods described in this FIREMON documentation. Collected data are then entered into a Microsoft® Access database. These data can be summarized, analyzed, and evaluated using the set of integrated programs developed specifically for FIREMON.

FIREMON has a flexible structure that allows the modification of sampling methods and local code fields to allow the sampling of locally important fire effects evaluation criteria.

## INTRODUCTION

We have developed a comprehensive ***Fire Effects Monitoring and Inventory System***, called ***FIREMON***, that integrates new and current ecological field sampling methods with remote sensing of satellite imagery to assess the effects of fire on important ecosystem components. The primary objective of FIREMON is to measure the immediate and long-term effects of a planned or unplanned fire on critical ecosystem characteristics so that fire managers can evaluate the impact of that fire on ecosystem health and integrity. This information can be used to refine fire management plans and prescriptions. This system is NOT used to document the behavior of the fire, but rather it is used to record the consequences of the fire on the landscape.

We used the National Park Service Fire Monitoring Handbook (FMH) (USDI NPS 2001) and the ECODATA Handbook (Hann and others 1988) as the framework for designing FIREMON sampling methods. However, we extended the utility of these protocols by providing nested levels of sampling intensity coupled with sampling flexibility. We designed FIREMON so that most of the data collected with FIREMON procedures will be compatible with other monitoring and inventory systems such as FMH and Natural Resource Information System (NRIS) databases. Additional sampling methods can be easily added to FIREMON as fire managers recognize their relevance in regard to inventorying and monitoring fire effects. A method to monitor water quality, for instance, would be a useful addition to the group of FIREMON sampling protocols.

*Monitoring* is the critical feedback loop that allows fire management to constantly improve prescriptions and fire plans based on the new knowledge gained from field measurements. *Inventory* is the description and quantification of important ecosystem and landscape elements and is critical to fire management activities for planning, prioritizing, and designing prescribed fire activities.

Monitoring the effects of wildland fire is critical for 1) documenting extent of fire effects, 2) assessing ecosystem damage and benefit, 3) evaluating the success or failure of a prescribed burn, 4) appraising the potential for future treatments, and 5) prioritizing stands for fire treatment. Objectives for monitoring depend on the type of fire. Wildfire monitoring is necessary to evaluate the possible need for rehabilitation or to assess a fire's potential impact to the ecosystem, while monitoring prescribed fires is invaluable for assessing the efficacy of the treatment. Monitoring data can have far-reaching applications in fire management because they provide the scientific basis for planning and implementing future burn treatments. Moreover, this information documents important fire effects, which can be used by other districts, agencies, and countries for their projects. Measuring postfire ecosystem response also allows us to understand the consequences of fire on important ecosystem components and share this knowledge in a scientifically based language.

Despite its importance, it is often a challenge for fire managers to install effective monitoring programs due to resource limitations inherent in time, money, people, and expertise. Also, often fire managers find themselves too busy with other essential duties to design and implement monitoring projects. And the perceived complexity of monitoring sampling designs has often overwhelmed or intimidated some fire managers. The issue of complexity is especially true when the fires to be monitored are large (greater than 1,000 acres), occur on diverse landscapes, and have complex severity patterns. Moreover, it is difficult to design a cost-efficient sampling strategy that will quantify stand- and landscape-level fire effects across an entire landscape using scientifically credible methods. But perhaps the main reason most fire monitoring projects never become implemented is the lack of standardized and comprehensive sampling methods and tools easily available to fire managers. Most fire management agencies do not have the scientifically based sampling protocols for inventorying pre- and postfire conditions to satisfy monitoring objectives. (The USDA Forest Service Monitoring and Evaluation Working Paper dedicates only one paragraph to data collection methods.) The major exception is the USDI National Park Service, which has extensive guidelines and protocols for sampling ecosystem characteristics that are important to monitoring fire effects (National Park Service 2001, http://www.nps.gov/fire/fire/fir_eco_firemonitoring.html). Collecting field data is easily the most expensive part of any monitoring and evaluation project, requiring extensive expertise in field sampling, fire and landscape ecology, and sampling methods design. Perhaps the single greatest challenge of designing a fire monitoring project is matching existing funding, personnel, and equipment with monitoring objectives to achieve scientifically credible evaluation data.

Monitoring is an extremely complex task that requires an extensive assessment of many ecosystem characteristics across multiple time and space scales. Fire effects monitoring, in this approach, does not include documentation of the behavioral characteristics of the fire, but rather the sampling of the ecosystem characteristics that are directly affected by the fire. These fire effects can be described at the plant level (mortality), at the stand level (fuel composition, species composition), and at the landscape level (patch dynamics, burn severity mosaic). Moreover, fire effects can be described over many timeframes including immediate (directly after fire), short (1 to 5 years postfire), or long (10 to 100 years postfire) term measurements. A valid sampling strategy for monitoring fire effects must provide for the integration and linkage of ecosystem response across these multiple time and space scales to provide meaningful data to fire management. Our intent in developing FIREMON is not to replace current systems of fire severity assessment, but rather to augment these efforts with a comprehensive and flexible set of recognized field and office methods.

It would be impossible, and probably inefficient, to design a fire monitoring program to include the measurement of all possible information a fire manager in any part of the United States would want to monitor. For instance, fire managers in the Western United States may not need a measurement, such as depth to water table, but this measurement might be absolutely critical to Eastern United States managers. Therefore, we have included local code fields in FIREMON that allow the manager to include other measurements that describe the macroplot. For example, hiding cover (horizontally projected plant cover) may be an important criterion in setting the objectives for a prescribed burn, so the manager could develop a coding system and use one of the FIREMON local code fields to assess hiding cover.

As managers attempt to oversee broader and broader areas for fire, fire effects information is increasingly difficult to obtain. Direct observation may be largely impeded by fire size, remoteness, and rugged terrain, and there may be little chance for sufficient reconnaissance on the ground. In some cases, the sheer number of areas to evaluate in one fire season is overwhelming. In others, managers with regional responsibilities may need to aggregate information from many districts to report their burn results, or to develop integrated plans. For circumstances such as these, FIREMON offers a section on Landscape Assessment (LA), which primarily addresses the need to identify and quantify fire effects over large areas, involving potentially many burns and covering tens of thousands of acres at a time. It incorporates remote sensing and GIS technologies that can produce a variety of derived products such as maps, images, and statistical summaries. The ability to compare results is emphasized, along with capacity to aggregate information across broad regions over time.

Landscape Assessment shows the spatial heterogeneity of burns, and how fire interacts with vegetation and topography, providing a *quantitative* picture of the whole burn as if viewed from the air. The quantity measured and mapped is "burn severity," defined here as a scaled index gauging the magnitude of ecological change caused by fire. In the process, two methodologies are integrated. One, the Normalized Burn Ratio (NBR), involves remote sensing using Landsat 30-m data and a derived radiometric value. The NBR is temporally differenced between pre- and postfire datasets to spatially determine the degree of change detected from burning. The other methodology, the Composite Burn Index (CBI), adds a complimentary field sampling approach. It entails a relatively large plot with independent severity ratings for individual strata within the community and a synoptic rating for the whole plot area. Plot sampling may be used to calibrate and validate remote sensing results, or it may be implemented as a stand-alone field survey for individual site assessment

# GENERAL DESCRIPTION

## What FIREMON Is...

FIREMON consists of a standardized set of sampling manuals, databases, field forms, analysis programs, and image analysis tools that will allow the manager to design and implement a fire effects monitoring project. To use FIREMON, a fire manager must first succinctly state the objectives of the proposed fire monitoring project. Then the manager must decide the amount of resources available to successfully conduct the project. Using this information, the manager goes to a series of FIREMON keys to decide which methods to use to accomplish the objectives, and the sampling strategy to employ to implement these methods across the landscape. Results from these keys are then used to design the fire monitoring project using FIREMON guidelines and procedures. Sampling equipment and plot forms are gathered and dispersed to sampling crews. The field crews then collect FIREMON data using the detailed methods described in this FIREMON publication. Collected data are then entered into a standardized database using Microsoft® Access software. These data are then summarized, analyzed, and evaluated using the set of FIREMON programs provided by this publication.

FIREMON is designed to be robust by being flexible. It allows fire managers to design a sampling strategy where only those ecosystem measurements of the greatest concern are measured. But FIREMON will still provide a myriad of comprehensive and detailed sampling schemes to measure the many important fire-related ecosystem elements. Sampling design focuses on wildland fire use objectives, rather than a shotgun approach where all ecosystem characteristics are measured to quantify ecosystem change. FIREMON is designed to be applicable for most land areas or ecosystems in the United States.

FIREMON is structured so that it can be easily learned. First, FIREMON resides on an Internet Web site so that it will be easily accessible to all. Second, the entire FIREMON system, including sampling methods, field forms, and databases, are available on CD so that it can be accessed from any computer with Microsoft Word and Access installed (versions 2000 and later). Finally, training courses have been developed to teach FIREMON to fire personnel with limited sampling experience.

## What FIREMON Is NOT…

To fully understand FIREMON, it is important to emphasize what FIREMON is NOT:

FIREMON is NOT intended to be a corporate database, although it surely could be at some point in the future.

FIREMON is NOT a replacement for FMH in the National Park Service or the NRIS protocols developed by the Forest Service. FIREMON can complement these systems and provide additional help with monitoring tasks.

FIREMON does NOT contain software for extensive data analysis. FIREMON software will provide a general report and statistical summary, but not extensive statistical analyses. More extensive analysis can be accomplished by exporting the data from FIREMON and using them in a statistical package. Also, additional statistical analysis can be added at a later date.

FIREMON is NOT used to document fire behavior; it is used to record the consequences of the fire on the landscape.

FIREMON is NOT just a fire monitoring package. Many procedures and the database within FIREMON are useful for other ecosystem inventory and monitoring. One inventory need we especially included in FIREMON is fuels. FIREMON contains the necessary components for sampling surface fuels for inventory, fuels mapping reference (ground-truth), and fuels summary for input to fire behavior and effects programs.

FIREMON does NOT include sampling methods for all important fire effects. For example, changes in water quality may be an important fire effects issue, but there is no water quality sampling protocol in FIREMON. The sampling methods in FIREMON were written using existing, recognized sampling methods. We were unable to find a standardized protocol for water quality sampling, so we did not include one. However, new sampling methods can be readily added into FIREMON in the future.

## The Four FIREMON Components

There are four major components to FIREMON:

1) **Integrated Sampling Strategy**—This is a set of step-by-step procedures for designing fire effects sampling projects. This component is composed of design keys, strategy descriptions, and guidelines for designing a successful fire monitoring project.

2) **Field Methods**—These are methods for sampling important ecosystem characteristics used to assess fire effects. There are currently 10 methods implemented into FIREMON: Plot Description (PD), Tree Data (TD), Fuel Load (FL), Species Cover (SC), Cover/Frequency (CF), Line Intercept (LI), Density (DE), Point Intercept (PO), Rare Species (RS), and Composite Burn Index (BI). These sampling methods provide a complete set of field sampling protocols to quantify changes in ecosystem characteristics due to fire to describe stand-level fire effects. Additionally, there are two database tables to record metadata (MD) information and fire behavior (FB).

   The Landscape Assessment component details how remotely sensed imagery can be used to design a spatially explicit strategy to locate, collect, and summarize field data across a burned landscape. These methods require extensive expertise in the processing of remotely sensed imagery.

3) **FIREMON Database**—Field data are stored in the Microsoft® Access-based FIREMON database. Data entry forms look like field forms, and drop down lists limit data entry errors.

4) **Analysis Tools**—These include queries in the FIREMON database for producing plot-level data summaries, and the FIREMON Analysis Tools (FMAT) software for analyzing collected field data. The FMAT program provides data summaries for either plot-level or grouped plots and statistical inference of grouped plots using Dunnett's procedure for multiple comparisons with a control. This test is designed to statistically compare pre- and posttreatment data.

The fire manager can choose to perform all or part of one or more components, but the real power of FIREMON is in the integration of all components to describe fire effects at multiple scales.

## Integrated Sampling Strategy

The Integrated Sampling Strategy (ISS) component provides the manager with step-by-step instructions on how to design a comprehensive, statistically valid field sampling effort for the purpose of quantifying fire effects over long periods across burned landscapes. This component describes how the detailed sampling procedures are selected, and how to place sample plots across project area. This will allow the fire manager to design a sampling procedure to implement on preburn or postburn areas for describing the effect of the wildfire or prescribed fire.

As in any project, there are three ways to get things done: good, fast, and cheap. But a fact of nature says we cannot accomplish these three goals simultaneously; one can only effectively manage for one and compromise on the remaining two. Therefore, the ISS has a three-level, hierarchically nested strategy for implementing each sampling method in the field assessment. This three-level strategy is geared toward a number of important sampling considerations that attempt to provide a compromise between good, fast, and cheap:

1. **Level I—Simple sampling scheme**. Fastest and cheapest while still collecting useful data in the context of the management objective. Use this scheme if little time, money, or personnel are available to complete the monitoring tasks.

2. **Level II—Recommended sampling scheme**. Somewhat fast, somewhat cheap, and somewhat good. Statistically valid data collected as efficiently as possible but with high levels of variability. Use this scheme if defensible numbers are needed from the monitoring effort, but there is limited time and/or resources.

3. **Level III—Detailed sampling scheme**. Real good but slow and somewhat costly. Statistically valid data with minimized levels of variation but with high collection costs. Use this scheme if the most statistically valid estimates are needed and time and money are not limiting.

These three sampling levels can be used at two spatial levels. The fire manager must pick the sampling level to assign to monitor the landscape and the sampling level to monitor the stands. For example, the land manager may not care about fire effects across the landscape, such as in a prescribed burn, but cares more about stand level changes across the burn unit. In this case, the fire manager would decide on Landscape Level I with Stand Level III. However, another fire manager may not care how a wildfire burned at the stand level, but wants to know general characteristics of how the fire burned across the landscape. In this case, Landscape Level II or III would be selected while Stand Level I or II might be selected, depending on time and resources.

## Field Assessment

The field assessment portion of FIREMON contains an extensive set of procedures for sampling important ecosystem characteristics before and after a prescribed or natural fire for ecosystems in the United States, including forests, grasslands, and shrublands. The design of FIREMON is such that the fire manager can tailor the field measurement procedures to match burn objectives or wildland fire use concerns. Moreover, the fire manager can scale the intensity of measurement to match resource and funding constraints. For example, to document tree mortality, the fire manager might choose one of three hierarchically nested sampling procedures, where the first procedure might provide general descriptions of tree mortality quickly at low cost (photopoints, walk-through), while the third procedure would document, in detail, individual tree health and vigor, to generate comprehensive data applicable to many analyses but costly to collect. A key has been developed to help fire managers decide the appropriate methods and sampling intensity for each.

The field assessment procedures are written into a handbook that can be taken into the field. The assessment is composed of 1) field methods, 2) plot forms, 3) cheat sheets, and 4) equipment lists. This assessment does not include details on how certain sampling procedures are selected; those details are in the ISS section.

FIREMON contains the following sampling procedures for monitoring ecosystem characteristics:

**Plot Description (PD)**—A generalized sampling scheme used to describe site characteristics on the FIREMON macroplot with biophysically based measurements.

**Tree Data (TD)**—Trees and large shrubs are sampled on a fixed-area plot. Trees and shrubs less than 4.5 ft tall are counted on a subplot. Live and dead trees greater than 4.5 ft tall are measured on a larger plot.

**Fuel Load (FL)**—The planar intercept (or line transect) technique is used to sample dead and down woody debris in the 1-hour, 10-hour, 100-hour, and 1,000-hour and greater size classes. Litter and duff depths are measured at two points along the along the base of each sampling plane. Cover and height of live and dead, woody and nonwoody vegetation is estimated at two points along each sampling plane.

**Species Composition (SC)**—Used for making ocular estimates of vertically projected canopy cover for all or a subset of vascular and nonvascular species by diameter at breast height (DBH) and height classes using a wide variety of sampling frames and intensities. This procedure is more appropriate for inventory than monitoring.

**Cover/Frequency (CF)**—A microplot sampling scheme to estimate vertically projected canopy cover and nested rooted frequency for all or a subset of vascular and nonvascular species.

**Point Intercept (PO)**—A microplot sampling scheme to estimate vertically projected canopy cover for all or a subset of vascular and nonvascular species. Allows more precise estimation of cover than the CF methods because it removes sampler error.

**Density (DE)**—Primarily used when the fire manager wants to monitor changes in plant species numbers. This method is best suited for grasses, forbs, shrubs, and small trees that are easily separated into individual plants or counting units, such as stems. For trees and shrubs over 6 ft tall the TD method may be more appropriate.

**Line Intercept (LI)**—Primarily used when the fire manager wants to monitor changes in plant species cover and height of plant species with solid crowns or large basal areas where the plants are about 3 ft tall or taller.

**Rare Species (RS)**—Used specifically for monitoring rare plants such as threatened and endangered species.

**Landscape Assessment (LA)**—Useful for mapping fire severity over large areas. Combines a ground-based burn severity assessment, the **Composite Burn Index (BI)** and a satellite derived remote sensing analysis method, the **Normalized Burn Ratio (BR)**. The LA methodology will assist in determining landscape level management actions where fire severity is a determining factor. See below for more information.

Each sampling method is discussed in detail in their respective sections. Additional sampling methods can be easily added to FIREMON as fire managers recognize their relevance.

## Landscape Assessment

The remote sensing of severity is captured by a new Landsat TM radiometric index we call the Normalized Burn Ratio, or NBR. The NBR evolved through sampling of TM band reflectance over burned surfaces, and was tested against three other TM measures appearing in the literature. Multitemporal differencing was employed to enhance contrast and detection of changes from before to after fire. Seasonal effects also were tested to determine the best time of year for TM data acquisition. Based on statistical and visual characteristics, NBR difference from early growing season dates was judged to be optimal, compared to other measures. Results clearly showed the extent of burning that represented a wide range of severity magnitude that was easily interpreted for each burn. Further, the full range of differenced NBR can be stratified into a finite number of ordinal severity levels, to facilitate summation of burns through mapping and tabular statistics. Those data provide a basis for monitoring burn impacts over large regions, and for comparing burns spatially and temporally.

Sensor characteristics make this approach suitable for moderate resolution (30-m) applications that require more extensive and precise information than rapid assessment techniques, and can be completed within a 1-year timeframe of the subject fire.

## FIREMON DOCUMENTATION STRUCTURE

FIREMON is presented using a series of sections to document the entire fire effects monitoring system. This set of documents is not necessarily designed to be read from front to back like a book, but rather it is designed for FIREMON users to read only those sections that are important to their sampling requirements. Every FIREMON user should read the Integrated Sampling Strategy (ISS) because it contains absolutely essential FIREMON sampling concepts and terminology that are used throughout all documents.

There is an obvious lack of citations in the bulk of FIREMON documentation. This was done on purpose to reduce clutter and improve readability. This does not mean that we didn't consult numerous sampling and monitoring texts during the development of FIREMON. The References sections contain citations for the journal articles, textbooks, reference books, and symposium proceedings used designing and developing FIREMON.

FIREMON also includes a glossary that defines common FIREMON terminology, and a How To... section that describes sampling techniques used in more than one of the FIREMON sampling methods.

We attempted to design FIREMON document structure so that major and minor headings describe critical monitoring tasks. This way, the FIREMON user can easily jump to a particular method or procedure instead of having to read the entire document. For this to work, each heading section must effectively stand on its own so the user does not have to read other sections to understand the section of interest. A side effect of this independent section treatment is that there is often redundant text across sections that may be annoying to those reading each section sequentially. We apologize for this repetition and hope you will recognize its purpose.

# Integrated Sampling Strategy (ISS) Guide

Robert E. Keane
Duncan C. Lutes

## SUMMARY

What is an Integrated Sampling Strategy? Simply put, it is the strategy that guides how plots are put on the landscape. FIREMON's Integrated Sampling Strategy assists fire managers as they design their fire monitoring project by answering questions such as:

- What statistical approach is appropriate for my sample design?
- How many plots can I afford?
- How many plots do I need?
- Where should I put my plots?
- What sampling methods should I use on my plots?

The Integrated Sampling Strategy (ISS) is used to design fire monitoring sampling projects by selecting the most appropriate sampling approach and the most efficient sampling strategy, then choosing the best sampling methods for a fire monitoring project. The first section of the ISS Guide introduces the FIREMON user to the terminology and inherent properties of sampling design in the FIREMON monitoring approach. The second section presents the preliminary information that must be collected or compiled for designing a monitoring project. The third section documents how a monitoring project is implemented. And the last section provides users with guides and keys to assist in developing the monitoring project. New users, especially those responsible for the design of monitoring programs, should read the third section in detail in order to gain the knowledge and understanding needed to implement an appropriate and successful FIREMON monitoring projects.

The ISS in FIREMON is critical to fire monitoring for several reasons. First, many fire managers do not have the background in ecosystem inventorying and sampling to design a statistically credible and efficient sampling strategy. Second, fire managers rarely have the time to learn sampling theory and concepts. Last, integrated sampling requires extensive experience in statistical sampling design and field implementation. FIREMON condenses this detailed information on sampling strategy into the ISS to guide the fire manager in planning and implementing an appropriate fire monitoring project.

## INTRODUCTION

The FIREMON Integrated Sampling Strategy (ISS) uses the best estimate of resources that the manager can provide to help design the plot level and landscape level sampling strategy of a fire effects monitoring project. A sampling strategy is different from a sampling method in that a sampling strategy describes where, when, and how the sampling methods (procedures for measuring things) are implemented across the landscape. This section allows the fire manager to match the appropriate sampling strategies with the scope and context of the project objectives.

The quickest way to design a fire effects monitoring project is to complete the set of sampling strategy and method keys provided in the FIREMON ISS. These keys provide guidance in the selection of various criteria needed to design a statistically credible and defensible monitoring sampling strategy. FIREMON provides methods for measuring fire effects at most levels of intensity and most any scale, and then provides guidance for data analysis that is appropriate for the data that have been collected. For example, a coarse sampling design that specifies pictures as the only data collected cannot be used to determine tree mortality, fuel consumption, or any other fire effect. Likewise, broad visual estimates of plant species canopy cover for a large area cannot be used to describe changes in plant composition.

Implementation of a FIREMON monitoring program is based on two components: objective(s) and sampling resources. The sampling objective or objectives provide the fundamental criteria for determining the sampling methods and, to a lesser extent, the sampling intensity that will be integrated into a FIREMON monitoring program. It is critical that the fire manager succinctly articulate the actual purpose of the sampling effort in the FIREMON sampling strategy and design process. Without an expression of the sampling purpose, the fire monitoring project is doomed to fail. The fire manager must explicitly state the reasons why a fire effects monitoring project is needed. These reasons provide the critical context to form the project objectives, which in turn drive the sampling methods. Sampling resources are less easily assessed as they are related to funds, time, personnel, and equipment, all of which can be somewhat dynamic throughout the course of the field season.

## Advanced Alternative to the FIREMON Integrated Sampling Strategy

The FIREMON ISS provides general guides and keys for you to use to determine the sampling strategy that best fits with the objectives and resources available for monitoring fire effects. Recently, new technology has been developed by Spatial Dynamics in cooperation with the USDI National Park Service Fire Monitoring Program that is an advanced alternative to the ISS presented in FIREMON. This new software is called FEAT or Fire Ecology Assessment Tool and it allows the user to interactively design sampling strategies with Geographic Information Systems (GIS) and integrated databases, and then implement the strategy on the landscape using GIS techniques and plot-level databases similar to FIREMON. FEAT is a complete fire monitoring software package that integrates the entire monitoring effort into one system. Users can use the FIREMON plot methods or they can use the FEAT plot methods for collecting data.

The system allows the user to examine a range of monitoring design applications and alternatives, such as:

- Random location of plots within an area using GIS techniques
- Identification of the sampling area and strata using any number of GIS layers
- Plot sampling methods linked to relational databases inside the GIS structure that allows plots to be shown on a GIS map and sampled attributes of the plot to be spatial displayed.
- Digital photo integration with plot and a GIS to allow point-and-click real-time information for each plot or sampling strata using photos or data.
- Ability to easily define new sampling protocols and modify existing protocols.
- Ability to manage tabular data using GIS.
- Designed to support efficient data entry into Personal Data Assistants (PDA).

FEAT is a comprehensive system that combines a number of software platforms to form an integrated fire effects monitoring package. The all-inclusive nature is the benefit of FEAT; however, some monitoring programs may find it difficult to meet the associate resource needs. For instance, there is a substantial initial financial commitment and ongoing maintenance overhead for the software needed to run FEAT (ARCMap/Spatial Analyst, Microsoft XP). To use the full capability and understand the underlying analysis within FEAT, specialized training is required. Also, there is a workload associated with updating and maintaining the GIS layers that FEAT requires. FEAT was developed to facilitate flexibility in sampling procedures and methods, so field methods can be extremely adaptable if required by your monitoring project. This is especially true if you want to modify a sampling procedure to measure a new entity or ecosystem characteristic; FEAT will allow one to easily modify or develop a new sampling protocol.

FIREMON users are encouraged to consider using FEAT for their monitoring system if they feel comfortable using the advanced features offered in FEAT, and have the financial commitment to obtain and keep the resources necessary to effectively apply the system. More information about FEAT can be obtained from the National Park Service, Fire Monitoring Program Web site: http://www.nps.gov/fire/fire/fir_ecology.html.

## Resampling Existing FIREMON Plots

If you are revisiting plots that have already been sampled then you do not necessarily need to read through the ISS at this time. Instead, carefully read through the FIREMON metadata (MD) information and/or FIREMON notebook to determine the methods and sampling intensity that were incorporated during the first sampling visit, and identify any optional fields or data variables that were developed at that time. Return to the FIREMON plots and sample using the same methods, intensity, and so forth, used during the original sampling. When reading the MD information you may also note any shortcomings identified by the previous sampling visit and modify the methods to make the sampling more effective. Use care when doing this so that the initial measurements can be used for analysis. For instance, changing the vegetation sampling method from cover to frequency would mean that the cover values could not be used in the analysis. Instead, the frequency method should be *added* to the list of methods applied at the plot and not used to replace the cover method.

Many studies examine change in vegetation attributes after a treatment. Generally, these attributes are related to the change in species numbers, the number of individual plants or vegetation cover as a result of the treatment. For instance, a manager might be interested in noting the difference in density of undesirable weed species after a prescribed fire. Or, a manager may want to study the difference in that same weed species in areas burned in the spring versus the fall in an attempt to identify an effective way to control its numbers. Whatever your reason for sampling it is important to recognize how plant attributes change during the season and take them into account with your sampling. Generally, this will mean sampling at the same time or times every year. It would be difficult to observe the effectiveness of treatment if, say, the first season the vegetation was sampled in late summer and the next year in early spring because plant growth during the year would influence plant attributes such as cover, density, and height. There is no hard and fast rule for timing the vegetation sampling, however, so it is up to the fire manager to determine the annual sampling schedule. The schedule will probably be set by date but could be set by some other attribute, such as the phenological stage of some species of interest. Recognize that rigid sampling schedules may make it difficult to finish all the sampling tasks each year. For example, if you decide that late season sampling is the most appropriate time for estimating species cover, some years you may not be able to sample because extreme fire danger keeps the monitoring crews out of the field or an early snowfall may make it impossible to sample fine and coarse woody debris.

## Terminology

There are a number of terms used in the FIREMON documentation that are either unique to FIREMON or imply a meaning that is specific to the FIREMON documentation. In general, these terms are used as shortcuts to reduce text and focus discussion. The more important FIREMON terms are stratified by subject area and put in context below. Complete definitions are located in the **Glossary**.

**The Project**—A *fire monitoring project* is a fire management activity used to evaluate the effects of a fire using field sampling and statistical analysis. A fire monitoring project that installs field plots AFTER a planned or unplanned fire (or other treatment or disturbance) is called a *postevent monitoring project*, whereas a project that establishes plots BEFORE and AFTER the burn is called a *complete fire monitoring project*. We recognize that disturbance occurs at many intensities and scales, so conceivably every monitoring project is both a postevent and complete monitoring project. However, we make the distinction based on the disturbance event that initiates the sampling program. *Sampling resources* are those assets available to the fire manager to accomplish the monitoring project, most frequently, funds, time, personnel, and equipment.

**The People**—A *FIREMON team* is the group of people involved in the planning and implementation of a fire monitoring project. This team is usually composed of a *FIREMON Project Leader* who oversees the project; a *FIREMON Architect* who plans and designs the appropriate fire monitoring methods and sampling strategies; and *field crews* who implement FIREMON methods in the field. The field crew is composed of the *crew leader* responsible for all logistics in the field and crew training; a *data recorder* who fills out the FIREMON plot sheets and *sampler* that does the actual collection of field data. There can be more than one sampler, and the crew leader and data recorder can also perform sampling duties. Be sure to let different members of the field crew try their hand at different tasks. In other words, if a person is a sampler on one plot let him or her switch jobs with the data recorder on the next. This will keep the field work from getting too monotonous and will let everyone become familiar with a number of field sampling procedures.

There may not always be a large number of people involved in a FIREMON project. For instance, in a small FIREMON sampling project, one person can be the FIREMON project leader, architect, crew leader, and data recorder. There should always be at least two people on the field crew, for safety sake. In the interest of good quality data it is useful to have one field crew member that has the expertise to overlook the sampler's observations, checking both the accuracy and precision of the recorded data. For instance, cover estimation can be quite difficult, especially for someone who is just starting out. It is important to have someone on the crew who is able to accurately estimate cover and to have that person check the cover estimations made by other crew members.

**The Sampling Procedure**—*Sampling strategies* are how, where, and why sampling methods are implemented on the landscape. *Sampling methods* are a set of procedures for measuring specific ecosystem attributes. The difference between strategies and methods can be somewhat vague. Think of measuring a tree's diameter with a diameter tape—that is a sampling method; then think of measuring tree diameters on all trees above 4.5 ft on a 0.25 acre circular plot randomly across a landscape—that is a sampling strategy. Finally, the *sampling approach* is the scheme used to drive the sampling strategy design process. Simply put, there are two sampling approaches in FIREMON, statistical and relevé. Each is discussed later in the ISS.

**The Sampling Unit**—The FIREMON *macroplot* defines the greater sampling area in which all of the sampling methods are nested. The size and shape of the macroplot is determined by sampling objectives and resources, but most macroplots will be rectangular or circular encompassing about 0.1 to 0.25 acres (0.04 to 0.1 ha).

Depending on the methods used, the FIREMON plot may be divided into *microplots*, also known as *quadrats*, *belts*, or *subplots*. Each one is a much smaller area used for measuring small-scale phenomena, such as ground cover or individual plant or species attributes. Microplots are usually located in a grid pattern within the macroplot. The size of the microplot depends on the size of the plant or species being measured, but typically it is about 3 ft square ($1 \text{ m}^2$). Some studies have found that certain types of vegetation are more effectively measured using belt transects. These belts are essentially elongated quadrats. In FIREMON we only associate subplots with the Tree Data (TD) methods where saplings and seedlings are sampled on a smaller plot—the microplot—nested within the larger plot used for sampling mature trees.

A *transect* is a one-dimensional line that is located within the macroplot. Ecological attributes that intersect or cross the transect are tallied or measured.

The vegetation sampling methods, in particular, use a macroplot to define the potential sampling area, with microplots located within, where data are actually collected. Microplot sampling allows macroplot scale attribute estimation using subplot sampling, and this can simplify sampling. For instance, determining plant density across a macroplot would be quite time consuming. However, by using microplots located within the macroplot, density can be sampled more quickly and with sufficient accuracy and precision. All site attributes such as slope, aspect, and elevation are recorded at the macroplot level.

Lastly, the FIREMON documentation uses some terms to describe spatial elements that need to be defined. *Stratifying factors* are defined by the project objectives and are the characteristics used to divide the treatment area or landscape into strata. *Polygons* are areas that exhibit unique characteristics in relation to the adjacent polygons and are usually defined by overlaying the different strata. Polygons can be defined by hand-drawn maps or electronically mapped in a Geographical Information System (GIS). A *Sampling stratum* is made up of the polygons that have similar attributes, as defined by all of the stratification factors. For example, if tree density and fuel load stratum were overlaid, a number of polygons would be defined; some polygons would have low tree density and low fuel load, some with high tree density and low fuel load, some with high tree density and high fuel load, and so on. All of the polygons that had low tree density and low fuel would be in the same sampling stratum; all of the polygons with low tree density and high fuel load would be in another sampling stratum, and so forth. Each polygon will belong to one of the sampling strata. A *landscape* is a large area that can be any size and shape but spatially defines stands and is composed of continuous polygons. The *sample landscape* is the area to be sampled in a FIREMON project and is often described by the prescribed burn map or wildfire map. In statistical terms, the sample landscape defines the population about which inferences will be made.

## Sampling Intensities

There are three ways to get things done: good, fast, and cheap. Unfortunately, we can only manage for one and compromise on the remaining two. The FIREMON ISS allows users to choose between three levels of sampling intensity based on the project objective(s) and sampling resources. This three-level strategy provides a context for striking a compromise between good, fast, and cheap:

Simple sampling intensity (Level I): Fastest and cheapest while still collecting useful data in the context of the management objectives. This scheme is used if there is limited time, money, or personnel available to complete the monitoring tasks. The data collected in this effort are usually qualitative and not suitable for statistical comparisons.

Alternative sampling intensity (Level II): Somewhat fast, somewhat cheap, and somewhat good. Statistically valid data collected as efficiently as possible but with poor estimates of variability. This scheme is used if defensible numbers are needed from the monitoring effort, but there is limited time and/or resources. Caution must be used in statistical inference due to the low number of samples that can be collected.

Detailed sampling intensity (Level III): Provides the most statistically defensible data, but most methods are slow and costly to implement. Data are statistically valid with appropriate estimates of variation but with high collection costs. Use this scheme if the most statistically valid estimates are needed, and time and money are not limiting.

These three sampling levels are implemented at two spatial levels—landscape and polygon. The fire manager must pick a sampling level to monitor landscape conditions and one level to monitor polygon-level conditions. This decision is based on the *sampling objectives* and the *sampling resources*. The sampling levels for each spatial scale may or may not be the same. For example, a land manager may not care about fire effects across the landscape, such as with a prescribed burn, but is more concerned with the polygon level changes across the burn unit. In this case, a fire manager may decide on Level I landscape sampling intensity and Level III polygon intensity. Another fire manager may not care how a wildfire burned at the polygon scale but wants to know general characteristics of how the fire burned across the landscape. In this case, landscape Level II or III would be selected while polygon Level I or II might be selected, depending on time and resources.

We refer to the sampling intensity levels frequently throughout the FIREMON documentation. However, they are intended as guidelines, not as rigid criteria. FIREMON allows the user to design sampling strategies at any level of intensity or complexity because the FIREMON procedures and methods have been constructed to be flexible and robust. For example, the Alternative Sampling Intensity, LEVEL II, may suggest that all trees above 4 inches DBH be measured individually using the TD method. However, the FIREMON architect can select any threshold DBH to accommodate the sampling objectives and the resources that are available. As long as the change is documented in the project records there would be no problem with dropping the diameter threshold from 4 to 2 inches, for instance. We have provided a metadata (MD) table in the FIREMON database so that changes to the sampling methods can be recorded and recovered easily.

## Sampling Approaches

There are two basic sampling approaches used in the FIREMON sampling strategy. The first is the *relevé approach*, used extensively in many ecological vegetation studies during the past 50 to 70 years. The relevé approach is used when documentation of important ecological characteristics is more important than statistically valid estimates of change. When using the relevé approach, one plot is placed in a representative portion of the stand or polygon "without preconceived bias," that is, the plots are not located to make the sampling results look good but, instead, are located with bias in order to represent the general conditions of the polygon or sampling stratum. Representativeness is based on stand history, vegetation composition, stand structure, and a host of other ecological attributes. The advantage of the relevé method is that the fire manager can choose where to locate plots based on past experience, management objectives, and crew safety. For example, the manager may wish to use a relevé approach if the restoration of an important plant community is the objective and the manager wants to make sure that the plots land inside this community. The disadvantage is that this approach is somewhat biased, and plot locations can be manipulated to influence monitoring results, making subsequent statistics highly suspect.

The next approach is the familiar *statistical approach* utilized in most natural resource inventories using systematic, random, or cluster plot establishment. Systematically established plots are distributed following a preset pattern, usually on a grid. Randomly established plots are located using some sort of random number routine. They are not regularly distributed across the sample site and will have some level of clumping. In FIREMON we describe how to purposely cluster plots in adjacent polygons or sampling stratum to allow less travel time between sampling locations. This is not the same as the traditional statistical method of *cluster sampling*, which can be quite complex and is outside the scope of FIREMON. The use of cluster plots can be problematic, statistically, because plots may not be distributed well enough to quantify variance, may not be independent, and samplers have an opportunity to place plots with bias. Despite these potential shortcomings, cluster plots have the advantage of allowing managers to sample a number of polygons relatively quickly.

Plots that follow a regular pattern are easier to relocate, so systematic sampling is recommended for sites that will be sampled multiple times. There is one cautionary note about systematic sampling. Ecologists have noted that some ecological variables have a periodic nature, that is, they vary across the landscape with some predictability. If fire managers develop a systematic plot design that happens to correspond to the periodicity of the attribute being sampled, the sampling results will be biased. The chances of this situation happening are quite small, however, and the convenience of being able to easily relocate sampling plots far outweighs the potential for biased results.

With the statistical approach, the emphasis is on gaining a statistically sound estimate of the sampling entities. It is assumed that the random or systematic establishment of macroplots across a landscape will adequately quantify the variability of sampled entities so that the entities can be compared using standard statistical tests. However, the only way to be certain that all characteristics a fire manager is interested in monitoring are sampled adequately is to design the sampling program with sufficient intensity to describe the variance of the most variable characteristic. Sampling at this intensity will probably lead to increased sampling effort.

A stratified approach describes how FIREMON plots are established across the landscape or sampling strata based on some land type stratification. The land stratification is based on the site characteristic or characteristics of interest. For instance, a fire manger may want to examine the effect of prescribed fire on exotic weed cover (one stratifying factor) and on sites with different fuel loads (the second stratifying factor). If there were three classes of weed cover and three of fuel load, the potential number of sampling strata would be $3^2$ or 9. Within the stratification, plots can be established using either a random or systematic approach. Both are well documented in the literature and have a proven track record, but most fire monitoring projects are designed using a stratified systematic plot approach for the reasons previously stated. Stratified sampling can reduce the overall cost of the project because stratification accounts for the within-stratification variability and that may reduce the total number of plots needed in the monitoring project. Stratification is especially useful if you are interested in examining treatment effects within the sampling strata.

## Sampling Design Keys

There are three sampling design keys in FIREMON. The first, the **Sample Approach Classification Key**, is designed to help the FIREMON architect determine whether a relevé or statistical sampling approach should be used. The second key, the **Sampling Intensity Key**, is designed to identify the sampling intensity level that is most applicable to the monitoring project. Last, the **Methods Classification Key** is used to guide the FIREMON architect to determine the sampling methods that should be used in the project. Each key uses the sampling objectives and resources to determine the keys' outcome. The FIREMON architect must determine the scale of the monitoring project—landscape or polygon—before using the keys.

Again, the FIREMON keys are not meant to be used as strict criteria on designing a fire monitoring project. They are meant only as guidelines for developing a locally relevant sampling design that optimizes available resources with the quality and quantity of data required to successfully accomplish the project objectives.

## Step-by-Step Procedures

If you are experienced with sampling methods and strategies or have previously implemented FIREMON fire monitoring projects you may not need to reread the detailed text in the next section. Instead, you can just refresh your memory on the steps needed to come up with a viable sampling design. In this case we have condensed the FIREMON ISS section into a series of step-by-step instructions to guide the design and implement your fire monitoring project (fig. ISS-1). These instructions are in the **Sampling Strategy Checklist** section and should be used as a quick reference for your monitoring project.

# PRELIMINARY SAMPLE DESIGN ACTIVITIES

In this section, the FIREMON sampling architect performs some preliminary tasks and analyses that will help design an integrated monitoring project using FIREMON sampling design strategies, techniques, and field methods. This section includes the most important design elements and should be sufficient for most managers when they are setting up their monitoring program. There are many texts and Web sites that give an indepth view of sampling design theory—more thorough than what we are presenting here. If you are interested in learning more, a good place to start is: http://statistics.fs.fed.us/checklists/checklists.html.

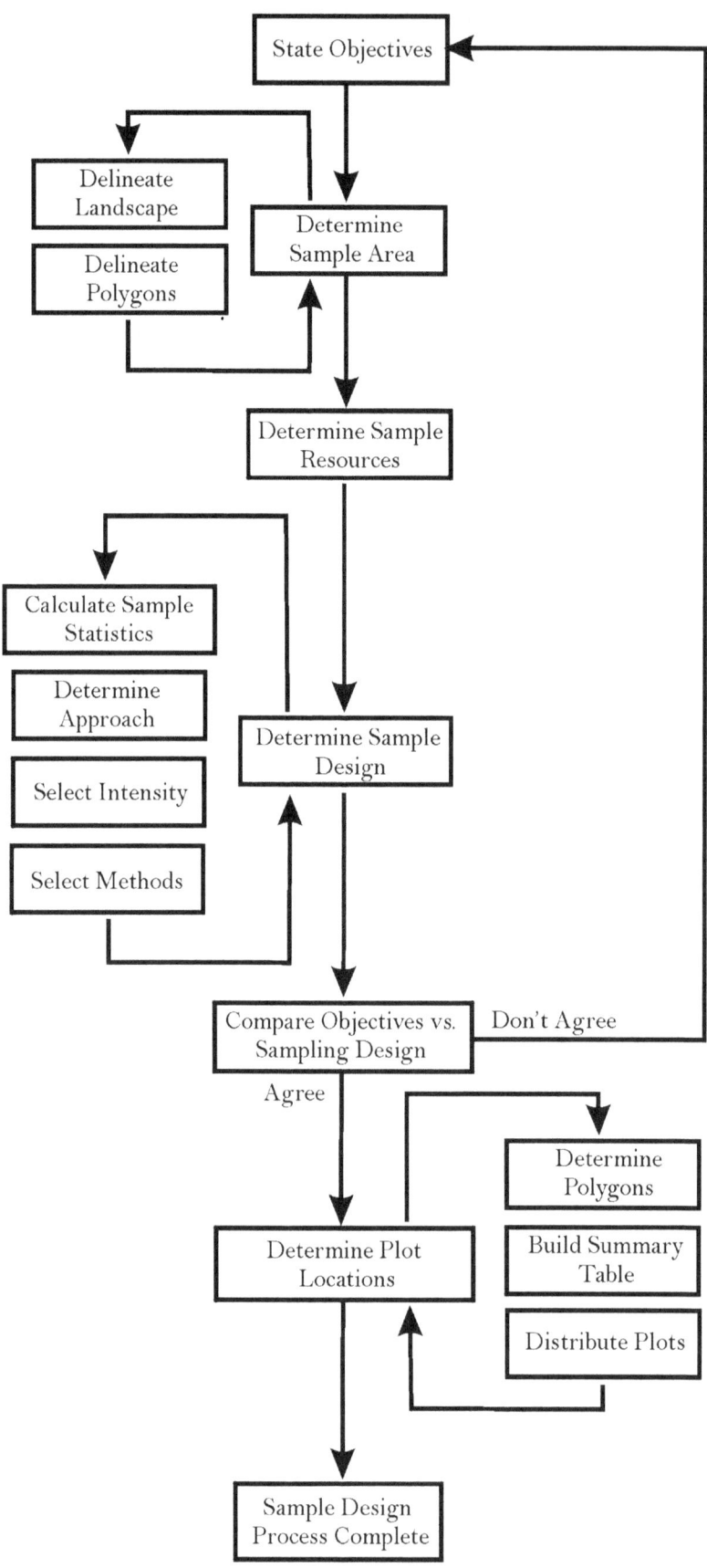

**Figure ISS-1**—Flowchart showing the general process for designing your FIREMON Sampling project.

We suggest that the FIREMON architect use the Metadata table to store important information for each fire monitoring sampling project. The Metadata table should contain a detailed listing of the project objectives, the resources available to the project, and the logic and reasoning used to design the sampling strategy and data analysis for the project. Also, the outcome of the FIREMON keys should be recorded in the notebook. Take special care to ensure that the decision process at each step is explained. Figures can be included in the MD table using the Document Link field.

The first, and most critical, step in a FIREMON sampling effort is to succinctly state the objectives of the monitoring project. This step should include a definitive description of the sample population. In other words, when completed the objectives should not just identify *what* the project will be accomplishing but also *where*. For instance, will they be applied across a watershed or just in one treatment unit? The next step is to identify the amount of resources available to accomplish the sampling task. Sample size is determined using the objectives and resources. Note that this is different than most scientific studies where objectives and variance determine the sample size. When developing FIREMON we recognized that, for most fire managers, resources determine sample size, not variance. Usually, the fire manager does not have the funds, time, or personnel to undertake a rigorous sampling program. The result is that the FIREMON approach may not always provide data for statistical inference—especially when sampling at the Simple or Alternative levels—or may do so at lower precision or certainty than typically used in rigorous research studies. In lieu of determining statistical significance, the manager may examine monitoring data, note the changes, identify how well the data represent what was seen in the entire treatment area, and then determine the apparent effectiveness of the treatments.

## Stating Monitoring Goals and Objectives

Succinct and comprehensive goals and objectives describing the purpose of the fire effects monitoring project set the tone for the remaining sample design and method decisions. It would be difficult to overemphasize the value of this step. To the person not dealing with them all of the time, goals and objectives can be difficult to differentiate. Briefly, goals are broad statements describing general intentions whereas objectives tend to be narrowly focused and precise. In terms of fire monitoring, goals generally explain the overall desired outcome of treatment while the objectives are the quantifiable measure used to evaluate the outcome.

Development of specific measurable objectives requires thoughtful reflection on what the FIREMON project manager wants from the monitoring effort. It may be intimidating to anticipate developing these objectives in light of the many diverse goals in fire management, but understanding exactly what questions the monitoring effort are supposed to answer will provide the context in which all other sampling design and implementation decisions are made. For instance, one manager may only want to qualitatively describe the general effects of a fire while another might want statistically valid estimates of change in vegetation and fuels across the landscape.

Many prescribed burns have a single goal of reducing fuel loading. Given that, a good objective statement would be: reduce dead and down woody debris biomass in the 3 inches and greater size class by at least 50 percent after the first burn. Or, if you have a specific desired future condition it could be: reduce dead and down debris in the 3 inches and greater size class to achieve an average fuel loading of 5 to 10 tons/acre. On the other end of the spectrum, if there were to be no actual measurements performed, the objective could be: complete a walk-through assessment to evaluate the reduction of dead and down debris within 6 weeks of the burn. As an example, a general goal coupled with specific objectives would be as follows:

Restore ecosystem processes and characteristics to pre-1900 conditions by:

1) Reducing fine woody fuel loadings by 80 percent or more after the first burn.
2) Reducing coarse woody debris by less than 50 percent after the first burn.
3) Killing 90 percent or more shade tolerant seedlings, saplings, and mature trees within 1 year of the second-entry prescribed burn.

4) Providing for at least 50 percent or greater survival in seral, shade-intolerant mature trees within 1 year of the second-entry prescribed.

5) Reducing duff depths by at least 10 percent for each prescribed burn entry.

6) Opening tree canopy by at least 50 percent after the first burn.

The FIREMON architect will have to decide if all the objectives can be accomplished with one burn and avoid conflicting objectives.

There is a downside to specifying detailed objective statements in that the monitoring project may become complex and expensive in order to monitor all the important characteristics. Additionally, it may be difficult to achieve all objectives with just one burn. The ecosystem characteristics important to evaluating the success or impact of a burn should be explicitly stated in the objective statement to guide sample design with the recognition that some objectives may be met earlier in the monitoring sequence than others. In other words, all objectives might not be met with one prescribed burn. Try to make objectives broad enough to facilitate an efficient sample design while being specific about the most important ecosystem attributes that must be treated.

Do not think of objective statements as static contracts of purpose and need. Objective statements should be modified and refined as the project proceeds in design and implementation. In fact, objectives should be altered as new information and resources become available—this is a basic tenet of adaptive management. Sometimes environmental factors can influence the sampling that can be done. If snow comes early or stays late on a sampling location then a survey of down dead fuel cannot be accomplished and objectives relating to down dead woody fuel would need to be postponed or eliminated. It is more desirable to add objectives than eliminate them, but you have to recognize that some circumstances are beyond your control. Lastly, understand that there might be parts of the objective so important to the project that they absolutely must be evaluated at any cost. For instance, say you are treating a Research Natural Area where an important plant population resides and you have an objective relating to identifying and tracking changes in the plant community. It would be critical to have a botanist on the crew to accurately identify all of the plant species so that the fire effects on the community can be determined. If the botanist leaves for another job you cannot just drop this objective, as it is critical to the project. Critical objectives like these should be noted and remain unchanged in the objective statement.

If you are planning on implementing a statistically based monitoring plan, then as you write the objectives, you should also consider remarking on the minimum amount of change you want to be able to detect and the confidence level that will be used for the analysis of the monitoring data. Both values affect the sampling intensity and should be indicated for the attributes most important to the project objectives. The minimum detectable change (MDC) parameter is an absolute value calculated by multiplying the mean of the attribute of interest by the percent change of that attribute you want to be able to detect. If you want to be certain that you are detecting a 10 percent reduction in down woody debris on a site that has 25 tons/acre the MDC is equal to 2.5 tons/acre. The confidence level is a measure of the certainty in which you state your statistical results. For example, a 95 percent confidence level means that you are 95 percent certain the change you identified in your statistical analysis really happened. Or conversely, one of 20 statistical tests will note a significant change in an attribute when actually the change did not occur. Most research studies set the confidence level at 95 or 99 percent. However, monitoring studies, especially those with limited resources, might not need to be as restrictive. The confidence level should never be set lower than 80 percent. As MDC decreases and as confidence level increases, sample size will increase so you may need to balance MDC and confidence level against the sampling resources. There is a further discussion of confidence level, detectable change, and sample size in the Implementing the Statistical Approach section below.

## S.M.A.R.T. Objectives

While objectives are critical to a well-written project plan, it is clear that writing "good" objectives can be difficult. You don't want a project to be determined a failure simple because the objectives were poorly written. The acronym S.M.A.R.T. relates to five properties of well-written objectives. As you write your project objectives refer to this list to make sure they are S.M.A.R.T.:

1) Objectives must be *Specific*. They must provide a description of the precision required for the objective and link it to a rate, percentage, or some other value. See the list of six objectives listed above for examples.
2) Objectives must be *Measurable*. There must be a system in place that can measure attributes of interest. In FIREMON we have provided a number of sampling procedures. However, for some attributes, such as water quality, we do not provide a method. In such cases you must be able to determine your own sampling procedures and apply them appropriately.
3) Objectives must be *Achievable*. Make sure what you are proposing can be and should be accomplished. For example, an objective that states, "Eliminate 100 percent of the exotic, invasive plant species after 1 year of treatment," is not valid, realistically.
4) Objectives must be *Relevant*. There is no point in making an objective that your treatment will have little or no influence over. Say the agency you work for has a goal of improving air quality in the watershed where your treatment unit is placed. You may have the ability to burn on a day that will reduce the negative impacts on the air quality across the watershed, but an objective that states, "Reduce PM2.5 emissions across the watershed," is not relevant to your treatment because, through your treatment, you cannot effectively control the other sources of PM2.5 emissions across the watershed.
5) Objectives must be *Time Based*. This one is simple—you must have a date or timeframe for completion of the objective. The start time is usually intuitive because generally it begins with the application of a treatment; however, if it is not obvious, clearly state the start date or timeframe.

The subject of setting goals and objectives has been covered extensively in other texts. A quick search of the World Wide Web will help you locate them. The National Park Service and U.S. Fish and Wildlife Service fire monitoring guides are also available online and provide information for fire related projects.

## Determining the Sample Area and Spatial Stratification

Perhaps the most important element of a monitoring program is *where* the treatments and subsequent monitoring project will be implemented. A detailed geographical description of the area to be sampled is an absolute necessity because it will also provide context for design descriptions. In statistical terms this description provides the *scope* of the treatments and, in most monitoring programs, the scope of inferences made by the statistical tests. Boundaries of the entire sample area should be explicitly stated and diagrammed on an appropriate map. In most cases, large scale maps (such as National Forest maps) will not provide the detail needed for a fire monitoring effort. Maps with a scale less than 1:30,000 will do a better job of accurately delineating the project area.

The entire sampling area must be spatially divided into sampling stratifications that match the sampling objective. Most resource managers delineate areas of homogeneous vegetation (stands), but fire monitoring can be stratified by other classifications such as aspect, slope, fuel condition, or land ownership. FIREMON presents procedures for mapping areas of similar fire severity from satellite imagery (see Landscape Assessment section), and the manager can also use severity as a stratifying factor.

As you define your strata be sure to match the mapping criteria with your sampling objectives. For example, if ponderosa pine restoration is a primary objective, then be sure the strata mapping guidelines delineate various successional stages of ponderosa pine communities. The sampling design can incorporate more than one stratification factor. For instance, a possible design might be to install FIREMON plots in all old-growth ponderosa pine stands that have slopes less than 50 percent and are on National Forest lands.

Figure ISS-2 shows three ecological characteristics mapped on a sample landscape: A) three levels of tree density, B) two levels of dead and down fuel load, and C) a corridor of exotic weed invasion along

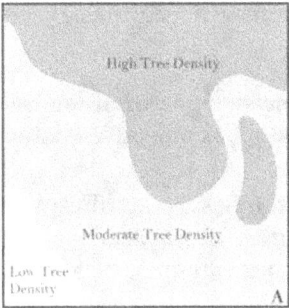 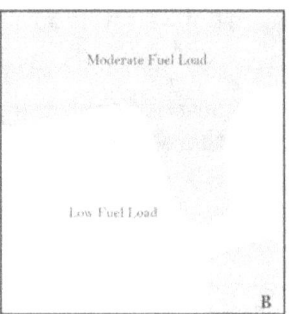 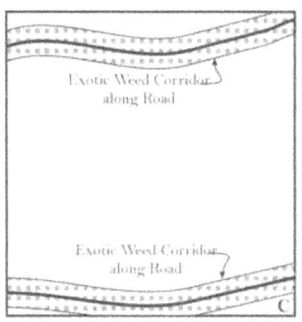 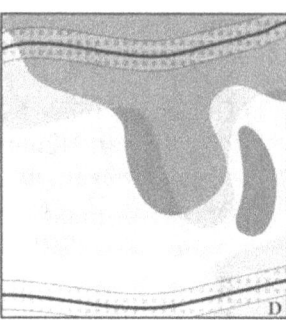

**Figure ISS-2**—Overlay maps of strata defined by the stratifying factors in your monitoring project to identify the different polygons on the landscape. Once a sampling design has been determined, the polygons will be sampled with FIREMON plots. In this figure the strata of A) tree density, B) fuel load, and C) exotic weed invasion are overlaid to identify the sample polygons in D. Each shade and/or patten combination represents a specific sampling strata. There are 17 polygons grouped into 9 sampling strata.

the roads. The levels would be determined by the FIREMON architect based on project objectives. In D, characteristics A, B, and C are combined to identify nine sampling strata divided into 17 polygons. One stratum has low tree density and low fuel without exotic weeds, another has low tree density and low fuel with exotic weeds, another has moderate tree density and low fuel without exotic weeds, and so on. Potentially there could have been 12 strata in this example (3 x 2 x 2 = 12) but not all of the combinations occurred. Note how quickly adding ecological characteristics and levels increases the potential number of sampling polygons, which in turn increases the complexity of the monitoring project. This example landscape will be used for demonstration throughout the ISS.

The mapping of sampling entities across the landscape is greatly dependent on the type of fire: prescribed burns or wildfires (*postevent monitoring projects* versus *complete monitoring projects*). The difference is that for wildfires and wildland fire use, fire effects monitoring plots are installed after the fire, whereas prescribed fire monitoring plots are measured both before and after the burn. Most wildland fire use burns (previously called prescribed natural fires) fall into the postevent category because of the absence of preburn plots. In these cases, sample stands must be identified after the wildfires using remotely sensed images (aerial photos or satellite imagery) taken before the fire if fire effects measurements are to be summarized by vegetation type. If fire severity stratification is necessary the Landscape Assessment methodology can be used.

The mapping effort, and its integration with FIREMON sampling efforts, can be made much easier if the mapping and analysis are done within a Geographical Information System (GIS). A GIS allows complex queries on landscape and stand attributes that make design and subsequent implementation of a FIREMON sampling strategy efficient. A GIS can produce maps of the sample area for reference and navigation, and the sampled FIREMON field data can be linked to the GIS for many other applications (landscape pattern analysis, satellite imagery mapping).

Two statistics must be computed once the sample area has been mapped and the landscape divided into polygons. First, compute the total treatment area of the study site(s). Exclude all areas that will not be sampled (talus slopes, lakes, glaciers) from the estimate. Then, compute the number of polygons or stands within the area to be sampled. These statistics will be used to determine the resources needed to accomplish the sampling.

The sampling environment, like the project goal and objectives, provides the spatial and logistical sideboards for project planning. There are four attributes about the sampling area that must be known before sampling design can continue: 1) size of area, 2) topographic complexity, 3) transportation network, and 4) ecological characteristics. The size of the sampling project is often dictated by the boundary of the burn, and burn boundaries are notoriously coarse, so it is important that a precisely developed burn map is provided for monitoring. Topography will dictate many aspects of the sampling effort. Steep, dissected landscapes will be difficult and dangerous to navigate, so the sampling project should be designed to accommodate or avoid these troublesome conditions. The network of roads, trails, and navigable terrain will provide the means of transporting crews to sampling areas. Remote areas with only trail access will require another level of planning because crews will probably need backcountry supplies along with the already extensive sampling gear, and this may require packstock support (mule and horse packing). Last, the ecological characteristics of the sample area will dictate the sampling design and methods. Forested environments will probably require time-intensive individual tree surveys, while rangeland types can be sampled using standard vegetation surveys. Areas with thick vegetation or high fuel loadings will be difficult to traverse. And areas with abundant threatened and endangered species will require a high resolution sampling design to properly evaluate fire's impact in small but highly valuable habitats.

## Determining Sampling Resources

The details of the FIREMON monitoring project design are determined by striking a compromise between cost, personnel, time, logistics, and sampling environment within the context of the project goals and objectives. For example, say that monitoring on the Clear Creek burn is essential to determine tree mortality and subsequent potential for salvage logging. The project goal might read, "Determine tree mortality and salvage potential." This statement provides critical information to determine what and how to sample for monitoring and evaluation. Obviously, the project goals aren't related to weeds, grazing, or fuel consumption, so sampling techniques that measure plant cover, plant biomass, and fuel loadings are not needed. A tree population sampling method is most appropriate here. Next, say there is limited funding, and the only people available are the fire crew, and there are only 3 weeks to perform the monitoring tasks. This means that a stratified random sample across the entire burn is inappropriate because it would cost too much and take too much time. However, a relevé approach might be the right compromise between sampling resources and project desires. Because the data must be used for two purposes—to determine tree mortality and the amount of timber in those trees (salvage potential)—a detailed, individual tree sampling method is warranted.

The FIREMON architect should consult with the FIREMON project manager to determine the exact amount of resources available to conduct the fire monitoring project. There are four types of resources that should be evaluated: 1) funding, 2) personnel (number of people and their expertise), 3) logistics, and 4) time. All of these resources are somewhat related, but each resource should be carefully appraised to determine its contribution to the monitoring project in the context of the extent and complexity of the sampling area.

Funding is easily the most important sampling resource because it dictates the level of all other resources. It is critical that the FIREMON architect knows the exact amount of money dedicated to the monitoring effort. This will help determine the number of people to hire, the number of vehicles to acquire, and the quantity and number of supplies to purchase. In short, funding often determines sampling intensity.

The number and qualifications of people to use in the FIREMON project is an important resource for the monitoring project. It is essential that the skills of the FIREMON field crew match the level of detail of the data to be collected. For example, the monitoring of plant species cover change requires a field botanist who can consistently and comprehensively identify vascular and nonvascular plant species. It is also important that the field crews have sufficient training in FIREMON methods and techniques. A poorly trained crew will invariably spend excessive amounts of time and money collecting questionable data that will be useful to no one. As funding often dictates sampling intensity, the experience and capabilities of the field crew will determine the quality of sampled data.

It is important that the logistic capacity of a FIREMON project be identified prior to designing a sampling project. Critical elements are 1) the number of available vehicles, 2) the amount of sampling equipment, 3) the amount of camping gear (if needed), and 4) computer equipment. Often, available vehicles and equipment can limit the staffing of monitoring projects. Required sampling equipment (compasses, clinometers, GPS units) must be available or rapidly and easily purchased. Maps of the sample area are absolutely essential for conducting a successful monitoring project. Laptop computers may also be used for data entry and reference to the FIREMON methods in the field. Logistical support determines the sampling ability.

The amount of time available to conduct the monitoring project can, in some circumstances, dictate the level of other resources. For example, it may be critical to establish monitoring plots across a large burn to determine appropriate levels of rehabilitation. To accomplish this objective, the sampling must take place directly after the burn and before the snow flies. This does not leave abundant time to mobilize extensive field crews and acquire new equipment and vehicles. Projects on fast timelines may need to forego extensive, statistically valid sampling designs in favor of relevé methods. The amount of time dictates the schedule of a sampling project.

The availability of funding, personnel, logistics, and time should be explicitly stated in the FIREMON field notebook. Obviously, the status of any of these resources can change; a good FIREMON architect will ensure there is plenty of flexibility in the sample design to accommodate changes in available resources, whether the changes are good or bad. There may be other resources or challenges to be included in the design of the sampling effort that are not mentioned here; for example, weather. Excessive rain or heat may hamper the sampling productivity of crews.

## Determining Sampling Design

When the major sampling resources have been identified and described they will be summarized into the FIREMON statistics that are used in the sample design keys. Be sure to document the calculation of the sampling resource statistics in the monitoring notebook. A number of these parameters will be hard to estimate when you first start your monitoring activities but will be easier to determine as you gain experience. When possible we have provided some guidelines for your initial values. When making your own estimates use your best judgment and be realistic about the numbers you chose. Remember, monitoring almost always takes longer and costs more than you think it will.

### Calculating FIREMON sample statistics

The first FIREMON sample design statistic that you will be calculating is the *Sampling Potential (SP)*, which is used to indicate the number of standard plots that can be installed during the sampling effort. This statistic integrates most sampling resources into one index that can describe the capacity to perform the monitoring project. SP is a function of project funds, crew costs, and plot production rate. Crew costs and plot production rate will probably need to be estimated.

You should first determine the amount of money available to conduct the entire monitoring effort *(Project Funds or PF)*. This amount should include salaries of existing personnel available to work on this project, including the FIREMON project leader and architect.

Next, estimate the *Crew Costs (CC)*. If necessary, include the cost of renting a vehicle for the period of sampling. Assume one vehicle will transport two people for the project. Calculate how much it will cost to outfit each crew with supplies, and if this figure is unknown, use $250.00. You will need to estimate the number of 8-hour workdays available to finish the monitoring project. Provide an estimate even if there appears to be plenty of time to finish the project. Use a target start and end date as a guide; try to identify a realistic day for starting the project and ending the project, then count the number of working days in between. Estimate the number of days that could be lost to inclement weather, if that is a possibility, and days lost to organizational, administrative, logistical, or personnel problems. In lieu of this information, add an additional 10 percent time to the project. Using your estimate of workdays, estimate the salary of one crew person for the duration of the monitoring project and multiply by the number people you plan to have on the crew. Finally, determine the CC by adding the transportation, equipment, wages, and any other expenses together and dividing by the number of workdays.

Last, *Plot Production Rate (PPR)* is an estimate of the number of plots that can be sampled per day by one crew. Early in the monitoring process PPR is probably unknown so use the rate of four plots per day

Integrated Sampling Strategy (ISS) Guide

as an estimate. In our experience the major factor in PPR is crew transport to the sample sites, not the actual sampling.

When you have determined PF, CC, and PPR use the following formula to calculate SP:

Equation ISS-1
$$SP = \frac{(PF)(PPR)}{CC}$$

where SP is the sampling potential (plots per crew for the entire project), PF is project funds in dollars, CC is crew costs in dollars per day, and PPR is plot production rate in plots per day per crew.

Here is an example of estimating SP: Assume that a manager has $3,000.00 to spend on installing monitoring plots the first year of a project. He has identified four people that could work on the crew and, after reviewing the other tasks that need to be done during the season, apart from this monitoring project, he notes that there are 10 work days that can be spent on the project. A best guess from past experience tells him that the crew can get about six plots done per day. All equipment and transportation is on hand, so the only expense is the crews' wage. Adding all of the wages together he notes that the crew costs about $400.00 per day. Dividing $400.00 into $3,000.00 he finds he will get 7 days of sampling in before all of the available funds are spent. The final calculation of SP is:

$$45 \approx \frac{(3000)(6)}{400}$$

Thus, there are resources to sample about 45 plots. If there is a question about the ability to assess the effectiveness of treatments using data from 45 plots, then the revision of project objectives, sampling methods, crew size, and/or the scope of the area to be tested with statistical inference must be revisited to bring the sampling intensity in line with the project objectives. See the Considering Tradeoffs section for more information.

Note that in this example the number of available workdays was not the limiting factor in calculation of SP, so it was not taken into account (the manager potentially had 10 days for sampling but the project funds only allowed 7 days of sampling). If there were only five days available for sampling then the available workdays *would* enter into the calculation, further lowering the SP. Only about 30 plots could be sampled (5 days X 6 plots/day).

From this simple example it is clear that part of the art of monitoring is balancing all of the components in the monitoring program so that the data collected are useful for assessing the treatments. Almost every component, including the objectives, sampling crew, sampling approach, sample size, monitoring area, sample stratifications, and study area can be modified to bring the monitoring data in line with the project objectives.

When you start your fieldwork, divide the samplers into crews making sure that the number and expertise in each sampling group are appropriate to the monitoring tasks. For example, don't send out a crew and expect them to collect species level data accurately without a good botanist in the group.

*Calculating the number of polygons to be sampled*

The next set of statistics attempts to quantify the amount of sampling required for the area to be monitored. First, determine the total size of the *Sample Area (SA)* in acres. This is often all the area within the burn boundary. Next, compute the *Number of Polygons (NP)* that compose the sample area. If you have determined your polygons by overlaying the sampling strata as shown in figure ISS-2, then this is as simple as adding up the number of polygons. If you are interested in sampling by stands within a prescribed fire, they were probably mapped prior to the fire treatments being set up and just need to be summed across the sampling area. For wildfire situations, NP can be taken from stand maps, satellite-derived vegetation cover type maps, or burn severity maps created prior to the fire. However, this type of spatial data will not be known for many monitoring projects. If NP is not known for your project, you can estimate it using average stand size. In the Northern Rockies we have commonly estimated the average stand to be 25 acres (10 ha) and used that value to calculate NP (NP = SA/25). This method assumes that the sampling will be stratified in space by stand characteristics (tree species, diameter, height, and so on). If this is not the case for your project, divide the landscape into the appropriate homogeneous sampling units. For example, burn severity polygons mapped from satellite imagery may be your spatial stratification. Or it could be treatment blocks within stands. The NP statistic is the number of spatially explicit sampling polygons in your project.

Sometimes there is no need to sample all of the stands or polygons in the sampling area. For example, monitoring plots might only be needed on steep areas where rehabilitation efforts will be prevalent. Or perhaps only forested areas need to be sampled to monitor tree establishment after wildfire. In these cases, calculate SA or NP only for those areas that are targeted for monitoring.

The NP and SP statistics are used to determine the suggested sampling approach and sampling intensity level. However, these statistics, and the statistics used to calculate them (CC, PPR, SA), can be modified to refine sampling strategies to fit the monitoring objectives or to generate several sampling alternatives for strategy design. For example, the sampling rate (PPR) can be doubled or halved to produce best and worst case sampling scenarios. Or NP or SA can be reduced or increased to match the sampling potential.

## Determining the sampling approach

The next important step in designing a FIREMON sampling strategy is to decide on a sampling approach to collect fire effects data. The two basic FIREMON approaches are *relevé* and *statistical*. Both these approaches can be stratified by any landform, ecological attribute, or disturbance characteristic. The selection of the appropriate approach will dictate nearly all other sampling details.

The **Sample Approach Classification Key** provides the guidance to select the approach to match your project. Read down the list of statements for each approach answering "yes" or "no" to each item that is important to the sampling project with special reference to your monitoring objectives, sample area, and available resources as mentioned above. Simply count the number of yes answers and no answers in each list. The approach with the most yes answers is the approach that probably should be used in the monitoring project. However, this checklist does not include all the subtle advantages and disadvantages of each approach with respect to your unique sample area. There may be other important elements that will influence your final decision. Be sure to document these special conditions in the FIREMON notebook for future reference.

In general, the *relevé* approach is used when time, money, or personnel limitations require the sampling to be done quickly but without compromising the temporal aspects of monitoring at the stand level. However, it is important to recognize the limitations of the relevé approach. First, the within-polygon variation is not quantified, so a statistical comparison across polygons (comparing one polygon to another) is not valid. Next, the variation of ecosystem elements, such as fuels, trees, or plants, is not measured across space, so a statistically valid landscape comparison is not possible. The only possible statistically valid comparison using the relevé method would be the comparison of one single plot measured at two time periods using FIREMON methods that captured within-plot variation. The independence of microplot samples is suspect, however, so temporal inference may not be appropriate. There may be an inclination for managers to use relevé plots because of the difficulties in estimating the variance of one or a number attributes, but if statistical inferences are to be made, estimates of variance for the most important entities (at least) need to be made so that the statistical approach can be applied.

The *statistical* approach is used when the answers obtained must be compared using statistically valid procedures across space and time. This is the most useful and most commonly applied approach for monitoring projects.

## Selecting the level of sampling intensity

The next step in sample design is to determine the level of sampling intensity for the monitoring project. Sampling intensity is usually described by the number of plots located across the project area and is related to the amount of variance to be explained. The more plots established across the landscape, the more likely that the range of response and variance has been captured in measurements of fire effects. This translates to more accurate and defensible comparisons and evaluations.

FIREMON has three levels of sampling intensity integrated into the sampling strategy to facilitate sampling design. These levels are intended only as guides for the inexperienced designers and not as recommendations. The FIREMON architect can design a monitoring project at any intensity level, not just the three mentioned here. Refer to the **Sampling Intensity Key** to decide which intensity level best fits your situation.

## Choosing sampling methods

This section describes how the FIREMON architect selects the sampling methods to employ at each plot during the project. Many people think this is one of the most difficult parts of a monitoring study, but in fact, choosing sampling methods is straightforward because you simply match the monitoring objectives to the attributes that need to be measured. Many managers get confused in sampling methods selection because of the complexity and diversity of sampling procedures available, but the selection process becomes simple when the decision is put in context of the objectives.

Methods for measuring fire effects are selected from the **Methods Classification Key** provided in FIREMON. Read each bullet in the Methods Key then refer to each of the project objectives to see if the bullet is true, and if so, employ the suggested method. Again, this key is intended as a guide and not a prescription. Use your own intuition and experience to modify results from the key to fit your special circumstances. FIREMON has been developed using established methods. Occasionally you may find that there is not a method that will assess the success of some objective. For example, there is no water quality sampling method in FIREMON. Thus, methods may need to be developed to monitor some attributes. These should be explicitly described in the MD table so that the exact procedure can be applied at the next sampling visit. Optionally, you might be able to add fields to existing methods to meet the objectives. For instance, if the Wildlife Biologist is interested in the presence of snag cavities, a field could be added in the Tree Data (TD) table of the database. (Use caution when adding fields to the FIREMON database, as it will make it difficult to merge your data with other FIREMON data.)

You will find that most of the time and money spent on field campaigns are in transporting crews to sampling areas and not on actual sampling. Therefore, it is often prudent to sample additional attributes at the FIREMON plot to strengthen monitoring analyses and to widen the scope of the monitoring project. This is especially true if the FIREMON architect is wondering whether or not to sample a particular attribute. It is much better to spend an additional 10 to 20 minutes on the plot sampling another fire effect, than it is to be frustrated because some component wasn't measured at the end of the sampling effort. For example, measuring crown characteristics for every tree on the macroplot may seem excessive if the sampling objective is to assess tree mortality, but those crown characteristics (percent crown scorch, tree DBH, height) could be used to develop salvage guidelines from percent crown scorch or predict crown fire potential using NEXUS (Scott 1999).

The process described in this section provides the FIREMON user important information on the elements of a sampling project that can be modified to fit the monitoring objectives. It is best to first compute all sample statistics from real data and then key the sampling approach, intensity, and methods. Then compare the key results to the monitoring objectives again to evaluate if the key results are appropriate. If not, go back and modify one or more sample statistics to achieve a more realistic result. Of course, if you modify a statistic, you must then implement that modification in the sample design. For example, if you reduce NP then you must make sure that the correct number of polygons are mapped. Alternatively, the FIREMON architect may reconsider the scope of the sampling (sampling a different, usually smaller, area than originally proposed) or reconsider the project objectives. This may result in fewer plots and/or a smaller area being sampled, but doing so could improve the quality of the collected data. Experienced sampling crews will be able to determine the most sensitive and important statistics and ensure that these attributes are well represented in the sample design.

## Considering Tradeoffs

The design of the sampling strategy is a constant tradeoff between statistical significance and logistical feasibility. The only way to have both is with sufficient funds, experienced personnel, and ample time. Unfortunately these three factors rarely coincide; therefore, a compromise in sampling rigor is usually necessary. The compromises of your FIREMON sampling design should be recorded in the FIREMON notebook or MD information to ensure the data are never used for inappropriate purposes.

Careful consideration should be given to assess whether or not the approach identified in the sampling approach key will actually accomplish the sampling objectives. For example, say one sampling objective is to quantify significant reductions in fuel loadings after a prescribed burn, which absolutely requires a statistical approach, but limited funds and personnel lead you to identify a relevé approach in the FIREMON sampling strategy design. The relevé approach will only provide qualitative descriptions of changes in fuel loadings and will not detect statistically significant differences in loading after treatment. So either the sampling objective must change to remove the statistical requirement or the statistical approach must be implemented.

The main limitation of the relevé approach is that it does not provide for any analysis of statistical significance, even if multiple relevé plots are established on a sample site. The subjective location of the relevé in a representative portion of the stand biases the sample and does not fulfill the assumption of randomness needed for classical statistics. Relevé plots are used only for qualitative reasons or descriptive purposes and should never be used to quantify changes in ecosystem characteristics. However, relevé methods allow for efficient collection of complex data over large land areas with limited resources.

The main limitation of the statistical approach is its high expense in time, funds, and personnel. A good statistical sample requires multiple plots in homogeneous areas (landscape stratification) and that often necessitates extensive resources, especially if many entities are being measured such as fuel loadings, tree populations, and vegetation cover. However, the statistical approach is *required* if changes in ecosystem characteristics must be quantified with some test of significance.

There are many ways to compromise sampling rigor with logistic restrictions and settle on a tradeoff between statistical validity and general description. If a statistical approach is necessary but time and funds are limited, it may be possible to reduce the number of entities being sampled. For example, modify the design to measure only fuel loadings and do not sample tree populations and plant cover. Or if a large landscape is treated, it might be possible to statistically sample a small representative area within the large area and sample the remaining areas with a relevé approach. Recognize that the statistical sample could only be used to quantitatively describe changes in the representative area and that extrapolation of those results to the large area would be highly questionable, especially if not supported by data collected from the relevé plots. Optionally, it may be possible to aggregate the sampling strata to minimize the number of sampling areas. And of course, there is always the possibility of optimizing sampling efficiency by cluster sampling around accessible locations. In any case, be sure to document the limitations of your tradeoffs so that others will not use the data for inappropriate analyses.

Be aware that all analyses using the FIREMON software package requires multiple plots within a sample strata to calculate an acceptable measure of variation for statistical tests of significance. If only one or two plots are established in a stratum, then the data can be used only for descriptive purposes, which may not be compatible with the statistical sampling objective. In short, the usefulness of your monitoring data increases with statistical validity. The more samples you collect in a treatment area, the higher the value of that data to other resource efforts. The FIREMON software does not generate statistical tests of significance for one plot across multiple monitoring visits, even if multiple microplot, transect, or belt techniques were used to quantify variation at the plot level. In FIREMON, the only way to calculate a variance is with multiple plots.

## Monitoring Prescribed Burn Projects

Ideally, every sampling polygon will be monitored to sufficiently track treatment effects. Identifying the polygons should not be a difficult task for a number of reasons. First, conventional prescribed burn projects require intensive planning and public involvement, and as a result, there is plenty of documentation and data on the area to be treated, including maps, stand delineations, treatment block delineations, and supporting stand and historical fire data. Also, because the objective of most prescribed burns is contingent on preburn conditions and the burn boundary is known prior to the fire, postburn mapping is generally not needed. Lastly, prescribed burns are usually small in area compared to wildfires or wildland fire use. Use the following sample approach guidelines to monitor your prescribed fire project.

**Relevé Approach**—Establish one FIREMON plot in each sample polygon in the prescribed burn area. Locate the relevé in an area that displays the typical ecological conditions within the stand. If you find unique features that you think should be sampled, such as seeps, dense thickets, and pockets of snags, you should locate another relevé for sampling those areas. In FIREMON, we generally do not consider the relevé approach appropriate for monitoring prescribed fire treatments.

**Statistical Approach**—Sample size equations are provided in **Sample Size Determination** section. If you have a variance estimate, use these equations to determine the appropriate number of samples. If the variance is unknown, plan on establishing at least five plots in each sample polygon in the prescribed burn area. Sampling fewer than five plots results in variance estimates that are suspect. If the sample polygon exceeds 50 acres, then establish another plot for every 5 acres in the polygon.

## Comparing Objectives and Sampling Design

Make sure you haven't specified any inconsistencies in your sampling strategy criteria. For example, the use of the relevé method at the Detailed sampling intensity level (Level III) is not correct or logical. Why use a descriptive method when you have plenty of resources to quantify the range of response and variation in fire effects measures? If you specified Level III intensity, then use the statistical approach. As always, these recommendations are provided to help you work through your design process, not as strict rules. There may be an occasion where the project objectives lead you to intensively sample a landscape but with few measurements taken on every plot. This would lead you to a "low" level of appropriateness in table ISS-1. However, if lots of plots measuring few attributes let you assess your objectives and stay within budget, then you have picked the right sampling scheme.

Finally, take one last thorough look at the project objectives and compare them with the sampling design that you have developed. Ask yourself if you will really be able to assess the effectiveness of the treatments using the sampling design that has been developed. If possible, have others familiar with the project review the monitoring plan. They may identify some shortcoming that you have missed. When you have decided on the approach, methods, and intensity you believe are appropriate for your project, record that information in the FIREMON notebook or MD table.

## DESIGNING A FIRE MONITORING PROJECT

This section is designed to help the FIREMON architect develop the integrated sampling strategy for a monitoring project using FIREMON sampling design, techniques, and field methods. This section is organized according to the design criteria that were determined in the previous section. If you haven't done so already, use the **Sample Approach Classification Key** and **Sampling Intensity Key** to identify the most appropriate approach and intensity for your monitoring project. After reading **Determining Polygons and Building a Summary Table** proceed to the section below that best fits your strategy criteria selections—relevé or statistical.

Table ISS-1 Appropriateness of sampling intensity levels by sample strategy. High indicates the sampling scheme is highly appropriate; Moderate indicates moderately appropriate; Low means somewhat inappropriate; and Inappropriate means the sampling design is not suggested.

| Approach | Level I-Simple | Level II-Alternative | Level III-Detailed |
|---|---|---|---|
| Relevé | High | Moderate | Inappropriate |
| Statistical | Low | Moderate | High |

## Determining Polygon Locations and Building a Summary Table

Regardless of the sampling approach you use, monitoring plots must be located across the project area so that the data collected will be useful for assessing the treatments applied. In FIREMON we propose using the project objectives to determine sampling strata then overlaying the strata to identify polygons. By ordering the sampling sequence of the polygons based on the most important project objectives, fire managers will have the best chance of collecting useful data for the project.

Each polygon on the sample landscape must be described by one or more ecological attributes for the prioritization method to work. *It is essential that the attributes used to describe the polygons and/or strata be consistent with the monitoring objectives.* Using the example ISS landscape, the attributes would be tree density, fuel load, and weed invasion. It is best if the attributes you use are stored in a GIS for digital map analysis, but it is possible to do the entire exercise using spreadsheets or simple pen-and-paper analyses.

Some stratification attributes may be secondary to the objectives but important to the project. Any site characteristic that influences fire behavior—for instance, slope—may be important to factor into the monitoring project because the fire could influence the monitored attributes in different ways depending on the behavior.

A good way to determine the stratification criteria is to create a summary table (table ISS-2). The columns in the table are the stratifying factors with polygons ranked by factor levels in the rows. The example summary table was developed by overlaying the strata, numbering the polygons, and grouping them by stratum (fig. ISS-3).

Next, determine the prioritization attributes so that the important polygons get sampled. The prioritization attributes will probably be related to the list of stratification attributes. You can make the prioritization as simple or as complicated as you like. For instance, you could decide that only stands with ponderosa pine cover types will be sampled, or you could decide to sample mature stands of shade-intolerant cover types on steep slopes for each habitat type. Be sure to keep your prioritization flexible enough so you can easily modify the selection criteria.

In the **Implementing the Relevé Approach** and **Implementing the Statistical Approach** sections below, tree density, fuels, and weeds are the prioritization attributes. The prioritizations are described further in each example.

**Table ISS-2** Develop a summary table to identify the number of polygons in each stratum. Label the first columns using the stratifying factors and list each combination of levels in the rows below. Give each stratum a unique name or code and in the last column list all that polygons that belong in that stratum. Include a table like this in the FIREMON project folder for future reference.

| Tree density | Fuel load | Exotic weeds | Stratum code | Polygon numbers |
|---|---|---|---|---|
| Low | Low | No | LLN | 1, 3 |
| Low | Low | Yes | LLY | 2 |
| Low | Moderate | No | LMN | N.A. |
| Low | Moderate | Yes | LMY | N.A. |
| Moderate | Low | No | MLN | 11 |
| Moderate | Low | Yes | MLY | 4 |
| Moderate | Moderate | No | MMN | 5, 7, 9, 12 |
| Moderate | Moderate | Yes | MMY | 6, 13 |
| High | Low | No | HLN | 8, 10, 16 |
| High | Low | Yes | HLY | N.A. |
| High | Moderate | No | HMN | 15, 17 |
| High | Moderate | Yes | HMY | 14 |

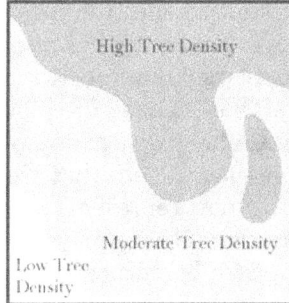 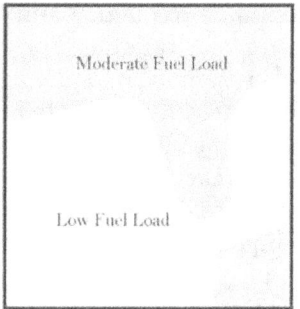 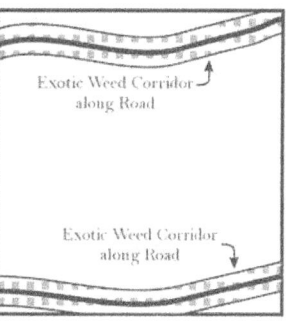 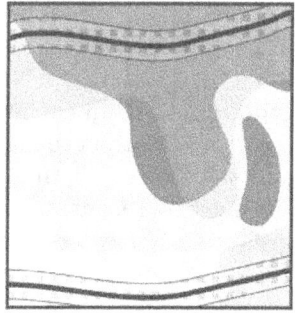

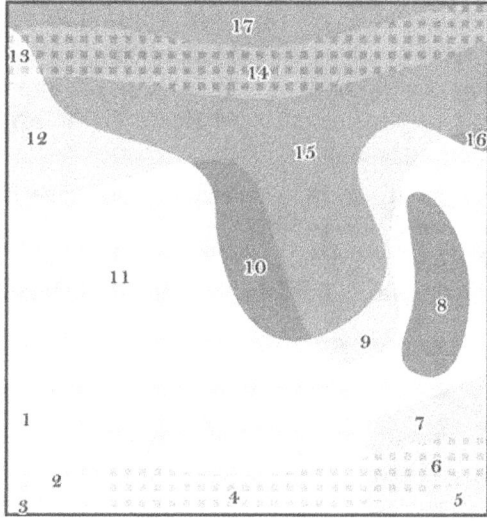

**Figure ISS-3**—Build a summary table by overlaying the stratifying factors to develop polygons, then number each polygon and note the stratum it belongs in. You should include a figure like this one in your FIREMON notebook for future reference.

## Implementing the Relevé Approach

### Background

The relevé approach requires that plots be located in a representative portion of the sample stand or polygon without preconceived bias. Data measured on relevé plots are not used to quantify the variation across the stands or polygons, but rather to provide a general description of the polygon and provide a baseline measurement of monitoring ecosystem characteristics for that polygon. The assumption in relevé sampling is that the plot is representative of a larger area (stand or polygon), and therefore conditions measured at the plot can be used to describe the stand or polygon as a whole. Thus, any fire effects measured on a plot can be used to describe fire effects across the entire polygon or sampling stratum. Two drawbacks of the relevé method are 1) the measured effects cannot be statistically compared, spatially, between polygons on a landscape and 2) extrapolation of relevé data across a polygon or stratum is controversial due to the subjective placement of the relevé and inherently high variability of ecological attributes.

### Polygon selection

Ideally, each stand or polygon on the landscape will have at least one relevé. This is typically not possible because some polygons may be inaccessible, or resources limit the ability to sample the entire landscape.

Therefore, a compromise must be struck between sample frequency and logistics so that the most important polygons can be sampled given the resources available. For relevé sampling the most important factor is ensuring that there is adequate representation in plot frequency so that important conditions within each sampling stratum can be summarized in reports or databases. Base the plot frequency on the area in each of the sampling stratum. For instance, if stratum A has twice as many acres as stratum B, then stratum A should have twice as much sampling. Sample the polygons that are the most representative of the sampling stratum and locate the relevé plots in the most representative area of each polygon. Remember that if you locate unique or unusual characteristics within a polygon, such as a site containing rare plants, they should also be sampled and noted as being not really representative of the entire polygon but used for monitoring unique attributes.

Figure ISS-4 shows how relevé plots might be distributed if put into the most representative portions of each polygon of the sample ISS landscape at the Level I-Simple sampling intensity. Note that not every polygon or even sampling stratum is being sampled. Each weed corridor is sampled with only one relevé, even though there are five possible strata that could have been sampled within the exotic weed stratification. Thus, this example assumes that the FIREMON architect determined that for some objective-based reason weed sampling was not a priority.

Sometimes it is more efficient to group sample locations close together, usually around an easily accessible point, rather than randomly selected throughout the landscape (fig. ISS-5). These "cluster plots" minimize transport time, thus they can be more efficient than using random or systematic sample selection of polygons, especially across large landscapes. The main disadvantage is the introduction of

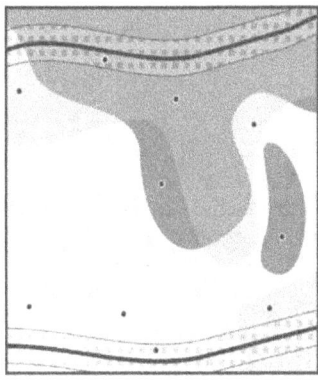

**Figure ISS-4**—this illustration shows plots located using the relevé approach at the Level 1-Simple sampling intensity. There are 10 plots distributed across the sample area. The exotic weed strata are not well sampled.

Relevé Simple

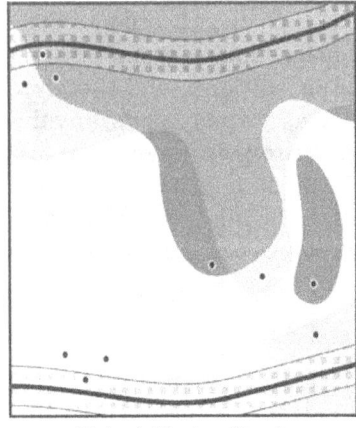

Relevé Cluster: Simple

**Figure ISS-5**—These 10 relevé sampling sites are clustered into three groups to increase sampling efficiency while at the same time getting a good spatial representation across the sample area.

bias because you are using only a small portion of the polygon to find a representative location for the relevé—there might be a more representative location, but it is outside the of area where you want to locate you cluster. Beyond the benefit of reduced transport times, cluster plots allow you note juxtaposition relationships of neighboring polygons.

Make an attempt to distribute clusters geographically around the landscape using transportation routes (roads, trails, rivers) as cluster centers. Cluster plots do have an element of subjectivity in the placement of cluster centers, but when sampling resources are low, cluster selection is a valuable alternative to random or systematic selection of sample stands.

## Relevé establishment

Relevé plots are established by navigating to a sample polygon and then visiting various parts of the polygon to find the range of vegetation and biophysical conditions. After examining the polygon the FIREMON crew leader will determine the location of the relevé. The vegetation and biophysical conditions inside the relevé must comprehensively describe conditions across the entire polygon. Representative conditions should be assessed from a wide range of ecological attributes. First and foremost, the relevé should represent the conditions of the polygon that are important to the project objectives. In the example illustrations (fig. ISS-4 and ISS-5) each of the relevé plots were placed in a spot that represents the tree density, fuel load, and weed conditions of the polygon (the stratifying factors presented in fig. ISS-2). Secondly, relevé plots should be located in areas of the polygon that reflect the characteristics of the entire polygon for attributes not related to the stratifying factors. For instance, if the majority of your polygon is gently sloped you would want to avoid locating your relevé in a steep draw because the fire behavior and fire effects there would not be typical of the polygon. Most of the secondary considerations, such as topography, fuel conditions, and disturbance history, will be related to fire. More specific examples are slope, slope position, aspect, elevation, fuel load by size class, fuel condition, fuel model, insect and disease damage, and past fire effects. Note that there may be more than one of location inside the polygon that is appropriate for sampling, and if so, the FIREMON crew leader needs to choose the one that will lead to the most representative sampling given the resources.

The procedure used to establish relevé plots has always been embroiled in controversy. Locating a plot in a representative portion of the polygon without preconceived bias is part fantasy, part science, and part guesswork. The result is that most plot locations will contain some element of sampler bias. However, in complex ecosystems with high spatial and temporal variability the relevé method is generally a simple, efficient, and tenable sampling approach. A good mitigation measure to minimize bias and subjectivity is to mark a plot location, then randomly choose a direction (you can use the second hand on your watch) and place the plot center 50 feet away along the randomly selected direction. Of course, this procedure could lead you to establish the plot outside the representative portion of the polygon you are interested in sampling, in which case you would need to try another random offset.

Crews may encounter a wide diversity of ecological conditions within one sample polygon, making it difficult to locate the relevé plot in an area of representative conditions. In these cases, if it is possible, divide the polygon and put a plot in each division. If the resources are available, crews should also establish plots in areas that are atypical of the polygon as a whole so that those unique sites an also be monitored over time. For example, small seeps, blowdowns, or benches may be included in a polygon because of the coarseness of stand mapping (the fine scale attributes were not discriminated out due to the scale of the mapped attributes). However, these special features should be sampled if there are enough resources. Crews should give sampling priority to features that are important to the project objectives.

The main concern with using relevé plots is to know their weaknesses, strengths, and applications. Relevé plots do not allow a statistical comparison across polygons or strata because of the lack of a spatial measure of variability for sampled ecosystem characteristics. Relevé plots are best used as descriptions of polygons that compose a landscape. Monitoring results from relevé plots cannot be extrapolated across space for statistically valid comparisons because of the missing variability measure.

However, relevé plots are appropriate for broad descriptions of ecological attributes within sampling strata and polygons.

## Using the Simple sampling intensity (Level I)

This sampling level assumes the number of FIREMON plots that can potentially be established in the monitoring effort (SP) is one-half or less than the number of plots needed to sample the entire landscape (NP). This level is often used with the relevé approach when monitoring is needed but there are few resources to complete the project. Level II will give you information about more ecological attributes (you can sample more polygons), but commonly Level I is the most realistic due to resource constraints.

The goal of this sampling scheme is to sample those stands or polygons that are the most important to fire effects monitoring. This can be difficult because often there isn't enough time, personnel, or funds to sample all of the important polygons. The key to a successful monitoring effort for this scheme is to prioritize those stands on the landscape that need sampling and sample them in order of priority. This means that the FIREMON architect must balance distribution of the plots across the landscape and accessibility with importance to management—a difficult task. Detailed below is a method to select sample polygons using the polygons identified in table ISS-2 above. The architect can vary the theme or strategy to fit local circumstances. Again, this is not a rigid procedural step, but rather a flexible framework for sampling design modification. It is important that a stratification system be explicitly stated and recorded in the FIREMON notebook.

Create a list of polygons to sample, ordering your sampling polygons based on your stratification and prioritization criteria. This list can be generated from a GIS or from a spreadsheet. Remember that this list is spatial in nature and does not take into account proximity and adjacency to other stands unless specifically designed and sorted.

If we use the ISS example landscape stratification described in the **Determining Polygons and Building a Summary Table** section then select two prioritization attributes—1) fuel load, especially in moderately loaded areas, and 2) exotic weed invasion, in that order—we can make up an ordered list of polygons for sampling (table ISS-3). In this example we will assume that SP is equal to nine. Figure ISS-6 shows how the plots might be located across the project area. For clarity, in this example tree density has also been used to order the polygons. However, tree density is not a prioritization attribute, so polygons that have the last two letters of the stratum code in common ("MN" in "MMN") could be sampled in any order. Sampling the polygons that are closer together first, will lower the sampling time.

We recommend that you have a list or map of polygons by stratification attribute in order of priority for field crews so that they can alter sampling if obstacles are encountered in the field. Remember to allow enough flexibility in the prioritization and selection process so that the field crews can modify sampling

Table ISS-3  Sites with moderate fuel load and weeds were used as prioritization attributes, then the strata in table ISS 2 were ordered to identify the polygons for sampling.

| Prioritization attribute 1 | Prioritization attribute 2 | Stratum code | Polygon number | Order of priority |
|---|---|---|---|---|
| Fuel load moderate to low | Exotic weeds | HMY | 14 | 1 |
| | | MMY | 6 | 2 |
| | | MMY | 13 | 3 |
| | | LMY | N.A. | |
| | | HLY | N.A. | |
| | | MLY | 4 | 4 |
| | | LLY | 2 | 5 |
| | | HMN | 15 | 6 |
| | | HMN | 17 | 7 |
| | | HLN | 8 | 8 |
| | | HLN | 10 | 9 |

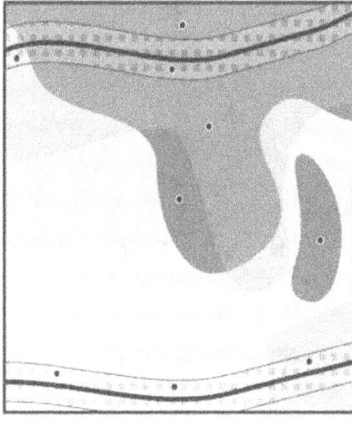

Relevé, Level I

**Figure ISS-6**—Nine relevé plots have been established in the most representative portion of the highest priority polygons in this example of the Level 1-Simple sampling scheme.

if problems arise. Examples could be high elevation snowfall or dangerous snags falling in a sample polygon. A table like ISS-3 should be included in the FIREMON notebook.

The low intensity of this sampling approach begs a cluster selection tactic to minimize transport times and maximize efficiency. In other words, the next prioritization criteria will often involve stand proximity. Those who know how to use GIS analysis can use GIS software routines to assign proximity measures to each stand based on transportation and prioritization attributes. For others, a color or grayscale map detailing prioritization and stratification attributes for each polygon overlaid on road, trail, and transportation route layers can be used to determine cluster centers and sample polygons.

The spatial prioritization criteria may be as simply stated as "sample all stands within 1 km of a road junction" or as complexly detailed as, "Sample all pole and mature stands within $\frac{1}{2}$ mile of a road junction that contain oak and hickory on slopes greater than 10 percent." Again, the selection of sample stands based on accessibility, adjacency, and attributes involves a great deal of subjectivity and sampling bias. We recommended that you have a list or map of polygons by stratum in order of priority for field crews so that they can alter sampling if obstacles are encountered in the field.

Using the prioritized polygons from table ISS-3 and then clustering the plots only marginally improves the sampling efficiency because many of the polygons that need to be sampled are not close to one another (fig. ISS-7). This is a good example of the realities of sampling.

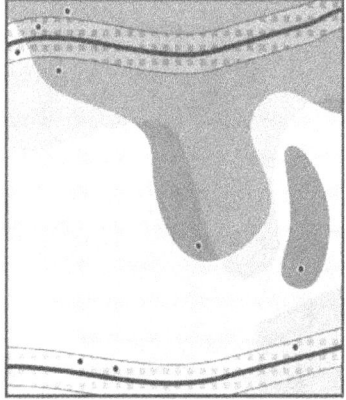

Relevé, Cluster, Simple

**Figure ISS-7**—Attempting to cluster the prioritized polygons from table ISS-3 does not substantially improve the sampling efficiency because the polygons that need to be sampled are widely distributed on the landscape.

Remember to allow enough flexibility in the prioritization and selection process so that the field crews can modify sampling if problems arise. Examples could be high elevation snowfall or dangerous snags falling in a sample polygon.

## Using the Alternative sampling intensity (Level II)

This sampling level assumes SP is close to NP. This level is often used with the relevé approach when monitoring is needed but there are not enough funds to implement a statistical approach to complete the project. The primary goal of this scheme is to describe all of the important conditions that need to be monitored on the sample landscape. This sampling intensity is recommended if a statistical approach is not warranted for the monitoring objective.

Sampling with the Level II approach assumes that you have enough resources to sample most or all of the polygons with one relevé. However, it is still important to prioritize the sampling so that the most critical polygons get sampled. The key to a successful monitoring effort for this scheme is to prioritize those stands on the landscape that need sampling and sample the most important ones first. This can be difficult to do because the FIREMON architect must balance distribution of the plots across the landscape and accessibility with importance to management. Detailed below is a method to select sample polygons using the polygons identified in table ISS-3 above. The architect can vary the theme or strategy to fit local circumstances. Again, this is not a rigid procedural step but rather a flexible framework for sampling design modification. It is important that a stratification system be explicitly stated and recorded in the FIREMON notebook.

Create a list of polygons to sample, ordering your sampling polygons based on your stratification and prioritization criteria. This list can be generated from a GIS or from a spreadsheet. Remember that this list is spatial in nature and does not take into account proximity and adjacency to other stands unless specifically designed and sorted.

If we use the ISS example landscape stratification described in the **Determining Polygons and Building a Summary Table** section then select two prioritization attributes—1) fuel load, especially in moderately loaded areas, and 2) exotic weed invasion, in that order—we can make up an ordered list of polygons for sampling (table ISS-4). In this example we will assume that SP is equal to 17. Figure ISS-8 shows how the plots might be located across the project area. For clarity, in this example tree density has also been used to order the polygons. However, tree density is not a prioritization attribute, so polygons that

Table ISS-4  Sites with moderate fuel load and weeds were used as prioritization attributes, then the strata in table ISS 2 were ordered to identify the polygons for sampling.

| Prioritization attribute 1 | Prioritization attribute 2 | Stratum code | Polygon number | Order of priority |
|---|---|---|---|---|
| Fuel load moderate to low | Exotic weeds | HMY | 14 | 1 |
| | | MMY | 6 | 2 |
| | | MMY | 13 | 3 |
| | | LMY | N.A. | |
| | | HLY | N.A. | |
| | | MLY | 4 | 4 |
| | | LLY | 2 | 5 |
| | | HMN | 15 | 6 |
| | | HMN | 17 | 7 |
| | | MMN | 5 | 8 |
| | | MMN | 7 | 9 |
| | | MMN | 9 | 10 |
| | | MMN | 12 | 11 |
| | | LMN | N.A. | 12 |
| | | HLN | 8 | 13 |
| | | HLN | 10 | 14 |
| | | HLN | 16 | 15 |
| | | LLN | 1 | 16 |
| | | LLN | 3 | 17 |

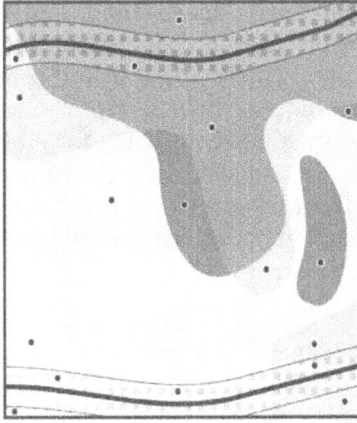

Relevé, Level II

Figure ISS-8—All 17 polygons on the example landscape are sampled with one releve when the Level II-Alternative sampling scheme is applied.

have the last two letters of the stratum code in common ("MN" in "MMN") could be sampled in any order. Sampling the polygons that are closer together first will lower the sampling time.

We recommend that you have a list or map of polygons by stratification attribute in order of priority for field crews so that they can alter sampling if obstacles are encountered in the field. Remember to allow enough flexibility in the prioritization and selection process so that the field crews can modify sampling if problems arise. Examples could be high elevation snowfall or dangerous snags falling in a sample polygon. A table like ISS-4 should be included in the FIREMON notebook.

## Using the Detailed sampling intensity (Level III)

There is rarely a situation that would match the relevé approach with a Detailed sampling intensity because the two are incompatible (see table ISS-1). Usually, if you have the resources to intensively sample a landscape, then a statistical approach is more appropriate to get more power from your monitoring results. However, if you want to use the relevé sampling approach with Level III sampling, then we suggest you follow the methods described in the previous section (Using the Alternative Sampling Intensity—Level II) with a small change. In the relevé Level II sampling scheme, plots are put in representative portion of most polygons on the sample landscape. With Level III sampling, we suggest that all special features be sampled within each polygon. The description or classification of special features must be explicitly stated in the sample design. For example, you might want to use a field on the Plot Description (PD) form, such a Landform, to detect special features. The goal of this sampling scheme is to sample all the polygons as well as the atypical conditions within the polygons, that represent conditions most important to fire effects monitoring. In this case, prioritization and stratification are not important; the only subjective element is the determination of special features. Cluster sampling is not needed because all polygons will be sampled.

## Implementing the Statistical Approach

Use the statistical approach when it is important to compare the differences across polygons or sampling strata using statistically valid techniques. The statistical approach in FIREMON attempts to quantify the variance in a wide variety of sampled entities within a sampled polygon. Two or more sampled polygons can be compared to ascertain whether they are significantly different before and after a treatment. Moreover, before and after measurements of fire effects can be compared using standard statistical techniques to obtain a measure of change for the entire polygon rather than for one representative plot within the polygon such as used in relevé sampling. The statistical approach has strong interpretative power, but it comes at a cost. It is often resource-intensive to implement a statistical approach in fire monitoring because multiple plots per polygon are needed to quantify the variance of the myriad of ecosystem characteristics that are measured to evaluate fire effects.

The complex challenges posed by field sampling, such as steep slopes, dense stands and wildlife (snakes, bears, and so on), coupled with the extensive challenges of statistical sampling mean designing a statistically valid fire effects sampling scheme can be an extremely difficult and complex task that requires extensive expertise in statistical sampling techniques, field sampling, and operational management. As a result, this section in FIREMON is only a starting point for statistically based sampling and is not intended as a complete reference on the subject. We beleive that the material presented here provides an adequate start for a statistically valid sampling effort; however, if you are doing Level III sampling, then we strongly recommend that you have your sampling scheme designed or, at least, reviewed by an agency statistician or sampling expert. This is especially true if you are setting up a monitoring project on a large area, such as a watershed.

The FIREMON statistical approach assumes the fire manager wants a statistically relevant estimate of ecosystem characteristics for each polygon sampled. Since multiple plots are needed to quantify the variance in ecosystem characteristics within a polygon, the most difficult task for the manager is determining which polygons will be sampled. The only difference between sampling intensities Level I, II, and III is the number of polygons that will be sampled with the resources available.

The statistical approach may involve sampling at two spatial scales, which can make the sampling design difficult. The monitoring objectives might call for a statistical test of significance in fire effects at the landscape scale and the polygon scale. For example, the purpose of a FIREMON project may be to test if the entire landscape experienced a 50 percent reduction in fuels. Here, every polygon on the landscape, or groups of polygons on the landscape, must be sampled to test for statistical significance. However, only polygon-level changes may be important in another FIREMON project. For example, did the sampled polygons achieve 50 percent duff reduction? The FIREMON architect must decide whether landscape level, polygon level, or both levels of statistical testing are relevant to the monitoring objective. Record this information in the FIREMON Metadata table.

### Sample size determination

The first important step in the statistical approach is to determine the number of plots needed to adequately sample each polygon. Most statistical sampling techniques recommend that sample size be determined by the amount of variability in the characteristic being sampled using standard formulae as determined from a pilot study—a small-scale sample collected simply to identify attribute variability. The two most often used statistical measures of variability are variance and standard deviation. Both are related to the difference between observed values in a group of numbers and the mean of those values. Standard deviation is simply the square root of the variance. Although either measure is appropriate, standard deviation is used more often because the units are the same as the units of the mean, whereas the units of variance are squared. For example, the standard deviation of 20 coarse woody debris (CWD) estimates might be 10 tons/acre or, equivalently, the variance would be 100 (tons/acre)$^2$. Variance and standard deviation estimates are easily made in spreadsheets, statistical software programs, and even on some handheld calculators. If resources allow, a pilot study of the characteristics that are important to the monitoring project should be done for each sampling stratum/sampling characteristic combination in the sampling project.

### Coefficient of variation for assessing variability

Coefficient of variation (CV) is a third measure of variability but one that is not used often. However, it is a good measure to use when comparing variability estimates. Its benefit lies in the fact that it relates the attribute's standard deviation to its mean and, since most ecological attributes exhibit increasing variability with increasing mean, CV provides a variability measure that is somewhat standardized for comparison among attributes. For example, assume that you have a study and note that the standard deviation of fine woody debris (FWD) load is 1.0 tons/acre and the standard deviation of CWD is 3.0 tons/acre. In absolute terms the variability of the CWD is higher, but if the mean load of FWD is 0.5 tons/acre and the mean load of CWD is 3.0 tons/acre then, relatively, FWD is more variable because the standard deviation is twice the mean, while the standard deviation of CWD equals the mean. The coefficient of variation is expressed as a percentage and is calculated using the formula:

Equation ISS-2 $$CV = \frac{s}{\bar{x}}(100)$$

where, s is the standard deviation estimate of an attribute $\bar{x}$ and is the estimated mean of the same attribute.

When trying to identify the attribute with the greatest variability we recommend using the coefficient of variation.

### *Determining the attribute variability used to calculate sample size*

Sample size determination can be confounding in FIREMON because many fire effects monitoring projects sample more than one characteristic. For example, it is common for changes in fuels, tree mortality, and vegetation cover to be monitored before and after fire. This is quite different from conventional forest inventory techniques that use only timber volume to compute the required number of plots. The question is, which attribute should be used to represent the variability to compute the requisite number of plots? In FIREMON, we recommend that you use either the standard deviation of the most important characteristic or the standard deviation of the characteristic that has the greatest coefficient of variation.

The selection of the characteristic or characteristics to use to determine the number of plots ultimately depends on the importance of the sampled characteristic in successfully completing the monitoring objectives. If fuel reduction is the highest priority, then select fuel loadings as the variable to compute number of plots. If there is more than one characteristic important to the sampling objectives, then select the variable with the largest variance to ensure adequate plot representation for the other characteristics. For instance, if fuel reduction, tree mortality, and plant succession share equal weight in the monitoring effort, then select the variable, probably fuel loading, that has the highest within-sampling strata variance and calculate the sample size using that characteristic.

If, after calculating the number of required plots, you find that the sampling intensity is too high compared to the sampling resources, you will need to reduce the number of plots to a manageable level. One way to reduce the sample size is to sample the most important characteristic rather than the one with the highest variability. If that doesn't sufficiently reduce the number of required plots, estimate the number of plots needed after removing the least important polygons from the study. The worst-case scenario is that you start eliminating sampling methods—for instance, in the previous example, eliminating the vegetation sampling—or need to replace some intensive methods with less intensive methods, such as substituting photos for vegetation measurements. Remember, you want to be collecting the best quality data you can, given the objectives and resources. It is better to have a few plots with useful data than have lots of plots with data that doesn't let you assess how well you met the project objectives.

This is usually the point where most sampling designs come to a dead halt. The FIREMON architect can easily choose the characteristics to sample, but then must obtain some measure of variation for that selected characteristic. Typically, fire managers do not have the field data to quantify variation in sampled characteristics although, occasionally, data collected from past sampling efforts in similar terrain and ecosystems can be used. If field or pilot data are available for your FIREMON project, analyze them to determine the variation of a particular characteristic across a stand or mapped classification category, and then use the standard deviation to compute the number of required plots using the equation presented below.

If you don't have an idea of variability, the easiest and fastest way to find a variability estimate is to contact a local expert knowledgeable about the location and variables you are interested in sampling, who may be able to provide a good estimate of variability. The local expert will be able to state the variability in terms of the mean, and this will help put the variability in perspective. For instance, "On that site, the standard deviation of CWD is about one and a half times the mean." If local experts cannot help, then examine research studies and reports to see if measures of variation for your characteristic of interest, on a similar landscape, are available. This can be time consuming and thus not possible for some projects. You may also be able to analyze previously collected FIREMON data by polygon or sampling stratum and identify an estimate of variability from the information. As a last resort, use your own experience or a best guess from what information you have been able to locate, and pick the standard deviation level that seems most appropriate—0.5, 1.0, or 1.5 times the mean. If you simply have no other information, calculate the required number of plots using a standard deviation that is equal to the mean of the attribute of interest.

There should be a measure of variability for the characteristic of interest for each sampling stratum in the monitoring project. For example, if the landscape is divided into polygons that are named according to cover type (ponderosa pine, Douglas-fir, lodgepole pine) and the selected variable is 1,000-hour fuel loading, then a measure of variability must be obtained for each cover type stratum on the landscape. This is often difficult because the data required to quantify variability are typically not available; therefore, it may be necessary to use the same standard deviation for a larger aggregation of sampling strata (all pine cover types).

## Calculating sample size

The number of required plots (NRP) per polygon per sampling stratum can be computed a number of different ways to meet different statistical objectives. Monitoring projects designed using the FIREMON protocol can use the following equation with reasonable assurance the sample size will be appropriate:

Equation ISS-3
$$NRP = \frac{s^2 \left(Z_{\alpha/2} + Z_\beta\right)^2}{MDC^2}$$

where: $s$ is the standard deviation of the difference of the first and second sampling visit.

$Z_\alpha$ is the Z-coefficient for the type I error rate from table ISS-5.

$Z_\beta$ is the Z-coefficient for the type II error rate from table ISS-5.

MDC is the Minimum Detectable Change of the difference of sampled values, in absolute terms.

The confidence level for your monitoring project should be indicated in the project plan or the project objectives. If so, then the false-change error rate is calculated from the confidence level ($\alpha$ = 1– (confidence level/100)). If not, choose an $\alpha$ level that you feel is appropriate given the project objectives. In most research level studies it is 0.05 or lower. However, fire monitoring projects may not have to be as restrictive. Never set the confidence level below 80 percent (error rate = 0.20). The higher the confidence level the greater the number of samples needed. Most monitoring studies are less concerned with making a Type II error during the statistical analysis so use a missed-change error rate of 0.20 unless you have a reason to use a lower rate. Calculate NRP using the $Z_\alpha$ and $Z_\beta$ values that correspond to the a and b error rates. When using this equation be sure to divide $\alpha$ and $\beta$ by 2 before selecting the

Table ISS-5—Determination of sample size is dependent on the acceptable error rate. Use the appropriate z-values in this table for your sampling project.

| False-change (Type I) error rate ($\alpha$) | $Z_{\alpha/2}$ | Missed-change (Type II) error rate ($\beta$) | $Z_\beta$ |
|---|---|---|---|
| 0.40 | 0.84 | 0.40 | 0.25 |
| 0.20 | 1.28 | 0.20 | 0.84 |
| 0.10 | 1.64 | 0.10 | 1.28 |
| 0.05 | 1.96 | 0.05 | 1.64 |
| 0.01 | 2.58 | 0.01 | 2.33 |

z-value. For example, if you are using the error rates α = 0.10 and β = 0.20, then $Z_\alpha$ and $Z_\beta$ would be 1.96 and 1.28, respectively.

Determining the MDC parameter can be confusing because it assumes the mean value of the attribute of interest is known, and generally it is not. Usually, you will be able to make an estimate that is sufficiently accurate for use in the NRP equation. If not, use either a pilot study or get information from an expert or the literature. Once the mean is known (or estimated) calculate MDC by multiplying the mean by the percent change you want to be able to detect. For instance, if you want to be able to detect a 20 percent change in a down woody debris load and you estimate the mean at 25 tons/acres, then MDC = 0.20(25) or 5 tons/acre. The lower the amount of change you want to detect the greater the number of plots you will need. Well-written objectives will give you some feeling for the detection level that is required in the study.

The NRP calculation must be completed for each polygon or sampling stratum in the project, then the NRP is summed across all polygons or sampling stratum on the landscape to compute the total NRP for the FIREMON sampling effort.

A good source for more information about determining NRP is: Measuring and Monitoring plant populations (Elzinga and others 1998 or Elzinga and others 2001). The entire 1998 publication is available as a PDF on the BLM library Web site: http://www.blm.gov/nstc/library/techref.htm. Select T.R. number 1730-1. Appendix Seven has a complete discussion about the determination of NRP.

### Using the Simple sampling intensity (Level I)

The Simple sampling intensity level is used when the number of required plots (NRP) is much greater (more than two times) than the sampling potential (SP). It is inappropriate to match the statistical approach with the Level I-Simple sampling intensity level because they are in conflict. If the number of plots to establish is limited, then it will be difficult to achieve a statistically valid measure of variability at the polygon and landscape scale as required by the statistical approach. If the statistical approach is most important for the monitoring objectives, then the landscape level statistical validity must be given up to achieve statistical validity at the polygon level. If landscape statistical validity is important, then it will be difficult to sample the required number of polygons. If statistical validity is important reevaluate the project objectives to reduce the number of polygons that need to be sampled, so that NRP is roughly the same as SP, then use a polygon prioritization process.

There are several methods for reducing the number of sample polygons under the statistical approach. As stated in the previous **Considering Tradeoffs** section, one method is to revisit the project objectives to see if there are some polygons that can be sampled less intensively in terms of the number of plots or the methods applied. Another method is to prioritize the sampling polygons. Base the prioritization on the project objectives so that the most important attributes are sampled, then eliminate sampling in the least important polygons. Using a GIS can simplify the prioritization task. It is recommended that the priority list include some measure of accessibility to optimize the number of sampled polygons with the number of plots possible. For example, stand selection might include the following criteria: 1) within a mile of a road, 2) contain ponderosa pine, and 3) on slopes less than 50 percent.

If we use the ISS example landscape stratification described in the **Determining Polygons** and **Building a Summary Table** section (table ISS-2), then select two prioritization attributes—1) fuel load, especially in moderately loaded areas, and 2) exotic weed invasion—in that order, we can make up an ordered list of polygons for sampling (table ISS-6). For clarity, in this example tree density has also been used to order the polygons. However, tree density is not a prioritization attribute, so polygons that have the last two letters of the stratum code in common ("MN" in "MMN") could be sampled in any order. Sampling the polygons that are closer together first will lower the sampling time. We recommended that you have a list or map of polygons by stratification attribute in order of priority for field crews so that they can alter sampling if obstacles are encountered in the field. After selecting the polygons for sampling, reassess the your decision to use the Level I sampling scheme with the statistical approach. If the sampling intensity does not match the information required for the project objectives,

**Table ISS-6**—Sites with moderate fuel load and weeds were used as prioritization attributes, then the strata in table ISS 2 were ordered to identify the polygons for sampling.

| Prioritization attribute 1 | Prioritization attribute 2 | Stratum code | Polygon number | Order of priority |
|---|---|---|---|---|
| Fuel load moderate to low | Exotic weeds | HMY | 14 | 1 |
| | | MMY | 6 | 2 |
| | | MMY | 13 | 3 |
| | | LMY | N.A. | |
| | | HLY | N.A. | |
| | | MLY | 4 | 4 |
| | | LLY | 2 | 5 |
| | | HMN | 15 | 6 |
| | | HMN | 17 | 7 |
| | | MMN | 5 | 8 |
| | | MMN | 7 | 9 |
| | | MMN | 9 | 10 |
| | | MMN | 12 | 11 |
| | | LMN | N.A. | 12 |
| | | HLN | 8 | 13 |
| | | HLN | 10 | 14 |
| | | HLN | 16 | 15 |
| | | LLN | 1 | 16 |
| | | LLN | 3 | 17 |

then adjust the methods and intensity you plan to use on each plot, move to the Level II sampling intensity, or use the relevé approach.

We recommended that you have a list or map of polygons by stratification attribute in order of priority for field crews so that they can alter sampling if obstacles are encountered in the field. Remember to allow enough flexibility in the prioritization and selection process so that the field crews can modify sampling if problems arise. Examples could be high elevation snowfall or dangerous snags falling in a sample polygon. A table like ISS-6 should be included in the FIREMON notebook.

## Using the Alternative sampling intensity (Level II)

The Level II-Alternative sampling intensity level is used when the number of required plots (NRP) is between one and two times greater than the sampling potential (SP). If NRP is greater than the number of polygons (NP) to be measured, then NP must be reduced to make NRP roughly equal to SP. The reduction will probably not compromise the statistical validity of the sample design if a landscape level measure of variability is not as important to the monitoring objective as a polygon level measure of variability.

There are several methods for reducing the number of sample polygons under the statistical approach. As stated in the previous **Considering Tradeoffs** section, one method is to revisit the project objectives to see if there are some polygons that can be sampled less intensively in terms of the number of plots or the methods applied. Another method is to prioritize the sampling polygons. Base the prioritization on the project objectives so that the most important attributes are sampled, then eliminate sampling in the least important polygons. Using a GIS can simplify the prioritization task. It is recommended that the priority list include some measure of accessibility to optimize the number of sampled polygons with the number of plots possible. For example, stand selection might include the following criteria: 1) within a mile of a road, 2) contain ponderosa pine, and 3) on slopes less than 50 percent.

If we use the ISS example landscape stratification described in the **Determining Polygons** and **Building a Summary Table** sections (table ISS-2), then select two prioritization attributes—1) fuel load, especially in moderately loaded areas, and 2) exotic weed invasion—in that order, we can make up an ordered list of polygons for sampling (table ISS-7). For clarity, in this example tree density has also been used to order the polygons. However, tree density is not a prioritization attribute so polygons

Integrated Sampling Strategy (ISS) Guide

Table ISS-7  Sites with moderate fuel load and weeds were used as prioritization attributes, then the strata in table ISS 2 were ordered to identify the polygons for sampling.

| Prioritization attribute 1 | Prioritization attribute 2 | Stratum code | Polygon number | Order of priority |
|---|---|---|---|---|
| Fuel load moderate to low | Exotic weeds | HMY | 14 | 1 |
| | | HLY | N.A. | |
| | | MMY | 6 | 2 |
| | | MMY | 13 | 3 |
| | | MLY | 4 | 4 |
| | | LMY | N.A. | |
| | | LLY | 2 | 5 |
| | | HMN | 15 | 6 |
| | | HMN | 17 | 7 |
| | | HLN | 8 | 8 |
| | | HLN | 10 | 9 |
| | | HLN | 16 | 10 |
| | | MMN | 5 | 11 |
| | | MMN | 7 | 12 |
| | | MMN | 9 | 13 |
| | | MMN | 12 | 14 |
| | | MLN | 11 | 15 |
| | | LMN | N.A. | |
| | | LLN | 1 | 16 |
| | | LLN | 3 | 17 |

that have the last two letters of the stratum code in common ("MN" in "MMN") could be sampled in any order. Sampling the polygons that are closer together first will lower the sampling time.

We recommended that you have a list or map of polygons by stratification attribute in order of priority for field crews so that they can alter sampling if obstacles are encountered in the field. Remember to allow enough flexibility in the prioritization and selection process so that the field crews can modify sampling if problems arise. Examples could be high elevation snowfall or dangerous snags falling in a sample polygon. A table like ISS-7 should be included in the FIREMON notebook.

If it is important to test for statistical significance at the polygon and landscape scale, then a random selection of polygons is warranted. This random selection should be weighted by some factor important to the monitoring objective. For example, it may be important to sample the greatest area, so selection should be weighted by area. This can be accomplished in a GIS by using a random selection algorithm linked to stand weights, or it can be done outside a GIS using a random number generator or list. Be sure enough stands will be sampled across the landscape to obtain a statistically valid estimate of variability.

## Using the Detailed sampling intensity (Level III)

The Detailed sampling intensity level is used when the number of required plots (NRP) is much less than the number of plots that are possible with the available resources (SP).

Developing a sampling design at this level can be quite complex and beyond the scope of the FIREMON ISS. We recommend that you employ an expert in statistical sampling to design your sampling project. It will cost a small fraction of what will be spent sampling and will reap great rewards. Contact a statistical expert at your local research institution or university for assistance.

## Plot distribution when using the statistical approach

In FIREMON we suggest plots be distributed using either systematic or random placement. When properly applied each will give you a statistically valid sample. The number of plots in each polygon can be weighted by area of each polygon in a sampling stratum or divided evenly among all of the polygons in the sampling stratum. For example, say NRP = 100 plots and there are two polygons in one sampling stratum with sizes of 20 and 80 acres. You could either weight the number of plots per polygon by area, resulting in 20 plots in the first and 80 plots in the second polygon. Or, you could divide the plots evenly and have 50 plots in each polygon. Unless all the polygons are close to the same size the first option is probably the most appropriate.

When using the systematic approach, plots are distributed by developing a grid spacing that will give you the appropriate number of plots within each of your polygons in each sampling stratum. The base corner of the grid should be located at random. Choose randomly distributed plot locations by developing a system to locate plots within a polygon with a list of random numbers. Depending on the variability of the attributes you are monitoring, one sampling stratum may need more or fewer plots per polygon than another stratum.

A third plot distribution technique is to locate the sampling plots in clusters. If you are establishing cluster plots, the location of plot clusters is dependent on landscape features such as roads and topography in order to reduce the sampling effort. The best cluster design allows the maximum number of plots to be established and sampled with the least bias in plot location. Cluster sampling may not result in a statistically valid sample if plots are placed with bias, are not independent samples, or do not allow sampling across the entire range of conditions in each stratum.

Figure ISS-9 shows how plots could be distributed on the ISS example landscape using each of the plot location methods at the Detailed level of sampling. Illustrations E, F, and G each show 66 sampling locations. You could easily make this the Alternative intensity (Level II) by limiting the methods on each plot to only those critical to the objectives. For example, at the Detailed level, you may plan to use cover/

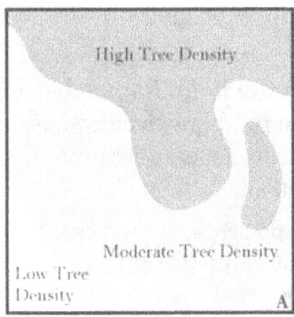 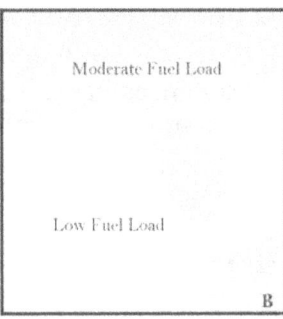  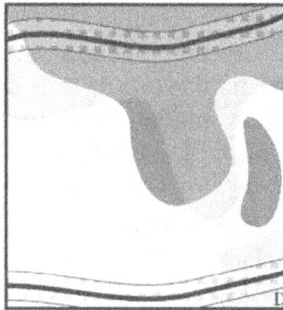

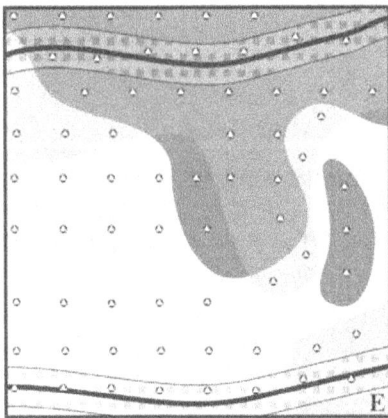 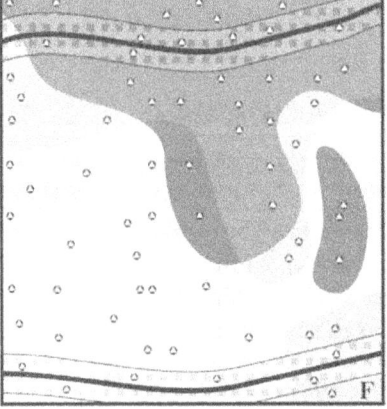 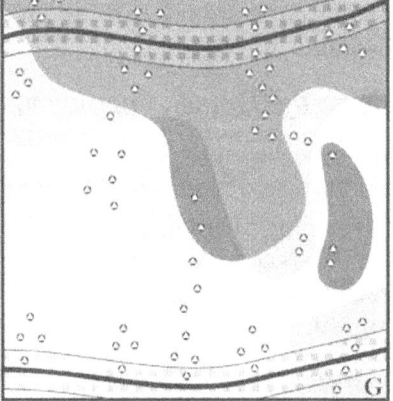

Statistical, Systematic, Detailed    Statistical, Random, Detailed    Statistical, Cluster, Detailed

**Figure ISS-9**—The three stratificaitons of the example ISS landscape (A, B, and C) are combined to identify the sampling polygons (D). Then, using the statistical, detailed approach, 66 sampling plots are distributed on the landscape map. Illustration (E) shows the systematic approach, (F) the random approach, and (G) the cluster approach.

frequency (for weed sampling), fuel load, and tree data methods on each plot in your study. To apply the Alternative level you could eliminate the cover/frequency methods when a plot falls out of the weed corridor. You would still be establishing 66 plot locations, but the sampling time would be reduced because you would not be using all three sampling methods on all of the plots.

## Establishing Control Plots

Control plots are established in areas outside the perimeter of your treatment unit in order to collect reference data that will be used to compare against your posttreatment data. Control plots are especially important in rangeland ecosystems where year-to-year variation in weather can mask changes caused by fires. For example, say a prescribed burn unit in a sagebrush-steppe ecosystem has 10 plots established within burn boundaries and five plots established outside burn boundaries. When the unit is burned, the postburn measurement of grasses shows a doubling of biomass. The inference is that the burn has increased grass production. However, remeasurement of the control plots (plots that were not burned) shows that the grass biomass on these plots has doubled as well. So the increase in grass biomass was actually a result of the some other factor, a wet spring, for instance, that occurred that year, not from the fire. Control plots allow you to assess the effects of factors other than those you applied with your treatments (such as weather or wildlife) on the characteristics you are monitoring.

While control plots are valuable they are not required in many monitoring efforts. The decision whether or not to establish controls depends on the ecosystem, monitoring objectives, and available resources. Control plots may not be necessary in ecosystems that are relatively unaffected by annual or decadal weather variations, such as forest and alpine environments. The establishment of controls is usually warranted if the monitoring objective is to statistically determine significant changes in ecosystem characteristics in environments where vegetation is substantially affected by annual fluctuations in weather. Examples are grasslands, shrublands, and some forest understories. If project resources are limiting, then, depending on objectives, control plot establishment is probably the first task to be trimmed.

As when determining the sampling intensity for monitoring plots, the number of control plots to establish outside the treatment boundary depends on the environment, objectives, available resources, and the availability of appropriate sites. At a minimum, at least one control plot should be established if you are using the relevé approach. At least three plots should be established when using the statistical approach. A sampling objective that specifies statistical significance will probably require more sampling units to adequately capture the variance of sampled entities and, thus, a greater number of control plots. The most complete sampling design will have at least one control plot for every sampling stratum in the study. Some areas have diverse mosaics of stand conditions within burn boundaries so more than one control area may be warranted to adequately capture the within-burn prefire heterogeneity. This is especially true of landscapes with highly diverse patch characteristics.

Controls should be established adjacent to or near the treatment area. They should be located in an area that represents the characteristics important to the monitoring objective, found inside the treatment area. Controls should be established at the same time as the pretreatment measurements, and they should be installed using the same methods as you used on the monitoring plots. The only difference between control plots and monitoring plots is their location. When using the relevé approach, establish control plots in an area outside the treatment area that best represents the area within the treatment area. Use random or systematic control plot location, when using the statistical approach. If there are no areas outside treatment boundaries that are suitable for control establishment, establish control plots in the next best area with similar aspect, slope, and elevation. Sometimes more than one control area will be needed because of diverse characteristics (slope, aspect, vegetation, and so on) within the burn. Remember, control plots are used to determine temporal variation in ecosystem characteristics caused by factors other than the treatments. They do not need to be placed on locations that have identical characteristics to the treatment area, but they should be similar. The use of potential vegetation types or site types can be helpful to stratify control sample areas.

Monitoring plots established inside treatment boundaries that do not burn or are not impacted by other treatment activities have intrinsic value as fire effects control plots. For instance, a prescribed fire, for some reason such as high fuel moisture, may not actually visit all the monitoring plots laid out inside a burn perimeter, and these unburned plots could supplement your other control plots. Care must be taken, however, because the reason that the plot didn't burn may be related to the attribute you are trying to control. For instance, if a plot didn't burn because the fuel moistures were high, it might mean that a seep is subirrigating the area around the unburned plot, which in turn would increase vegetation cover and height. Using this unburned plot as a control plot to compare vegetation would be inappropriate. If there are monitoring plots that will be used as control plots, sample them at the same time and with the same methods as your monitoring plots.

Controls are sometimes useful for monitoring fire effects on plots when you were not able to install monitoring plots before an event, such as a wildfire. Control plots can approximate preburn conditions and these pseudo-preburn conditions can be used as reference. Simply establish the controls outside of the burn boundary in areas that best approximate the characteristics found prior to the burn as observed inside the fire perimeter. Use snags and downed logs to help visualize what the stand looked like before the burn. Many people do not establish controls after a fire because the subsequent information is somewhat suspect, as the preburn stand conditions within the fire can never be truly determined. However, when they are appropriately used, postfire control plots can provide useful information. You may need to establish controls on many site types to cover the wide range of environmental conditions within the burned area.

Ideally, each control plot should be remeasured at the same time as the burned monitoring plots. It is important that the controls be measured at least twice—once to establish the controls and once at the end of the monitoring effort.

## FIREMON GUIDES AND KEYS

### Sampling Strategy Checklist

*Suggested step-by-sep procedure for designing a firemon monitoring sampling effort*

*See figure ISS-1*

1. **State the monitoring objective(s).** Describe, in detail, the reasons why this monitoring effort is being implemented.
2. **Determine the sample area.** Create a map that explicitly identifies the boundaries of the landscape to be sampled.
3. **Determine the sample stratification.** Create a map that delineates each polygon on the sample landscape, and then describe each polygon by one or more attributes that will be used to stratify monitoring results.
4. **Determine the sampling resources.** Record all resources that can be used in this sampling effort including personnel, monies, time, and vehicles.
5. **Calculate sampling resource statistics.** Compute the sampling statistics that are used throughout the FIREMON sampling design process (SP, NRP, and so on).
6. **Determine the sampling approach.** Select the most appropriate sampling approach, statistical or relevé, using the keys and text presented in FIREMON.
7. **Determine the sampling intensity level.** Select the most appropriate level of sampling intensity from the keys and text presented in FIREMON.
8. **Choose the most appropriate sampling methods.** Select the measurement methods that will be used to sample ecological characteristics using the keys and the text presented in FIREMON.

9. **Compare monitoring design and project objectives**. If the design is compatible with the objectives, distribute plots and begin sampling. Otherwise, review the project and go through the checklist again.
10. **Determine plot locations**. Use stratification factors to determine polygons, then distribute plots over the sampling area.

*Suggested step-by-step procedure for implementing a FIREMON monitoring sampling effort*

1. **Locate a FIREMON sampling polygon**. Use the sample design to select, then navigate to a stand on the sample landscape.
2. **Locate a FIREMON sampling plot**. Use the directions in How to Locate a Plot to go to an area within a stand that will be sampled.
3. **Establish a FIREMON sampling plot**. Use the directions in How to Establish a Plot to permanently or semipermanently mark the area to sample.
4. **Follow the procedures for each selected sampling method**. Refer to the Sample Methods discussions for the sampling methods and protocols selected for the project and follow those instructions for the appropriate sampling strategy and intensity level.
5. **Record measured field information on FIREMON plot forms**. Write down all information and measured entities on the plot forms provided in FIREMON. You may also record the measurements directly onto a field laptop computer.
6. **Check recorded information**. Double check your entries on the plot forms and make sure all fields are completed and appear correct.
7. **Enter data on plot form into FIREMON databases**. Enter the recorded field data into the Microsoft Access Databases provided by FIREMON.

## Field Assessment

The field assessment portion of FIREMON contains an extensive set of procedures for sampling important ecosystem characteristics before and after a prescribed or natural fire for terrestrial ecosystems. The field assessment is composed of 1) field methods, 2) plot forms and cheat sheets, and 3) equipment lists. FIREMON has been designed so that the fire manager can tailor the field measurement procedures to match burn objectives or wildland fire use concerns. Additionally, the fire manager can scale the intensity of measurement to match resource and funding constraints. For example, to document tree mortality, the fire manager would choose one of three hierarchically nested sampling procedures, where the first procedure would provide general descriptions of tree mortality quickly at low cost (photopoints, walk-through), while the third procedure would document, in detail, individual tree health and vigor, to generate comprehensive data applicable to many analyses but costly to collect. A key has been provided help the fire manager decide the appropriate methods and sampling intensity.

*Sampling protocols*

FIREMON contains the following sampling procedures for monitoring many ecosystem characteristics:

**Plot Description (PD)**—A generalized sampling scheme used to describe site characteristics on the FIREMON macroplot with biophysically based measurements.

**Species Composition (SC)**—Used for making ocular estimates of vertically projected canopy cover for all or a subset of vascular and nonvascular species by DBH and height classes using a wide variety of sampling frames and intensities. This procedure is more appropriate for inventory than monitoring.

**Cover/Frequency (CF)**—A microplot sampling scheme to estimate vertically projected canopy cover and nested rooted frequency for all or a subset of vascular and nonvascular species.

**Point Intercept (PO)**—A microplot sampling scheme to estimate vertically projected canopy cover for all or a subset of vascular and nonvascular species. Allows more precise estimation of cover than the CF methods because it removes sampler error.

**Line Intercept (LI)**—Primarily used when the fire manager wants to monitor changes in plant species cover and height of plant species with solid crowns or large basal areas where the plants are about 3 feet tall or taller.

**Density (DE)**—Primarily used when the fire manager wants to monitor changes in plant species numbers. This method is best suited for grasses, forbs, shrubs, and small trees, which are easily separated into individual plants or counting units, such as stems. For trees and shrubs over 6 feet tall the TD method may be more appropriate.

**Rare Species (RS)**—Used specifically for monitoring rare plants such as threatened and endangered species.

**Tree Data (TD)**—Trees and large shrubs are sampled on a fixed-area plot. Trees and shrubs less than 4.5 feet tall are counted on a subplot. Live and dead trees greater than 4.5 feet tall are measured on a larger plot.

**Fuel Load (FL)**—The planar intercept (or line transect) technique is used to sample dead and down woody debris in the 1-hour, 10-hour, 100-hour, and 1,000-hour and greater size classes. Litter and duff depths are measured at two points along the base of each sampling plane. Cover and height of live and dead, woody, and nonwoody vegetation is estimated at two points along each sampling plane.

**Landscape Assessment (LA)**—Useful for mapping fire severity over large areas by combining satellite-derived Normalized Burn Ratio (BR) with a ground-based indicator of fire severity, Composite Burn Index (BI). The LA methodology will assist in determining landscape-level management actions where fire severity is a determining factor.

**Composite Burn Index (BI)**—The BI methodology is a subset of the LA methods. It provides users with a ground-based fire severity index derived from a number of plot measurements.

**Normalized Burn Ratio (BR)**—The BR method is the subset of the LA methods. It describes how to derive remotely sensed spatial information on burn severity, using Landsat satellite data.

We used the Western Region Fire Monitoring Handbook (FMH) (National Park Service 2001, see http://www.nps.gov/fire/fire/fir_eco_firemonitoring.html), ECODATA handbook (Hann and others 1988; Jensen and others 1993; Keane and others 1990) and the USDA Forest Service Natural Resources Information System (NRIS) protocols as the framework for selecting and designing FIREMON sampling methods. We modified these protocols so that there are now nested levels of sampling intensity coupled with nested levels of sampling flexibility.

## Sample Approach Classification Key

Answer each bullet with a "yes" or "no" and add up the answers. More "yes" answers suggest using the statistical approach, and more "no" answers suggest using the relevé approach.

1. ***Sufficient sampling resources are available***. There is plenty of funding, ample time, and sufficient personal (with necessary skills) to complete a detailed monitoring effort.
2. ***NP < SP***. There are sufficient resources to sample the entire landscape. Sampling resources allow more than one plot in each stand.
3. ***An estimate of across and within stand variation is important***. The project objectives require an estimate of variability in ecosystem characteristics or the statistical comparisons of sampled attributes.
4. ***A statistician or statistics expert is available for consultation***. Someone can easily be contacted to answer questions about your sampling design. There is sufficient expertise for designing a valid statistical sampling scheme.

5. ***Navigation across the sample landscape is relatively easy***. Steep, dangerous terrain, long travel distances, or other features that limit plot access are not present on major portions of the landscape.
6. ***Few ecosystem components are being measured for assessing fire effects***. The monitoring objectives are concerned with just one or two ecosystem attributes whose variation must be quantified.

## Sampling Intensity Key

Answer each bullet under each intensity level with a "yes" or "no." Count up the number of "yes" votes for each intensity level, and the level with the most "yes" votes would suggest that this may be the most likely intensity level for your monitoring study. The bullets are listed in order of importance with the first few bullets most important in sample design.

### Simple sampling intensity (Level I)

1. ***Little funding is available.*** This project has little financial support and must be done with existing personnel and equipment.
2. ***SP << NP***. There are many more polygons to be sampled than there are potential plots to sample them. A good rule-of-thumb is that NP is more than twice the number of potential plots, then a simple sampling approach is appropriate.
3. ***There is little time to conduct the project***. There are only a few weeks or months to sample, and the project must be completed as quickly as possible.
4. ***Description is more important than comparison or evaluation***. The monitoring objectives can be accomplished by establishing enough plots to generally describe fire effects without quantifying the variability of the sampled attributes.
5. ***There are few people available***. It will be difficult to hire or obtain a crew of experienced field technicians, or it will be difficult to train inexperienced crews to collect the fire effects data
6. ***Travel across the landscape is difficult and restrictive***. The polygons being sampled are so difficult to traverse that establishing multiple plots in each stand would be laborious and time consuming. The landscape may be steep and dangerous, composed of thick vegetation and deep fuels, or contain many dangerous obstacles such as deep rivers, cliffs, ice, and so forth.

### Alternative sampling intensity (Level II)

1. ***Sampling resources are available but limited***. Funding and other resources (people, vehicles) are available but not abundant. One or more categories of sampling resources is limiting such as few people, money, or time, but overall, there appear to be resources available.
2. ***SP = > NP***. There are about the same number of sample polygons as there are potential plots to sample them. There is a possibility that some polygons will be sampled with more than one plot. Or it is possible that some polygons may not be important to the objectives and could be removed from the list of polygons to be sampled.
3. ***An estimate of across and within stand variation is important but not essential***. An estimate of error in comparing polygon conditions is desired but not essential. It is more important that conditions within each polygon on the landscape be described so that management can proceed.
4. ***Many ecosystem components are being measured for assessing fire effects***. The monitoring objective is concerned with describing fire effects for many ecosystem elements (fuels, trees, plant species) so that an integrated stand-level evaluation of fire effects is possible.
5. ***A compromise is desired between the simple and detail methods***. Level I intensity is not enough to accomplish monitoring objectives and there are not enough sampling resources to implement a project at Level III intensity.

6. ***The sample area is complex, rugged, or remote***. Having to sample at a high intensity is not desirable. The size, topography, and limited roads/trails within the sampling area may limit the possibility of establishing enough plots to quantify polygon conditions across the entire area.

### *Detailed sampling intensity (Level III)*

1. ***Sampling resources are abundant***. There are sufficient resources to conduct the monitoring project. None of the sampling resource categories is limiting.
2. ***SP >> NP***. The sampling potential is high enough that multiple plots can be established in each polygon and should allow *at least* two plots per polygon.
3. ***An estimate of stand variation is important***. An estimate of error across the sampling polygons is essential for describing fire effects. It is important that the error in fire effects be quantified to make management decisions.
4. ***A statistically defensible comparison is important***. It is important that results from this study capture sampled variations so statistical comparisons can be made with error estimates. This implies that the sampling approach is required to get the most statistically defensible results to back up any management action.
5. ***Only a few ecosystem components are being measured for assessing fire effects***. The monitoring objective is only concerned with one or two ecosystem attributes whose variability can be easily quantified for computing the number of plots to establish.
6. ***The sampling environment allows an intensive sampling effort***. There are no foreseeable dangers or restrictions in the area to be sampled; the area is safely and easily accessible. Crews are sufficiently trained and available to conduct this extensive sampling.

## Methods Classification Key

Use this key to help you decide what methods should be used in your monitoring project. Start at the top of the key and, for every statement that is true for your situation, record the suggested methods or fields in your FIREMON notebook. Once finished, review the methods you have identified. Compare them with the sampling resources available and project objectives to determine if they are right for your project.

1. A description of the ***physical environment*** is important for providing context to monitoring results or for providing stratification for the data analyses. Physical descriptions include elevation, aspect, slope, soils, landform, and slope position. These measurements provide a general description of biophysical processes that might influence fuel, fire, and vegetation dynamics. It is strongly recommended that these variables be recorded at each plot.

   1.1. **Complete Biophysical Settings Fields in PD method**

2. A general description of ***vegetation characteristics*** is important for understanding monitoring results, stratifying analyses, or validating satellite imagery. This description includes lifeform (trees, shrubs, and grasses) cover by size class and an estimation of cover type and potential vegetation type. These measurements are especially helpful in describing general vegetation conditions and for relating plot-level stand-related vegetation characteristics to satellite imagery analysis and mapping.

   2.1. **Complete Vegetation Fields in PD method**

3. A general description of ***stand fuel characteristics*** is important for summarizing monitoring and interpreting monitoring results, stratifying analyses, or correlating with satellite imagery. These descriptions include ground cover (ash, bare soil, rock), fuel model, and crown fuel characteristics. These measurements are especially helpful in describing general fire-related fuel conditions and for relating plot-level fuels characteristics to satellite imagery analysis and mapping.

   3.1. **Complete the Fuels Fields in PD method**

4. ***Documentation of plot conditions and location*** is important to the monitoring objective. This includes photo-documentation, written notes, maps, and so on. This method is used to strengthen the

documentation of the location of the plot and to document spatial characteristics of the plot using visual tools.

### 4.1. Complete the Common and Comments Fields in the PD method

5. A general description of *fire behavior and effects* at the plot level is important for describing fire conditions in the interpretation of monitoring results, stratification of analyses, or presentations with the public. Descriptions include photo-documentation of fire behavior and effects, and quantification of overall fire effects. These methods are used to generate a pictorial assessment of fire behavior and effects, and are especially effective for relating fire effects to those unfamiliar with fire ecology.

### 5.1. Complete the Fire Behavior and Effects Fields in the PD method

6. A general description of *ambient weather conditions, fuel moistures, and fire behavior* for a prescribed fire is important for assessing fire effects. This information is collected for the entire fire event and is more complete than the plot level descriptive information collected using the PD methods.

### 6.1. Complete the Fire Behavior (FB) methods

7. *Changes in plant species cover and/or height* is important in assessing fire effects. Objectives state monitoring of individual species presence, cover, and/or height is important to the project. Possible applications include changes in species cover and height of threatened and endangered species, important forage species, and reductions in tree understory. This method is used mostly to track succession development in vegetation over time and to quantify changes in species cover due to disturbance.

7.1. A statistically valid comparison of changes in species cover over time is not important. Simple Sampling Intensity Level I. *Descriptive changes in species cover will fulfill monitoring objectives*. Species measurements include cover and height estimates without error estimates.

7.1.1. **Complete the Species Composition (SC) methods**. (See the Vegetation Sampling Overview for more information.)

7.2. *A statistical comparison of changes in species occurrence is important* for the successful completion of objective. Species measurements include plant frequency, cover, and height with error estimates.

7.2.1. *Height of >50 percent vegetation cover less than 3 feet*. Majority of plot is composed of plants that are less than 6 feet tall.

7.2.1.1. *A quadrat based examination of plant frequency or subjectively determined plant cover* is important to the project.

7.2.1.1.1. **Complete Cover/Frequency (CF) methods**. (See the Vegetation Sampling Overview for more information.)

7.2.1.2. *Objective determination of plant cover* is important for the project. Mostly fine-textured vegetation.

7.2.1.2.1. **Complete the Point Cover (PO) methods.** (See the Vegetation Sampling Overview for more information.)

7.2.1.3. *It is important to monitor changes in the number of plant species and/or the number of individual plants*.

7.2.1.3.1. **Complete the Density (DE) methods**. (See the Vegetation Sampling Overview for more information.)

7.2.2. *Height of >50 percent vegetation greater than 3 feet*.

7.2.2.1. **Complete Line Intercept (LI) method**. (See the **Vegetation Sampling Overview** for more information.)

8. Changes in *fuel loadings* are important to successfully completing monitoring objectives. Fuel measurements include loadings of all woody fuel size classes, duff, litter, live and dead shrub and herbaceous; duff and litter depths; and coarse woody debris description with rot classes. This method

is used mostly for estimating fuel consumption and smoke generation, and to describe stand-level fuel characteristics.

8.1. **Complete the Fuel Load (FL) method**

9. Changes in *tree or stand characteristics* (mortality, survival, damage) are important for describing fire effects for monitoring objective. Tree measurements include health, insect and disease evidence and damage, crown characteristics, diameter, height, and fire damage. This method is used mostly to quantify tree and stand mortality and to describe stand-level tree characteristics.

9.1. **Complete the Tree Data (TD) method**

10. Documentation of *aggregate fire effects within strata* is important to determine cumulative burn severity on the community. Coverage of large areas and diverse conditions is important, with minimal time spent per plot. Also, a means to calibrate or validate remote sensing data is sought for moderate resolution applications. This method integrates independent ratings of severity by strata to determine understory, overstory, and overall severity.

10.1. **Read the Landscape Assessment (LA) section, and complete the Composite Burn Index (BI) method.**

11. Spatial representation of *fire effects is desired over large areas using GIS*, targeting burns exceeding about 100 ha (250 ac). Analysis of historic burns (back to about 1983) is desired. A need exists to monitor burns over long periods, as in a regional fire atlas, or to relate fire effects to environmental variables continuously across a landscape. Also, mapping is important, to display or analyze burn results at a landscape resolution of 30 meters.

11.1. **Read the Landscape Assessment (LA) section, and complete the Normalized Burn Ratio (BR) method.**

## Vegetation Sampling Overview

The FIREMON system uses five vegetation sampling procedures that are useful for sampling vegetation for most monitoring situations. Each method has its strengths and weaknesses, and the descriptions below are provided to help you determine which sampling method is best for your project.

**Species Composition (SC) Method**: This method is primarily used to acquire inventory data over large areas using few examiners. The SC method is useful for documenting important changes in plant species cover and composition over time. However, this method is not designed to monitor statistically significant changes in vegetation over time. The SC sampling method primarily addresses individual plant species canopy cover and height for vascular and nonvascular plants. Canopy cover and average height may be recorded by size classes for plant species. Size class data can provide important structural information about the stand such as the vertical distribution of plant species cover.

**Cover/Frequency (CF) Method**: This method is primarily used for monitoring changes in plant species cover, height, and frequency. The CF sampling method primarily addresses individual plant species canopy cover, height, and frequency for vascular and nonvascular plants less than 3 feet (1 m) in height. The FIREMON line intercept (LI) method is better suited for estimating cover of shrubs greater than 1 m in height (Western United States shrub communities, mixed plant communities of grasses, trees, and shrubs, and open grown woody vegetation). The CF methods can also be used to estimate ground cover. However, the FIREMON point intercept (PO) method is better suited for estimating ground cover. We suggest that if you are primarily interested in monitoring changes in ground cover, you use the PO method because it is not a subjective measure. The PO method is also better suited for sampling fine-textured herbaceous communities (dense grasslands and wet meadows). However, if rare plant species are of interest, the CF methods are preferred because it is easier to sample rare species with quadrats than with points or lines.

**Line Intercept (LI) Method**: This method is primarily used for monitoring changes in plant species cover and height. This method is primarily designed to sample plant species with solid crowns or large basal areas. The LI method works best in open grown woody vegetation (Western United States shrub communities), especially shrubs greater than 3 feet (1 m) in height. The CF method is generally preferred for sampling herbaceous plant communities with vegetation less than 3 feet (1 m) in height. However, the LI method can be used in combination with the CF method if shrubs greater than 3 feet (1 m) exist on the plot. This is probably the best method of sampling canopy cover in mixed plant communities with grasses, shrubs, and trees. This method is not well suited for sampling single stemmed plants or dense grasslands. The PO method is better suited for sampling fine-textured herbaceous communities such as dense grasslands and wet meadows. Canopy cover measured with line intercept is less prone to observer bias than ocular estimates of cover in quadrats such as in the CF method. However, if rare plant species are of interest, the CF methods are preferred because it is easier to sample rare species with quadrats than with points or lines.

**Point Intercept (PO) Method**: This method is primarily used when the fire manager wants to monitor changes in plant species cover and height or ground cover. This sampling method is best suited for sampling ground cover and grasses, forbs, and shrubs less than 3 ft (1 m) in height. The point cover method works best for fine leaved plant species (grasslands and wet meadows) and species with open canopies (pastures and grasslands), which can be more difficult to estimate with the LI method. The PO method can provide a more accurate estimate of cover than the ocular estimates used in the CF sampling method because sampler error is removed. Examiners only decide if the sampling pole contacted a plant species or ground cover class. It can be difficult to detect rare plants unless many points are used for sampling. If rare plant species are of interest the CF methods are preferred since it is easier to sample rare species with quadrats than with points or lines. We suggest you use the PO method if you are primarily interested in monitoring changes in ground cover (bare ground, herbaceous cover, and so forth).

**Density (DE) Method**: This method is primarily used when the fire manager wants to monitor changes in plant species numbers. This method is primarily suited for grasses, forbs, shrubs, and small trees, which are easily separated into individual plants or counting units, such as stems. However, we recommend using the FIREMON TD sampling method for estimating tree density. The DE sampling method uses density to assess changes in plant species numbers over time. The quadrat size and belt width varies with plant species or item and size class, allowing different size sampling units for different size plants or items. Quadrat size and belt width should be adjusted according to plant size and distribution. This method is particularly useful for sampling rare plants where monitoring an increase or decrease in numbers is important. This method also provides useful information on seedling emergence, survival, and mortality.

**Rare Species (RS) Method**: This method is primarily used to monitor uncommon grass, forb, shrub, and tree species of special interest, including Threatened and Endangered species. Individual plants are monitored for changes in plant survivorship, growth, and reproduction over time. Individual plants are spatially located using distance along and from a permanent baseline and permanently tagged. Data are collected for status (living or dead), stage (seedling, nonreproductive, or reproductive), size (height and diameter), and reproductive effort (number of flowers and fruits).

## Analysis Tools

Fire effects monitoring is defined by two tasks: field data collection and evaluation. Field data collection has been discussed in detail in the Integrated Sampling Strategy and Field Assessment documentation. Discussed here are the software and procedures for the evaluation of the field data to assess fire effects.

The FIREMON Analysis Tools component encompasses two major tasks: 1) data entry and 2) data analysis. Data entry is accomplished in FIREMON by physically entering the collected field data into a set of standardized Microsoft Access databases. Data analysis is accomplished using a set of database queries and computer programs developed specifically for FIREMON.

The collected data are stored in Microsoft Access FIREMON databases that have nested data entry screens with error checking capabilities. The database structure is similar to that of ECODATA (Hann and others 1988; Keane and others 1990). Linked to these databases are database queries and computer programs that summarize the sampled data into the reports required by fire management agencies.

The database entry screens are designed so that all methods are present on the home screen, and the user need only enter data in the screens describing the selected methods. For instance, if only fuels were measured on a monitoring plot, then the user only enters information in the Required fields of the Plot Description form and in the Fuel Load table. In standardized fields, users select the code from a dropdown list. Most lists can be modified be the user by adding, removing, and modifying the codes.

The analysis programs perform a variety of tasks. First, is a program that scans the entire database and locates empty fields. The workhorse program imports the database and performs common calculations on the data. For example, fuels data are entered by transect, but what is needed is an estimate of fuel loading, which is the summarization of the transect data. The base code computes fuel loadings and then summarizes results in a table. The analysis program summarizes all database entries into one report for storage as a computer file or paper report. The database also contains built-in queries for computations at the plot level.

The statistics program performs the temporal monitoring analyses using Dunnett's t-test and provides output reports showing the results of statistical tests. The summary can be stratified by any macroplot-level field or by the user.

The analysis package can also produce the necessary files to run the Forest Vegetation Simulator and the associated Fire and Fuels Extension. Last, data can be exported for input into spreadsheets or statistical programs.

# Plot Description (PD)

## Sampling Method

**Robert E. Keane**

## SUMMARY

The Plot Description (PD) form is used to describe general characteristics of the FIREMON macroplot to provide ecological context for data analyses. The PD data characterize the topographical setting, geographic reference point, general plant composition and cover, ground cover, fuels, and soils information. This method provides the general ecological data that can be used to stratify or aggregate fire monitoring results. The PD method also has comment fields that allow for documentation of plot conditions and location using photos and notes. The key for the FIREMON database—made up of the Registration Key, Project ID, Plot Number, and Date—is part of the PD form.

## INTRODUCTION

The Plot Description (PD) methods were designed to describe important ecological characteristics of the FIREMON macroplot. The macroplot is the area where the other FIREMON methods will be applied. All fields in the PD method pertain to the entire macroplot and should be estimated and recorded so that they describe the macroplot as a whole.

The seven general categories of data in the PD method are 1) required, 2) plot information, 3) biophysical settings, 4) vegetation, 5) ground cover, 6) fire, and 7) common/comment. Only the required fields must be completed. However, within each category, there are some groups of fields that belong together and must be completed as a group. These will be evident on the PD data form and discussed in detail in this chapter.

All fields in the required category must be completed regardless of the sampling methods employed. These fields uniquely identify the plot data within the FIREMON database.

## SAMPLING PROCEDURE

This method assumes that the sampling strategy has already been selected and the macroplot has already been located. If this is not the case, then refer to the **FIREMON Integrated Sampling Strategy** for further details.

The PD sampling methods described here are the recommended procedures for this method. Later sections will describe how the FIREMON three-tier sampling design can be used to modify the recommended procedure to match resources, funding, and time constraints.

Plot Description (PD) Sampling Method

The sampling procedure is described in the order of the fields that need to be completed on the PD data form, so it is best to reference the data form when reading this section.

If there are data that you would like to collect but cannot due to broken equipment or other unforeseen circumstances, record each instance in the Comments field for the plot. For instance, if you cannot measure the slope because the clinometer was broken, leave the Slope field empty and note in the Comments field, "No slope measurements were taken because the clinometer was broken." This will explain empty fields to future users of the data. Do not enter 0 (zero) in a field that could not be assessed. Either leave the field blank or enter the code that denotes you were not able to assess the attribute.

See **How To Locate a FIREMON Plot, How To Permanently Establish a FIREMON Plot, and How to Define the Boundaries of a Macroplot** in the **How-To Guide** chapter for more information on setting up your macroplot.

## Required PD Fields—Database Key

These four fields constitute the key for your FIREMON database. If you are entering data these fields *must* be entered.

The FIREMON Analysis Tools program will allow summarization and comparison of plots only if they have the same Registration and Project Codes. This restriction is set because typically each monitoring project has unique objectives with the sample size and monitoring methods developed for specific reasons intimately related to each project. Comparisons made between projects with dissimilar methods may not be appropriate.

**Registration Code**—The Registration Code is a four-character code determined by you or assigned to you. The Registration Code should be used to identify a large group of people, such as all the people at one District of a National Forest or the people working under one monitoring leader. You are required to use all four characters. Choose your Registration Code so that the letters and numbers are related to your business or organization. For example:

MFSL = Missoula Fire Sciences Lab

MTSW = Montana DNRC, Southwest Land Office

CHRC = Chippewa National Forest, Revegetation Crew

RMJD = Rocky Mountain Research Station, John Doe

**Project Code**—The Project Code is an eight-character code used to identify project work that is done within the group. You are not required to use all eight characters. Some examples of Project Codes are:

TCRESTOR = Tenderfoot Creek Restoration

BurntFk = Burnt Fork Project

SCF1 = Swan Creek Prescribed Fire, Monitoring Crew 1

BoxCkDem = Box Creek Demonstration Project

It will be easier to read the sorted results if you do not include digits in the left most position of the project code. For instance, if two of your projects are 22Lolo and 9Lolo, then when sorted 22Lolo will come before 9Lolo. The preferred option would be to name the projects Lolo09 and Lolo22, although Lolo9 and Lolo22 will also sort in the proper order.

**Plot Number**—Identifier that corresponds to the site where sampling methods are applied. Integer value.

**Sampling Date**—Enter the date of sampling as an eight-digit number in the MM/DD/YYYY format where MM is the month number, DD is the day of the month, and YYYY is the current year. For example, April 01, 2001, would be entered 04/01/2001.

Plot Description (PD) Sampling Method

## Organization Code Fields

These four fields are provided so that users can sort and summarize data using agency location codes—for instance, USFS Region, Forest, and District. All four fields allow alphanumeric characters.

Field 1: Organization Code 1—four-character field.

Field 2: Organization Code 2—two-character field.

Field 3: Organization Code 3—two-character field.

Field 4: Organization Code 4—two-character field.

## Plot Information Fields

**Field 5: Examiner Name**—The name of the FIREMON crew boss or lead examiner should be entered up to eight-characters. This is a nonstandardized field so anything can be entered here, but we suggest the name follow the convention of first letter in first name followed by a dot followed by the entire last name. So, Smokey Bear would be s.bear and John Smith would be j.smith. We strongly suggest that there are no blanks in the text—for example, don't enter Smokey Bear as s. bear.

**Field 6: Units**—Enter "E" if you will be collecting data using English units or "M" if you using metric units. These units are used for all measurements in the sampling. The only exception is the Error Units field associated with the GPS location. GPS error may be in English or metric units regardless of what is entered in Field 6.

The macroplot is the area where you will be applying the FIREMON methods. The size of the macroplot ultimately dictates the representative area to be sampled (table PD-1). If vegetation is dense, large plot sizes usually take longer to sample because it is difficult to traverse the plot. However, some ecosystems have large trees scattered over large areas so that large plot sizes are needed to obtain realistic estimates. Studies have attempted to identify the optimum plot size for different ecosystems but have only provided mixed results. We offer the following table to help determine the plot size that matches the fire monitoring application. Plot size and shape selection should be determined by the FIREMON project leader prior to entering the field.

Usually, the 0.1-acre circular plot will be sufficient for most ecosystems, and this size should be used if no other information is available. A general rule of thumb is that the plot should be big enough to capture at least 20 trees above 4 inches diameter at breast height (DBH) on average (across all plots in your project). It is important that the plot size stay constant across all plots in a sampling project. For example, if a FIREMON project contains shrublands, grasslands, and forests, don't change the plot size when you sample each one. Select the largest plot size (forests, in this example) and use it for all ecosystems. In general we suggest using a circular PD macroplot.

Two fields in the PD method are used to describe plot shape and size. If the plot shape is circular, then enter plot radius/length in Field 7 and enter 0 (zero) in Field 8. If a rectangular plot shape is required, the length of the macroplot is entered in Field 7 and the width is entered in Field 8. No other plot shapes are used in FIREMON.

Table PD-1  Suggested FIREMON macroplot plot sizes.

| Average plant height | Pant cover | Suggested plot size | Plot radius | Suggested plot size | Plot radius |
|---|---|---|---|---|---|
| ft | % | acres | ft | $m^2$ | m |
| X < 15 | <50 | 0.10 | 37.2 | 400 | 11.3 |
|  | >50 | 0.05 | 26.3 | 200 | 8.0 |
| 15 < X < 100 | <50 | 0.10 | 37.2 | 400 | 11.3 |
|  | >50 | 0.08 | 33.3 | 300 | 9.8 |
| X > 100 | <50 | 0.40 | 74.5 | 1,000 | 17.8 |
|  | >50 | 0.13 | 42.5 | 500 | 12.6 |

## Plot size

**Field 7: Plot Radius** (ft/m)—If the macroplot is circular enter the radius of the macroplot. Enter the length of the macroplot if it is rectangular.

**Field 8: Plot Width** (ft/m)—Enter the width of the plot if it is rectangular, or enter zero (0) or leave the field blank if the macroplot shape is circular.

## Sampling information

FIREMON data can be collected on "Monitoring" plots or "Control" plots. Monitoring plots are located inside the treatment area so that you can compare the effects of different treatments on the sampled attributes. Control plots are placed outside the treatment area and used to check that any changes in the sampled attributes were actually due to the treatments and not some unrelated factor. This topic is discussed more in the Integrated Sampling Strategy document.

**Field 9: Plot Type**—Enter "M" if you are sampling a monitoring plot or "C" if you are sampling a control plot.

**Field 10: Sampling Event**—Monitoring requires that sampling be stratified by space and time. Since monitoring is a temporal sampling of repeated measures, it is essential that you record the reason for sampling to provide a context for analysis. The Sampling Event field is used to document why the plot is being measured at this particular time (as recorded by Date). The Sampling Event field will help you track changes at the plot level more easily than if you used only the sampling date. The codes used for this field are: 1) P is the pretreatment measurement of the plot, 2) R is the posttreatment, remeasurement of the plot, and 3) IV indicates an Inventory plot that is not permanently monumented and won't be resampled (table PD-2). The codes P and R are followed by a numeric value that indicates the sampling visit of the current sampling. For instance, if you sample a plot once before a prescribed fire the code would be P1, then when you sample after the fire, the code will be R1 for the first sampling, R2 for the second sampling, and so on. When you change event codes, from P to R, you should start the sequential sample number over at 1. The FIREMON database will accept data for up to three pretreatment measurements. When you are sampling a plot that has been sampled once or more before you will have to consult previously collected FIREMON data so that you use the appropriate sequential sample number. For simplicity we have only provided standardized codes for pre- and posttreatment measurements. This may be a problem if, for instance, you plan on three measurements: one preharvest, one postharvest/preburn, and one postburn. We suggest using the P before any treatments are applied then using R codes after the first treatment. In the previous example the codes would be: P1 for the preharvest sample, R1 for the postharvest/preburn sample, and R2 for the postburn sample. Be sure to note the sampling event numbering scheme in the Metadata table. You can make up your own codes if you chose. However, the FIREMON Analysis Tools program will not recognize codes other than those listed in table PD-2 and won't be able to do any analysis for you. If you are doing inventory sampling (you will not be resampling the plots) code them IV.

## Linking fields

**Field 11: Fire ID**—Enter a Fire ID of up to 15 characters. The ID number or name that relates the fire that burned this plot to the same fire described in the Fire Behavior (FB) table. This field links this plot

**Table PD-2** Sampling Event codes.

| Code | Event |
|---|---|
| P$n$ | Pretreatment measurement, sequential sample number. |
| R$n$ | Posttreatment remeasurement of a plot, sequential sample number. |
| IV | Inventory plot, not a monitoring plot. |

scale data with the fire scale data in the FB method. There may be many FIREMON plots referencing one fire. This field will be empty until after the burn has been completed.

**Field 12: Metadata ID**—Enter code of up to 15 characters that links the plot data to the MD table. The Metadata (MD) table is used to store information on the sampling intensity and methods used in the monitoring project. This field is highly recommended so that important information will be recorded for future reference.

## Georeferenced plot positions

The next set of fields is important for relocating FIREMON sample plots and for using FIREMON plot data in mapping and map validation of remote sensing projects. These fields fix the geographic location of the plot center.

Geographic coordinates are nearly always obtained from a Geographic Positioning System (GPS). GPS technology uses data from at least four orbiting satellites to triangulate your position in three dimensions (X, Y, Z, or North, East, Elevation) to within 3 to 50 meters of accuracy. GPS receivers are available from many sources, and there are a wide range of GPS models to choose from depending on various sampling criteria. GPS selection and training are not part of the FIREMON sampling methods. However, a number of resources provide advice on purchasing the right GPS for your sampling needs. A wide variety of public and private agencies also provide excellent training. We recommend that the georeferenced coordinates for FIREMON plots be taken from a GPS receiver and not from paper maps such as USGS quadrangle maps because of the high degree of error. Average the plot location over at least 200 readings to reduce the location error.

Many map projections are available to record FIREMON plot georeferenced coordinates. Users can use either latitude-longitude (lat-long) or the UTM (Universal Transverse Mercator) coordinate system. If you are using UTM coordinates, record easting and northing to the nearest whole meter. If you are using lat-long coordinates, record latitude and longitude to the sixth decimal place using decimal degrees (this corresponds to about 1 meter of ground distance at 45 degrees latitude). The down side of lat-long coordinates is that it is difficult to visualize the measurements on the ground (how far is 0.05 degrees latitude). Be especially alert because units of degrees-min-seconds look similar to decimal degrees. If using lat-long coordinates, enter data in Fields 14, 15, 19, 20, and 21. If using UTM coordinates enter data in Fields 16 to 21

**Field 13: Coordinate System**—Record the coordinate system being used. Latitude and longitude (lat-long) or Universal Transverse Mercator (UTM).

**Field 14: Latitude**—If using the lat-long system, enter the latitude, in decimal degrees to six decimal places.

**Field 15: Longitude**—If using the lat-long system, enter the longitude, in decimal degrees to six decimal places.

**Field 16: Northing**—If using the UTM system, enter the UTM northing to the nearest whole meter.

**Field 17: Easting**—If using the UTM system, enter the UTM easting to the nearest whole meter.

**Field 18: Zone**—If using the UTM system, enter the UTM zone of the plot center.

**Field 19: Datum**—If using the UTM system, enter the datum used in conjunction with the UTM coordinates.

**Field 20: Position Error**—Enter the position error value provided by the GPS unit. This should be entered regardless of whether you are using lat-long or UTM coordinates.

**Field 21: Error Units (E/M)**—Enter the units associated with the GPS error. May be different than the units listed in Field 6.

Fields 5 through 21 make up the information that is critical to have for every FIREMON macroplot, regardless of the sampling intensity or methods you will be using to collect data.

Plot Description (PD) Sampling Method

The following sections describe the measurement or estimation of various ecosystem characteristics that are important to fire effects monitoring.

## Biophysical Setting Fields

The biophysical setting describes the physical environment of the FIREMON plot relative to the organisms that grow there. Many site characteristics can be included in a description of biophysical setting, but only topography, geology, soils, and landform fields are implemented in FIREMON.

### Topography

**Field 22: Elevation (ft/m)**—Enter the elevation above MSL (mean sea level) of the FIREMON plot in feet (meters) to the nearest 100 feet (30 m). Elevation can be estimated from three sources. Most GPS readings include an estimate of elevation, and these estimates are usually fairly accurate. Elevation can also be estimated from an altimeter. There are many types of altimeters, but most are barometric, estimating elevation from atmospheric pressure. Altimeters are notoriously fickle and need calibration nearly every day. When there are frequent weather systems passing the area, altimeters should be calibrated every 4 hours. Finally, elevation can be taken from USGS topographic maps.

**Field 23: Plot Aspect**—Enter the aspect of the FIREMON plot in degrees true north to the nearest 5 degrees. Aspect is the direction the plot is facing. For example, a slope that faces exactly west would have an aspect of 270 degrees true north. Be sure to record the aspect that best represents the macroplot as a whole and not just the point where you are standing. Also, be sure you check your compass reading with your knowledge of the area to be sure that the aspect indicated is really correct. Often, metal on sampling equipment, or iron rebar plot center, can influence the estimation of aspect. For information about using a compass see **How to Use a Compass—Sighting and Setting Declination** in the **How-To-Guide** chapter.

**Field 24: Slope**—Record the plot slope using the percent scale to the nearest 5 percent. The slope is measured as an average of the uphill and downhill slope from plot center. See How To Measure Slope in the How-To Guide chapter for more information. Be sure the recorded slope reflects the slope of the entire plot and not just the line where you are standing. Slope values should always be positive.

**Field 25: Landform**—Enter up to a four-character code that best describes the landform containing the FIREMON macroplot from table PD-3. See **Appendix C: NRIS Landform Codes** for a complete list.

**Field 26: Vertical Slope Shape**—Enter up to a two-character code using the classes in table PD-4 that best describes the general contour of the terrain upslope and downslope from plot center. As you look up and down the slope, estimate a shape class that best describes the horizontal contour of the land (fig. PD-1).

**Field 27: Horizontal Slope Shape**—Enter up to a two-character code using the classes in table PD-4 that best describes the general contour of the terrain upslope and downslope from plot center. This is

Table PD-3 Landform codes.

| Code | Landform |
|------|----------|
| GMF | Glaciated mountains foothills |
| UMF | Unglaciated mountains foothills |
| BRK | Breaklands river breaks badlands |
| PLA | Plains rolling planes plains w/breaks |
| VAL | Valleys swales draws |
| HIL | Hill low ridges benches |
| X | Did not assess |

Table PD-4 Slope shapes.

| Code | Slope shape |
|------|-------------|
| LI | Linear or planar |
| CC | Depression or concave |
| PA | Patterned |
| CV | Rounded or convex |
| FL | Flat |
| BR | Broken |
| UN | Undulating |
| OO | Other shape |
| X | Did not assess |

Plot Description (PD) Sampling Method

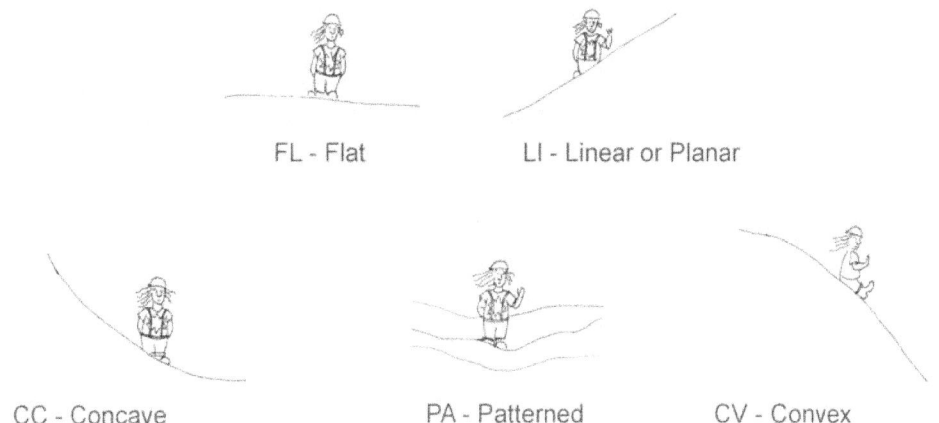

**Figure PD-1**—These illustrations depict the different types of vertical slope shapes. Horizontal slope shapes use the same classification but are determined by examining the across slope profile, rather than up and down the slope.

an estimate of the general shape of the slope parallel to the contour of the slope. As you look across the slope along the contour, estimate a shape class that best describes the horizontal contour of the land (fig. PD-1).

## Geology and soils fields

**Field 28: Primary Surfical Geology**—This is the first of five fields used to describe geology and soils. Determine the geological rock type composing the parent material at the plot and enter the appropriate code from table PD-5 into the field. Generally, identification of surficial geology requires someone with specialized training and experience.

**Field 29: Secondary Surfical Geology (Field 29-SGEOLOGY)**—Use this field only if you have coded a primary surficial geology type. Determine the secondary geological rock type composing the parent material at the plot and enter the appropriate code from table PD-6 into the field. Generally,

**Table PD-5** Common primary surficial geology codes.

| Primary code | Rock type 1 |
|---|---|
| IGEX | Igneous extrusive |
| IGIN | Igneous intrusive |
| META | Metamorphic |
| SEDI | Sedimentary |
| UNDI | Undifferentiated |
| X | Did not assess |

**Table PD-6** Common secondary surficial geology codes. Additional codes are listed in Appendix B.

| Secondary code | Rock type 2 | Secondary code | Rock type 2 |
|---|---|---|---|
| ANDE | Andesite | CONG | Conglomerate |
| BASA | Basalt | DOLO | Dolomite |
| LATI | Latite | LIME | Limestone |
| RHYO | Rhyolite | SANS | Sandstone |
| SCOR | Scoria | SHAL | Shale |
| TRAC | Trachyte | SILS | Siltstone |
| DIOR | Diorite | TUFA | Tufa |
| GABB | Gabbro | MIEXME | Mixed extrusive and metamorphic |
| GRAN | Granite | MIEXSE | Mixed extrusive and sedimentary |
| QUMO | Quartz monzonite | MIIG | Mixed igneous (extrusive and intrusive) |
| SYEN | Syenite | MIIGME | Mixed igneous and metamorphic |
| GNEI | Gneiss | MIIGSE | Mixed igneous and sedimentary |
| PHYL | Phyllite | MIINME | Mixed intrusive and metamorphic |
| QUAR | Quartzite | MIINSE | Mixed intrusive and sedimentary |
| SCHI | Schist | MIMESE | Mixed metamorphic and sedimentary |
| SLAT | Slate | X | Did not assess |
| ARGI | Argillite | | |

identification of surficial geology requires someone with specialized training and experience. Table PD-6 is an abridged list of common surficial types. A complete list is included in Appendix B: NRIS Lithology Codes.

**Field 30: Soil Texture Class**—The description of soil on the FIREMON plot is limited to a general description because fire effects are not influenced by fine-scale soil characteristics. Generally, identification of soil texture requires someone with specialized training and experience. Many fire effects can be described by general soil characteristics, and soil texture is one of those general characteristics. Enter the code that best describes the texture of the soil on the FIREMON macroplot (table PD-7). These soil textures are described in many soils textbooks. If you are unsure of how to evaluate soil texture or have no confidence in your estimates, then use the X code or leave the field blank. We have only included the codes for soil texture required by FOFEM; if additional codes are desired you may design them on your own and note them in the MD table.

**Field 31: Erosion Type**—Erosion is an important second order fire effect that needs to be documented. We have based the Erosion Type on the classification used by the Natural Resources Conservation Service Soil Survey Handbook (table PD-8). See www.nrcs.usda.gov/technical/references/ for more information. If your macroplot is on a site that has moved in its entirety through landslip, include that information in the Comments field of the PD form, then code Field 31 with the code that identifies the erosion conditions you are seeing on the surface. Be sure to record erosion on preburn plots in order to provide the reference conditions. The types of erosion are listed along with the codes in table PD-8. Enter the code that best describes the erosion occurring on the plot.

**Field 32: Erosion Severity**—The severity of the erosion event is extremely difficult to assess and is best estimated by those who have some experience with erosion processes. We have based the Erosion Severity on the classification used by the USDA Natural Resources Conservation Service Soil Survey Handbook (table PD-9). The severity codes use the depth and extent of erosion to quantify severity. Enter the code that best fits the severity of the erosion on the plot in this field. Severity codes do not apply to tunnel erosion. If you have tunnel erosion on your plot enter –1 in this field.

## Vegetation Fields

These PD fields describe general aspects of the vegetation using percent canopy cover as the measurement unit. All vegetation fields require an estimate of the percent vertically projected canopy cover recorded by class (table PD-10). Cover estimation methods are described in the **How To Estimate Cover** section of the **How-To Guide** chapter.

The seasonal timing of cover estimates can lead to substantially different cover estimations especially for the shrub and herbaceous components. Critically consider how and when cover should be estimated based on project objects, resources, and the sampling experience of the crew. One option may be to attempt to estimate what the cover would be at the peak of the growing season. Doing so can remove

**Table PD-7** Soil texture codes.

| Code | Description | Code | Description |
|---|---|---|---|
| C | Clay | S | Sand |
| CL | Clay loam | SC | Sandy clay |
| COS | Coarse sand | SCL | Sandy clay loam |
| COSL | Coarse sandy loam | SI | Silt |
| FS | Fine sand | SIC | Silty clay |
| FSL | Fine sandy loam | SICL | Silty clay loam |
| L | Loam | SIL | Silt loam |
| LCOS | Loamy coarse sand | SL | Sandy loam |
| LFS | Loamy fine sand | VFS | Very fine sand |
| LS | Loamy sand | VFSL | Very fine sandy loam |
| LVFS | Loamy very fine sand | X | Did not assess |

**Table PD-8** Erosion type codes.

| Code | Erosion type |
|---|---|
| S | Stable, no erosion evident |
| R | Water erosion, rill |
| H | Water erosion, sheet |
| G | Water erosion, gully |
| T | Water erosion, tunnel |
| W | Wind erosion |
| O | Other type of erosion |
| X | Did not assess |

**Table PD-9** Erosion severity codes.

| Code | Erosion severity |
|---|---|
| 0 | Stable, no erosion is evident. |
| 1 | Low erosion severity; small amounts of material are lost from the plot. On average less than 25 percent of the upper 8 inches (20 cm) of soil surface have been lost across the macroplot. Throughout most of the area the thickness of the soil surface layer is within the normal range of variability of the uneroded soil. |
| 2 | Moderate erosion severity; moderate amounts of material are lost from the plot. On average between 25 and 75 percent of the upper 8 inches (20 cm) of soil surface have been lost across the macroplot. Erosion patterns may range from small, uneroded areas to small areas of severely eroded sites. |
| 3 | High erosion severity; Large amounts of material are lost from the plot. On average 75 percent or more of the upper 8 inches (20 cm) of soil surface have been lost across the macroplot. Material from deeper horizons in the soil profile is visible. |
| 4 | Very high erosion severity; Very large amounts of material are lost from the plot. All of the upper 8 inches (20 cm) of soil surface have been lost across the macroplot. Erosion has removed material from deeper horizons of the soil profile throughout most of the area. |
| 1 | Unable to assess. |

**Table PD-10** Cover codes. Use these codes to record vegetation cover in the fields that call for cover estimation.

| Code | Cover class |
|---|---|
| 0 | Zero percent cover |
| 0.5 | >0  1 percent cover |
| 3 | >1  5 percent cover |
| 10 | >5  15 percent cover |
| 20 | >15  25 percent cover |
| 30 | >25  35 percent cover |
| 40 | >35  45 percent cover |
| 50 | >45  55 percent cover |
| 60 | >55  65 percent cover |
| 70 | >65  75 percent cover |
| 80 | >75  85 percent cover |
| 90 | >85  95 percent cover |
| 98 | >95  100 percent cover |

some of the seasonal variation in vegetation sampling. However, it can also lead to error in the cover estimates.

Cover of herbaceous plants often appears greater when they are dormant because they fall over and lie flat on the ground. To get accurate values for these species, estimate cover as if they were erect.

Vegetation cover in these PD fields is stratified by lifeform and size class. This makes determining canopy cover difficult because estimations require quite a bit of experience to arrive at consistent assessments of lifeform and size class cover when lifeforms and classes are unevenly distributed in all three dimensions. If you are unable to make an estimation for any reason, leave the field blank and note the reason in the comments section (Field 81). Always enter the code 0 (zero) when there is no cover for that ground element.

Vegetation cover does not need to sum to 100 percent by lifeform because there will probably be overlapping cover across all lifeforms. However, the total cover for each lifeform must always be greater than any of the covers estimated for the size classes within that lifeform.

### Vegetation—trees

The following fields provide an estimate of tree cover by size class.

**Field 33: Total Tree Cover**—Enter the percent canopy cover of all trees using the canopy cover codes presented in table PD-10. This estimate includes cover of ALL tree species from the smallest of seedlings to the tallest of old growth stems. It includes all layers of canopy vertically projected to the ground.

**Field 34: Seedling Tree Cover**—Enter the percent canopy cover of all trees that are less than 4.5 feet (1.4 m) tall using the codes in table PD-10. This cover estimate includes only the small seedlings.

**Field 35: Sapling Tree Cover**—Enter the percent canopy cover of all trees that are greater than 4.5 feet (1.4 m) tall and less than 5.0 inches (13 cm) DBH using the codes in table PD-10.

**Field 36: Pole Tree Cover**—Enter the percent canopy cover of all trees that are greater than 5 inches (13 cm) DBH and less than 9 inches (23 cm) DBH using FIREMON cover codes in table PD-10.

**Field 37: Medium Tree Cover**—Enter the percent canopy cover of all trees that are greater than 9 inches (23 cm) DBH up to 21 inches (53 cm) DBH using the codes in table PD-10.

**Field 38: Large Tree Cover**—Enter the percent canopy cover of all trees that are greater than 21 inches (53 cm) DBH up to 33 inches (83 cm) DBH using the FIREMON codes in table PD-10.

**Field 39: Very Large Tree Cover**—Enter the percent canopy cover of all trees that are greater than 33 inches (83 cm) DBH using the codes in table PD-10.

*Vegetation—shrubs*

The next set of fields allows the FIREMON sampler to estimate shrub cover in three height size classes.

**Field 40: Total Shrub Cover**—Enter the percent canopy cover of all shrubs on the plot into using the FIREMON canopy cover in table PD-10. This cover estimate includes vertically projected cover of all shrub species of all heights.

**Field 41: Low Shrub Cover**—Enter the percent canopy cover of all shrubs that are less than 3 feet (1 m) tall on the plot using the codes in table PD-10.

**Field 42: Medium Shrub Cover**—Enter the percent canopy cover of all shrubs that are greater than 3 feet (1 m) tall and less than 6.5 feet (2 m) tall on the plot using the codes in table PD-10.

**Field 43: Tall Shrub Cover**—Enter the percent canopy cover of all shrubs that are greater than 6.5 feet (2 m) tall on the plot using the codes in table PD-10.

*Vegetation—herbaceous*

Cover of grasses, forbs, ferns, mosses, and lichens are entered in the next set of vegetation fields. If you feel uncomfortable distinguishing between species within and across lifeforms, try to get some additional training from the ecologist, forester, or other resource specialists at your local office. Phenological adjustments must be made for many herbaceous species because most cure during the dry season, making cover estimation difficult. Follow the suggestions in the **How To Estimate Cover** section of the **How-To Guide** chapter to get the correct cover estimates.

**Field 44: Graminoid Cover**—Enter the percent canopy cover of all graminoid species on the plot into using the codes in table PD-10. Graminoid cover includes all grasses, sedges, and rushes in all stages of phenology. This cover is for all sizes and species of graminoids.

**Field 45: Forb Cover**—Enter the percent canopy cover of all forbs on the plot using the FIREMON cover codes in table PD-10.

**Field 46: Fern Cover**—Enter the percent canopy cover of all ferns on the plot using the FIREMON cover codes in table PD-10.

**Field 47: Moss and Lichen Cover**—Enter the percent canopy cover of all mosses and lichens on the plot using the codes in table PD-10. These mosses and lichens can be on the ground or suspended from plants in the air (arboreal).

*Vegetation—composition*

The following fields document the dominant plant species in each of three layers or strata on the FIREMON plot. These fields are used to describe the existing vegetation community based on dominance in cover. These descriptions are especially useful in satellite classification for mapping vegetation, developing existing vegetation community classifications, and for stratifying FIREMON fire effects results.

For a species to be dominant it has to have at least 10 percent canopy cover in that stratum, and the species must have higher cover than any other species in that stratum. In the PD method, two species per stratum are used to describe dominance. The first species (Species 1) is the most dominant in terms of canopy cover, and the second species (Species 2) is the second most dominant. Use the NRCS plant code or local species code to record the species.

There are three strata for stratifying dominant existing vegetation. The first stratum is called the Lower Stratum and is the cover of all plants less than 3 feet (1 m) tall. The Mid Stratum is for plants 3 to 10 feet (1 to 3 m) tall, while the Upper Stratum is for plants taller than 10 feet tall (3 m). Only species cover within the stratum is used to assess dominance. Many shade tolerant tree species can be dominant in all three strata.

If there are no species above 10 percent cover in a stratum, enter the code N indicating that there are no species that qualify for dominance. The same applies if there is no secondary species for dominance.

**Field 48: Upper Dominant Species 1**—Enter the species code of the most dominant species in the upper level stratum of the FIREMON plot. This is the stratum that is greater than 10 feet (3 m) above ground level.

**Field 49: Upper Dominant Species 2**—Enter the species code of the second most dominant species in the upper level stratum of the FIREMON plot. This is the stratum that is greater than 10 feet (3 m) above ground level.

**Field 50: Mid Dominant Species 1**—Enter the species code of the most dominant species in the mid level stratum of the FIREMON plot. This is the stratum that is greater than 3 feet and less than 10 feet (1 to 3 m) above ground level.

**Field 51: Mid Dominant Species 2**—Enter the species code of the second most dominant species in the mid level stratum of the FIREMON plot. This is the stratum that is greater than 3 feet and less than 10 feet (1 to 3 m) above ground level.

**Field 52: Lower Dominant Species 1**—Enter the species code of the most dominant species in the lowest level stratum of the FIREMON plot. This is the stratum that is less than 3 feet (1 m) above ground level.

**Field 53: Lower Dominant Species 2**—Enter the species code of the second most dominant species in the lowest level stratum of the FIREMON plot. This is the stratum that is less than 3 feet (1 m) above ground level.

## Potential vegetation

An important characteristic for describing biotic plant communities, especially in the Western United States, is the potential vegetation type. Potential vegetation generally describes the capacity of a site or FIREMON plot to support unique vegetation species or lifeforms. Potential vegetation is evaluated by describing the vegetation that would eventually occupy a site in the absence of disturbance over a long time. For example, an alpine site can only support herbaceous communities because these sites are too cold for shrubs or trees, whereas a clearcut cedar-hemlock site has the potential to support coniferous forest ecosystems. Potential vegetation classifications are highly ecosystem specific and are locally developed for certain regions, so a standardized potential vegetation classification for the entire United States does not currently exist. In FIREMON, potential vegetation is evaluated to broad lifeforms to aid in the interpretation of FIREMON results.

**Field 54: Potential Vegetation Type ID**—Potential vegetation types are the foundation of many management decisions. Many forest plans and project designs stratify treatments by potential vegetation type to achieve better results. Unfortunately, there is no national standard list of potential vegetation types in the United States. Instead, we have provided a generic field for the user to enter his or her own PVT code to stratify FIREMON results. This field is not standardized and any combination of alpha or numeric characters can be used. Do not use spaces in the text (enter ABLA/VASC). Be sure you document your codes in the FIREMON MD table. There are 16 characters available in this field.

**Field 55: Potential Lifeform**—Enter the potential lifeform code that best describes the community lifeform that would eventually inhabit the FIREMON plot in the absence of disturbance (table PD-11).

## Ground Cover Fields

This next set of PD fields describes the fuels complex on the FIREMON plot. The first group of fuels fields characterizes ground cover by various characteristics important for evaluating fire effects. The

**Table PD-11**  Potential lifeform codes.

| Code | Potential lifeform |
|---|---|
| AQ | Aquatic   Lake, pond, bog, river |
| NV | Nonvegetated   Bare soil, rock, dunes, scree, talus |
| CF | Coniferous upland forest   Pine, spruce, hemlock |
| CW | Coniferous wetland or riparian forest   Spruce, larch |
| BF | Broadleaf upland forest   Oak, beech, birch |
| BW | Broadleaf wetland or riparian forest   Tupelo, cypress |
| SA | Shrub dominated alpine   Willow |
| SU | Shrub dominated upland   Sagebrush, bitterbrush |
| SW | Shrub dominated wetland or riparian   Willow |
| HA | Herbaceous dominated alpine   Dryas |
| HU | Herbaceous dominated upland   grasslands, bunchgrass |
| HW | Herbaceous dominated wetland or riparian   ferns |
| ML | Moss or lichen dominated upland or wetland |
| OT | Other potential vegetation lifeform |
| X | Did not assess |

standard FIREMON percent cover class codes (PD-10) are used to quantify ground cover. Ground cover is critical for describing fuel continuity and cover, but it is also used for evaluation of erosion potential and for classification of satellite imagery.

A group of generalized fuel attributes are used to describe biomass characteristics for the entire FIREMON plot. The first fields describe surface fuel characteristics through standardized fuel models, while the last fields describe crown fuel characteristics important for fire modeling.

## Ground cover

Ground cover attempts to describe important attributes of the forest floor or soil surface. Ground cover is estimated into 10 categories, with each category important for calculating subsequent or potential fire effects. Ground cover is another difficult sampling element. Cover within a category is evaluated as the vertically projected cover of that category that occupies the ground. *Only elements that are in direct contact with the ground are considered in the estimation of ground cover.* Ecosystem components suspended above the ground, such as branches, leaves, and moss, are not considered in the estimation of ground cover.

Ground cover is described by a set of 10 fields where the sum *must add to 100 percent* (unlike the PD vegetation cover fields) plus or minus 10 percent. We suggest the following strategy for making these cover estimates. First, estimate ground cover for those categories with the least ground cover. These categories are the easiest to estimate with high accuracies. Be sure you scan the entire FIREMON plot to check for mineral soil, moss/lichen, and rock ground cover. Next, estimate the basal vegetation field to the cover codes 0.5, 3, or 10 (basal vegetation rarely exceeds 15 percent ground cover). Lastly, use the ground cover fields with the most cover (this is often only one or two fields, such as duff/litter) to make your estimate add to 100 percent. See **How to Estimate Cover** in the **How-To Guide** chapter for more information. If you are unable to make an estimation for any reason, leave the field blank and note the reason in the **Comments** section (Field 81). Always enter the code 0 (zero) when there is no cover for that ground element.

**Field 56: Bare Soil Ground Cover**—Estimate the percent ground cover of bare soil using the codes in table PD-10. Bare soil is considered to be all those mineral soil particles less than $\frac{1}{16}$ inch (2 mm) in diameter. Bare soil does not include any organic matter. The bare soil can be charred or blackened by the fire.

**Field 57: Gravel Ground Cover**—Estimate the percent ground cover of gravel using the codes in table PD-10. Gravel is those mineral soil particles greater than $\frac{1}{16}$ inch (2 mm) in diameter to 3 inches (80 mm) in diameter. Again, gravel does not include any organic soil colloids. The gravel can be charred or blackened by the fire.

**Field 58: Rock Ground Cover**—Estimate the percent ground cover of rock using the codes in table PD-10. Rock ground cover is considered to be all those mineral soil particles greater than 3 inches (8 cm) in diameter, including boulders. Rocks can be blackened by the fire.

**Field 59: Litter and Duff Ground Cover**—Estimate the percent ground cover of all *uncharred* litter and duff on the soil surface using the codes in table PD-10. Litter and duff cover is mostly organic material, such as partially decomposed needles, bark, and leaves, deposited on the ground. Do not include any woody material into this ground cover category unless it is highly decomposed twigs or logs that appear to be part of the duff. Sometimes after a fire the litter and duff will be charred and the cover of this charred litter/duff is estimated into the Charred Ground Cover field and not here. Other ground cover elements that are included in this category include plant fruits, buds, seeds, animal scat, and bones. If human litter appears on the FIREMON plot, pick it up, throw it away, and do not include it in the ground cover estimate.

**Field 60: Wood Ground Cover**—Estimate the percent ground cover of all *uncharred* woody material using the codes in table PD-10. Woody ground cover is only those wood particles that are recognizable as twigs, branches, or logs. Do not include cover of suspended woody material, such dead branches connected on shrub or tree stems, into this field.

**Field 61: Moss and Lichen Cover**—Enter the percent canopy cover of all mosses and lichens on the plot using the codes in table PD-10. These mosses and lichens can be on the ground or suspended from plants in the air (arboreal). This is the same estimate as in Field 43. The duplication is because some people consider moss and lichens ground cover and some consider it vegetation.

**Field 62: Charred Ground Cover**—Estimate the percent ground cover of all *charred organic* material using the codes in table PD-10. Char is the blackened charcoal left from incomplete combustion of organic material. Char can occur on any piece of organic matter, such as duff, litter, logs, and twigs, and cover of all char is lumped into this category. Do not include ash into the charred ground cover. If it is difficult to distinguish char and black lichen, try to scrape the black area with your fingernail and then rub your nail on your plot sheet. Char will usually leave a mark.

**Field 63: Ash Ground Cover**—Estimate the percent ground cover of all ash material using the codes in table PD-10. Ash can sometimes look like mineral soil, but mineral surface feels sandy or gritty when touched, while ash will often feel like a powder. Ash can occur in a variety of colors (red, gray, white), but light gray is often the primary shade.

**Field 64: Basal Vegetation Ground Cover**—Estimate the percent ground cover of basal vegetation using the codes in table PD-10. Basal vegetation is the area of the cross-section of the stem where it enters the ground surface expressed as a percent of plot cover. This category is extremely difficult to estimate, but fortunately, it has some repeatable characteristics. First, basal vegetation rarely exceeds 15 percent cover, so it will only get four valid FIREMON cover codes: 0, 0.5, 3, or 10. Next, it is highly ecosystem specific. Usually only forested ecosystems have high basal vegetation ground covers. This field is only used for vascular plant species. All nonvascular species are estimated in the Moss/Lichen Ground Cover field.

**Field 65: Water Ground Cover**—Estimate the percent ground cover of standing water using the codes in table PD-10. Water ground cover includes rainwater puddles, ponding, runoff, snow, ice, and hail. Do not include wet surfaces of other ground cover categories in this estimate. Although water is often only ephemeral, its cover must be recorded to make cover estimates sum to 100.

### General fuel characteristics

These fields are designed to describe general, plot-level fuel attributes for mapping and modeling fuel characteristics to predict fire behavior and effects. For instance, these fields could provide the information needed to run the FARSITE model. Estimation of fuel characteristics is highly subjective and dependent on the experience of the FIREMON crew. If more objective, repeatable, and accurate fuel estimates are needed, then use the Fuel Load (FL) and the Tree Data (TD) methods to more accurately

Plot Description (PD) Sampling Method

and objectively measure information on surface and crown fuels. The crown fuel description fields (Fields 68-70) are often used as model inputs to determine crown fire spread rates, especially in the FARSITE fire growth model. Because these fields only pertain to crown fuels, they should only be completed if there is a significant tree canopy layer (greater than 10 percent canopy cover) above the surface fuel layer (>6 feet [2 m] tall) on the plot. The canopy layer can extend into the surface fuel layer (below 6 feet [2 m]); however, canopy layer must extend above the surface fuel layer to be considered canopy fuels instead of surface fuels.

**Field 66: Surface Fire Behavior Fuel Model**—Choose the appropriate fire behavior fuel model from the Anderson 1983 publication, Aids for Determining Fuel Models for Estimating Fire Behavior, or a custom fire behavior fuel model

**Field 67: Fuel Photo Series ID**—Many areas in the United States have associated photo series guides. The guides use photos to describe typical fuel loadings by major cover types and geographical area. Each picture is linked to intensively sampled fuel loadings. These series are used to visually estimate fuel loadings by matching a picture from the guide with the current conditions of different fuel classes on the macroplot. If used as described in the guides, you would record a photo number for each component. For instance, you would record a picture number for the photo that best correlates to the 1-hour fuels on the macroplot, record another picture for the 10-hour fuels, and so on. However, often only one picture is recorded per plot. This is for two reasons. First, many people don't know that each fuel component should be matched to a photo. Second, in many of the guides it is difficult to see the fine woody debris or make an accurate assessment of the duff and litter from the photographs. It is important to note that this method is highly subjective and notoriously inaccurate, but it is often the only means available for quantifying the fuelbed loadings.

In FIREMON we provide only one field for photo guide information. Compare the current fuel conditions on the macroplot with the pictures in a photo series, and record the photo number of the picture that most closely matches the plot conditions, using a locally designed code. You can use the publication number combined with the picture number to uniquely identify the photo. For instance, if you are using the photo series for estimating natural fuels in the Lake States (Ottmar and Vihnanek 1999) you could combine the NFES publication number, 2579, and the plot number of the photo that best describes your fuels conditions. In this case you would enter NFES2579MP04 in Field 63. You can use up to 12 characters. Design this field to best suit your needs, but document your code conventions in the FIREMON MD table. If you want to record more than one photo number, they can be recorded in the Comments section.

**Field 68: Stand Height (ft/m)**—Estimate the height of the highest tree stratum that contains at least 10 percent canopy cover. This value is used to model crown fire spread. Estimate to the nearest 3 feet (1 m).

**Field 69: Canopy Fuel Base Height (ft/m)**—The lowest point above the ground at which there is a sufficient amount of tree canopy fuel to propagate a fire vertically into the canopy. Canopy fuel base height (CFBH) is a stand level measurement that provides an index for crown fire initiation and should account for dense dead vertical fuels (lichens, needle-drape, dense dead branches) that could provide a conduit for entrance of a surface fire into the crown. Estimate canopy base height to the nearest foot (0.3 m). This is a macroplot-based assessment. Take into account the dead fuels attached to standing trees that on individual trees might not be sufficient to move flames up the tree, but when intermingled with the branches from other trees would. A trick to estimating canopy base height for the entire FIREMON plot is to envision a plastic sheet on the ground with a hole for each tree. Then, mentally try to lift the plastic sheet to the first dense section of the crown (part of crown having burnable biomass that could catch fire). The average height of the imaginary plastic sheet is the CFBH for the plot.

Because the CFBH assessment is subjective some crews may not be comfortable making it. Optionally, estimate the live canopy base height for the stand by imagining a plastic sheet lifted to the live crown on each tree and record the average height of the sheet. This assessment is somewhat less subjective than CFBH but does not capture the dead canopy fuels. If you collect these data rather than CFBH be sure to note it in the Metadata table.

**Field 70: Canopy Cover**—Estimate the percent canopy cover of the forest/tree canopy above 6 feet (2 m) using the codes in table PD-10. This value is used to estimate crown bulk density for crown fire spread modeling. Be sure you estimate cover as percent vertically projected canopy cover that includes the cover for all species.

## Fire Behavior and Effects Fields

These FIREMON fields are used to identify the fire event and to describe the fire behavior and the subsequent fire effects. Fire behavior is a physical description of the fire, whereas fire effects are assessed from observations of the ecosystem after the fire has burned the area. Fire behavior data will generally be collected at two scales: the plot scale and the fire scale. Plot scale data are collected on the FIREMON macroplot and are contained in just two fields on the PD field form: flame length and fire spread rate. There is also one field to enter the file name of a fire behavior photo. There will probably never be a fire where samplers are able to collect these data on every macroplot, but the information can be useful in determining relationships between fire behavior and fire effects. Recording flame length and spread rate, as well as taking a fire behavior photo, on even a subset of the total plots will be to your advantage. You will be collecting only flame length and spread rate data during a fire event. Any other fields on the PD form that are important to your project will be completed before the fire. Fire scale data—things such as fuel moistures, plume behavior, and spotting observations—are recorded in the FIREMON Fire Behavior (FB) table.

### Fire behavior

Enter the plot scale estimates of flame length and fire spread in the following two fields. This information will be collected during the fire event but using the data sheets from the most recent sampling before the fire. For example, if there were two preburn sampling visits, record fire behavior data in Fields 71 and 72 on the field forms where P2 was coded in the Sampling Event field (Field 10). This may lead to some confusion because you will be doing most of your sampling before the fire, then waiting until the weather allows you to burn at a later date. At that time you will have to relocate the field forms and fill in additional fields—Fire ID, Flame Length, Spread Rate, and Fire Behavior Picture. Remember, you can also use the Date field to identify the most recent forms.

**Field 71: Flame Length (ft/m)**—Flame length is the length of the flames from the center of the combustion zone to the end of the continuous flame. It is more highly correlated with fire intensity than flame height (fig. PD-2). Estimate flame length as an average within the FIREMON macroplot boundaries to the nearest 0.5 feet (0.2 m).

**Field 72: Spread Rate (ft/min or m/min)**—Estimate the average speed of the fire as it crosses the macroplot in feet per minute to nearest 1 foot per minute (meters per minute to nearest 0.3 meter). Estimate spread rate by noting the number of minutes it takes for the flaming front to pass two points separated by a known distance.

**Field 73: Fire Behavior Picture**—Enter the picture code—up to 15 characters—for a picture that best shows fire behavior as the flaming front crosses the FIREMON plot. This code will link to a digital

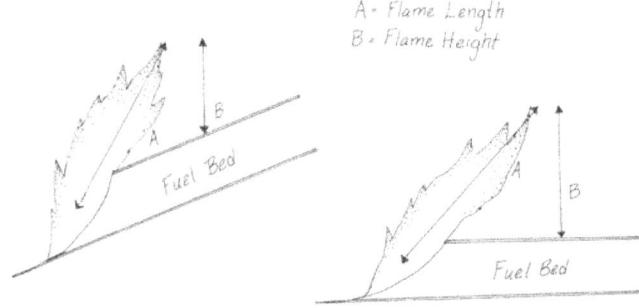

**Figure PD-2**—Illustration showing flame length versus flame height measurement. Enter your flame length estimate (A) into Field 71.

picture placed into the FIREMON database. The picture code could be something like R01P02 for Roll 1, picture number 2 for film cameras, or it could be a filename (for example, file0001.jpg) for digital cameras. Scan slides or paper photographs into JPEG files for entry into the FIREMON database.

## Fire effects

Fire effects must be evaluated from the burned evidence left on the FIREMON plot after the fire has passed. The fire severity classification used in the PD method is based on the NPS Fire Monitoring Handbook. Fire severity on larger areas (30 X 30 m) can be obtained by completing the Composite Burn Index methods (see the Landscape Assessment methods).

**Field 74: Fire Severity Code**—Enter the number (0 to 5) corresponding to the fire severity observed in the substrate and then in the overstory on the FIREMON plot using the descriptions in table PD-12. This fire severity classification is based on that used in the NPS Fire Monitoring Handbook

**Table PD-12** Use these fire severity class to determine the fire severity across the FIREMON macroplot.

| Fire severity code | Substrate | Forest vegetation | Shrubland vegetation | Grassland vegetation |
|---|---|---|---|---|
| Unburned (5) | Not burned | Not burned | Not burned | Not burned |
| Scorched (4) | Litter partially blackened; duff nearly unchanged; wood/leaf structures unchanged. | Foliage scorched and attached to supporting twigs. | Foliage scorched and attached to supporting twigs. | Foliage scorched |
| Lightly burned (3) | Litter charred to partially consumed; upper duff layer may be charred but the duff is not altered over the entire depth; surface appears black; where litter is sparse charring may extend slightly into soil surface but soil is not visibly altered; woody debris partially burned; logs are scorched or blackened but not charred; rotten wood is scorched to partially burned. | Foliage and smaller twigs partially to completely consumed; branches mostly intact. | Foliage and smaller twigs partially to completely consumed; branches mostly intact; typically, less than 60 percent of the shrub canopy is consumed. | Grasses with approximately two inches of stubble; foliage and smaller twigs of associated species partially to completely consumed; some plant parts may still be standing; bases of plants are not deeply burned and are still recognizable. |
| Moderately burned (2) | Litter mostly to entirely consumed, leaving coarse, light colored ash (ash soon disappears, leaving mineral soil); duff deeply charred, but not visibly altered; woody debris is mostly consumed; logs are deeply charred, burned out stump holes are evident. | Foliage twigs and small stems consumed; some branches still present. | Foliage twigs and small stems consumed; some smaller branches (0.25 0.50 inches) still present; typically, 40 to 80 percent of the shrub canopy is consumed. | Unburned grass stubble usually less than 2 inches tall, and mostly confined to an outer ring; for other species, foliage completely consumed, plant bases are burned to ground level and obscured in ash immediately after burning. |
| Heavily burned (1) | Litter and duff completely consumed, leaving fine white ash (ash disappears leaving mineral soil); mineral soil charred and/or visibly altered, often reddish; sound logs are deeply charred, and rotten logs are completely consumed. | All plant part consumed, leaving some or no major stems or trunks; any left are deeply charred. | All plant parts consumed leaving only stubs greater than 0.5 inch in diameter. | No unburned grasses above the root crown; for other species, all plant parts consumed. |
| Not applicable (0) | Only inorganic material on site before burn. | None present at time of burn. | None present at time of burn. | None present at time of burn. |

(http://www.nps.gov/fire/fire/fir_eco_science_monitoring_FMH.html). You will make two assessments of severity. First, examine the fire severity of the substrate component across the macroplot, select severity code from the table that most closely matches the effects you see, and record the code. Second, examine the overstory for the appropriate type—forest, shrubland, or grassland—then select the severity code from the table that most closely matches the effects you see, and record the code. You will enter both these numbers in the same field in the database. For instance, a fire resulting in a moderately burned substrate and lightly burned overstory would be recorded as 23. Be sure the Fire Severity Code is determined only from observations made inside the macroplot.

## Common Fields

Photographs—conventional or digital—are a useful means to document the FIREMON plot a number of ways. They provide a unique opportunity to visually assess fire effects and document plot location in a database format. Previously established FIREMON plots can be found by orienting the landmarks in photos to visual cues in the field. Photos can be compared to determine important changes after a fire. Last, photos provide excellent communication tools for describing fire effects to the public and forest professionals.

Document the FIREMON macroplot location using two photographs taken facing north and east. For the north-facing photo move about 10 feet (3 m) south of the FIREMON macroplot center, then take the photo facing north, being sure that the plot center stake or rebar will be visible in the picture (fig. PD-3). Then, move west of the plot center about 10 feet (3 m) and take a photo facing east, again being sure that the plot center stake or rebar will be visible in the picture. For these pictures be sure that the camera is focused on the environment surrounding the plot, not the distance or foreground, and that the camera is set for the correct exposure and aperture for existing light conditions. A flash might be needed in low-light conditions.

**Figure PD-3**—Take your plot photos so that they show the plot center and the general plot conditions.

Enter an identifier in Field 75 for the north-facing photo and Field 76 in the east-facing field. Photos taken with conventional film can be identified by assigning a code that integrates the roll number or name (John Smith Roll 1) and the picture number (number shown on the camera). For example, John Smith Roll 1 and picture 8 might be assigned JSR01P08 on the PD data form. You must label the roll so that you will be able to find the correct photos after the film has been developed. One way is to take a picture of a card with the roll information on it, as your first photo. Or you could write the roll information on the film canister before you load it into the camera. The first method is the more foolproof. For digital cameras, enter the file name of the digital picture. Film photos will need to be scanned once they are developed and stored on your computer in digital format. The file names in Fields 75 and 76 will be linked to the plot photos when you enter your data into the FIREMON database.

**Field 75: North Digital Photo**—Enter a code of up to 15 characters that uniquely describes the location of the photo taken in the direction of due north. This field in the PD database will be linked to the actual digital photo when you enter data into the FIREMON database.

**Field 76: East Digital Photo**—Enter code of up to 15 characters that uniquely describes the location of the photo taken in the direction of due east. This field in the PD database will be linked to the tactual digital photo when you enter data into the FIREMON database.

There are many methods for documenting the before and after plot conditions using a series of photos. Rather than describe these procedures in FIREMON, we recommend you use the methods of Hall (2002) for photo point documentation. Hall's guide establishes and analyzes photo points over time, and it is useful for fire monitoring. You can download Hall's publication at: www.fs.fed.us/pnw/pubs/gtr526/. We have provided fields for two photo points per FIREMON plot. We strongly recommend a comprehensive photo documentation of the plot conditions. These two additional photo fields will provide you with the opportunity to record important changes on the FIREMON plot.

Enter an identifier in Fields 77 and 78 for the first and second photo points, respectively. The file names in these fields will be linked to the plot photos when you enter your data into the FIREMON database.

**Field 77: Photo Point 1**—Enter a code of up to 15 characters that uniquely describes the first photo taken at a point in or near this FIREMON plot. This field in the PD database will be linked to the actual digital photo when you enter data into the FIREMON database.

**Field 78: Photo Point 2**—Enter a code of up to 15 characters that uniquely describes the first photo taken at a point in or near this FIREMON plot. This field in the PD database will be linked to the actual digital photo when you enter data into the FIREMON database.

## Comments Fields

It is impossible for any standardized sampling methodology to estimate all ecosystem characteristics that are important to fire effects monitoring. There may be attributes that are locally important but of limited value in a nationwide fire effects sampling system such as FIREMON. A sampling method design that accounts for all ecological variables across North America would be so large and complex it would be difficult to use and apply. We have tried to reduce complexity in FIREMON, but as a result, we probably missed some variables that describe important ecological conditions for your region. The Comments Fields allow locally important observations to be included into standardized and nonstandardized fields.

### Local codes

We included some unstandardized fields so that plot level ecological data that do not fit in any standardized field can be recorded for later use. For example, you will notice that there is no PD field for structural stage, which is an important vegetation attribute for many land management applications. We omitted structural stage because there are many unstandardized classifications of structural stage across the country that are applicable for only local conditions and for a limited number of management objectives. However, some FIREMON users may have developed structural stage classes that they want to use, and the Local Code and Comments fields allow them a place to store and document that information.

**Field 79: Local 1**—Enter a user-designed code that is up to 10 characters in length and uniquely describes some condition on the FIREMON plot. To avoid confusion and database problems, do not embed blanks in your codes. Document your coding method in the Comments field.

**Field 80: Local 2**—Enter a user designed code that is up to 10 characters in length, and uniquely describes some condition on the FIREMON plot. To avoid confusion and database problems, do not embed blanks in your codes. Document your coding method in the Comments field.

## Comments

The Comments field is provided so that the field crew can record any information associated with the macroplot that cannot be recorded elsewhere on the PD form. For example, you can record ecological conditions on the plot, directions for plot location, sampling conditions that might affect data quality, and/or other attributes important for management objectives.

It is important that field samplers accurately describe ecological characteristics on the FIREMON plot so that these can be integrated into the monitoring analysis. Important ecological attributes include: wildlife utilization (browsing, grazing), human use (clearcutting, logging, mining), fire characteristics (abnormalities, burn coverage), topographic characteristics (seeps, swales), and/or disturbances (insects, disease).

The notetaker should provide detailed notes for relocating the plot for future remeasurements including succinct, short directions such as "proceed 140 degrees azimuth from junction of roads 432 and 543 exactly 200 meters to a blazed 100 cm spruce." Write the directions clearly, so it will be easy for others to use them when the plot needs to be resampled.

It is important that observations of any factor that might affect the quality and integrity of the collected data be recorded. An often-recorded sampling condition is the weather—"cold, rainy, windy day," for instance—but many other factors can be entered, such as "high stand density that precluded accurate measurement of diameter and canopy cover."

Comments should directly address the purpose of FIREMON sampling. For example, a sampling objective might be an evaluation of coarse woody debris, so a useful comment might be "many large logs consumed by fire; most were rotten."

**Field 81: Comments**—Enter up to a 256-character comment. Try to use shorthand and abbreviations to reduce space as long as the comments are still understandable. You might try to organize comments in a standard order with appropriate punctuation. For example, you might describe weather first and use only colons to separate the next major category of comments.

## Precision Standards

Use these standards for the PD method (table PD-13).

# SAMPLING DESIGN CUSTOMIZATION

This section will present several ways that the PD sampling method can be modified to collect more detailed information or streamlined to collect only the most important tree characteristics. First, the suggested or recommended sample design is detailed, then modifications are presented.

## Recommended PD Sampling Design

The recommended PD sampling design follows the Alternative FIREMON sampling intensity where the optimal number of fields are sampled to achieve a strong, but limited field sample. We suggest that besides the Required PD field set, you complete all fields in the Biophysical Setting field set, the Vegetation field set, and the Comments field set. This leaves the Fuels and Fire field sets empty.

**Table PD-13**  Precision guidelines for TD sampling.

| Component | Standard |
|---|---|
| Latitude | ±0.000001 degree |
| Longitude | ±0.000001 degree |
| Northing | ±1 meter |
| Easting | ±1 meter |
| Elevation | ±100 ft/30 m |
| Aspect | ±5 degrees |
| Slope | ±5 percent |
| All cover estimates | ±1 class |
| Stand hight | ±3 ft/1 m |
| Canopy fuel base height | ±1 ft/0.3 m |
| Flame length | ±0.5 ft/0.2 m |
| Spread rate | ±1 ft/min. or 0.3 m/min. |
| Severity class | ±1 class |

However, completion of both of these field sets would require less than 5 minutes per plot, even under the worst conditions. So it probably would be prudent to complete all PD fields, even if you are working under the Alternative FIREMON sampling intensity.

## Streamlined PD Sampling Design

The streamlined PD sampling design follows the Simple FIREMON sampling intensity where only the minimal set of fields are measured. For the PD method, the minimal set of fields are simply those in the Required field set. No other fields need be completed. However, completion of the Comments and the two plot pictures would add great detail to the simple structure.

## Comprehensive PD Sampling Design

The comprehensive PD sampling design follows the Detailed FIREMON sampling intensity and is quite easy to implement. Simply complete all fields in the PD data form and leave none blank.

## User-Specific PD Sampling Design

There are three ways to create user-designed fields for describing local ecological conditions. The two local fields in the Comments field set each allow up to a 10-character code in the database. This means the user can design a complex code to describe some important ecological characteristic critical to fire management. For example, the presence of weeds may be a significant management concern, so these fields might describe the cover and species, respectively, of the dominant weed.

Creative approaches can be used to enter local data if more than two fields are needed. Using the weed example, the cover and weed species can be integrated in one field by making the first six characters the local species code and the next two characters the FIREMON cover code. A third attribute, say plant height, could be added as a two-character code.

The 256-character comments field also can contain mixes of locally designed fields. Some people create search engines within a database query that look for certain combinations of special characters and numbers to link to a locally created standard field. For example, the term $SRF could be entered in the comments field to indicate the dominant fire regime (Stand-Replacement Fire).

## Sampling Hints and Techniques

Field sampling can become quite complicated, especially when visiting complex ecosystems with many canopy strata and high biodiversity. It can be easy for the field crew to become overwhelmed by all the heterogeneity on the landscape. It is important that the field crew concentrate their evaluation of the PD fields to those ecosystem characteristics inside the FIREMON macroplot.

Plot Description (PD) Sampling Method

# PLOT DESCRIPTION (PD) FIELD DESCRIPTIONS

## Required PD Fields—Database key
**Registration Code.** A four-character code determined by you or assigned to you. All four characters must be used.
**Project Code.** An eight-character code used to identify project work that is done within the group. You are not required to use all eight characters.
**Plot Number.** Identifier that corresponds to the site where sampling methods are applied. Integer value.
**Sampling Date.** Eight-digit number in the MM/DD/YYYY format where MM is the month number, DD is the day of the month, and YYYY is the current year.

## Organization Codes
Field 1: **Organization Code 1**. Four-character field used to identify part of the agency location code.
Field 2: **Organization Code 2**. Two-character field used to identify part of the agency location code.
Field 3: **Organization Code 3**. Two-character field used to identify part of the agency location code.
Field 4: **Organization Code 4**. Two-character field used to identify part of the agency location code.

## Plot Information Fields
Field 5: **Examiner Name**. Eight-character field used to identify the crew boss or lead examiner.
Field 6: **Units**. (E/M). Units of measure use on the plot—English or metric.
Field 7: **Plot Radius** (ft/m). Radius of the macroplot. If the macroplot is rectangular, plot length.
Field 8: **Plot Width** (ft/m). Width of the plot if it is rectangular. Enter 0 (zero) or blank if the plot is circular.

### Sampling information
Field 9: **Plot Type** (M/C). Plot type—Monitoring or Control.
Field 10: **Sampling Event** (P$n$/R$n$/IV). Treatment relative sampling identification. Valid codes are in table PD-2 of the sampling method.

### Linking fields
Field 11: **Fire ID**. Fire ID of up to 15 characters. The ID number or name that relates the fire that burned this plot to the same fire described in the Fire Behavior (FB) table.
Field 12: **Metadata ID**. Metadata ID of up to 15 characters that links the plot data to the MD table.

### Georeferenced plot positions
Field 13: **Coordinate System**. Identifies whether lat-long or UTM coordinates were used. This field is automatically filled based on the data entered in Fields 14 to 21. The user does not see this field.
Field 14: **Latitude**. Latitude. Precision: $\pm 0.000001$ decimal degree.
Field 15: **Longitude**. Longitude. Precision: $\pm 0.000001$ decimal degree.
Field 16: **Northing**. UTM northing. Precision: $\pm 1$ m.
Field 17: **Easting**. UTM easting. Precision: $\pm 1$ m.
Field 18: **Zone**. UTM zone of the plot center.
Field 19: **Datum**. Datum used in conjunction with the UTM coordinates.
Field 20: **GPS Position Error**. Position error value provided by the GPS unit
Field 21: **GPS Error Units** (E/M). Enter the units associated with the GPS error. May be different than the units listed in Field 6.

## Biophysical Setting Fields

### Topography
Field 22: **Elevation** (ft/m). Plot elevation. Precision: $\pm 100$ ft/30 m.
Field 23: **Plot Aspect** (degrees). Aspect measured in degrees true north. Precision: $\pm 5$ degrees.
Field 24: **Slope** (percent). Average plot slope. Precision: $\pm 5$ percent.
Field 25: **Landform**. Four-letter landform code. Valid codes are in table PD-3 of the sampling method.
Field 26: **Vertical Slope Shape**. Two-letter slope shape code. Valid codes are in table PD-4 of the sampling method.
Field 27: **Horizontal Slope Shape**. Two-letter slope shape code. Valid codes are in table PD-4 of the sampling method.

### Geology and soils fields
Field 28: **Primary Surficial Geology**. Four-letter code describing the geological rock type composing the parent material. Valid codes are in table PD-5 of the sampling method.
Field 29: **Secondary Surficial Geology**. Four-letter code describing the secondary geological rock type composing the parent material. Use this field only if you have coded a primary surficial geology type. Valid codes are in table PD-6 of the sampling method. Table PD-6 is an abridged list of common surficial types. A complete list is included in the Lithology Codes Appendix.
Field 30: **Soil Texture Class**. Up to four-letter code that describes the soil texture. Valid codes are in table PD-7 of the sampling method.
Field 31: **Erosion Type**. One-letter code describing the erosion on the plot. Valid codes are in table PD-8 of the sampling method.
Field 32: **Erosion Severity**. One-number code describing the severity of the soils erosion. Valid codes are in table PD-9 of the sampling method.

## Vegetation Fields

### Vegetation—trees
Field 33: **Total Tree Cover**. Vertically projected canopy cover of all the trees on the macroplot. Valid codes are in table PD-10 of the sampling method. Precision: $\pm 1$ class.
Field 34: **Seedling Tree Cover**. Vertically projected canopy cover of all trees that are less than 4.5 feet (1.4 m) tall. Valid codes are in table PD-10 of the sampling method. Precision: 1 class.
Field 35: **Sapling Tree Cover**. Vertically projected canopy cover of all trees that are greater than 4.5 feet (1.4 m) tall and less than 5.0 inches (13 cm) DBH. Valid codes are in table PD-10 of the sampling method. Precision: $\pm 1$ class.
Field 36: **Pole Tree Cover**. Vertically projected canopy cover of all trees that are greater than 5 inches (13 cm) DBH and less than 9 inches (23 cm) DBH. Valid codes are in table PD-10 of the sampling method. Precision: $\pm 1$ class.
Field 37: **Medium Tree Cover**. Vertically projected canopy cover of all trees that are greater than 9 inches (23 cm) DBH up to 21 inches (53 cm) DBH. Valid codes are in table PD-10 of the sampling method. Precision: $\pm 1$ class.
Field 38: **Large Tree Cover**. Vertically projected canopy cover of all trees that are greater than 21 inches (53 cm) DBH up to 33 inches (83 cm) DBH. Valid codes are in table PD-10 of the sampling method. Precision: $\pm 1$ class.
Field 39: **Very Large Tree Cover**. Vertically projected canopy cover of all trees that are greater than 33 inches (83 cm) DBH. Valid codes are in table PD-10 of the sampling method. Precision: $\pm 1$ class.

# Plot Description (PD) Sampling Method

*Vegetation—shrubs*

Field 40: **Total Shrub Cover**. Vertically projected canopy cover of all shrubs on the plot. Valid codes are in table PD-10 of the sampling method. Precision: ±1 class.
Field 41: **Low Shrub Cover**. Vertically projected canopy cover of all shrubs that are less than 3 feet (1 meter) tall. Valid codes are in table PD-10 of the sampling method. Precision: ±1 class.
Field 42: **Medium Shrub Cover**. Vertically projected canopy cover of all shrubs that are greater than 3 feet (1 meter) tall and less than 6.5 feet (2 meters) tall. Valid codes are in table PD-10 of the sampling method. Precision: ±1 class.
Field 43: **Tall Shrub Cover**. Vertically projected canopy cover of all shrubs that are greater than 6.5 feet (2 meters) tall. Valid codes are in table PD-10 of the sampling method. Precision: ±1 class.

*Vegetation—herbaceous*

Field 44: **Graminoid Cover**. Vertically projected canopy cover of all graminoid species on the plot. Valid codes are in table PD-10 of the sampling method. Precision: ±1 class.
Field 45: **Forb Cover**. Vertically projected canopy cover of all forbs on the plot. Valid codes are in table PD-10 of the sampling method. Precision: ±1 class.
Field 46: **Fern Cover**. Vertically projected canopy cover of all ferns on the plot. Valid codes are in table PD-10 of the sampling method. Precision: ±1 class.
Field 47: **Moss and Lichen Cover**. Vertically projected canopy cover of all mosses and lichens on the plot. Valid codes are in table PD-10 of the sampling method. Precision: ±1 class.

*Vegetation—composition*

Field 48: **Upper Dominant Species 1**. Either the NRCS plants species code or the local code of the most dominant species in the upper level stratum (greater than 10 feet [3 m] above ground level). Plant code is either the NRCS plant code or locally defined code.
Field 49: **Upper Dominant Species 2**. Either the NRCS plants species code or the local code of the second most dominant species in the upper level stratum (greater than 10 feet [3 m] above ground level). Code is either the NRCS plant code or locally defined code.
Field 50: **Mid Dominant Species 1**. Either the NRCS plants species code or the local code of the most dominant species in the mid level stratum (greater than 3 feet and less than 10 feet [1 to 3 m] above ground level). Code is either the NRCS plant code or locally defined code.
Field 51: **Mid Dominant Species 1**. Either the NRCS plants species code or the local code of the second most dominant species in the mid level stratum (greater than 3 feet and less than 10 feet [1 to 3 m] above ground level). Code is either the NRCS plant code or locally defined code.
Field 52: **Lower Dominant Species**. Either the NRCS plants species code or the local code of the most dominant species in the lowest level stratum (less than 3 feet [1 m] above ground level).
Field 53: **Lower Dominant Species**. Either the NRCS plants species code or the local code of the second most dominant species in the lowest level stratum (less than 3 feet [1 m] above ground level).

*Potential vegetation*

Field 54: **Potential Vegetation Type ID**. A 10-character, unstandardized code used to identify locally determined potential vegetation type.
Field 55: **Potential Lifeform**. Two-letter potential lifeform code. Valid codes are in table PD-11 of the sampling method.

## Ground Cover Fields

*Ground cover*

Field 56: **Bare Soil Ground Cover**. Percent ground cover of bare soil. Valid codes are in table PD-10 of the sampling method. Precision: ±1 class.
Field 57: **Gravel Ground Cover**. Percent ground cover of rock. Valid codes are in table PD-10 of the sampling method. Precision: ±1 class.
Field 58: **Rock Ground Cover**. Percent ground cover of rock. Valid codes are in table PD-10 of the sampling method. Precision: ±1 class.
Field 59: **Litter and Duff Ground Cover**. Percent ground cover of all uncharred litter and duff on the soil surface. Valid codes are in table PD-10 of the sampling method. Precision: ±1 class.
Field 60: **Wood Ground Cover**. Percent ground cover of all uncharred woody material. Valid codes are in table PD-10 of the sampling method. Precision: ±1 class.
Field 61: **Moss and Lichen Cover**. Percent canopy cover of all mosses and lichens on the plot. Valid codes are in table PD-10 of the sampling method. Precision: ±1 class.
Field 62: **Charred Ground Cover**. Percent ground cover of all *charred organic* material. Valid codes are in table PD-10 of the sampling method. Precision: ±1 class.
Field 63: **Ash Ground Cover**. Percent ground cover of all ash material. Valid codes are in table PD-10 of the sampling method. Precision: ±1 class.
Field 64: **Basal Vegetation Ground Cover**. Percent ground cover of basal vegetation using the codes. Valid codes are in table PD-10 of the sampling method. Precision: ±1 class.
Field 65: **Water Ground Cover.** Percent ground cover of standing water. Valid codes are in table PD-10 of the sampling method. Precision: ±1 class.

*General fuel characteristics*

Field 66: **Surface Fire Behavior Fuel Model**. Fire behavior fuel model from the Anderson 1983 publication, Aids for Determining Fuel Models for Estimating Fire Behavior. Or custom fire behavior fuel model.
Field 67: **Fuel Photo Series ID**. A 12-character, unstandardized field to enter a photo guide publication number and photo number that is similar to fuel characteristics seen on the plot.
Field 68: **Stand Height** (ft/m). Height of the highest stratum that contains at least 10 percent vertically projected canopy cover. Precision: ±3 ft/1 m.
Field 69: **Canopy Fuel Base Height** (ft/m). Lowest point above the ground at which there is a sufficient amount of canopy fuel to propagate a fire vertically into the canopy. Precision: ±1 ft/0.3 m.
Field 70: **Canopy Cover**. Percent canopy cover of the forest canopy above 6 feet (2 m). Valid codes are in table PD-10 of the sampling method. Precision: ±1 class.

## Fire Behavior and Effects Fields

*Fire behavior*

Field 71: **Flame Length** (ft/m). Length of the flames from the center of the combustion zone to the end of the continuous flame. Precision: +0.5 ft/ 0.2 m.
Field 72: **Spread Rate** (ft/min. or m/min.). Average speed of the fire across the macroplot. Precision: ±1 ft/min. or 0.3 m/min.
Field 73: **Fire Behavior Picture**. Up to a 15-character filename used to identify the location of a digital photo showing the fire behavior on the plot.

*Fire effects*

Field 74: **Fire Severity Code**. A two-number code describing the fire severity on the plot. Range for 1 (heavily burned) to 5 (unburned). Valid codes are in table PD-12 of the sampling method. Precision: ±1 class.

## Common Fields

Field 75: **North Digital Photo**. Up to a 15-character filename used to identify the location of a digital photo showing general plot conditions facing north.
Field 76: **East Digital Photo**. Up to a 15-character filename used to identify the location of a digital photo showing general plot conditions facing east.
Field 77: **Photo Point 1**. Up to a 15-character filename used to identify the location of a digital photo for general use.
Field 78: **Photo Point 2**. Up to a 15-character filename used to identify the location of a digital photo for general use.

## Comments Fields

*Local codes*

Field 79: **Local 1**. User designed code that is up to 10 characters in length.
Field 80: **Local 2**. User designed code that is up to 10 characters in length.

## Comments

Field 81: **Comments**. A 256-character, unstandardized comment field.

# FIREMON PD Cheat Sheet

### Sampling event codes

| Code | Event |
|---|---|
| P*n* | Preburn measurement, sequential sample number. |
| R*n* | Postburn remeasurement of a plot, sequential sample number. |
| C*n* | Control plot measurement, sequential sample number. |
| IV | Inventory plot, not a monitoring plot. |

### Vertical and horizontal slope shape

| Code | Slope shape |
|---|---|
| LI | Linear or planar |
| CC | Depression or concave |
| PA | Patterned |
| CV | Rounded or convex |
| FL | Flat |
| BR | Broken |
| UN | Undulating |
| OO | Other shape |
| X | Did not assess |

### Primary geologic codes

| Primary code | Rock type 1 |
|---|---|
| IGEX | Igneous extrusive |
| IGIN | Igneous intrusive |
| META | Metamorphic |
| SEDI | Sedimentary |
| UNDI | Undifferentiated |
| X | Did not assess |

### Soil types

| Code | Description | Code | Description |
|---|---|---|---|
| C | Clay | S | Sand |
| CL | Clay loam | SC | Sandy clay |
| COS | Coarse sand | SCL | Sandy clay loam |
| COSL | Coarse sandy loam | SI | Silt |
| FS | Fine sand | SIC | Silty clay |
| FSL | Fine sandy loam | SICL | Silty clay loam |
| L | Loam | SIL | Silt loam |
| LCOS | Loamy coarse sand | SL | Sandy loam |
| LFS | Loamy fine sand | VFS | Very fine sand |
| LS | Loamy sand | VFSL | Very fine sandy loam |
| LVFS | Loamy very fine sand | X | Did not assess |

### Secondary geologic codes

| Secondary code | Rock type 2 |
|---|---|
| ANDE | Andesite |
| BASA | Basalt |
| LATI | Latite |
| RHYO | Rhyolite |
| SCOR | Scoria |
| TRAC | Trachyte |
| DIOR | Diorite |
| GABB | Gabbro |
| GRAN | Granite |
| QUMO | Quartz monzonite |
| SYEN | Syenite |
| GNEI | Gneiss |
| PHYL | Phyllite |
| QUAR | Quartzite |
| SCHI | Schist |
| SLAT | Slate |
| ARGI | Argillite |
| CONG | Conglomerate |
| DOLO | Dolomite |
| LIME | Limestone |
| SANS | Sandstone |
| SHAL | Shale |
| SILS | Siltstone |
| TUFA | Tufa |
| MIEXME | Mixed extrusive and metamorphic |
| MIEXSE | Mixed extrusive and sedimentary |
| MIIG | Mixed igneous (extrusive & intrusive) |
| MIIGME | Mixed igneous and metamorphic |
| MIIGSE | Mixed ineous and sedimentary |
| MIINME | Mixed intrusive and metamorphic |
| MIINSE | Mixed intrusive and sedimentary |
| MIMESE | Mixed metamorphic and sedimentary |
| X | Did not assess |

### Precision

| Component | Standard |
|---|---|
| Latitude | ±0.000001 degree |
| Longitude | ±0.000001 degree |
| Northing | ±1 meter |
| Easting | ±1 meter |
| Elevation | ±100 ft/30 m |
| Aspect | ±5 degrees |
| Slope | ±5 percent |
| All cover estimates | ±1 class |
| Stand height | ±3 ft/1 m |
| Canopy fuel base height | ±1 ft/0.3 m |
| Flame length | ±0.5 ft/0.2 m |
| Spread rate | ±1 ft/min. or 0.3 m/min. |
| Severity class | ±1 class |

### Cover classes

| Code | Cover class |
|---|---|
| 0 | Zero percent cover |
| 0.5 | >0 1 percent cover |
| 3 | >1 5 percent cover |
| 10 | >5 15 percent cover |
| 20 | >15 25 percent cover |
| 30 | >25 35 percent cover |
| 40 | >35 45 percent cover |
| 50 | >45 55 percent cover |
| 60 | >55 65 percent cover |
| 70 | >65 75 percent cover |
| 80 | >75 85 percent cover |
| 90 | >85 95 percent cover |
| 98 | >95 100 percent cover |

### Common landforms

| Code | Landform |
|---|---|
| GMF | Glaciated mountains foothills |
| UMF | Unglaciated mountains foothills |
| BRK | Breaklands river breaks badlands |
| PLA | Plains rolling planes plains w/breaks |
| VAL | Valleys swales draws |
| HIL | Hill low ridges benches |
| X | Did not assess |

### Erosion severity codes

| Code | Erosion severity |
|---|---|
| 0 | Stable, no erosion is evident. |
| 1 | Low erosion severity; small amounts of material are lost from the plot. On average less than 25 percent of the upper 8 in. (20 cm) of soil surface have been lost across the macroplot. Throughout most of the area the thickness of the soil surface layer is within the normal range of variability of the uneroded soil. |
| 2 | Moderate erosion severity; moderate amounts of material are lost from the plot. On average between 25 and 75 percent of the upper 8 in. (20 cm) of soil surface have been lost across the macroplot. Erosion patterns may range from small, uneroded areas to small areas of severely eroded sites. |
| 3 | High erosion severity; large amounts of material are lost from the plot. On average 75 percent or more of the upper 8 in. (20 cm) of soil surface have been lost across the macroplot. Material from deeper horizons in the soil profile is visible. |
| 4 | Very high erosion severity; very large amounts of material are lost from the plot. All of the upper 8 in. (20 cm) of soil surface have been lost across the macroplot. Erosion has removed material from deeper horizons of the soil profile throughout most of the area. |
| 1 | Unable to assess. |

# FIREMON PD Cheat Sheet (cont.)

## Plot description (PD) equipment list

| | |
|---|---|
| Camera with film and flash | Maps, charts, and directions |
| Clear plastic ruler (2) | Map protector or plastic bag |
| Clinometer (2) | Logger's tape (2 plus steel tape refills) |
| Clipboard | Magnifying glass |
| Cloth tape (2) | Pocket calculator |
| Compass (2) | Plot sheet protector or plastic bag |
| Flagging | Previous measurement plot sheets |
| Geographic Positioning System or GPS receiver | Field notebook |
| Indelible ink pen (Sharpie, Marker) | PD data forms and cheat sheet |
| Lead pencils with lead refills | |

## Potential lifeform codes

| Code | Potential lifeform |
|---|---|
| AQ | Aquatic   Lake, pond, bog, river |
| NV | Nonvegetated   Bare soil, rock, dunes, scree, talus |
| CF | Coniferous upland forest   Pine, spruce, hemlock |
| CW | Coniferous wetland or riparian forest   Spruce, larch |
| BF | Broadleaf upland forest   Oak, beech, birch |
| BW | Broadleaf wetland or riparian forest   Tupelo, cypress |
| SA | Shrub dominated alpine   Willow |
| SU | Shrub dominated upland   Sagebrush, bitterbrush |
| SW | Shrub dominated wetland or riparian   Willow |
| HA | Herbaceous dominated alpine   Dryas |
| HU | Herbaceous dominated upland   grasslands, bunchgrass |
| HW | Herbaceous dominated wetland or riparian   ferns |
| ML | Moss or lichen dominated upland or wetland |
| OT | Other potential vegetation lifeform |
| X | Did not assess |

## Erosion types

| Code | Erosion type |
|---|---|
| S | Stable, no erosion evident |
| R | Water erosion, rill |
| H | Water erosion, sheet |
| G | Water erosion, gully |
| T | Water erosion, tunnel |
| W | Wind erosion |
| O | Other type of erosion |
| X | Did not assess |

## Plot level fire severitty codes

| Fire severity code | Substrate | Forest vegetation | Shrubland vegetation | Grassland vegetation |
|---|---|---|---|---|
| **Unburned (5)** | Not burned | Not burned | Not burned | Not burned |
| **Scorched (4)** | Litter partially blackened; duff nearly unchanged; wood/leaf structures unchanged. | Foliage scorched and attached to supporting twigs. | Foliage scorched and attached to supporting twigs. | Foliage scorched |
| **Lightly burned (3)** | Litter charred to partially consumed; upper duff layer may be charred but the duff is not altered over the entire depth; surface appears black; where litter is sparse charring may extend slightly into soil surface but soil is not visibly altered; woody debris partially burned; logs are scorched or blackened but not charred; rotten wood is scorched to partially burned. | Foliage and smaller twigs partially to completely consumed; branches mostly intact. | Foliage and smaller twigs partially to completely consumed; branches mostly intact; typically, less than 60 percent of the shrub canopy is consumed. | Grasses with approximately two inches of stubble; foliage and smaller twigs of associated species partially to completely consumed; some plant parts may still be standing; bases of plants are not deeply burned and are still recognizable. |
| **Moderately burned (2)** | Litter mostly to entirely consumed, leaving coarse, light colored ash (ash soon disappears, leaving mineral soil); duff deeply charred, but not visibly altered; woody debris is mostly consumed; logs are deeply charred, burned out stump holes are evident. | Foliage twigs and small stems consumed; some branches still present. | Foliage twigs and small stems consumed; some smaller branches (0.25 0.50 inches) still present; typically, 40 to 80 percent of the shrub canopy is consumed. | Unburned grass stubble usually less than 2 inches tall, and mostly confined to an outer ring; for other species, foliage completely consumed, plant bases are burned to ground level and obscured in ash immediately after burning. |
| **Heavily burned (1)** | Litter and duff completely consumed, leaving fine white ash (ash disappears leaving mineral soil); mineral soil charred and/or visibly altered, often reddish; sound logs are deeply charred, and rotten logs are completely consumed. | All plant part consumed, leaving some or no major stems or trunks; any left are deeply charred. | All plant parts consumed leaving only stubs greater than 0.5 inch in diameter. | No unburned grasses above the root crown; for other species, all plant parts consumed. |
| **Not applicable (0)** | Only inorganic material on site before burn. | None present at time of burn. | None present at time of burn. | None present at time of burn. |

# Plot Description (PD) Form

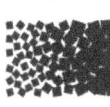

|  | | | | | | | | |
|---|---|---|---|---|---|---|---|---|
| | Registration ID | | Project ID | | Plot ID | | Date | PD Page __ of __ |
| Org Info | Field 1: Organization Code 1 | Field 2: Organization Code 2 | Field 3: Organization Code 3 | Field 4: Organization Code 4 | | | | |
| Plot Info | Field 5: Examiner | Field 6: Units (M / C) | Field 7: Plot Radius ft / m | Field 8: Plot Width ft / m | | | | |
| | Field 9: Plot Type | Field 10: Sampling Event | Field 11: FireID | Field 12: MetadataID | | | | |
| Georeferenced Position | Field 13: Coordinate System (Lat-Long / UTM / Albers) | Field 14: Latitude | Field 15: Longitude | | | | | |
| | | Field 16: Northing | Field 17: Easting | Field 18: Zone | | | | |
| | | Field 19: Datum | Field 20: GPS Position Error (E / M) | Field 21: GPS Error Units (E / M) | | | | |
| Biophysical Settings | Field 22: Elevation | Field 23: Plot Aspect | | | | | | |
| | Field 24: Slope | Field 25: Landform | | | | | | |
| | Field 26: Vertical Slope Shape | Field 27: Horiz. Slope Shape | | | | | | |
| Geology & Soils | | Field 28: Surf. Geo. 1 | Field 29: Surf. Geo. 2 | | | | | |
| | | Field 30: Soil Texture | Field 31: Erosion Type | | | | | |
| | | Field 32: Erosion Severity | | | | | | |
| Tree Cover | Field 33: Total Tree Cover | Field 34: Seedling Tree Cover | | | | | | |
| | Field 35: Sapling Tree Cover | Field 36: Pole Tree Cover | | | | | | |
| | Field 37: Medium Tree Cover | Field 38: Large Tree Cover | | | | | | |
| | Field 39: Very Lg. Tree Cover | | | | | | | |
| Shrub Cover | | Field 40: Total Shrub Cover | Field 41: Low Shrub Cover | | | | | |
| | | Field 42: Medium Shrub Cover | Field 43: Tall Shrub Cover | | | | | |
| Herb Cover | Field 44: Graminoid Cover | Field 45: Forb Cover | | | | | | |
| | Field 46: Fern Cover | Field 47: Moss & Lichen Cover | | | | | | |
| Composition | | Field 48: Upper Dom. Spp. 1 | Field 49: Upper Dom. Spp. 2 | | | | | |
| | | Field 50: Mid Dominant Spp. 1 | Field 51: Mid Dominant Spp. 2 | | | | | |
| | | Field 52: Lower Dom. Spp. 1 | Field 53: Lower Dom. Spp. 2 | | | | | |
| | | Field 54: Potential Veg. ID | Field 55: Potential Lifeform | | | | | |
| Ground Cover | Field 56: Bare Soil Cover | Field 57: Gravel Ground Cover | | | | | | |
| | Field 58: Rock Ground Cover | Field 59: Litter & Duff Cover | | | | | | |
| | Field 60: Wood Ground Cover | Field 61: Moss & Lichen Cover | | | | | | |
| | Field 62: Charred Ground Cover | Field 63: Ash Ground Cover | | | | | | |
| | Field 64: Basal Veg. Cover | Field 65: Water Ground Cover | | | | | | |
| Fuels | | Field 66: Surface Fire Behavior Fuel Model | Field 67: Fuel Photo ID | | | | | |
| | | Field 68: Stand height ft / m | Field 69: Canopy Fuel Base Ht ft / m | | | | | |
| | | Field 70: Canopy Cover | | | | | | |
| Behavior & Effects | Field 71: Flame length ft / m | Field 72: Spread Rate ft/min m/min | | | | | | |
| | Field 73: Fire Behavior Picture | Field 74: Fire Severity Code | | | | | | |
| Common Fields | | Field 75: North Photo | Field 76: East Photo | | | | | |
| | | Field 77: Photo Point 1 | Field 78: Photo Point 2 | | | | | |
| | | Field 79: Local Code 1 | Field 80: Local Code 2 | | | | | |

Write comments on back

# Tree Data (TD)

## Sampling Method

Robert E. Keane

## SUMMARY

The Tree Data (TD) methods are used to sample individual live and dead trees on a fixed-area plot to estimate tree density, size, and age class distributions before and after fire in order to assess tree survival and mortality rates. This method can also be used to sample individual shrubs if they are over 4.5 ft tall. When trees are larger than the user-specified breakpoint diameter the following are recorded: diameter at breast height, height, age, growth rate, crown length, insect/disease/abiotic damage, crown scorch height, and bole char height. Trees less than breakpoint diameter and taller than 4.5 ft are tallied by species-diameter-status class. Trees less than 4.5 ft tall are tallied by species-height-status class and sampled on a subplot within the macroplot. Snag measurements are made on a snag plot that is usually larger than the macroplot. Snag characteristics include species, height, decay class, and cause of mortality.

## INTRODUCTION

The Tree Data (TD) methods were designed to measure important characteristics about individual trees so that tree populations can be quantitatively described by various dimensional and structural classes before and after fires. This method uses a fixed area plot sampling approach to sample all trees that are within the boundaries of a circular plot (called a macroplot) and that are above a user-specified diameter at breast height (called the breakpoint diameter). All trees or shrubs below the breakpoint diameter are recorded by species-height-status class or species-diameter-status class, depending on tree height. The TD method is appropriate for both inventory and monitoring.

The TD sampling methods were developed using fixed-area sampling procedures that have been established and accepted by forestry professionals. Field sampling and the data collected are similar to that in ECODATA.

The fixed area plot used to describe tree characteristics in the TD methods is different from the standard timber inventory techniques that use plotless point sampling implemented with the prism. We use the fixed-area technique for a number of reasons. First, the plotless method was designed to quantify stand characteristics using many point samples across a large area (stand). This means that the sampling strategy is more concerned with conditions across the stand than within the plot. Second, plotless sampling was designed for inventorying large, merchantable trees and is not especially useful for describing tree populations—especially within a plot—because the sampling distribution for the plotless methods undersamples small and medium diameter trees. These small trees are the ones many fire managers are interested in monitoring, for instance, in a restoration project. Similarly, canopy fuels are not adequately sampled using plotless techniques because there is insufficient number of trees in all age classes to obtain realistic vertical fuel loadings and distributions. Finally, there are many ecosystem characteristics recorded at each FIREMON macroplot, and the origin and factors that control these characteristics are highly interrelated. For example, shrub cover is often inversely related to tree cover on productive sites. Plotless sampling does not adequately allow the sampling of tree characteristics that influence other ecosystem characteristics, at the plot level, such as loading and regeneration. The expansion of trees per area estimates from fixed-area plots is much less variable than density estimation made using plotless methods The contrast between point and fixed area plot sampling is really a matter of scale. Point or prism sampling is an efficient means to get stand-level estimations, but it is inadequate for describing tree characteristics within a plot.

Many characteristics are recorded for each tree. First, species and health status are recorded for each tree. Then, structural characteristics of diameter breast height (DBH), height, live crown percent, and crown position are measured to describe physical dimensions of the trees. Age and growth rate describe life stage and productivity. Insect and disease evidence is recorded in the damage codes. A general description of dead trees is recorded in snag codes. Fire severity is assessed by estimates of downhill bole char height and percent of crown scorched. There is one column to build user-defined codes for each tree, if needed. Each tree above the breakpoint diameter gets a tree tag that permanently identifies it for further measurements.

Besides being used to inventory general tree characteristics on the macroplot, the TD method can be used to determine tree survival or fire-caused tree mortality after a burn and used to describe the pre- and postburn tree population characteristics by species, size, and age classes. Values in many TD fields can be used to compute a host of ecological characteristics of the tree. For example, the tree's DBH, height, and live crown percent can be used to compute stand bulk density for modeling crown fire potential.

There are many ways to streamline the TD procedure. First, the number of measured trees can be lowered by raising the breakpoint diameter. A large breakpoint diameter will exclude the individual measurement of the many small trees on the macroplot. Next, age estimates of individual trees can be simplified by taking age in broad diameter and species classes. Last, the FIREMON three-tier sampling design can be employed to optimize sampling efficiency. See **User Specific TD Sampling Design** in the section below on **Sampling Design Customization.**

## SAMPLING PROCEDURE

This method assumes that the sampling strategy has already been selected and the macroplot has already been located. If this is not the case, then refer to the FIREMON **Integrated Sampling Strategy** for further details.

The sampling procedure is described in the order of the fields that need to be completed on the **TD data form**. The sampling procedure described here is the recommended procedure for this method. Later sections will describe how the FIREMON three-tier sampling intensity classes can be used to modify the recommended procedure to match resources, funding, and time constraints.

### Preliminary Sampling Design

There is a set of general criteria recorded on the TD data form that are the basis of the user-specified design of the TD sampling method. Each general TD field must be designed so that the sampling captures the information needed to successfully satisfy the management objective within time, money, and personnel constraints. These general fields must be decided before the crews go into the field and should reflect a thoughtful analysis of the expected problems and challenges in the fire monitoring project.

Tree Data (TD) Sampling Method

See **Populating the Plant Species Codes Lookup Table** in the **FIREMON Database User Manual** for more details.

## Plot size selection

There may be as many as three nested fixed plots in the FIREMON TD methods. First, the macroplot is the primary sampling plot, and it is the plot where all live tree population characteristics are taken. Next is the snag plot, which may be larger than the macroplot, and it is used to record a representative sample of snags. Often the incidence of snags on the landscape is so low that the macroplot area is not large enough to reliably describe snag populations. The snag macroplot allows more snags to be sampled. Last is the subplot where all seedlings are counted. This is smaller than the macroplot and is provided to streamline the counting of densely packed seedlings. All of the plots are concentrically located around the macroplot center.

## Macroplot area

The size of the macroplot ultimately dictates the number of trees that will be measured, so large plot sizes usually take longer to sample because of the large number of trees on the plot. However, some ecosystems have widely spaced trees scattered over large areas so that large plot sizes are needed to obtain statistically significant estimates. There have been many studies to determine the optimum plot size for different ecosystems with mixed results. We offer table TD-1 to help determine the plot size that matches the fire monitoring application. Unless the project objectives dictate otherwise, use the median tree diameter and median tree height—for trees greater than breakpoint diameter—to determine the size of your macroplot. Enter the macroplot size for the TD method in Field 1 of the TD data form

In general, the 0.1 acre (0.04 ha) circular plot will be sufficient for most forest ecosystems and should be used if no other information is available. A general rule-of-thumb is the plot should be big enough to capture at least 20 trees above breakpoint diameter (see definition in next section) on the average across all plots in your project. Though it is not absolutely necessary, extra measures should be taken so that plot sizes are the same for all plots in a project.

## Subplot area

All seedlings—trees less than 4.5 ft (1.37 m) tall—are measured on a subplot nested within the macroplot (fig. TD-1). Use seedling density to determine the subplot size (table TD-2) unless the project objectives dictate otherwise.

Again, make an effort to keep the subplot radius constant across all plots in the FIREMON project. Subplot sampling is discussed in later sections of this document. The area of the subplot is entered in Field 2 on the TD data form.

## Snag plot area

Snags are dead trees greater than breakpoint diameter. Snags can be measured within the macroplot, but often their numbers are so low that a larger plot is needed to detect changes in snag populations. A suitable snag plot size is difficult to determine because snags are nonuniformly distributed across the landscape. A good rule-of-thumb for sizing the snag plot is to double the macroplot diameter, which will

**Table TD-1** Use the median tree diameter and median tree height to determine the size of the sampling macroplot.

| Median tree diameter (trees greater than breakpoint diameter) | Median tree height | Suggested plot radius | Suggested plot size |
|---|---|---|---|
| < 20 inches (<50 cm) | < 100 ft (<30 m) | 37.24 ft (11.28 m) | 0.1 ac (0.04 ha) |
| 20 to 40 inches (<100 cm) | 100 to 130 ft (<40 m) | 52.66 ft (12.61 m) | 0.2 ac (0.05 ha) |
| > 40 inches (<200 cm) | > 130 ft (<50 m) | 74.47 ft (17.84 m) | 0.4 ac (0.1 ha) |

**Figure TD-1** — The subplot is nested inside the macroplot. The recommended circular plot shape is shown, but rectangular plots may also be used.

**Table TD-2** Use seedling density to determine the subplot radius.

| Seedling density | Subplot radius | Area |
| --- | --- | --- |
| Typical | 11.77 ft (3.57 m) | 0.01 ac (0.004 ha) |
| <2 seedlings per species | 18.62 ft (5.64 m) | 0.025 ac (0.01 ha) |
| >100 seedlings per species | 5.89 ft (1.78 m) | 0.0025 ac (0.001 ha) |

increase the snag plot area by a factor of four. Enter the snag plot area in Field 3 of the TD form even if it is the same size as the macroplot. If snags are important to the project objectives, you should choose a plot radius that will provide you with at least 20 snags, meaning the plot could be quite large.

## Breakpoint diameter

Choose a breakpoint diameter that allows at least 20 trees to be measured on each FIREMON macroplot (fig. TD-2). The same breakpoint diameter should be used for all of the macroplots in the study.

The breakpoint diameter is the tree diameter at breast height (DBH) above which all trees are tagged and measured individually and below which trees are tallied to species-DBH classes. Selection of the breakpoint diameter must account for fire monitoring objectives as well as sampling limitations and efficiency. For example, a large breakpoint diameter (>8 inches) will exclude many small trees from individual measurements and reduce the sampling time. As long as the project objectives are broad (70

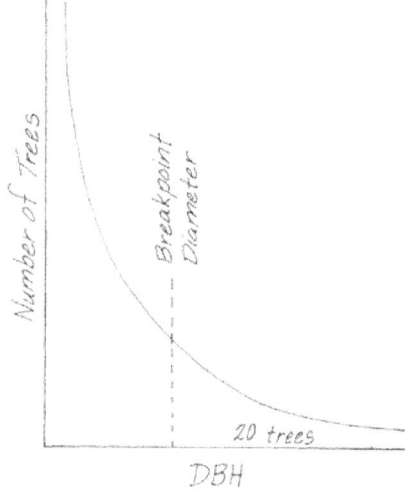

**Figure TD-2** — Choose a breakpoint diameter that leaves at least 20 trees in the tail of the distribution. The diameter distribution can be derived from a pilot study or estimated from previous experience.

to 80 percent mortality of saplings) then a tally of trees below breakpoint will be sufficient. However, if the objectives are specific (determine the effect of crown scorch and char height on fire caused mortality of trees 2 inches and greater DBH) then a lower breakpoint diameter, and the associated increase in sampling effort, will be necessary. Individual measurement of small trees might be unrealistic if there are high sapling densities (>1,000 trees) on the macroplot. Selection of an appropriate breakpoint diameter requires some field experience and knowledge of the resources available to complete the fire monitoring project. In FIREMON we suggest using a 4-inch (10-cm) breakpoint diameter if no other information is available. The breakpoint diameter is entered in Field 4 of the TD data form.

In the **How-To Guide** chapter of this manual, see **How To Locate a FIREMON Plot**, **How To Permanently Establish a FIREMON Plot**, and **How to Define the Boundaries of a Macroplot** for more information on setting up your macroplot.

## Preliminary Sampling Tasks

Before setting out for your field sampling, lay out a practice macroplot and subplot in an area with easy access. Even if there are just a few trees on your practice plot, getting familiar with the plot layout and the data that will be collected before heading out will make the first day or two of field sampling less frustrating. It will also let you see if there are any pieces of equipment that will need to be ordered.

When you are ready to go into the field, consult the **TD Equipment list** and gather the necessary materials. You will probably spend most of your day hiking from plot to plot, so it is important that supplies and equipment are placed in a comfortable daypack or backpack. Be sure you pack spare equipment so that an entire day of sampling is not lost if something breaks. Spare equipment can be stored in the vehicle rather than the backpack. Be sure all equipment is well maintained and there are plenty of extra supplies such as data forms, map cases, and pencils.

All TD data forms should be copied onto waterproof paper because inclement weather can easily destroy valuable data recorded on standard paper. Data forms should be transported into the field using a plastic, waterproof map protector or plastic bag. The day's sample forms should always be stored in a dry place (office or vehicle) and not be taken back into the field for the next day's sampling.

If the sampling project is resampling previously installed FIREMON plots, then it is highly recommended that plot sheets from the first measurement be copied and brought to the field for reference. These data can be extremely valuable for help in identifying sample trees that have lost their tree tags or have fallen over, and the data can provide an excellent reference for verifying measurements.

We recommend that one person on the field crew, preferably the crew boss, have a waterproof, lined field notebook for recording logistic and procedural problems encountered during sampling. This helps with future remeasurements and future field campaigns. All comments and details not documented in the FIREMON sampling methods should be written in this notebook. For example, snow on the plot might be described in the notebook, which would be helpful in plot remeasurement.

It is beneficial to have plot locations for several days of work in advance in case something happens, such as if the road to one set of plots is washed out by flooding. Locations and/or directions to the plots you will be sampling should be readily in order to reduce travel times. If the FIREMON plots were randomly located within the sampling unit, it is critical that the crew is provided plot coordinates before going into the field. Plots should be referenced on maps and aerial photos using pin-pricks or dots to make navigation easy for the crew and to provide a check of the georeferenced coordinates. It is easy to misrecord latitude and longitude coordinates, and marked maps can help identify any erroneous plot positions.

A field crew of three people is probably the most efficient for implementation of the TD sampling method. For safety reasons there should never be a one-person crew. A crew of two people will require excessive walking and trampling on the plot. More than three people will probably result in some people waiting for critical tasks to be done and unnecessary physical damage to the plot.

For simplicity, we will refer to the people in the three-person crew as the crew boss, a note taker, and a technician. The crew boss is responsible for all sampling logistics including the vehicle, plot directions, equipment, supplies, and safety. The note taker is responsible for recording the data on data forms or onto the laptop. The technician will perform most individual tree measurements while the note taker estimates tree heights. Of course, the crew boss can be the note taker, and probably should be for most situations. The initial sampling tasks of the field crew should be assigned based on field experience, physical capacity, and sampling efficiency. Sampling tasks should be modified as the field crew gains experience and tasks should also be shared to limit monotony.

## Sampling Tasks

### Define macroplot boundary

The first task to be completed is to define the boundary of the TD sampling plot. The TD macroplot should be established so that the sampling plots for all of the methods overlap as much as possible. In the **How-To Guide** chapter, see **How To Establish Plots with Multiple Methods**.

If you are sampling using only the TD methods, it is not so important that flags are placed exactly at the fixed distance; rather, it is more important that the trees inside the macroplot are clearly identified for sampling. This means that borderline trees (questionable trees on the boundaries of the plot) be flagged so that the tree sampler knows if the tree is inside or outside the macroplot. This is done by tying the flag on a branch on the outside of the tree (away from plot center) if the tree is inside the macroplot, or tying the flag on a branch that faces toward plot center if the tree is outside the boundaries. Avoid tying flagging to tree boles as it can lead to some confusion about whether trees are "in" or "out." If you must flag around a bole, face the knot toward plot center if the tree is "out" or face the knot away from plot center if the tree is "in." See **How To Define the Boundaries of a Macroplot** in the **How-To Guide** chapter for more information. With either method you will need to adjust for slope as you go around the plot (see **How to Adjust for Slope** in the **How-To Guide**).

The snag and subplot boundaries probably don't need the extensive marking and flagging as required for the macroplot. For the snag plot, we suggest fixing a cloth or fiberglass tape to the macroplot center and proceeding out to the distance of the Snag Plot Radius. We suggest a cloth or fiberglass tape because your diameter tape will probably not be long enough. Leave the tape pulled out and start traversing the snag plot to search for snags. You will be able to determine whether most of the snags you find are "in" or "out" just by looking. However, you will have to use the tape to measure snags that are on the border. If you have a second cloth or fiberglass tape, this task will go quicker. You probably won't need to flag the seedling subplot because it is so small. Seedling sampling is described in detail below.

### Define initial sample position

Once the macroplot has been defined and all perimeter flags have been hung, flag the tree inside the macroplot, farthest from plot center and greater than breakpoint diameter, that is closest to due north (360 degrees azimuth) from plot center. Mark it near the base so that flagging won't be confused with the plot boundary marking. This tree will be the first tree measured and will also indicate when the sampling has been completed. If tree density is high, then you may want to flag the closest tree to the left of your "first" tree as the "last" tree (you will be sampling clockwise around the plot). Tie a flag at the base of the tree(s) so that it will not be confused with plot boundary trees. You may even want to use another flag color, or you may want to use tree chalk instead of a flag (fig. TD-3).

### Sampling seedlings

In FIREMON we call trees less than 4.5 ft (1.37 m) tall seedlings. Seedlings are sampled using a small subplot within the larger macroplot. This is done because seedlings are more numerous than larger trees so they can be sampled on a smaller area and still allow a representative sample. In most cases, attempting to sample seedlings on the entire macroplot would be inefficient and time consuming.

# Tree Data (TD) Sampling Method

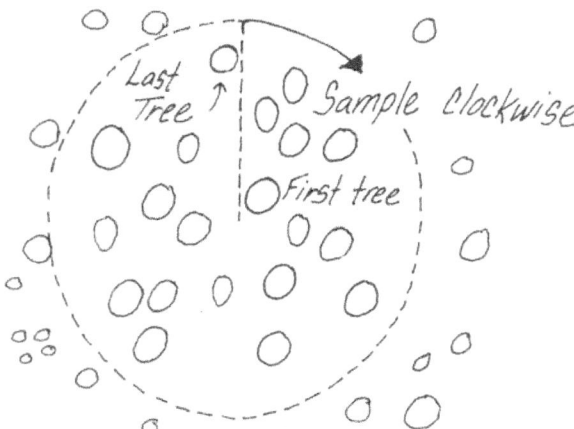

**Figure TD-3**—Sample trees clockwise around the plot starting with the first tree located clockwise from an azimuth of 360 degrees true north.

Subplot sampling is done before individual tree measurements because repeated walking around the macroplot center by field crews may trample some seedlings and bias the sample.

To start your seedling counts attach the zero end of a logger's tape at plot center and walk away headed due north until you are at the distance (corrected for slope) you selected for you subplot radius from table TD-2. Once at the required distance, hold the tape just above the seedlings, then sweep clockwise around the plot, tallying seedlings by species into their respective status and height classes as you go. Use the following status classes:

**H—Healthy** tree with little biotic or abiotic damage.

**U—Unhealthy** tree with some biotic or abiotic damage, and this damage will reduce growth. However, it appears the tree will not immediately die from the damage.

**S—Sick** tree with extensive biotic or abiotic damage, and this damage will ultimately cause death within the next 5 to 10 years.

**D—Dead** tree or snag with no living tissue visible.

The height classes in table TD-3 are suggested, but you may use any classes you choose. The Analysis Tools program assumes that the height class value recorded is the midpoint of the class. Make sure you note any size class changes in the Metadata table. Estimating height using a stick marked with the height classes will make counting quicker and easier.

We recommend that you sample seedlings using two crew members, with one person tallying all of the seedlings on the inner part of the plot and the other person counting all the seedlings on the outer part. Also, this way, when the tape encounters a tree as it is swept clockwise, the sampler closest to the plot center can keep track of which seedlings have been counted so that the sampler holding the tape can move around the obstructing tree and return the tape to the proper point before the tally is resumed.

**Table TD-3** Size classes for seedlings.

| | Midpoint height class | | |
|---|---|---|---|
| Class | Height range | Class | Height range |
| | ft | | m |
| 0.2 | >0.0  0.5 | 0.1 | >0.0  0.2 |
| 1 | >0.5  1.5 | 0.3 | >0.2  0.5 |
| 2 | >1.5  2.5 | 0.6 | >0.5  0.8 |
| 3 | >2.5  3.5 | 0.9 | >0.8  1.0 |
| 4 | >3.5  4.5 | 1.2 | >1.0  1.4 |

Be sure to carefully search the subplot for all seedlings, no matter how small. Use the dot tally method for seedlings (fig. TD-4) and when you have finished, enter the counts in Seedlings table of the TD data form by midpoint class (Fields 31 to 34).

Sampling seedlings by height class allows a compromise between sampling efficiency and the detail required for describing fire effects.

If a tree is broken below 4.5 ft (1.37 m), but you believe that the tree would be taller than 4.5 ft if unbroken, you should still sample it as a seedling.

## Saplings and mature trees

In FIREMON we divide trees taller than breast height (4.5 ft) into two groups: saplings and mature trees. Both groups are measured on the macroplot. Saplings are trees taller than 4.5 ft (1.37 m) but smaller than the breakpoint diameter, and mature trees are greater than breakpoint diameter. Trees above the breakpoint diameter are tagged, measured, and recorded in table 1 of the TD data form. Trees below the breakpoint diameter, but taller than 4.5 ft (1.37 m), are measured and tallied in table 2 of the TD data form. It is most efficient if both sampling tasks are done at the same time as the sampling proceeds around the macroplot because, especially in dense stands, it will be difficult determining which trees have been measured.

As with seedlings, sample saplings and mature trees in a clockwise direction starting with your "first" tree. The best way to do a repeatable sample is to measure trees in the order that a second hand would hit them as it moved around the plot (fig. TD-5). Sometimes this method means that you travel back and forth between the middle and outer portions of the plot, which may seem inefficient, but the benefit is

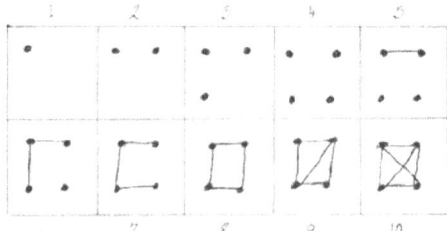

**Figure TD-4**—Use the dot tally method to make counts quickly with little chance of error.

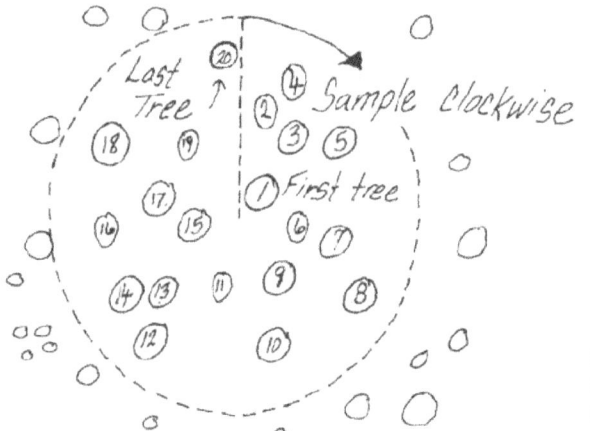

**Figure TD-5**—Sample trees by measuring the tree you flagged due north of the plot center, first, then move in a clockwise direction. It is easiest to sample all trees above DBH at the same time.

that you will be able to relocate trees easier at a later date if something happens to the trees or the tree tags. For example, if tagged trees are blown down by heavy winds or knocked down by other falling trees, you will be able to account for their mortality by locating them, in order, around the plot.

Once a tree is identified for measurement, the sampler must decide if it is above or below the breakpoint diameter. Although, usually you will be able to do this visually, use a ruler or diameter tape if the tree is borderline. We have found that it is usually less time consuming to initially estimate DBH using a clear plastic ruler rather than with a diameter tape (see **How To Measure Diameter with a Ruler** in the **How-To Guide** chapter). However, if the ruler estimate is within 1 inch (2 cm) of breakpoint diameter, we recommend using a diameter tape because the diameter measurement will be more reliable.

## Measuring DBH

The diameter of a tree or shrub is conventionally measured at exactly 4.5 ft (1.37 m) above the ground surface, measured on the uphill side of the tree if it is on a slope. Wrap a diameter tape around the bole or stem of the plant, without twists or bends, and without dead or live branches caught between the tape and the stem (fig. TD-6). Pull the tape tight and record the diameter. If you are only determining the diameter class then the measurement only needs to identify the class that the tree is in. If you are measuring the diameter of a mature tree then measure to the nearest 0.10 inch (0.5 cm). Large diameter trees are difficult to measure while standing at one point, so you have to hook the zero end of the tape to the plant bole (bark) and then walk around the tree, being sure to keep the tape exactly perpendicular to the tree stem and all foreign objects from under the tape. In the **How-To Guide** chapter, see **How To Measure DBH** for more information.

Forked trees should be noted in the damage codes discussed in later sections. Forked trees often occur in tall shrublands or woodlands (pinyon juniper, whitebark pine) where trees are clumped due to bird caches or morphological characteristics. Many tree-based sampling techniques suggest that diameter be measured at the base rather than at breast height, but we feel this may bias an estimate of tree mortality by not counting survival of individual stems. Moreover, basal diameters may not adequately portray canopy fuels for fire modeling.

Trees are "live" if they have any green foliage on them regardless of the angle at which they are leaning. If a tree has been tipped or is deformed by snow, make all the measurements but use the damage codes described below to record the nature of the damage.

**Figure TD-6**—DBH measurement. The diameter tape should carefully be pulled tight around the tree without twists or bends.

## Sampling saplings

Saplings are recorded in species-diameter class-status groups rather than as individuals. As mentioned above we suggest using a 4-inch (10-cm) breakpoint diameter. Tally trees by the diameter classes in table TD-4. The Analysis Tools program assumes that the diameter class value you record is the midpoint of the class. Make sure you note any size class changes in the Metadata table.

Note that no age or growth characteristics are recorded for trees below breakpoint diameter. The only fire severity attribute assessed from these data is tree mortality by species-diameter-status class.

Once a tree is determined to be below the breakpoint diameter the sampler must 1) estimate diameter to the appropriate diameter class, 2) determine the species of the tree, 3) identify the tree status, and 4) estimate height to nearest 1 ft (0.3 m). Tree diameter is most easily estimated using a clear plastic ruler (see **How-To Guide** chapter, **How To Measure Diameter with a Ruler**). A diameter tape can be used, but it is often cumbersome when diameters are less than 4 inches (10 cm). Once the diameter class, species, status, counts, and average heights have been determined, record them in Fields 24 to 28, respectively. Optionally, average live crown percent can be recorded in Field 29 using the classes in table TD-5. Record the class based on the percent of the tree that has live foliage growing from it, from the top of the live foliage to the ground.

Height and live crown percent are challenging to estimate because their values must represent all trees within the species-diameter-status class combination. The best way to do this is to look at the first tree, assign it to a species-diameter-status class, visually estimate height to the nearest 1 ft (0.5 m), estimate live crown percent class, and record the estimates in the Saplings table of the TD data form. As more trees in the same species-diameter-status class are found, the note taker must adjust the average height and average crown ratio of the class. Methods for measuring height and live crown percent are discussed in detail in the next section.

## Sampling mature trees

Mature trees are those with a diameter at breast height that is greater than or equal to the breakpoint diameter.

**Tagging the trees**—If you are doing a monitoring project you should tag all of the mature trees on your macroplot so that they can be identified in the future. (If you are simply doing an inventory, they do not need to be tagged. In this case, sequentially number the trees as you sample. The tree numbers are needed in the database.) There are two methods for tagging the sample trees depending on whether you are using steel or aluminum tags. The tag number is recorded in Field 5 on the TD data form.

**Table TD-4** Diameter classes for saplings.

| Midpoint diameter class | | | |
|---|---|---|---|
| Class | Diameter range | Class | Diameter range |
| | inches | | cm |
| 0.5 | >0  1 | 1.2 | >0  2.5 |
| 1.5 | >1  2 | 3.8 | >2.5  5 |
| 2.5 | >2  3 | 6.2 | >5  7.5 |
| 3.5 | >3  4 | 8.8 | >7.5  10 |

**Table TD-5** Use these classes to record live crown and crown scorch.

| Code | Live crown |
|---|---|
| | Percent |
| 0 | Zero |
| 0.5 | >0  1 |
| 3 | >1  5 |
| 10 | >5  15 |
| 20 | >15  25 |
| 30 | >25  35 |
| 40 | >35  45 |
| 50 | >45  55 |
| 60 | >55  65 |
| 70 | >65  75 |
| 80 | >75  85 |
| 90 | >85  95 |
| 98 | >95  100 |

We suggest using steel casket tags because they will not melt during the heat of the fire. Nail the casket tags to the tree using high-grade nails that also won't melt during fires. Each tag should be tightly nailed to the tree bole with just enough pressure to prevent it from moving, but not so tight that the tag is driven into the bark. If the tag is allowed to move or twirl in the wind, the movement might wear through the nail and the tag could fall off. Nail the tags at breast height with the tag facing toward plot center. This is done so that each tree can be identified while standing in one place and will make relocating the plot center easier. It will also reduce macroplot travel, which can cause compaction and seedling trampling. If the trees are going to be cut, nail the tag less than 1 ft from the ground, facing plot center. This will leave the tags available for posttreatment sampling and will keep them out of the sawyer's way if the trees are cut with a chain saw.

Unfortunately, casket tags are expensive and the steel tags and nails can damage saws, so they should not be used if the trees you are tagging will eventually be harvested. As an alternative to steel tags, nail aluminum tags at DBH on the downhill side of all the mature trees using aluminum nails. Putting tags on the downhill side of the trees will help keep them out of the hottest part of the flame and help keep the tags and nails from melting. Pound the nails in at a downward angle, until the tags are tight but not driven into the bark. Typically, some of the tags will be melted by the fire, but by using tree characteristics (species and DBH, primarily) and the clockwise sampling scheme you should be able to relocate all the mature trees. Occasionally, the head of the nail will melt off and the wind will blow the tag off the angled nail, so you may be able to find the tag lying at the base of the tree.

Although it is not recommended, sometimes it may be necessary to remove some branches with a hatchet or bow saw so that the tag can be firmly attached to the tree. If needed, remove just the problem branches. Removing too many branches may influence the tree's health and/or modify the fuelbed around the tree.

**Measuring General Attributes**—Record the general characteristics of the tree (species and health) in the next set of fields. These characteristics allow the stratification of results and provide some input values needed to compute crown biomass and potential tree mortality.

Enter the species of the tagged tree in Field 6 and the tree status in Field 7 of the TD data form. Tree status describes the general health of the tree. Use the following tree status codes:

**H—Healthy** tree with little biotic or abiotic damage.

**U—Unhealthy** tree with some biotic or abiotic damage, and this damage will reduce growth. However, it appears the tree will not immediately die from the damage.

**S—Sick** tree with extensive biotic or abiotic damage and this damage will ultimately cause death within the next 5 to 10 years.

**D—Dead** tree or snag with no living tissue visible.

Tree status is purely a qualitative measure of tree health, but it does provide an adequate characteristic for stratification of preburn tree health and for determining postburn survival. Remember that trees marked as dead (code D) indicate the tree was sampled on the snag plot.

**Measuring Structural Characteristics**—Five important structural characteristics are measured for each mature tree: DBH, height, live crown percent, crown fuel base height, and crown class. These structural characteristics are used to assess a number of fire-related properties such as canopy bulk density, vertical fuel ladders, height to the base of the canopy, and potential fire-caused mortality. These characteristics can also be used to compute the input parameters needed by the fire growth model FARSITE (Finney 1998), FVS (USDA Forest Service), and FOFEM (Reinhardt and others 1997).

Tree DBH is measured using a diameter or logger's tape. Be sure to measure DBH on the uphill side of the tree with the diameter tape perpendicular to the tree stem and directly against the bark. Measure DBH to nearest 0.1 inch (0.2 cm) and record in Field 8 on TD data form. (See **How To Measure DBH** in the **How-To Guide** chapter.)

Measure tree height from the ground level to the top of the bole or highest living foliage, whichever is higher, to the nearest 1 ft (0.5 m), and enter the height in Field 9 of the TD data form. Tree height is commonly measured with a clinometer, but it can be measured with laser technology or other surveying techniques if that equipment is available and the crew has adequate training in using that technology (see **How to Measure Plant Height** in the **How-To Guide** chapter).

Live crown percent (LCP) is the percent of the tree bole that is supporting live crown. The illustrations in figure TD-7 show four trees with different crown characteristics. Tree A has a crown that should be the most common crown form you will see. Tree B is missing live foliage from the upper part of the crown (the area may be filled with dead branches and needles), and tree C is missing foliage along one side of the stem. Estimate LCP by visually redistributing the live tree crown evenly around the tree so the branches are spaced at about the same branch density as seen along the bole and form the typical conical crown. In the instance of recent abiotic or biotic damage, use the damage and severity codes described below to improve the crown fuel estimates. Record LCP in Field 10 on TD data form using the classes in table TD-5.

Sometimes, you'll see a lone live branch at the bottom of the crown that doesn't appear as part of the crown, and in these cases you can ignore the branch's contribution to LCP.

Crown Fuel Base Height (CFBH) is important for assessing the risk of crown fire. In Field 11 record either the height of the dead material that is sufficient to carry a fire from the lower to the upper part of the tree crown or, if the dead fuel is insufficient, the height of the lowest live foliage. The dead material may include dead branches associated with mistletoe infection, lichens, dead needles, and so forth. This is probably the most subjective field of the TD assessment. We suggest that this information should not be collected unless there is a knowledgeable crew member available that can consistently estimate CFBH. He or she should also record in the Metadata table the characteristics used for the assessment so that it can be made the same way in future sampling.

Optionally, users can record height to live crown (ignore the dead component) to limit some subjectivity. Height to live crown is defined as the height of the lowest branch whorl that has branching in two quadrants—excluding epicormic branches and whorls not part of the main crown—measured from the ground line on the uphill side of the tree. If you decide to estimate height to live crown instead of CFBH, be sure to note that in the Metadata information.

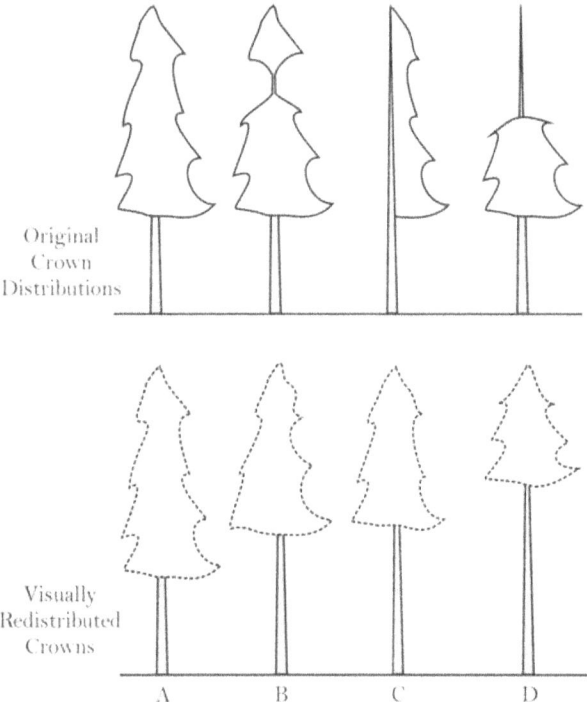

**Figure TD-7**—To estimate LCP visually rearrange the tree branches so that they are distributed around the tree bole, then estimate the percent length of the total tree stem that those branches would occupy. In this illustration the LCP class of trees A, B, C, and D are 70, 50, 50, and 30, respectively.

# Tree Data (TD) Sampling Method

The tree and live crown base height measurement protocols presented in FIREMON were designed to be consistent with the standard inventory techniques used by forest management for timber inventories. These inventories attempt to determine the amount of wood volume that can be harvested from a stand and provide an estimate of potential down woody debris in temporal simulations such as the Forest Vegetation Simulator (FVS). In most instances the FIREMON tree sampling methods will be sufficient. However, in stands where dead topped trees predominate and you need to calculate potential fire-caused tree mortality (for salvage tables, for example), you must make some modifications to your tree sampling.

The most important variable for determining fire-caused mortality is the percent of crown volume that is scorched by the fire. To make this calculation you need to know flame height, live crown base height, and live crown top height. These values are used in the mortality equations to determine the reduction in photosynthetically active material caused by fire scorch. Fire simulations, such as FVS, compute flame height. The crown measurements need to come from field data.

Make the following changes to the standard FIREMON tree measurements if you will be estimating fire-caused mortality of dead top trees. First, do not estimate crown fuel base height. Instead, estimate the live crown base height (the height of the lowest branch whorl that has branching in two quadrants—excluding epicormic branches and whorls not part of the main crown—measured from the ground line on the uphill side of the tree). Next, either measure and record the height to the top of the live foliage or make crown measurements so that you can calculate the height to the top of the live foliage. If you are measuring it directly, enter the height to the top of the live foliage in the Local Code field. If you want to calculate the height to the top of the live foliage, record the total height (height to the top of the tree bole) and record a damage and severity code to indicate the extent of the dead top. For example, the Forest Service Natural Resource Information System code for dead topped trees is 99002. The severity code is the percent of the total height that is dead. (Live foliage top height is calculated by multiplying the total bole height by the percent dead top, divided by 100, then subtracting that value from total bole height.) Using the second method—where you calculate the top of the live crown from bole height and percent tree height that is dead—gets you a more complete set of tree data. Remember, if trees do not have dead tops, then the top of the tree (bole) and the top of the live foliage are essentially the same, so there is no need to calculate live foliage top height. Be sure to document any methods modifications in the Metadata table.

Crown classes (Field 12) represent the position in the canopy of the crown of the tree in question and describe how much light is available to the crown of that tree (fig. TD-8).

There are six categories describing crown class:

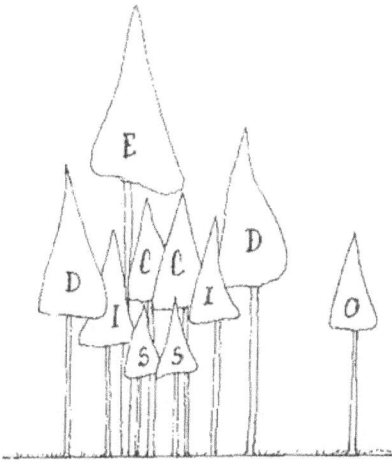

**Figure TD-8**—Use this illustration of crown classes to help you describe the crown class of the tree you are measuring.

**O—Open grown**, the tree is not taller than other trees in the stand but still receives light from all directions.

**E—Emergent**, the crown is totally above the canopy of the stand.

**D—Dominant**, the crown receives light from at least three to four directions.

**C—Codominant**, the crown receives light from at least one to two directions.

**I—Intermediate**, the crown only receives light from the top.

**S—Suppressed**, the crown is entirely shaded and underneath the stand canopy.

Crown class may be used for data report stratification, but it is also an important variable in the computation of tree biomass and leaf area.

**Measuring Tree Age Characteristics**—Age characteristics allow better interpretation of the fire monitoring data by identifying important age and growth classes and using them for results stratification. In addition, growth data allows direct comparison of the change in growth rate as a result of the fire.

Tree age is estimated by extracting a core from the tree at stump height (about 1 ft above ground-line) on the downhill side of the tree using an increment borer. Rings on the increment core can be counted in the field or counted later in the office.

Coring trees to determine tree age is a time-consuming procedure that often requires more than 50 percent of the sampling time. There are many reasons for this. First, there are a lot of time and effort involved in coring the tree, especially large trees, and often you will need to take more than one core per tree in order to get a usable core because of rot or because you missed the pith. Next, it is difficult to get comfortable when drilling at stump height because the sampler is stooped over in an unpleasant position. Certainly, you do not want to have to core every mature tree on your FIREMON macroplot, so the question is how many samples are enough? We suggest that one tree per species per 4 inch (10 cm) diameter class be cored if you are sampling at the Level III - Detailed intensity. The FIREMON project manager can increase the diameter class width to 8 or 12 inches (20 or 30 cm) to further limit coring time on the plot. Enter the tree age in Field 13 of the TD data form for each tree that was cored.

Growth rate should be determined for those trees that are cored for age. However, the growth rate core should be taken at DBH because the age core, taken at stump height, is not a good indicator of growth rate. Remove a core that is deep enough to allow you to measure the last 10 years growth. Growth rate is the distance between the cambium and 10th growth ring in from the cambium, measured to the nearest 0.01 inch (0.1 mm) (fig. TD-9).

If changes in growth rates are a critical part of the monitoring effort (check sampling objectives), then more trees need to be cored for growth rate. If growth rate is an important facet of the fire monitoring project, we suggest that all tagged trees be cored to determine the 10-year growth increment. Enter the growth rate in Field 14 on the TD data form.

**Measuring Damage and Severity**—This information may not be essential for determining the effects of prescribed and wildland fires. However, it can be useful for describing, interpreting, and stratifying monitoring results. We recommend that these fields be completed for most fire monitoring applications.

Biotic/Abiotic Damage and associated Severity Codes are entered in Fields 17 to 20 in table 1 of the TD data form. These codes describe and then quantify the degree of damage by biotic (insect, disease,

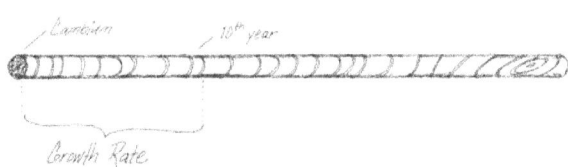

**Figure TD-9**—Tree growth rate is determined by measuring the last 10 years' growth.

## Tree Data (TD) Sampling Method

browsing) and abiotic (wind, snow, fire) agents for the mature trees. Additionally, you will be able to use the abiotic codes for snags. The extensive list of damage and severity codes is presented in **Appendix A: NRIS Damage Categories, Agent, Severity Ratings, and Tree Parts**. These codes are the same ones that are used in the USDA Forest Service Common Stand Exam Guide. For your FIREMON project we suggest that you go through the appendix, select codes that are important to your location and project objectives, and take a printed list with you in the field. If your project requires you to collect damage and severity codes, as a minimum you should use the ones presented in table TD-6. You can add your own codes if needed but be sure to document them in the FIREMON Metadata table. When trees exhibit multiple forms of damage, record the damage with the greatest severity first, then the damage with the second greatest severity.

It takes a great deal of field experience to identify each of the damaging agents present in a tree. Classes at local colleges or USDA Forest Service offices teach insect and disease identification. If the FIREMON crew does not have experience or training in the identification of biotic and abiotic damage sources, then do not fill out these fields.

**Measuring Fire Severity**—Two fire severity measurements are specific to the TD sampling method and apply to the mature trees. The first is bole char height, which is the height of continuous char measured above the ground, on the downhill side of the tree, or on flat ground the height of the lowest point of continuous char (fig. TD-10). It is used to quantify potential tree mortality, and some mortality prediction equations use bole char height as an independent variable. Measure char height with the logger's tape or cloth tape, holding the tape on the downhill side of the tree, and measure to the top of charred part of the bole keeping the tape exactly vertical to the tree. Be sure to measure vertical height. Do not measure along the tree bole, which might be tempting if the tree is leaning. Record bole char height to the nearest 0.1 ft (0.3 m) in Field 21 of the TD data form.

The second fire severity measurement is percent of crown scorched (PCS), which directly relates to tree mortality. It is extremely difficult to estimate because the sampler often does not know if all the charred branches were alive prior to the fire. However, estimates of PCS to the nearest 10 percent class are usually adequate for use in mortality equations, so the bias introduced by dead branches should not cause major problems.

Table TD-6  Record tree and snag damage using these codes or select a code from the set listed in the **Appendix A: NRIS Damage Categories, Agents, Severity Ratings, and Tree Parts Damage and Severity Code Appendix**.

| Damage code | Description | Severity code |
|---|---|---|
| 00000 | No damage | No Damage |
| 10000 | General Insects | 101 Minor  Bottlebrush or shortened leaders or <20% of branches affected, 0 to 2 forks on stem, or <50% of the bole with larval galleries. |
|  |  | 102 Severe  3 or more forks on bole, or 20% or more branches affected or terminal leader dead or 50% or more on the bole with visible larval galleries. |
| 19000 | General diseases | 191 Minor  Short term tree vigor probably not affected. |
|  |  | 192 Severe  Tree vigor negatively impacted in the short term. |
| 25000 | Foliage diseases | 251 Minor  <20% of the foliage affected or <20% of crown in brooms. |
|  |  | 252 Severe  >20% of the foliage affected or >20% of the crown in brooms. |
| 50000 | Abiotic damage | 501 Minor  <20% of the crown affected, bole damage is <50% circumference. |
|  |  | 502 Severe  >20% of the crown affected bole damage is >50% circumference. |
| 90000 | Unknown | 900  0 9% affected |
|  |  | 901  10 19% affected |
|  |  | 902  20 29% affected |
|  |  | 903  30 39% affected |
|  |  | 904  40 49% affected |
|  |  | 905  50 59% affected |
|  |  | 906  60 69% affected |
|  |  | 907  70 79% affected |
|  |  | 908  80 89% affected |
|  |  | 909  90 99% affected |

**Figure TD-10**—Measure char height on the downhill side of the tree.

Estimate PCS by trying to rebuild the tree crown in your head, and then estimate the amount of live crown volume that was damaged or consumed by the fire. The foliage could have been entirely consumed, or scorched black or brown with needles/leaves still attached to the branches. Enter percent crown scorched in Field 22 of the TD data form using the classes in table TD-5.

## Sampling snags

Two fields on the TD field form are specific to snags: snag decay class and primary mortality agent. Snags progress through a series of stages after the tree dies, from standing with red, dead needles still attached to a well decayed stump. As snags pass through these stages, they function differently in the ecosystem, thus it is important to record snag decay class so the ecosystem characteristics can be quantified. Enter the appropriate snag decay class code from table TD-7 into Field 15 of the TD Field Form.

Snag characteristics, for example, wildlife preference and snag persistence, can be greatly influenced by the mortality agent, so it is important to try and identify what killed the tree. This can be difficult for snags that have been standing for a few years, as it is generally accepted that after 5 years you cannot determine the cause of mortality with any certainty. Also, typically there is more than one mortality agent. In FIREMON we suggest recording the primary cause of mortality. For instance, a fire may injure a tree enough to cause stress, which will, in turn, reduce its resistance so that it cannot effectively protect itself from beetle attacks, and then the beetles introduce a fungus that eventually kills the tree. In this case the primary (first) agent of mortality is fire. If you can determine the primary cause of mortality for the tree, record it in Field 16 using the codes in table TD-8. If you cannot determine the primary cause of mortality, use the U code.

## Precision Standards

Use these precision standards for the TD sampling (table TD-9).

**Table TD-7**  Determine snag decay class using these descriptive characteristics.

| Snag code | Limbs | Top of bole | Bark | Sapwood | Other |
|---|---|---|---|---|---|
| 1 | All present | Pointed | 100% remains | Intact | Height intact |
| 2 | Few, limbs | May be broken | Some loss, variable | Some decay | Some loss in height |
| 3 | Limb stubs only | Usually broken | Start of sloughing | Some sloughing | Broken top |
| 4 | Few or no limb stubs | Always broken some rot | 50% or more loss of bark | Sloughing evident | Loss in height always |
| 5 | No limbs or limb stubs | Broken and usually rotten | 20% bark remaining | Sapwood gone | Decreasing height with rot |

Tree Data (TD) Sampling Method

**Table TD-8** Use these mortality codes to identify the primary (first) cause that killed the tree.

| Mortality code | Description |
| --- | --- |
| F | Fire caused |
| I | Insect caused |
| D | Disease caused |
| A | Abiotic (flooding, erosion) |
| H | Harvest caused |
| U | Unable to determine |
| X | Did not assess |

**Table TD-9** Precision guidelines for TD sampling.

| Component | Standard |
| --- | --- |
| DBH | ±0.1 inch/0.25 cm |
| Height | ±1 ft/0.3 m |
| Live crown percent | ±1 class |
| Live crown base height | ±1 ft/0.3 m |
| Crown class | ±1 class |
| Age | ±10 percent of total years |
| Growth rate | ±0.01 inch/0.1 mm |
| Decay class | ±1 class |
| Mortality code | Best guess |
| Damage code | Appropriate category |
| Severity code | ±1 severity class |
| Char height | ±1 ft/0.3 m |
| Crown scorch | ±1 class |
| Count | ±10 percent of total count |
| Average height (saplings) | ±1 ft/0.3 m |
| Average live crown percent | ±1 class |

# SAMPLING DESIGN CUSTOMIZATION

This section presents several ways that the TD sampling method can be modified to collect more detailed information or streamlined to collect only the most important tree characteristics. First, the suggested or recommended sample design is detailed, then modifications are presented.

## Recommended TD Sampling Design

The recommended TD sampling design follows the Alternative FIREMON sampling intensity and is listed below:

**Breakpoint diameter**: 4 inches (10 cm).

**Macroplot size**: 0.1 acre (0.04 ha) [37.24 ft (11.28 m) radius].

**Subplot radius**: 11.77 ft (3.57 m).

Measure all fields EXCEPT Growth Rate, and only measure Age for one tree for each species-4 inch (species-10 cm) diameter class.

## Streamlined TD Sampling Design

The streamlined TD sampling design follows the Simple FIREMON sampling intensity and is designed below:

**Breakpoint diameter**: 6 inches (15 cm).

**Plot size**: 0.1 acre (0.04 ha) [37.24 ft (11.28 m) radius].

**Subplot radius**: 5.89 ft (1.78 m).

Only measure Fields 1-11, Fields 20-21, Fields 21-32.

Do not measure table 1 fields Age, Growth Rate, Damage, Damage Severity, Snags.

## Comprehensive TD Sampling Design

The comprehensive TD sampling design follows the Detailed FIREMON sampling intensity and is detailed below:

**Breakpoint diameter**: 4 inches (10 cm).

**Plot size**: 0.1 acre (0.04 ha) [37.24 ft (11.28 m) radius].

**Subplot radius**: 5.89 ft (1.78 m).

Measure all fields, but measure Age for one tree in each species-4 inch (species-10 cm) diameter class. Measure growth rate on all tagged trees, but only drill deep enough to measure 10-year record.

## User-Specific TD Sampling Design

There are optional fields in each of the TD Field Form tables called Local Code. These will allow the sampler to record some local characteristic that is important to evaluating fire severity and effects. For example, a sampler might be concerned about the height of dead branches above the ground to describe the vertical fuel ladder. Or maybe the sampler wishes to describe the value of the timber in this tree (sweep, crook, broken top) for salvage reasons. The Local Code field allows the sampler to design the TD data form to local situations and objectives. These codes must be recorded in the FIREMON metadatabase to ensure that future users know their meaning.

There are many ways the user can adjust the TD sample fields to make sampling more efficient and meaningful for local situations. First, be sure the plot is big enough and the breakpoint diameter is large enough to get an ecologically valid sample. Use the 20 tree threshold for most cases (each macroplot should contain at least 20 trees above breakpoint DBH), but that should be adjusted based on the advice of local experts. Next, be sure to measure age and growth rate often enough to get a meaningful description for the plot, but not so detailed that it subsumes all sample time.

Use the Local Code for any other characteristic that is of interest in fire monitoring. Note that there is only one field so measuring two tree characteristics will take some creative thinking. For example, you could combine both measurements into one code. For example, the code 11 might indicate a code 1 for tree taper and a code 1 for timber value.

## Sampling Hints and Techniques

Many times the sampler will encounter more trees than the Mature Trees table can accept. Another plot sheet can be started, and the note taker can record at the top of the form that this is the second page of two (or maybe more?). It would be better if the TD plot sheet were copied to both sides of the waterproof paper. That way, the two or more plot sheets do not need to be attached for organization. The same situation happens when there are more species below breakpoint diameter than the Seedling or Sapling tables on the TD data form can accept. Again, if this happens, start another TD data form.

Tree Data (TD) Sampling Method

# TREE DATA (TD) FIELD DESCRIPTIONS

Field 1: **Macroplot Size**. Macroplot size (acre/ha).
Field 2: **Microplot Size**. The size of the microplot where trees less the 4.5 ft tall are sampled (acre/ha).
Field 3: **Snag plot size**. The size of the plot where dead trees greater than breakpoint diameter are measured (acre/ha).
Field 4: **Breakpoint Diameter**. DBH above which trees are measured individually and below which trees are tallied by diameter-species-status classes (inch/cm).
**Table 1: Mature Trees**. Trees greater than breakpoint diameter at breast height
Field 5: **Tag Number**. Tag number attached to mature trees. The tagged numbers need not be in sequence.
Field 6: **Species Code**. Either the NRCS plants species code or the local code for that species. Precision: No error.
Field 7: **Tree Status**. Tree status code that best describes the current health of the tree. Precision: $\pm 1$ class.
**H—Healthy** tree with little biotic or abiotic damage.
**U—Unhealthy** tree with some biotic or abiotic damage, and this damage will reduce growth. However, it appears the tree will not immediately die from the damage.
**S—Sick** tree with extensive biotic or abiotic damage, and this damage will ultimately cause death within the next 5 to 10 years.
**D—Dead** tree or snag with no living tissue visible.
Field 8: **DBH**. The diameter of the tree at breast height (inch/cm). Precision: $\pm 0.1$ inch/0.25 cm.
Field 9: **Tree Height**. The vertical height of the tree (ft/m). Precision: $\pm 1$ ft/0.3 m.
Field 10: **Live crown percent**. The percent class that best describes the percent of the tree stem that is supporting live crown based on the distance from the ground to the top of the live foliage. Valid classes are in table TD-5 of the TD sampling method. Precision: $\pm 1$ class.
Field 11: **Crown Fuel Base Height**. Height above the ground of the lowest live and/or dead fuels that have the ability to move fire higher in the tree (ft/m). Precision: $\pm 1$ ft/0.3 m.
Field 12: **Crown Class**. Code that describes the tree crown's position in forest canopy. Precision: $\pm 1$ class.
**O—Open grown**, the tree is not taller than other trees in the stand but still receives light from all directions.
**E—Emergent**, the crown is totally above the canopy of the stand.
**D—Dominant**, the crown receives light from at least three to four directions.
**C—Codominant**, the crown receives light from at least one to two directions.
**I—Intermediate**, the crown only receives light from the top.
**S—Suppressed**, the crown is entirely shaded and underneath the stand canopy.
Field 13: **Tree Age**. Tree age taken from sample core. Precision: $\pm 10$ percent of total years.
Field 14: **Growth Rate**. The distance measured across the 10 most recent growth rings (inch/mm). Precision: $\pm 0.01$ inch/0.1 mm.
Field 15: **Decay Class**. Decay measure of snags. Valid classes are in table TD-7 of the sampling method. Precision: $\pm 1$ class.
Field 16: **Mortality Code**. Damage that initiated the tree mortality (the first damage to the tree causing its reduced vigor). Valid codes are in table TD-8 of the sampling method. Precision: best guess.
Field 17 to Field 20: **Damage and Severity Codes**. Enter the codes that describe evidence of a damaging agent on the tree and the severity of the damage in order of prevalence on the tree. See **Appendix A: NRIS Damage Categories, Agents, Severity Ratings, and Tree Parts**. Precision: Correct damage category, $\pm 1$ severity class.
Field 21: **Bole Char Height**. Enter the height of the highest contiguous char measured on the downhill side of the tree. Precision: $\pm 0.1$ ft/0.3 m.
Field 22: **Percent Crown Scorched**. Enter the percent of crown that has been killed by fire. Include both scorched and consumed foliage. Valid codes are in table TD-5 of the sampling method. Precision: $\pm 1$ class.
Field 23: **Local Variable**. User defined value or code.
**Table 2: Saplings**—Trees less than breakpoint diameter and taller than 4.5 ft.
Field 24: **Diameter Class**. Class of the trees being sampled. The Analysis Tools program assumes that the diameter class value in this field represents the midpoint of the DBH range of the trees being sampled. Precision: No error
Field 25: **Species Code**. Code of sampled entity. Either the NRCS plants species code or the local code for that species. Precision: No error.
Field 26: **Status Code**. Tree status code that best describes the current health of the tree. Codes presented above in the Field 7 description. Precision: $\pm 1$ class.
Field 27: **Count**. The number of trees tallied for the appropriate diameter-species-status class. Precision: $\pm 10$ percent of total count.
Field 28: **Average Height**. The average height of all trees tallied for this diameter-species-status class. Precision: $\pm 1$ ft/0.3 m.
Field 29: **Average Live crown percent**. Enter the average live crown percent of the trees tallied for this diameter-species-status class. Valid classes are in table TD-5 of the TD sampling method. Precision: $\pm 1$ class.
Field 30: **Local Code**. User defined value or code.
**Table 3: Seedlings**—Trees less than 4.5 ft tall.
Field 31: **Height Class**. Class of the trees being sampled. The Analysis Tools program assumes that the height class value in this field represents the midpoint of the height range of the trees being sampled. Precision: No error.
Field 32: **Species Code**. Code of sampled entity. Either the NRCS plants species code or the local code for that species. Precision: No error.
Field 33: **Status Code**. Tree status code that best describes the current health of the tree. Codes presented above in the Field 7 description. Precision: $\pm 1$ class.
Field 34: **Count**. The number of trees tallied for the appropriate height-species-status class. Precision: $\pm 10$ percent of total count.
Field 35: **Local Code**. User defined value or code.

# FIREMON TD Cheat Sheet

## Tree status

**H—Healthy** tree with very little biotic or abiotic damage.
**U—Unhealthy** tree with some biotic or abiotic damage, and this damage will reduce growth. However, it appears the tree will fully recover from this damage.
**S—Sick** tree with extensive biotic or abiotic damage and this damage will ultimately cause death within the next 5 10 years.
**D—Dead** tree or snag with no living tissue visible.

## Mortality codes

| Mortality code | Description |
|---|---|
| F | Fire caused |
| I | Insect caused |
| D | Disease caused |
| A | Abiotic (flooding, erosion) |
| H | Harvest caused |
| U | Unable to determine |
| X | Did not assess |

## Crown class

**O—Open grown**, or the tree is not near any other tree
**E—Emergent**, or the crown is totally above the canopy of the stand
**D—Dominant**, or the crown receives light from at least 3 4 directions
**C—Codominant**, or the crown receives light from at least 1 2 directions
**I—Intermediate**, or the crown only receives light from the top
**S—Suppressed**, or the crown is entirely shaded and underneath the stand canopy

## Live crown ratio and crown scorch classes

| Code | Live crown percent |
|---|---|
| 0 | Zero percent |
| 0.5 | >0 percent |
| 3 | >1 5 percent |
| 10 | >5 15 percent |
| 20 | >15 25 percent |
| 30 | >25 35 percent |
| 40 | >35 45 percent |
| 50 | >45 55 percent |
| 60 | >55 65 percent |
| 70 | >65 75 percent |
| 80 | >75 85 percent |
| 90 | >85 95 percent |
| 98 | >95 100 percent |

## Snag code descriptions

| Snag code | Limbs | Top of bole | Bark | Sapwood | Other |
|---|---|---|---|---|---|
| 1 | All present | Pointed | 100% remains | Intact | Height intact |
| 2 | Few limbs | May be broken | Some loss, variable | Some decay | Some loss in height |
| 3 | Limb stubs only | Usually broken | Start of sloughing | Some sloughing | Broken top |
| 4 | Few or no limb stubs | Always broken some rot | 50% or more loss of bark | Sloughing evident | Loss in height always |
| 5 | No limbs or limb stubs | Broken and usually rotten | 20% bark remaining | Sapwood gone | Decreasing height with rot |

## Sapling classes

| | Midpoint diameter class | | |
|---|---|---|---|
| Class | Diameter range | Class | Diameter range |
| | inches | | cm |
| 0.5 | >0 1 | 1.2 | >0 2.5 |
| 1.5 | >1 2 | 3.8 | >2.5 5 |
| 2.5 | >2 3 | 6.2 | >5 7.5 |
| 3.5 | >3 4 | 8.8 | >7.5 10 |

## Seedling classes

| | Midpoint height class | | |
|---|---|---|---|
| Class | Height range | Class | Height range |
| | ft | | m |
| 0.2 | >0.0 0.5 | 0.1 | >0.0 0.2 |
| 1 | >0.5 1.5 | 0.3 | >0.2 0.5 |
| 2 | >1.5 2.5 | 0.6 | >0.5 0.8 |
| 3 | >2.5 3.5 | 0.9 | >0.8 1.0 |
| 4 | >3.5 4.5 | 1.2 | >1.0 1.4 |

## Precision

| Component | Standard |
|---|---|
| DBH | ±0.1 inch/0.25 cm |
| Height | ±1 ft/0.3 m |
| Live crown ratio | ± 1 class |
| Crown fuel base height | ±1 ft/0.3 m |
| Crown class | ±1 class |
| Age | ± 10 percent of total years |
| Growth rate | ±0.01 inch/0.1 mm |
| Decay class | ±1 class |
| Mortality code | Best guess |
| Damage code | Appropriate category |
| Severity code | ±1 severity class |
| Char height | ±1 ft/0.3 m |
| Crown scorch | ±1 class |
| Count | ± 10 percent of total count |
| Average height (saplings) | ±1 ft/0.3 m |
| Average live crown percent | ±1 class |

# FIREMON TD Cheat Sheet (cont.)

**Damage and severity codes—short list**

| Damage code | Description | Severity code |
|---|---|---|
| 00000 | No damage | No damage |
| 10000 | General insects | 101 Minor Bottlebrush or shortened leaders or <20% of branches affected, 0 2 forks on stem or <50% of the bole with larval galleries |
| | | 102 Severe 3 or more forks on bole, or 20% or more branches affected or terminal leader dead or 50% or more on the bole with visible larval galleries. |
| 19000 | General diseases | 191 Minor Short term tree vigor probably not affected. |
| | | 192 Severe Tree vigor negatively impacted in the short term. |
| 25000 | Foliage diseases | 251 Minor <20% of the foliage affected or <20% of crown in brooms. |
| | | 252 Severe >20% of the foliage affected or >20% of the crown in brooms. |
| 50000 | Abiotic damage | 501 Minor <20% of the crown affected, bole damage is <50% circumference. |
| | | 502 Severe >20% of the crown affected bole damage is >50% circumference. |
| 90000 | Unknown | 900 0 9% affected |
| | | 901 10 19% affected |
| | | 902 20 29% affected |
| | | 903 30 39% affected |
| | | 904 40 49% affected |
| | | 905 50 59% affected |
| | | 906 60 69% affected |
| | | 907 70 79% affected |
| | | 908 80 89% affected |
| | | 909 90 99% affected |

**Tree data (TD) equipment list**

| | |
|---|---|
| Camera with film | Increment corer (2) |
| Clear plastic ruler (2) | Indelible ink pen (for example, Sharpie, Marker) |
| Clipboard | Lead pencils with lead refills |
| Clinometer (2) | Maps, charts, and directions |
| Cloth tape (2) | Map protector or plastic bag |
| Compass (2) | Masking tape |
| Diameter tape (2) | Mount boards or plastic straws |
| Flagging | Logger's tape (2 plus steel tape refills) |
| Hammer (2) | Magnifying glass |
| Hard hat | Pocket calculator |
| Hatchet (1) | |

# Tree Data (TD) Form

## TD Table 1 - Mature Trees
Trees >Breakpoint Diameter

(Field 1) Macroplot Size : _ _._ _ (ac./ha)
(Field 2) Microplot Size : _ _._ _ (ac./ha)
(Field 3) Snag Plot Size : _ _._ _ (ac./ha)
(Field 4) Breakpoint Diameter : _ _._ _ (in/cm)

RegistrationID: _ _ _ _ _ _
ProjectID: _ _ _ _ _ _
PlotID: _ _ _ _
Date: _ _ / _ _ / _ _

Plot Key

TD Page _ _ of _ _

Notes

Crew

| Field 5 | Field 6 | Field 7 | Field 8 | Field 9 | Field 10 | Field 11 | Field 12 | Field 13 | Field 14 | Field 15 | Field 16 | Field 17 | Field 18 | Field 19 | Field 20 | Field 21 | Field 22 | Field 23 |
|---|---|---|---|---|---|---|---|---|---|---|---|---|---|---|---|---|---|---|
| Tag Number | Species | Tree Status | DBH (in/cm) | Height (ft/m) | Live Crown Percent | CrwnFuel Base Ht (ft/m) | Crown Class | Age | Growth Rate | Decay Class | Mortality Code | Damage Code 1 | Severity Code 1 | Damage Code 2 | Severity Code 2 | Char Height (ft/m) | Crown Scorch % | Local Code |
| | | | | | | | | | | | | | | | | | | |
| | | | | | | | | | | | | | | | | | | |
| | | | | | | | | | | | | | | | | | | |
| | | | | | | | | | | | | | | | | | | |
| | | | | | | | | | | | | | | | | | | |
| | | | | | | | | | | | | | | | | | | |
| | | | | | | | | | | | | | | | | | | |
| | | | | | | | | | | | | | | | | | | |
| | | | | | | | | | | | | | | | | | | |
| | | | | | | | | | | | | | | | | | | |
| | | | | | | | | | | | | | | | | | | |
| | | | | | | | | | | | | | | | | | | |
| | | | | | | | | | | | | | | | | | | |
| | | | | | | | | | | | | | | | | | | |
| | | | | | | | | | | | | | | | | | | |

# Tree Data (TD) Form

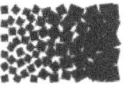

(Field 1) Macroplot Size: _ _ . _ _ (ac / ha)
(Field 2) Microplot Size: _ _ . _ _ (ac / ha)
(Field 3) Snag Plot Size: _ _ . _ _ (ac / ha)
(Field 4) Breakpoint Diameter: _ _ . _ _ (in / cm)

TD Page _ _ of _ _

RegistrationID: _ _ _ _ _
ProjectID: _ _ _ _ _
PlotID: _ _ _ _
Date: _ _ / _ _ / _ _

## TD Table 2 - Saplings
Trees <Breakpoint Diameter & >4.5 ft

| Field 24 | Field 25 | Field 26 | Field 27 | Field 28 | Field 29 | Field 30 |
|---|---|---|---|---|---|---|
| Diameter Class (in/cm) | Species | Status | Count | Average Height (ft/m) | Average Live Crown % | Local Code |
| | | | | | | |
| | | | | | | |
| | | | | | | |
| | | | | | | |
| | | | | | | |
| | | | | | | |
| | | | | | | |
| | | | | | | |
| | | | | | | |
| | | | | | | |
| | | | | | | |
| | | | | | | |

## TD Table 3 - Seedlings
Trees <4.5 ft

| Field 31 | Field 32 | Field 33 | Field 34 | Field 35 |
|---|---|---|---|---|
| Height Class (ft/m) | Species | Status | Count | Local Code |
| | | | | |
| | | | | |
| | | | | |
| | | | | |
| | | | | |
| | | | | |
| | | | | |
| | | | | |
| | | | | |
| | | | | |
| | | | | |
| | | | | |

Plot Key

Notes

Crew

# Fuel Load (FL)

## Sampling Method

**Duncan C. Lutes**
**Robert E. Keane**

## SUMMARY

The Fuel Load method (FL) is used to sample dead and down woody debris, determine depth of the duff/litter profile, estimate the proportion of litter in the profile, and estimate total vegetative cover and dead vegetative cover. Down woody debris (DWD) is sampled using the planar intercept technique based on the methodology developed by Brown (1974). Pieces of dead and down woody debris are tallied in the standard fire size classes: 1-hour (0 to 0.25 inches or 0 to 0.6 cm), 10-hour (0.25 to 1.0 inches or 0.6 to 2.5 cm), 100-hour (1.0 to 3.0 inches or 2.5 to 8 cm). Pieces greater than 3 inches (8 cm) in diameter are recorded by diameter and decay class. Duff and litter depth are measured at two points along each 60-ft (20-meter) sampling plane. Litter depth is estimated as a proportion of total duff and litter depth. Cover of live and dead vegetation is estimated at two points along each 60-ft (20-meter) sampling plane. Biomass of DWD, duff, litter, and vegetation is calculated using the Analysis Tools software.

## INTRODUCTION

The Fuel Load (FL) methods are used to quantify three general components of the fuel complex: dead and down woody debris (DWD), duff and litter, and understory vegetation. Biomass estimates of dead and down woody debris are collected for the size classes that fire scientists have found important for predicting fire behavior and effects—1-hour, 10-hour, 100-hour, and 1,000-hour and greater. DWD measurements are based on the planar intercept methods published by Brown (1974). The sampling area is an imaginary plane extending from the ground, vertically from horizontal (not perpendicular to the slope) to a height 6 ft (2 m) above the ground. Pieces that intercept the sampling plane are measured and recorded. Frequently the term "line transect sampling" is used when discussing the planar intercept method. As far as the FL methodology is concerned the two terms can be interchanged as long as samplers recognize that the "line" is really the measuring tape laid on the litter layer while the "plane" extends above and below the tape, from the top of the duff layer to a height of 6 ft (2 m). Duff and litter are assessed by measuring the depth of the duff/litter profile down to mineral soil, and estimating the percent of the total duff/litter depth that is litter. The biomass of live and dead, woody and nonwoody understory vegetation is estimated using cover and average height estimations. The data collected using the FL methods are used to model fire behavior or to indicate potential fire effects. Forest managers often prescribe fuel treatments, at least partially, on the data collected using the FL methodology. The load of DWD can also be used to estimate the total carbon pool that is stored in the dead material, or DWD data can be used as an indicator of habitat for wildlife. Standing dead trees (snags) are sampled using the FIREMON Tree Data (TD) methods.

The FL methods allow data collection for a wide number of fuel characteristics on each plot. However, field crews are not required to sample every characteristic represented on the field form. In fact, FIREMON was developed specifically so that crews only sample the characteristics they are interested in, as determined by the goals and objectives of the project. In most cases the data collected from plot to plot will be the same although there are situations when some characteristic may be sampled on a subset of the sampling plots.

Dead wood is important in many forest processes. Fire managers need to have an estimate of down dead fuel because it substantially influences fire behavior and fire effects. Smaller pieces of DWD are generally associated with fire *behavior* because they reach ignition temperature more readily than larger pieces. The time it takes for a flaming front to move across a fuel complex is an example of fire behavior influenced by the smaller DWD. Larger pieces of DWD, on the other hand, are usually associated with fire *effects* because, once ignited, these large pieces generally burn longer in both the flaming and smoldering phases of combustion. Soil heating and emissions from combustion are two fire effects closely tied to large DWD. Fire intensity and duration are directly related to fuel load and influence fire *severity* (a general term used to describe the amount of change in the floral and faunal components of a burned site). Logs contribute to forest diversity by providing important nutrient and moisture pools in forest ecosystems. These pools support microfauna and provide sites for the regeneration of understory plants. Logs are frequently used by animals for food storage and cover as well as feeding and nesting sites. Duff and litter are rich in nutrients and microfauna, both of which are intrinsically related to the overall vigor of herbaceous and woody species. Disturbance that substantially reduces the amount of DWD, duff and litter, and understory vegetation can increase soil movement and cause siltation into streams. Duff and litter also provide a layer of insulation during fire, which reduces heat transfer to the soils below. In the absence of an insulating layer of duff and litter, the high levels of soil heating can reduce soil nutrients, and kill microfauna and underground living plant tissues.

A full description of the FL method is provided in the **Sampling Procedure** section below. However, to help the sampling crew understand the research behind and the uses for the FL sampling there is a brief overview provided here.

Two specific components of dead woody fuel are measured using the FL methods: fine woody debris (FWD) and coarse woody debris (CWD). Ecologists often refer to FWD and CWD independently because they function differently in forest ecosystems. FWD are pieces less than 3 inches (8 cm) diameter, and include 1-hour, 10-hour, and 100-hour fuels. CWD includes pieces 3 inches (8 cm) or greater in diameter and at least 3 ft (1 m) in length, also called 1,000-hour and greater fire fuels (table FL-1).

Pieces of DWD are sampled if they pass through the 6-ft (2-meter) high sampling plane. Fine woody pieces are recorded as simple counts. Diameter and decay class are recorded for each piece of CWD. DWD

Table FL-1 Ecologists and fire managers often use different terms to define the same dead woody debris. Typically 1, 10 and 100 hour fuels are grouped together by ecologists and called "fine woody debris." They term 1,000 hour fuels and larger, "coarse woody debris."

| Dead woody class | | | Piece diameter | Piece diameter |
|---|---|---|---|---|
| | | | inches | cm |
| DWD | | | | |
| | FWD | 1 hr | 0  0.25 | 0  0.6 |
| | | 10 hr | 0.25  1.0 | 0.6  2.5 |
| | | 100 hr | 1.0  3.0 | 2.5  8.0 |
| | CWD | 1,000 hr and greater | 3.0 and greater | 8.0 and greater |

biomass estimation is made using equations published in Brown (1974). FIREMON provides six optional assessments for CWD: 1) diameter of the large end of the log, 2) log length, 3) distance along the tape where the piece intercepts the plane, 4) the percent of diameter lost to decay in hollow logs, 5) the percent of log length lost to decay in hollow logs, and 6) percent of the surface of CWD that is charred.

At two points along the base of each sampling plane, measurements are made of duff/litter depth and estimations of the percent of the duff/litter profile that is litter. At these same locations the sampling crew will also estimate the cover of live and dead herbs and shrubs as well as average height of herbs and shrubs.

The planar intercept sampling methodology used in the FL protocol was originally developed by Warren and Olsen (1964) for sampling slash. Brown (1974) revised the original sampling theory to allow for more rapid fuel measurement while still capturing the intrinsic variability of forest fuels. Brown's method was developed strictly to provide estimates of fuel load in the size classes important to fire behavior. He determined the length of the sampling plane needed for each size class and, for FWD, determined quadratic mean diameter for several species. Planar sampling has been reduced to its most fundamental and efficient level while still providing good estimates of DWD.

The planar intercept technique assumes that DWD is randomly oriented directionally on the forest floor. Typically, this assumption does not hold true (for instance in areas of high wind, trees tend to fall with the prevailing winds). FIREMON uses a sampling scheme that reduces bias introduced from nonrandomly oriented pieces by orienting the DWD sampling planes in different directions. This sampling design greatly reduces or eliminates the bias introduced by nonrandomly oriented DWD (Howard and Ward 1972; Van Wagner 1968).

The planar intercept method also assumes that pieces are lying horizontal on the forest floor. Brown (1974) developed a nonhorizontal correction for FWD and noted that a correction for CWD would not substantially improve biomass estimates; therefore, samplers do not record piece angle as part of the FL methodology.

DWD is notoriously variable in its distribution within and between forest stands. Frequently, the standard deviation of DWD samples exceeds the mean. This variability requires large numbers of samples for statistical tests.

There are many ways to streamline or customize the FL sampling method. The FIREMON three-tier sampling design can be employed to optimize sampling efficiency. See the sections on **Optional Fields** and **Sampling Design Customization** in this chapter.

## SAMPLING PROCEDURE

This method assumes that the sampling strategy has already been selected and the macroplot has already been located. If this is not the case, then refer to the FIREMON **Integrated Sampling Strategy** for further details.

The FL sampling procedure is presented in the order of the fields that need to be completed on the **FL data form**, so it is best to reference the data form when reading this section. The sampling procedure described here is the recommended procedure for this method. Later sections will describe how the FIREMON three-tier sampling design can be used to modify the recommended procedure to match resources, funding, and time constraints.

### Preliminary Sampling Tasks

Before using the FL methods in the field we suggest that you find a place close by where you can lay out at least one plot of three transects. This will give the field crew an opportunity to practice and learn the FL methods in a controlled environment where they are not battling steep slopes and tall vegetation. Even if the spot you chose does not have DWD, you can find some branches to lie on the ground to simulate the sampling environment. Be sure to pick a spot where you will be able to make estimates of vegetation cover, vegetation height, and depth of the duff/litter profile. Use the FL Equipment List to determine the materials you will need.

Preparations need to be made before proceeding into the field for FL sampling. First, all equipment and supplies in the **FL Equipment List** must be purchased and packed for transport into the field. Since travel to FIREMON plots is usually by foot, it is important that supplies and equipment be placed in a comfortable daypack or backpack. Be sure you pack spare equipment so that an entire day of sampling is not lost if something breaks. Spare equipment can be stored in the vehicle rather than the backpack. Be sure all equipment is well maintained and there are plenty of extra supplies such as data forms, map cases, and pencils.

All FL data forms should be copied onto waterproof paper because inclement weather can easily destroy valuable data recorded on standard paper. Data forms should be transported into the field using a plastic, waterproof map protector or plastic bag. The day's sample forms should always be stored in a dry place (office or vehicle) and not be taken back into the field for the next day's sampling.

If the sampling project is to resample previously installed FIREMON plots, then it is recommended that plot sheets from the first measurement be copied and brought to the field for reference. These data can be valuable for help in relocating the FIREMON plot.

It is recommended that one person on the field crew, preferably the crew boss, have a waterproof, lined field notebook for recording logistic and procedural problems encountered during sampling. This helps with future remeasurements and future field campaigns. All comments and details not documented in the FIREMON sampling methods should be written in this notebook.

Plot locations and/or directions should be readily available and provided to the crews in a timely fashion. It is beneficial to have plot locations for several days of work in advance in case something happens, such as the road to one set of plots is washed out by flooding. Plots should be referenced on maps and aerial photos using pinpricks or dots to make navigation easy for the crew and to provide a check of the georeferenced coordinates. If possible, the spatial coordinates should be provided if FIREMON plots were randomly located.

Three people allow the most efficient sampling of down debris. There should never be a one-person field crew for safety reasons, and any more than three people will probably result in some people waiting for tasks to be done and cause unnecessary trampling on the plot. Assign one person as data recorder and the other two as samplers. Samplers count FWD and measure CWD pieces that intercept the sampling plane, make duff/litter measurements, and make cover and height estimates along each sampling plane. One sampler should count the 1-hour, 10-hour, and 100-hour size classes while the other measures the CWD. The remainder of the sampling tasks—duff/litter measurements and vegetation cover and height estimates—can be divided between the samplers after they have completed their first tasks.

The crew boss is responsible for all sampling logistics including the vehicle, plot directions, equipment, supplies, and safety. The initial sampling tasks of the field crew should be assigned based on field experience, physical capacity, and sampling efficiency, but sampling tasks should be modified as the field crew gains experience and shared to limit monotony.

## Determining Piece Size

An important task when sampling fuels is to properly determine whether each piece is in the 1-hour, 10-hour, 100-hour, or 1,000-hour and greater size class. Often it will be clear by examining which size class the pieces belong in. This is especially true as field crews gain experience sampling fuels. However, while samplers are calibrating their eyes or when pieces are clearly on the boundary between two size classes, samplers need to take the extra effort to measure pieces and assign them to the proper class. Each sampling crew should have at least one set of sampling dowels for this task. The set is made up of two dowels. One measures 0.25 inch (0.6 cm) in diameter and 3 inches (8 cm) long. Use this dowel to determine whether pieces are in the 1-hour or 10-hour class. The second dowel is 1 inch (2.5 cm) in diameter and 3 inches (8 cm) long. Use this dowel to separate the 10-hour from the 100-hour fuels. Cutting the dowels into 3 inch (8 cm) lengths makes them useful to discern 100-hour and 1,000-hour fuels. The go/no-go gauge is a tool that can speed up the sampling process (fig. FL-1). The gaps in the tool correspond to the 1-hour and 10-hour fuel sizes, and they allow quick assessment of fuel size. Make it out of sheet aluminum (about 0.06 inch thick) so that it is lightweight and durable. Or make one out of an old plastic card (such as the ones you get at grocery stores); while it won't be as durable as an aluminum one, it is easier to make because you can cut the openings using scissors.

FL sampling requires 12 tasks for each sampling plane:

1) Layout the measuring tape, which defines the sampling plane.
2) Measure the slope of the sampling plane.
3) Count FWD.
4) Measure CWD.
5) Measure depth of the duff/litter profile.
6) Estimate the proportion of the profile that is litter.
7) Estimate cover of live woody species.
8) Estimate cover of dead woody species.
9) Estimate average height of live and dead woody species.
10) Estimate cover of live nonwoody species.
11) Estimate cover of dead nonwoody species.
12) Estimate average height of live and dead nonwoody species.

Tasks 5 through 12 are made at two points along each line. Data are recorded on the FL data forms after completing each of steps 2 through 12. You will learn that sampling in order 1 through 12 is not the fastest way to sample a plot. Instead, use task list provided in table FL-2 as a general guide for sampling, and modify it as needed to make for the most efficient sampling.

## Modifying FL Sampling

In the FL method we suggest sampling over a 60-ft (20-m) distance with an addition 15 ft (5 m) of buffer provided to keep from disturbing fuels around the plot center (fig. FL-2). The 60-ft (20-m) plane is the shortest recommended for sampling CWD. However, there are instances of high fuel loads, in slash for instance, where shorter planes for DWD may be justified. If the FIREMON architect wants to use shorter (or longer) sampling planes based on research or expert knowledge, the database can accommodate that data. This write-up assumes that the FIREMON crew is using the suggested FL method.

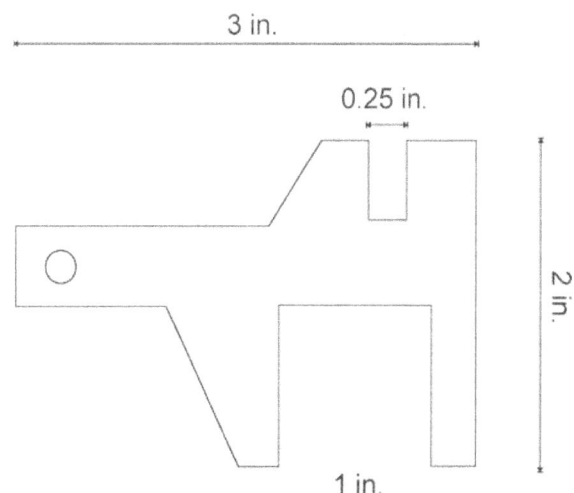

**Figure FL-1**—A go/no-go gauge helps samplers tally 1-hour, 10-hour, and 100-hour fuels quickly and accurately.

Fuel Load (FL) Sampling Method

**Table FL-2**—General task list for sampling with the FL method.

| Task | Recorder | Sampler 1 | Sampler 2 |
|---|---|---|---|
| Organize materials | 1 | | |
| Layout tape | | 1 (guider) | 1 (guidee) |
| Measure slope | 2 (record data) | 2 | 2 |
| Count FWD | 3 (record data) | 3 | |
| Measure duff/litter and veg. at 75 ft mark | 4 (record data) | | 3 |
| Measure CWD | 5 (record data) | | 4 |
| Measure duff/litter and veg. at 45 ft mark | 6 (record data) | 4 | |
| Check for complete forms | 7 | | |
| Collect equipment | | 5 | 5 |

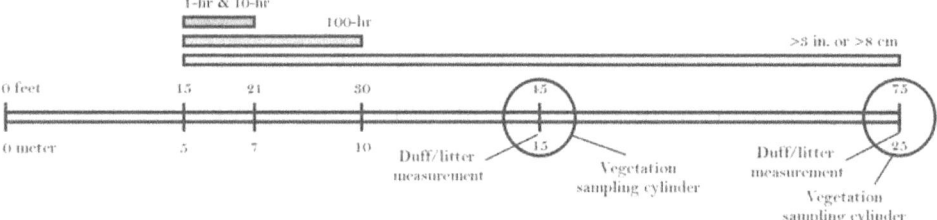

**Figure FL-2**—Dead fuels, duff/litter, and vegetation data are recorded at specific locations on or along each sampling plane. The 1-hour and 10-hour fuels are sampled from the 15-ft (5-m) to the 21-ft (7-m) marks along the plane, the 100-hr fuels are sampled from the 15-ft (5-m) to the 30-ft (10-m) marks, and pieces 3 inches (8 cm) and larger are sampled between the 15-ft (5-m) and 75-ft (25-m) marks along the plane. Duff/litter measurements are made in a representative area within a 6-ft (2-m) diameter circular area at the 45-ft (15-m) and 75-ft (25-m) marks. The cover of live and dead vegetation is estimated within an imaginary 6-ft (2-m) diameter by 6-ft (2-m) high sampling cylinder at the 45-ft (15-m) and 75-ft (25-m) marks.

Additionally, the field crew does not have to use the suggested locations for sampling duff/litter and vegetation. As long as they are thoughtfully placed (for instance, do not sample duff/litter in an area where you will be sampling FWD) these measurements can be made elsewhere along the sampling plane. Record any sampling modifications in the FIREMON Metadata table.

## Laying Out the Measuring Tape

A measuring tape laid close to the soil surface defines the sampling plane. The sampling plane extends from the top of the duff layer to a height of 6 ft (2 m). When laying out the tape, crew members need to step carefully to minimize trampling and compacting fuels—DWD, duff/litter and vegetation—especially along the sampling plane. While the data recorder is arranging field forms and so forth, the other two crew members can lay out the measuring tape for the first sampling plane. Have one crew member stand at plot center (see **How to Locate a FIREMON Plot** in the **How-To Guide** chapter) holding the zero end of the tape, then, using a compass, he or she will guide second crew member (see **How to Use a Compass—Sighting and Setting Declination** in the **How-To Guide** chapter) on an azimuth of 090 degrees true north. The second sampler will move away from plot center, following the directions of the first crew member, until he or she reaches the 75-ft (25-m) mark on the tape. The process of laying out the tape is typically more difficult than it sounds because the tape needs to be straight, not zigzagging around vegetation and trees (fig. FL-3). It pays to sight carefully with the compass and identify potential obstructions before rolling out the tape.

Fuel Load (FL) Sampling Method

**Figure FL-3**—The measuring tape, which represents the lower portion of the sampling plane, should be as straight as possible. If the tape is not straight it needs to be offset left or right until it can be established without kinks or bends.

The second crew member must follow the directions given by the first in order to stay on line and that can take him or her under low branches of trees and shrubs, through thick brush…or worse. The smallest crew member generally has the greatest success at this task, but be sure everyone gets an opportunity. Once the second crew member is at the appropriate location, the first crew member will hold the zero end of the tape over plot center while the second crew member pulls the tape tight. Together, they will move the tape down as close to the ground as possible without struggling to get it so close to the ground that the debris to be measured is disturbed. In most cases, the tape will end up resting on some of the DWD and low vegetation but below the crowns of shrubs, seedlings, and so forth. It is not unusual to get to this point and realize that a large tree, rock, or other obstruction won't allow the tape to be laid straight; instead there is a kink where it hits an obstruction. DWD shouldn't be sampled over a tape that isn't straight so crew members need to lift the tape above the vegetation, move both ends of the line left or right (keep it oriented at the same azimuth) until the tape won't be influenced by any obstructions, then place it back down and straight on the soil surface. Usually this offset won't need to be more than a few feet left or right, but on sites with even moderate amounts of tall vegetation, offsetting the tape can mean considerable work.

Once established, anchor the tape and do not move its position until all sampling is finished for the sampling plane. Most tapes have a loop on the zero end that a spike can be placed through to keep it anchored, and a spike or stick though the handle on the other end of the tape will hold it in place. Roll-up tapes (fig. FL-4) usually have a winding crank that can be flipped so that the knob points toward the reel. In this position the knob will lock the reel so the tape won't unwind when it is pulled tight.

Mark the 0-ft and 75-ft (25-m) marks along the tape so that the plane can be easily re-established. This is especially true when sampling will be done both pre- and post-treatment. Bridge spikes, 8 to 10 inches long, work well because they are relatively permanent when driven completely into the ground and can be relocated with a good metal detector, if needed. Animals such as deer and elk tend to pull survey flags out of the ground, so the flags should not be used as the only indicator of tape position. If spikes and flags are used together, do not wrap the survey flag wire around the spike.

## Determining the Slope of the Measuring Tape

Once the tape has been secured, use a clinometer to measure the percent slope of the line. Aim the clinometer at the eye level of sampler at the other end of the line (fig. FL-5). If there is a height difference of the samplers, adjust the height where you are aiming so that the slope reading is accurate. Carefully, read the percent slope from the proper scale in the instrument and report to the data recorder who will enter it in Field 7 on the FL Field Form.

## What Are "Woody," "Dead," and "Down" Debris?

Before sampling any DWD the terms "woody," "dead," and "down" need to be understood so data gathered with the FL methods are consistent between field crews. "Woody" refers to a plant with stems,

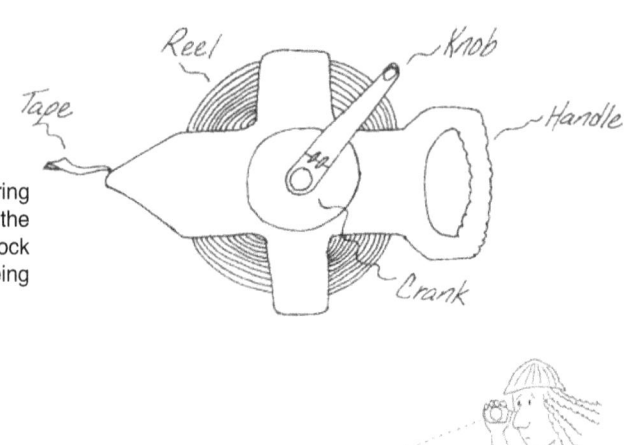

**Figure FL-4**—Parts of a roll-up measuring tape. The crank can usually be flipped in the opposite direction allowing the knob to lock the reel. This will keep more tape from being pulled off.

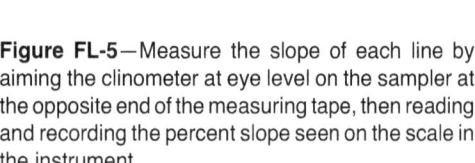

**Figure FL-5**—Measure the slope of each line by aiming the clinometer at eye level on the sampler at the opposite end of the measuring tape, then reading and recording the percent slope seen on the scale in the instrument.

branches, or twigs that persist from year to year. The structural parts support leaves, needles, cones, and so forth, and it is these structural components that are tallied along the sampling plane.

"Dead" DWD has no live foliage. Sampling deciduous species in the dormant season can be a challenge and should only be done by crews with the expertise to identify dormant versus dead trees and shrubs.

## CWD

CWD includes pieces 3 inches (8 cm) or greater in diameter. Some authors suggest CWD must also be at least 3 ft (1 m) in length. Because this may lkead to unrealistic CWD biomass values—especially in logging slash where many pieces may not meet the length criteria—we do not suggest defining CWD using a length component. CWD at an angle of greater than 45 degrees above horizontal where it passes through the sampling plane should only be considered "down" if it is the broken bole of a dead tree where at least one end of the bole is touching the ground (not supported by its own branches, or other live or dead vegetation). If CWD is at an angle of 45 degrees or less above horizontal where it passes through the sampling plane, then it is "down" regardless of whether or not it is broken, uprooted, or supported in that position (fig. FL-6 and FL-7).

Do not sample a piece of CWD if you believe the central axis of the piece is lying in or below the duff layer where it passes through (actually, under) the sampling plane (fig. FL-8). These pieces burn more like duff, and the duff/litter methodology will allow field crews to collect a representative sample of this material.

## FWD

FWD are pieces less than 3 inches (8 cm) diameter. Pieces of FWD that are "woody," "dead," and" down" fall into three general categories: 1) pieces that are not attached to the plant stems or tree boles where

# Fuel Load (FL) Sampling Method

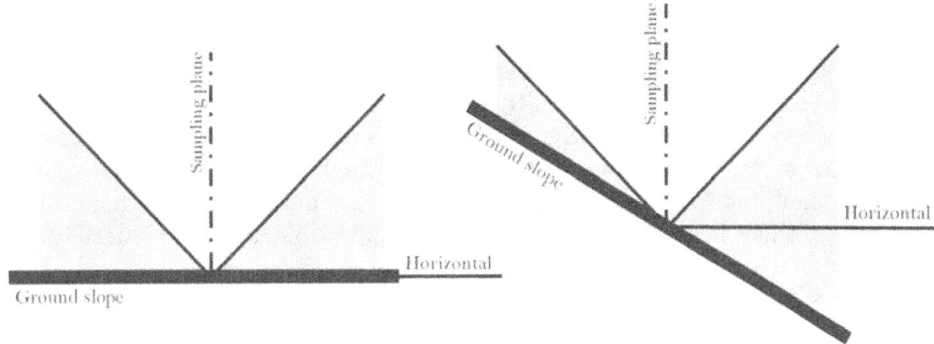

**Figure FL-6**—CWD pieces crossing through the sampling plane at an angle less than 45 degrees from horizontal (represented by the shaded areas in the figure) are always considered to be "down." Some CWD leaning greater than 45 degrees may be considered "down." See the text for details.

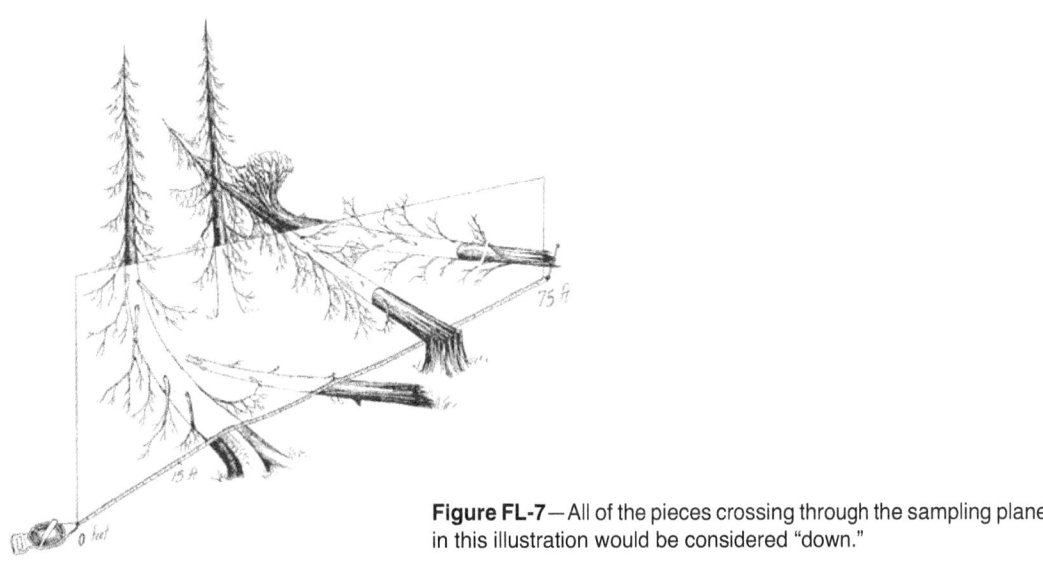

**Figure FL-7**—All of the pieces crossing through the sampling plane in this illustration would be considered "down."

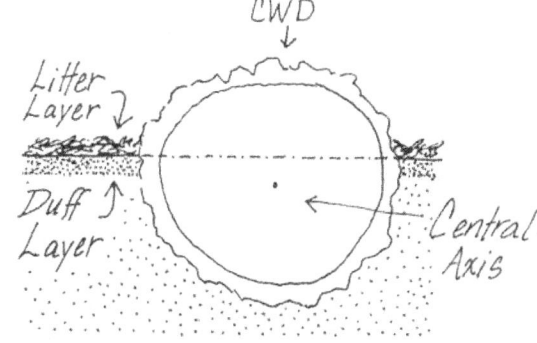

**Figure FL-8**—Do not sample CWD when the central axis of the piece lies in or below the duff layer.

they grew and have fallen to the ground, 2) pieces that are not attached to the plant stems or tree boles where they grew but are supported above the ground by live or dead material, and 3) pieces attached to stems or boles of shrubs or trees that are themselves considered "dead" and "down." Note that it is possible for FWD to be considered "dead" even though it has green foliage attached because the rules consider any piece severed from the plant where it grew to be both "dead" and "down." Fresh slash and broken branches are examples of green material considered "dead." Sample dead pieces only when they are still attached to "dead" and "down" trees and shrubs. Do not sample dead branches attached to live trees and shrubs even if those branches are broken but hanging from the plant where they grew. Piece angle of FWD is not critical in determining whether or not it is "down." Do not tally needles, grass blades, pine cones, cone scales, bark pieces, and so forth, as they are not "woody" in nature. This material is considered litter and is measured as part of the duff/litter profile.

## DWD Sampling Distances

DWD is sampled along a certain portion of the sampling plane based on the size of the piece (fig. FL-2). The 1-hour and 10-hour fuels are sampled from the 15-ft (5-m) to the 21-ft (7-m) marks along the plane, the 100-hr fuels are sampled from the 15-ft (5-m) to the 30-ft (10-m) marks, and pieces 3 inches (8 cm) and larger are sampled between the 15-ft (5-m) and 75-ft (25-m) marks along the plane. The distances for sampling FWD are shorter than for CWD because pieces of FWD are more numerous, so a representative sample can be obtained with a shorter sampling distance. DWD is not measured along the first 15 ft (5 m) of the tape because fuels are usually disturbed around plot center by the activity of the sampling crew as they get organized to lay out the tape. The Analysis Tools program will accept different sampling plane lengths from the ones suggested here. If you use different lengths, record the reason for changing them in the Metadata (MD) table. Enter the sampling plane length for 1-, 10-, 100-, and 1,000-hour fuels in Fields 1 through 4 of the FL field form. If you are using a predetermined number of sampling planes per plot, enter that value in Field 5, otherwise the field will be filled in at the end of the plot sampling. This issue is more completely covered in later sections.

## Sampling FWD

The crew member at the zero end of the tape should sample FWD to maximize sampling efficiency. Count the 1-hour and 10-hour fuels that pass through the sampling plane from the 15-ft (5-m) to the 21-ft (7-m) marks on the measuring tape. Remember the plane extends from the top of the duff layer vertically to a height of 6 ft (2 m). The best way to identify the pieces intercepting the plane is to lean over the tape so that your eye is positioned vertically a few feet over the measuring tape at the 15-ft (5-m) mark. Then, while looking at one edge of the tape, maintain your head in that same vertical position over the line and move ahead to the 21-ft mark while making separate counts for the 1-hour and 10-hour fuels that cross under or above the edge of the tape. Each piece needs to be classified as 1-hour or 10-hour fuel by the diameter where it intercepts the sampling plane, defined by one edge of the measuring tape. Samplers should use the dowels or the go/no-go gauge discussed earlier to classify fuels that are close to the size class bounds. Often pieces above will cover pieces below. It is important to locate all the pieces that intercept the plane in order to get accurate fuel load data (fig. FL-9). When finished tallying the 1-hour and 10-hour fuels, report the counts to the data recorder who will enter them in Fields 8 and 9 on the data sheet.

Use the same basic procedure to count the 100-hour fuels that pass through the sampling plane from the 15-ft (5-m) to the 30-ft (10-m) marks on the tape. Report the information to the data recorder who will enter the count in Field 10 on the data sheet.

## Sampling CWD

The CWD sampling plane is 6 ft (2 m) high and extends from the 15-ft (5-m) mark to the 75-ft (25-m) mark along the measuring tape. Sample CWD that intercepts the sampling plane and meets the dead, down, and woody requirements discussed above. In general, at least two fields are recorded for each piece of CWD: diameter and decay class. Percent char, log length, diameter of the large end, point of intersect, and

# Fuel Load (FL) Sampling Method

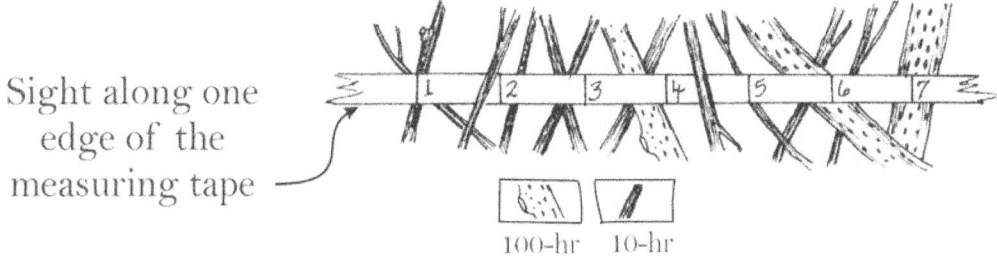

**Figure FL-9**—Tally pieces that intercept the sampling plane both above and below the measuring tape. Focus on one edge of the tape to make counting easier. Be sure to note any lower fuels that are hidden by pieces above. In this illustration there are 11 1-hour and 3 10-hour fuels.

estimations of volume lost to decay are additional data that may be collected for each piece of CWD. See the **Sampling Design Customization** section at the end of this document for more information. CWD sampling should be done by the crew member who is standing at the 75-ft (25-m) end of the tape while moving toward the zero end. This will keep him or her out of the way of the other sampler and will reduce the chances of the FWD being inadvertently disturbed before being sampled.

Diameter measurement and decay class are determined on each piece of CWD where it passes through the sampling plane. Measure diameter perpendicular to the central axis of each piece to the nearest 0.5 inch (1 cm) (fig. FL-10). If a piece crosses through the sampling plane more than once, measure it at each intersection. A diameter tape or caliper work best for diameter measurements, but a ruler can give good results if it is used so that parallax error does not introduce bias (See **How to Measure Diameter with a Ruler** in the **How-To Guide** chapter).

Use the descriptions in table FL-3 to determine the decay class for CWD at the same point where diameter measurement was made. Decay class can change dramatically from one end of a piece of CWD to the other, and often the decay class at the point where the diameter measurement was taken does not reflect the overall decay class of the piece. However, by recording the decay class at the point where diameter was measured, the field crew will collect a representative sample of decay classes along each sampling plane. The transect number, sequential piece number (log number), diameter, and decay class for each piece are entered in Fields 16 through 19, respectively.

## What Are, "Duff," "Litter," and the "Duff/Litter Profile"?

Duff and litter are two components of the fuel complex made up of small, woody, and nonwoody pieces of debris that have fallen to the forest floor. Technically, packing ratio, moisture content, and mineral content are used to discriminate the litter and duff layers. Samplers will find it easier to identify each layer by using the following, more general, criteria. "Litter" is the loose layer made up of twigs, dead

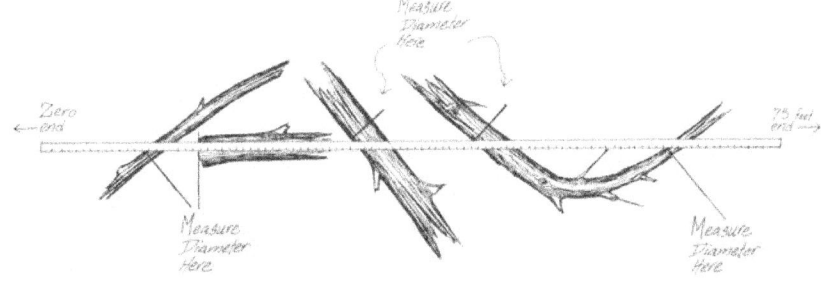

**Figure FL-10**—Measure the diameter of CWD crossing through the sampling plane perpendicular to the central axis of the piece. If a curved piece passes through the plane more than once measure its diameter at each intersection.

Fuel Load (FL) Sampling Method

**Table FL-3** Use these descriptions to determine the decay class where the log crosses the sampling plane.

| Decay class | Description |
|---|---|
| 1 | All bark is intact. All but the smallest twigs are present. Old needles probably still present. Hard when kicked. |
| 2 | Some bark is missing, as are many of the smaller branches. No old needles still on branches. Hard when kicked. |
| 3 | Most of the bark is missing, and most of the branches less than 1 inch in diameter also missing. Still hard when kicked. |
| 4 | Looks like a class 3 log but the sapwood is rotten. Sounds hollow when kicked, and you can probably remove wood from the outside with your boot. Pronounced sagging if suspended for even moderate distances. |
| 5 | Entire log is in contact with the ground. Easy to kick apart but most of the piece is above the general level of the adjacent ground. If the central axis of the piece lies in or below the duff layer then it should not be included in the CWD sampling, as these pieces act more like duff than wood when burned. |

grasses, recently fallen leaves, needles, and so forth, where the individual pieces are still identifiable and little altered by decomposition. The "duff" layer lies below the litter layer and above the mineral soil. It is made up of litter material that has decomposed to the point that the individual pieces are no longer identifiable. The duff layer is generally darker than the litter layer and is more aggregated because of the fine plant roots growing in the duff material.

The "duff/litter profile" is the cross-sectional view of the litter and duff layers. It extends vertically from the top of the mineral soil to the top of the litter layer. The FL methods use the depth of the duff/litter profile and estimation of the percent of the total duff/litter depth that is litter to estimate the load of each component.

Litter usually burns in the flaming phase of consumption because it is less densely packed and has lower moisture and mineral content than duff, which is typically consumed in the smoldering phase. Litter is usually associated with fire *behavior*, and duff with fire *effects*.

## Sampling Duff and Litter

Duff and litter are not sampled using the planar intercept method. Instead, duff/litter measurements are made using a duff/litter profile at two points along each sampling plane. The goal is to develop a vertical cross-section of the litter and duff layers without compressing or disturbing the profile. As samplers finish collecting DWD data, they can start making the duff/litter measurements.

Duff/litter depth measurements are made at a point within 3 ft (1 m) of the 45-ft (15-m) and 75-ft (25-m) marks along the tape. Follow the same instructions at both measurement locations. Select a sampling point within a 3-ft (1-m) radius circle that best represents the duff/litter characteristics inside the entire circle. Samplers can make the profile using a trowel or boot heel. Using a boot heel in deep duff and litter generally results in poor profiles, which in turn make measurement difficult. Use the blade of the trowel to lightly scrape just the litter layer to one side. Then return the blade to the point where the litter scrape was started, push the trowel straight down as far as possible through the duff layer and move the material away from the profile. Use the trowel to work through the duff layer until mineral soil is noted at the bottom of the profile. Mineral soil is usually lighter in color than the duff and more coarse in composition, often sandy or gravelly. If a boot is used, drive the heel down and drag it toward you. As with the trowel, continue working through the duff until mineral soil is noted. It is important to not disturb the profile by compacting or pulling it apart on successive scrapes. The profile that is exposed should allow an accurate measurement of duff/litter depth (fig. FL-11).

Use a plastic ruler to measure total depth of the duff/litter profile to the nearest 0.1 inch (0.2 cm). Place the zero end at the point where the mineral soil meets the duff layer, then move either your index finger or thumb down the ruler until it is level with or just touches the top of the litter (fig. FL-12). While keeping your finger in the same position on the ruler, lift the ruler out of the profile and note the duff/litter depth, indicated by your finger. If your ruler is not long enough to measure the duff/litter depth use the ruler to make marks on a stick, and measure the profile with the stick. If you use the stick measurement method often, get a longer ruler. Next examine the duff/litter profile and estimate the

# Fuel Load (FL) Sampling Method

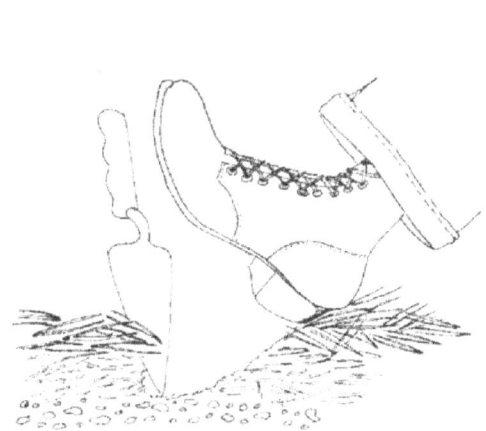

**Figure FL-11**—Use your boot to carefully pull the litter and duff layers away, until you are down to mineral soil.

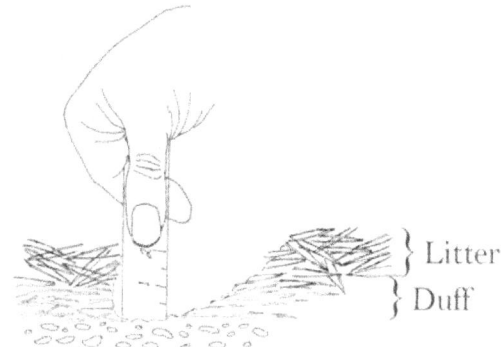

**Figure FL-12**—Use a plastic ruler to estimate duff and litter depth. Place the zero end at the intersection of the mineral soil and duff layer, then mark top of the litter layer using your thumb or finger. In this illustration the duff/litter depth is 2 inches (5 cm), and the proportion of that depth that is litter is about 50 percent.

percent of the total depth that is made up of litter, to the nearest 10 percent. Finally, report the duff/litter depth measurement and litter percent estimate to the data recorder who, depending on measurement point, will enter the data in Fields 11 and 12 or Fields 13 and 14 on the FL field form.

Duff and litter measurements are most easily and accurately made on the vertical portion of the profile as long as that portion of the profile is representative of the true duff/litter depth (it wasn't negatively impacted when the profile was developed). Sometimes the most vertical part is where the back of the trowel blade or boot heel went in, as depicted in figures FL-11 and FL-12, and sometimes it is along one side of the profile.

## What Is "Woody" and "Nonwoody" Vegetation?

The last fuel characteristics that field crews will sample along each sampling plane are the covers of trees, shrubs, and herbs. These can be divided into woody and nonwoody species. Both trees and shrubs are woody species. They are easily identified because their stems persist, and growth does not have to start at ground level each growing season. Trees generally have a single, unbranched stem near ground level, and shrubs generally have multiple stems near ground level. Woody species can be evergreen or deciduous. Deciduous species lose their foliage at the end of the growing season, but the aerial woody portions of the plant remain. Herbs are nonwoody plants whose aerial portions die back at the end of the growing season. Most experienced field samplers will have an intuitive idea of which vegetation is woody and which is not. One way to help identify nonwoody plants is to remember that, in general, weather factors, such as wind, rain, snow, and so forth, collapse herb foliage and stems back to or near the ground between growing seasons.

Small trees, shrubs, and herbs influence fire behavior because their branches and foliage are suspended above the ground allowing more efficient heating and burning of the parts. Dense, suspended fuels can lead to fires that are difficult or impossible to control. The fires in chaparral vegetation in the Western United States are an example. By estimating the cover and heights of woody and nonwoody vegetation, fire managers can estimate the volume, density, and biomass of vegetation. All three of the characteristics are strongly associated with fire behavior.

## Sampling Vegetation Cover and Height

Estimate vegetation cover and height at the 45-ft (15-m) and 75-ft (25-m) marks on the measuring tape. Field crews will estimate the vertically projected cover of vegetation within a 6-ft (2-m) tall by 6-ft (2-

m) diameter imaginary sampling cylinder. Use the marks on the measuring tape to help visualize the 6-ft (2-m) diameter. For instance, when standing at the 45-ft (15-m) mark, the 42-ft (14-m) and 48-ft (16-m) marks will identify the boundary of the cylinder along the tape. Use that measurement to get a good idea of the distance needed on each side, perpendicular to the tape, required to form the imaginary base of the cylinder. Many people have an arm's width spread that is about 6 ft (2 m). Each sampler should measure his or her arm span and use that measurement to help them visualize the sampling cylinder.

The extent of plant cover (foliage and supporting parts) is a function of phenological stage. Early in the season many plants may not have completely leaved out, in mid-season plant cover and height reaches a maximum, and then in late season plant material, especially herbaceous vegetation, moves from the live to dead class. So the question is: when should vegetation be sampled? Should you sample at the same time every year regardless of the growth stage of the vegetation, or should you sample at the peak of growth, or should you sample during the burning season…? Often an examination of the project objectives will help determine the best time for sampling. However, the reality is that sampling will typically occur when field crews have the time, and that may have little to do with the objectives or the plants' phenological stage during previous sampling visits.

In the FIREMON FL vegetation sampling we suggest *estimating* the peak cover and height regardless of the seasonal timing of the sampling visit. This adds a certain amount of error in the cover and height estimates, so it may not be advisable in some monitoring programs. For instance, if project objectives are specifically interested in monitoring fuel characteristics during the burning season, it will be better to actually sample during the burning season. The benefit of estimating peak cover is twofold. First, it allows some sort of standardization between vegetation assessments by partially eliminating the variation due to seasonal changes in vegetation characteristics. Second, it gives the manager some idea of the maximum biomass, maximum vertical distribution of the fuel complex, and maximum live component. Then he or she can use that information to estimate vegetation characteristics during other times of the season, say, when considering an end of the season prescribed fire. For example, if cover in a grassland is estimated at peak to be 70 percent cover, 3 ft tall, with 20 percent dead material you can figure that at the end of the growing season biomass will be the same, or nearly so, but with a majority closer to the ground and with a much higher dead component. (The biomass equations in FIREMON are based on oven-dry weight, so when cover and height are equal there is no difference in biomass between live and dead plants.) Given the same weather conditions, the resulting fire behavior and, possibly, fire effects will be more extreme at the end of the season than during the peak because of denser packing and lower fuel moistures.

Herbaceous species such as some grasses fall over when they become dormant. In order to get good biomass estimates, when sampling dormant material you need to estimate cover and height as if plants were erect. For height, simply lift the tops of a few plants up from the ground and measure the height. Cover can be a bit more challenging as the flattened grass makes cover look greater than it is. Usually lifting the grass and examining the basal distribution of the stems will lead to sufficient estimates. Sometimes cover estimation can be accurately made simply by recalling what was seen at the peak of the growing season in similar areas. Because of the woody component, shrub height does not change much between growing and dormant season. However, during the dormant season, shrub cover must be estimated by imagining the plants with their foliage. Getting accurate estimates this way may be difficult, but reasonable cover assessments are possible with practice.

Again, estimating cover and height at the peak of the growing season is a *suggested* sampling scheme for the vegetation component. You should sample vegetation using the methods you feel the most comfortable with and that meet the needs of the project. Be sure to record in the Metadata table the how, when, and why of your FL vegetation sampling.

Six attributes are measured at each vegetation sampling point. There are four cover estimations for vegetation: 1) live woody species (trees and shrubs), 2) dead woody species, 3) live nonwoody species (herbs), and 4) dead nonwoody species. There are two height estimations: 5) the woody component and 6) the nonwoody component. "Cover" is the vertically projected cover contributed by each of the four categories within the sampling cylinder. It includes plant parts from plants rooted in the sampling cylinder and plant parts that project into the sampling cylinder from plants rooted outside, for instance, live and dead branches. Estimate cover by imagining all the vegetation in the class being sampled, say live shrub cover, compressed straight down to the ground. The percent of the ground covered by the compressed vegetation inside the 6-ft (2-m) diameter sampling area what is being sampled. The cover of dead branches on a live plant should be included in the dead cover estimate. We recommend not including the cover of the cross-sectional area of vertically oriented single stemmed trees in the live or dead woody cover estimate. The stems don't really count as surface fuel because they do not contribute much to fire behavior or fire effects. Also, if the sampling cylinder was located on an area with an unusually high number of tree stems, the vertical projection of the foliage would probably be overlapping the area of the stems; thus the actual cover would be the same with or without the stems. See **How To Estimate Cover** in the **How-To Guide** chapter for additional hints on how to estimate cover accurately.

Two conditions make cover estimations difficult and, frequently, inaccurate. First, the equations used to estimate biomass assume that all of the plant parts for each species are included in the cover and height estimation. In other words, if looking at the cover of a woody shrub species, samplers need to estimate the cover of all the parts, even things like the foliage, which are not "woody." Second, estimating cover is not something people do often; it is only with practice and experience that good estimations of plant cover can be made. Fortunately, the cover classes used in FIREMON are typically 10 percent so the precision of cover estimates are secondary to accuracy (table FL-4).

In addition to the cover estimates, samplers will make two height estimates at each vegetation sampling location, one for the average height of the live and dead woody species and one for the average height of the live and dead nonwoody species. Make your height estimate by noting the maximum height of all the plants in the class and then recording the typical or average of all the maximum heights. Some people like to envision a piece of plastic covering just the plants in one class, then to estimate the average height of the plastic above the ground. Either method will work and give answers that are of adequate precision. Estimate height to the nearest 0.5 ft (0.2 m). Remember, for both the cover and height estimation, only include the vegetation that is within the sampling cylinder. A fast way to make accurate height assessments is for samplers to measure their ankle, knee, and waist heights then estimate vegetation height based on those points. See **How To Measure Plant Height** in the **How-To Guide** chapter for more information.

Record the vegetation cover classes and height data in Fields 22 through 33 on the FL Field Form.

Table FL-4 Cover of each of the four vegetation categories is recorded on the field form in one of the following classes.

| Code | Cover |
|------|-------|
|      | percent |
| 0    | No cover |
| 0.5  | >0 1 |
| 3    | >1 5 |
| 10   | >5 15 |
| 20   | >15 25 |
| 30   | >25 35 |
| 40   | >35 45 |
| 50   | >45 55 |
| 60   | >55 65 |
| 70   | >65 75 |
| 80   | >75 85 |
| 90   | >85 95 |
| 98   | >95 100 |

# Fuel Load (FL) Sampling Method

## Finishing Tasks

The most critical task before moving to the next sampling plane or plot is to make certain that all of the necessary data have been collected. This task is the responsibility of the data recorder. Also, the recorder should write down any useful comments. For instance, you might comment on some unique or unusual characteristic on or near the plot that will help samplers relocate the plot. Include notes about other plot characteristics, such as "evidence of deer browse" or "deep litter and duff around trees." Finally, collect the sampling equipment and move ahead to start sampling the next plane.

## Successive Sampling Planes

On each FL plot the field crew will collect data for at least three sampling planes. (If you are resampling an existing FL plot read the **Resampling FL Plots** section, below.) Follow the FL plot design in figure FL-13. The first sampling plane is always oriented at an azimuth of 090 degrees true north, the second is oriented 330 degrees, and the third at 270 degrees.

Planes are oriented in multiple directions to avoid bias that could be introduced by DWD pieces that are not randomly oriented on the forest floor. The DWD biomass estimate with the FL methods is an average across all of the sampling planes.

It is not necessary for one sampling plane to start at the exact 75-ft (25-m) mark of the previous one. In fact it is better if the start of the new line is 5 ft or so away from the end of the last so that the activity around the new start does not adversely impact the fuel characteristics at the end of the last one. The duff/litter layer and woody and nonwoody vegetation at the 75-ft (25-m) mark, in particular, could be disturbed by field crew traffic, which can bias the data when the plot is resampled. Make sure that no portion of the new sampling plane will be crossing fuels that were sampled on the previous plane. Once the start of the new sampling plane has been determined, collect the data as you did on the first plane. Look ahead and see which starting point will guarantee a straight line before you start laying out the next sampling plane. If the sample sampling planes are going to be remeasured be sure to carefully mark 0-ft and 75-ft end of each sampling plane.

## Determining the Number of Sampling Planes

After the crew has finished sampling three planes, the data recorder will sum up the counts of all the DWD pieces (1-hour, 10-hour, 100-hour, and 1,000-hour pieces), and if that number is greater than 100 then the crew is finished sampling DWD for the plot. If the count is less than 100 then the crew needs

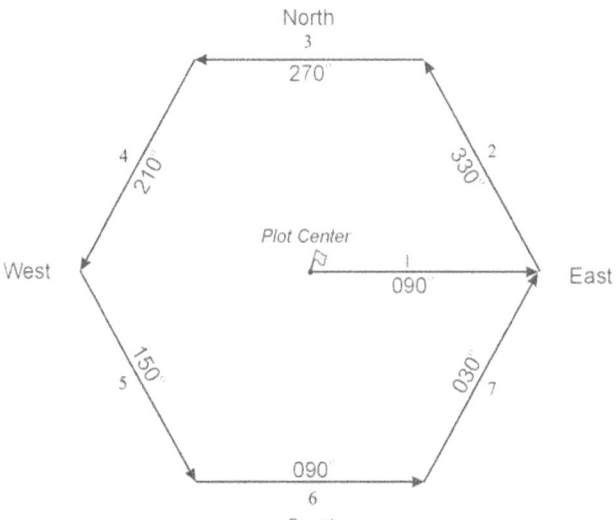

**Figure FL-13**—The FL plot design allows a representative sample of DWD to be obtained while reducing or eliminating the bias introduced by nonrandomly oriented pieces. Data are collected on and along three to seven sampling planes.

Fuel Load (FL) Sampling Method

to sample another line. If another line is needed, refer to the FL plot design (fig. FL-13), lay out the next sampling plane, and collect the FL data. When finished with that plane, recalculate the DWD piece count. Again, if the count is greater than then 100, the sampling is finished; if not, another plane needs to be sampled. Continue sampling until either total piece count is greater than 100 or seven planes have been sampled. Once sampling has begun on a sampling plane, data must be collected for the *entire* plane. When the sampling is completed record the number of planes that you sampled in Field 5 of the FL data form.

DWD is only one part of the surface fuels complex. The 100-piece rule is also meant to help guide the sample size for the duff, litter, and vegetation components of the complex. Thus, even if there is little DWD, the duff, litter, and vegetation should be sampled sufficiently. Conversely, if you are sampling numerous pieces of DWD, in principle that material should be carrying most of the fire, so a reduced number of litter, duff, and vegetation assessments should not be an issue. If this is not the case modify your plot level sampling so you get dependable estimates of all the FL components. Record any sampling modifications in the Metadata table for the project.

Two potential shortcomings can be encountered when using the 100-piece rule. First, the greater the clumping or aggregation of fuels on the forest floor the greater the opportunity of having a high number of piece counts on one or more sampling planes. These clumps can lead to an overestimation of DWD biomass. For example, say that as an experiment a field crew wants to compare biomass values using the 100-piece rule versus sampling with five sampling planes. The first three sampling planes are in the exact same location for the comparison. In the experiment it just happens that the third plane crosses over a spot where there is an accumulation of FWD, and when the third sampling plane was finished the crew had sampled 112 pieces—the end of sampling for the 100-piece rule data. They continued to sample two more planes for their five-plane comparison but recorded no more data because the planes crossed a small grassy area. Back at the office they ran their data through the FIREMON database and noted that they sampled 5.3 tons/acre of material using the 100-piece rule but only 3.2 tons/acre when five sampling planes were used, even though in the field they sampled exactly the same pieces. This is because the tons/acre value that comes from the planar intercept calculation is the average across all of the sampling planes. In the first case the denominator was 3 and in the second it was 5. The example presents an extreme case, but recognize that any aggregation of fuels can lead to overestimation—and always an overestimation—and the earlier in the sampling plane sequence that the aggregation is encountered, the greater the opportunity for overestimation. The second shortcoming of the 100-piece rule is that, for comparison, when plots are resampled, the number of sampling planes has to be the same as the first time the plot was sampled. It can be time consuming (and presents an opportunity for errors) to look up all of the original plots in the database and note the number of planes sampled at each. Despite these shortcomings, the 100-piece rule works well most of the time and frees the FIREMON architect from determining the number of planes that will need to be sampled on each plot of the project. Finally, the 100-piece rule is especially useful in inventory sampling where plots are sampled only once.

If the 100-piece rule is not used for the DWD sampling then the architect must determine the number of sampling planes that will be used throughout the project. The task is to sample with sufficient intensity to capture the variation while not wasting time sampling too intensively. This is made more difficult when fuels vary greatly across the project area. Assuming that the project funding is not limiting the sampling intensity, we suggest determining the number of sampling planes per plot using a pilot study. Install pilot plots in the study area or a similar ecosystem, and sample using the 100-piece rule (you don't need to measure any attributes, just count pieces of DWD). Be sure to put plots in areas representing the range of DWD piece densities you will be sampling in your study area. Depending on the variability of the fuels, after sampling 10 or 20 plots you will be able to identify a good number of sampling planes to use in your project. You should pick the number that lets you meet the 100-piece limit on at least 80 percent of your plots. For example, say that you had 20 plots in your pilot study, and the number of sampling planes needed to count 100 pieces at each plot was:

| | |
|---|---|
| 3 sampling planes | 2 plots |
| 4 sampling planes | 5 plots |
| 5 sampling planes | 10 plots |
| 6 sampling planes | 3 plots |

Then for your project you could use five sampling planes per plot and be getting sufficient estimates of DWD. (Be sure to enter this information in Field 5 on each data form and make a note of the methods used to determine the number of sampling planes for the project in the Metadata table.) We suggest an absolute minimum of three planes per plot be sampled for the DWD. Generally, DWD is the most variable of the FL attributes, so the duff, litter, and vegetation sampling intensity should be adequate when sampled at the DWD intensity.

## Resampling FL Plots

The FL methods are unique in FIREMON in that they allow a variable number of sampling planes on each FL plot, based on piece count. When resampling a FL plot, always sample the same number of planes as were sampled when the plot was sampled the *first* time. Never use the 100-piece rule when resampling. Instead, look through the FIREMON database and record the number of sampling planes that were used when each plot was first sampled, and then sample only that number in subsequent sampling.

## What If...

"...No matter where I start my next line, it runs off a cliff." There is no way that we can foresee every problem samplers will encounter in the field. The best way for a crew to deal with unique situations is to apply the FL methods as well as they can in order to sample the appropriate characteristics based on the project objectives, then make a record in the Comments section on the PD data form of what was encountered and how it was handled. For instance, if a crew initially planned to lay out a line that in the end headed off a cliff, then the crew could regroup and use the next azimuth from the FL plot design, and lay out the sampling plane in that direction.

## Precision Standards

Use these standards when collecting data with the FL methods (table FL-5).

# SAMPLING DESIGN CUSTOMIZATION

## Alternative FL Sampling Design

**Number of sampling planes:** Minimum 3, maximum 7. Continue sampling until count of pieces, across all sizes, is greater than 100. If you are resampling an existing plot, use the same number of planes as were used in the initial survey.

**Duff/Litter depth measurements per plane:** 2.

**Vegetation assessments per plane:** 2.

**Large debris piece measurements:** Diameter and decay class.

Table FL-5 Precision guidelines for FL sampling.

| Component | Standard |
|---|---|
| Slope | ±5 percent |
| FWD | ±3 percent |
| CWD diameter | ±0.5 inch/1 cm |
| CWD decay class | ±1 class |
| Duff/litter depth | ±0.1 inch/0.2 cm |
| Percent litter estimation | ±10 percent |
| Vegetation cover estimation | ±1 class |
| Vegetation height estimation | ±0.5 ft/0.2 m |

Fuel Load (FL) Sampling Method

## Streamlined FL Sampling Design

**Number of line sampling planes:** 3. If you are resampling an existing plot use the same number of planes as were used in the initial survey.

**Duff/Litter depth measurements per plane:** 2.

**Vegetation assessments per plane:** 2.

**Large debris piece measurements:** Diameter and decay class.

## Comprehensive FL Sampling Design

**Number of sampling planes:** 7. If you are resampling an existing plot use the same number of planes as were used in the initial survey.

**Duff/Litter depth measurements per plane:** 2.

**Vegetation assessments per plane:** 2.

**Large debris piece measurements:** Diameter and decay class.

## Optional Data

**Percent of log that is charred**—Measured to assess extent and severity of fire. Record the percent of the surface of each individual piece of CWD passing through the sampling plane that has been charred by fire using the classes in table FL-6.

**Diameter at large end of log**—Measured for wildlife concerns. Record the diameter of the large end of the log to the nearest inch (2 cm). If a piece is broken but the sections are touching, consider that one log. If the broken sections are not touching then consider them two logs and record the diameter of the large end of the piece that is passing through the sampling plane.

**Log length**—Important for wildlife concerns and useful for rough determination of piece density. Record length of CWD to the nearest 0.5 ft (0.1 m). If a piece is broken but the two parts are still touching then record the length end-to-end or sum the lengths for broken pieces not lying in a straight line (fig. FL-14). If piece is broken and the two parts are not touching, then measure only the length of the piece that intercepts the sampling plane.

**Distance from beginning of line to log**—This measurement makes relocation of specific logs easier, which is especially important when calculating fuel consumption on a log-by-log basis. Frequently, logs that were included in prefire sampling roll away from the sampling plane during a fire, while other logs

Table FL-6 Assign the amount of surface charred by fire, for each piece of CWD, into one of these char classes.

| Class | Char |
|---|---|
| | percent |
| 0 | No char |
| 0.5 | >0  1 |
| 3 | >1  5 |
| 10 | >5  15 |
| 20 | >15  25 |
| 30 | >25  35 |
| 40 | >35  45 |
| 50 | >45  55 |
| 60 | >55  65 |
| 70 | >65  75 |
| 80 | >75  85 |
| 90 | >85  95 |
| 98 | >95  100 |

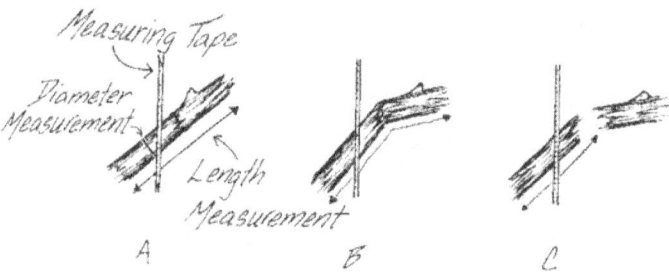

**Figure FL-14**—Log length and diameter measurement for optional CWD data. In each case the diameter measurement is made at the same point. In A and B the broken pieces are touching so length includes both pieces. In C, only the piece crossed by the measuring tape is measured for length.

not originally sampled will roll into the plane. Recording the distance from the start of the line, in addition to permanently marking the logs with tags, will make postfire sampling easier. Record the distance from the start of the measuring tape to the point where the diameter was measured.

**Percent of log that is hollow**—This characteristic is important for wildlife concerns but also allows more accurate estimates of carbon. Estimate the percent diameter and percent length that has been lost to decay. Record data using the classes in table FL-7.

**Table FL-7** Use these classes for recording the percent of diameter and length lost to rot in CWD.

| Class | Lost to decay |
|---|---|
| | percent |
| 0 | No Loss |
| 0.5 | >0   1 |
| 3 | >1   5 |
| 10 | >5   15 |
| 20 | >15   25 |
| 30 | >25   35 |
| 40 | >35   45 |
| 50 | >45   55 |
| 60 | >55   65 |
| 70 | >65   75 |
| 80 | >75   85 |
| 90 | >85   95 |
| 98 | >95   100 |

Fuel Load (FL) Sampling Method

# FUEL LOAD (FL) FIELD DESCRIPTIONS

Field 1: **1-Hr.** Sampling plane length for the 1-hour fuels (ft/m).
Field 2: **10-Hr.** Sampling plane length for the 10-hour fuels (ft/m).
Field 3: **100-Hr.** Sampling plane length for the 100-hour fuels (ft/m).
Field 4: **1,000-Hr.** Sampling plane length for the 1,000-hour fuels (ft/m).
Field 5: **Number of Planes.** Number of sampling planes/transects that the data were recorded along. (One "plane" includes data for the 1-hour through 1,000-hour fuels.)
Table 1. FWD, litter and duff.
Field 6: **Plane/Transect Number.** Sampling plane/transect number for data recorded in Fields 7 to 15.
Field 7: **Slope** (percent). Slope of the sampling plane. Precision: ±5 percent.
Field 8: **1-hour Count.** Count of pieces in the 1-hour size class (0–0.25 inch). Precision: ±3 percent total count.
Field 9: **10-hour Count.** Count of pieces in the 10-hour size class (0.25–1.0 inch). Precision: ±3 percent total count.
Field 10: **100-hour Count.** Count of pieces in the 100-hour size class (1.0–3.0 inches). Precision: ±3 percent total count.
Field 11: **D/L Depth 1** (inch/cm). Duff/litter depth measured at the first location. Precision: ±0.1 inch/0.2 cm.
Field 12: **Litter Percent 1.** Percent of the total duff/litter depth that is litter at the first location. Precision: ±10 percent.
Field 13: **D/L Depth 2** (inch/cm) Duff/litter depth measured at the second location. Precision: ±0.1 inch/0.2 cm.
Field 14: **Litter Percent 2.** Percent of the total duff/litter depth that is litter, at the second location. Precision: ±10 percent.
Field 15: **Local Code.** Optional data field.
Table 2. CWD
Field 16: **Plane/Transect Number.** Sampling plane/transect number for CWD data recorded in Fields 17 to 20.
Field 17: **Log Number.** Log number, numbered sequentially by transect.
Field 18: **Diameter** (inch/cm). Diameter of the piece measured perpendicular to the longitudinal axis. Precision: ±0.5 inch/1 cm.
Field 19: **Decay Class.** Decay class of the log where it crosses the plane. Valid classes are in table FL-3 of the sampling method. Precision: ±1 class
Field 20: **Local Code.** Optional data field.
Table 3. Vegetation.
Field 21: **Plane/Transect.** Sampling plane/transect number for vegetation measurements recorded in Fields 22 to 33.
Field 22: **Live Tree/Shrub Cover 1.** Cover class of live trees and shrubs at the first sampling location. Precision: ±1 class. Valid classes are in table FL-4 of the sampling method.
Field 23: **Dead Tree/Shrub Cover Class 1.** Cover class of dead trees and shrubs at the first sampling location. Precision: ±1 class. Valid classes are in table FL-4 of the sampling method.
Field 24: **Average Tree/Shrub Height 1** (ft/m). Average height of live and dead tree/shrub component at the first sampling location. Precision: ±0.5 ft/0.2 m.
Field 25: **Live Herb Cover 1.** Cover class of live herbs at the first sampling location. Precision: ±1 class. Valid classes are in table FL-4 of the sampling method.
Field 26: **Dead Herb Cover 1.** Cover class of dead herbs at the first sampling location. Precision: ±1 class. Valid classes are in table FL-4 of the sampling method.
Field 27: **Average Herb Height 1** (ft/m). Average height of live and dead herb component at the first sampling location. Precision: ±0.5 ft/0.2 m.
Field 28: **Live Tree/Shrub Cover 2.** Cover class of live trees and shrubs at the second sampling location. Precision: ±1 class. Valid classes are in table FL-4 of the sampling method.
Field 29: **Dead Tree/Shrub Cover 2.** Cover class of dead trees and shrubs at the second sampling location. Precision: ±1 class. Valid classes are in table FL-4 of the sampling method.
Field 30: **Average Tree/Shrub Height 2** (ft/m). Average height of live and dead tree/shrub component at the second sampling location. Precision: ±0.5 ft/0.2 m.
Field 31: **Live Herb Cover 2.** Cover class of live herbs at the second sampling location. Precision: ±1 class. Valid classes are in table FL-4 of the sampling method.
Field 32: **Dead Herb Cover 2.** Cover class of dead herbs at the second sampling location. Precision: ±1 class. Valid classes are in table FL-4 of the sampling method.
Field 33: **Average Herb Height 2** (ft/m). Average height of live and dead herb component at the second sampling location. Precision: ±0.5 ft/0.2 m.

# FIREMON FL Cheat Sheet

## Sampling plane layout

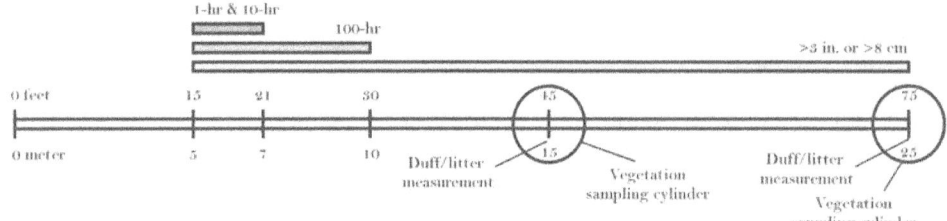

## Task list

| Task | Recorder | Sampler 1 | Sampler2 |
|---|---|---|---|
| Organize materials | 1 | | |
| Layout tape | | 1 (guider) | 1 (guidee) |
| Measure slope | 2 (record data) | 2 | 2 |
| Count FWD | 3 (record data) | 3 | |
| Measure duff/litter and veg. at 75 ft mark | 4 (record data) | | 3 |
| Measure CWD | 5 (record data) | | 4 |
| Measure duff/litter and veg. at 45 ft mark | 6 (record data) | 4 | |
| Check for complete forms | 7 | | |
| Collect equipment | | 5 | 5 |

## Precision

| Component | Standard |
|---|---|
| Slope | ±5 percent |
| FWD | ±3 percent |
| CWD diameter | ±0.5 inch/1 cm |
| CWD decay class | ±1 class |
| Duff/litter depth | ±0.1 inch/0.2 cm |
| Percent litter estimation | ±10 percent |
| Vegetation cover estimation | ±1 class |
| Vegetation height estimation | ±0.5 ft/0.2 m |

## CWD decay class

| Decay class | Description |
|---|---|
| 1 | All bark is intact. All but the smallest twigs are present. Old needles probably still present. Hard when kicked. |
| 2 | Some bark is missing, as are many of the smaller branches. No old needles still on branches. Hard when kicked. |
| 3 | Most of the bark is missing and most of the branches less than 1 inch in diameter also missing. Still hard when kicked. |
| 4 | Looks like a class 3 log but the sapwood is rotten. Sounds hollow when kicked and you can probably remove wood from the outside with your boot. Pronounced sagging if suspended for even moderate distances. |
| 5 | Entire log is in contact with the ground. Easy to kick apart but most of the piece is above the general level of the adjacent ground. If the central axis of the piece lies in or below the duff layer then it should not be included in the CWD sampling, as these pieces act more like duff than wood when burned. |

## Fuel load (FL) equipment list

0.25 inch diameter by 3 inch long dowel
1.0 inch diameter by 3 inch long dowel
75 foot tape
Bridge spikes
Clipboard
Compass
Clear plastic 6 inch ruler
 (w/0.1 inch gradations)
Clinometer (with percent scale)
Diameter tape, caliper, yardstick, or similar
 (w/0.1 inch gradations) for measuring large log diameter

Field notebook
FL cheatsheet
FL data forms
Hard hat
Pencils/pens
Sharp trowel or
 small shovel
Small stakes or rebar
Survey flags

### Optional
Go/No Go gauge (see FL Sampling Methods for details)
Hammer, hatchet, or big rock to pound in bridge spikes

## Plot layout

## Cover class

| Code | Canopy cover |
|---|---|
| | Percent |
| 0 | Zero |
| 0.5 | >0  1 |
| 3 | >1  5 |
| 10 | >5  15 |
| 20 | >15  25 |
| 30 | >25  35 |
| 40 | >35  45 |
| 50 | >45  55 |
| 60 | >55  65 |
| 70 | >65  75 |
| 80 | >75  85 |
| 90 | >85  95 |
| 98 | >95  100 |

## Piece sizes

| Dead woody class | | | Piece diameter |
|---|---|---|---|
| | | | inches (cm) |
| DWD | FWD | 1 hr | 0  0.25 (0  0.6) |
| | | 10 hr | 0.25  1.0 (0.6  2.5) |
| | | 100 hr | 1.0  3.0 (2.5  8.0) |
| | CWD | 1,000 hr and greater | 3.0 and greater (8.0 and greater) |

# Fuel Load (FL) Form

RegistrationID: _ _ _ _ _ _
ProjectID: _ _ _ _ _ _
PlotID: _ _ _ _ _
Date: _ _ / _ _ / _ _ _ _

FL Page _ _ of _ _ _

Plot Key

| Measurement Distances ft / m | Field 1 1-hour | Field 2 10-hr | Field 3 100-hr | Field 4 1000-hr | Field 5 # of Transects |
|---|---|---|---|---|---|
| | | | | | |

## FL Table 1 - Fine Woody Debris (<3in / <8cm) - Duff & Litter

| Field 6 Transect Number | Field 7 Slope % | Field 8 1-hr | Field 9 10-hr | Field 10 100-hr | Field 11 Duff/Litter Depth 1 (in/cm) | Field 12 Litter % 1 | Field 13 Duff/Litter Depth 2 (in/cm) | Field 14 Litter % 2 | Field 15 Local Code |
|---|---|---|---|---|---|---|---|---|---|
| 1 | | | | | | | | | |
| 2 | | | | | | | | | |
| 3 | | | | | | | | | |
| 4 | | | | | | | | | |
| 5 | | | | | | | | | |
| 6 | | | | | | | | | |
| 7 | | | | | | | | | |

## FL Table 2 - Coarse Woody Debris (>3in / >8cm)

| Field 16 Transect Number | Field 17 Log Number | Field 18 Diameter (in/cm) | Field 19 Decay Class | Field 20 Local Code |
|---|---|---|---|---|
| | | | | |
| | | | | |
| | | | | |
| | | | | |
| | | | | |
| | | | | |
| | | | | |
| | | | | |
| | | | | |
| | | | | |
| | | | | |
| | | | | |

| Field 16 Transect Number | Field 17 Log Number | Field 18 Diameter (in/cm) | Field 19 Decay Class | Field 20 Local Code |
|---|---|---|---|---|
| | | | | |
| | | | | |
| | | | | |
| | | | | |
| | | | | |
| | | | | |
| | | | | |
| | | | | |
| | | | | |
| | | | | |
| | | | | |
| | | | | |

Notes/Dot Tally Space:

Crew:

# Fuel Load (FL) Form

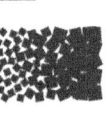

## FL Table 2 - Continuation
### Coarse Woody Debris - Duff & Litter (<3in / <8cm)

RegistrationID: _ _ _ _
ProjectID: _ _ _ _
PlotID: _ _ _
Date: _ _ / _ _ / _ _ _ _

Plot Key

Notes

FL Page _ _ of _ _

| Field 16 | Field 17 | Field 18 | Field 19 | Field 20 | Field 16 | Field 17 | Field 18 | Field 19 | Field 20 | Field 16 | Field 17 | Field 18 | Field 19 | Field 20 | Field 16 | Field 17 | Field 18 | Field 19 | Field 20 |
|---|---|---|---|---|---|---|---|---|---|---|---|---|---|---|---|---|---|---|---|
| Transect Number | Log Number | Diameter (in/cm) | Decay Class | Local Code | Transect Number | Log Number | Diameter (in/cm) | Decay Class | Local Code | Transect Number | Log Number | Diameter (in/cm) | Decay Class | Local Code | Transect Number | Log Number | Diameter (in/cm) | Decay Class | Local Code |
| | | | | | | | | | | | | | | | | | | | |
| | | | | | | | | | | | | | | | | | | | |
| | | | | | | | | | | | | | | | | | | | |
| | | | | | | | | | | | | | | | | | | | |
| | | | | | | | | | | | | | | | | | | | |
| | | | | | | | | | | | | | | | | | | | |
| | | | | | | | | | | | | | | | | | | | |
| | | | | | | | | | | | | | | | | | | | |
| | | | | | | | | | | | | | | | | | | | |
| | | | | | | | | | | | | | | | | | | | |
| | | | | | | | | | | | | | | | | | | | |
| | | | | | | | | | | | | | | | | | | | |
| | | | | | | | | | | | | | | | | | | | |
| | | | | | | | | | | | | | | | | | | | |
| | | | | | | | | | | | | | | | | | | | |
| | | | | | | | | | | | | | | | | | | | |
| | | | | | | | | | | | | | | | | | | | |
| | | | | | | | | | | | | | | | | | | | |

# Fuel Load (FL) Form

RegistrationID: _ _ _ _
ProjectID: _ _ _ _
PlotID: _ _ _ _
Date: _ _ / _ _ / _ _ _ _

FL Page _ _ of _ _

Plot Key

## FL Table 3 - Vegetation Data

| Field 21 | Field 22 | Field 23 | Field 24 | Field 25 | Field 26 | Field 27 | Field 28 | Field 29 | Field 30 | Field 31 | Field 32 | Field 33 |
|---|---|---|---|---|---|---|---|---|---|---|---|---|
| Transect | Live Tree/Shrub Cover 1 | Dead Tree/Shrub Cover 1 | Average Tree/Shrub Height 1 (ft/m) | Live Herb Cover 1 | Dead Herb Cover 1 | Average Herb Height 1 (ft/m) | Live Tree/Shrub Cover 2 | Dead Tree/Shrub Cover 2 | Average Tree/Shrub Height 2 (ft/m) | Live Herb Cover 2 | Dead Herb Cover 2 | Average Herb Height 2 (ft/m) |
| 1 | | | | | | | | | | | | |
| 2 | | | | | | | | | | | | |
| 3 | | | | | | | | | | | | |
| 4 | | | | | | | | | | | | |
| 5 | | | | | | | | | | | | |
| 6 | | | | | | | | | | | | |
| 7 | | | | | | | | | | | | |

Notes:

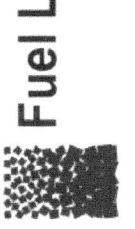

# Species Composition (SC)

## Sampling Method

John F. Caratti

## SUMMARY

The FIREMON Species Composition (SC) method is used to provide ocular estimates of cover and height measurements for plant species on a macroplot. The SC method provides plant species composition and coverage estimates to describe a stand or plant community and can be used to document changes over time. It is suited for a wide variety of vegetation types and is especially useful in plant communities with tall shrubs or trees. The method is relatively fast and efficient to conduct in the field and facilitates sampling many sites over large areas using few examiners. The SC method does not quantify the variability within a plot and cannot be used to detect statistically significant changes over time.

## INTRODUCTION

The Species Composition (SC) method is designed to provide plant species composition, and cover and height estimates to describe the plant community found on the FIREMON plot. This method uses a circular macroplot to record plant species characteristics. Cover, height, and optional user-specific attributes are recorded for each plant species or ground cover within the macroplot. Plant height is measured in feet or meters.

This method is primarily used when the user wants to acquire inventory data over large areas using few examiners. The SC method is useful for documenting important changes in plant species cover and composition over time. However, this method is not designed to monitor statistically significant changes in vegetation over time due to the subjective nature of the cover estimations. The SC sampling method primarily addresses individual plant species cover and height for vascular and nonvascular plants, by size class.

Cover is an important vegetation attribute that is used to determine the relative influence of each species on a plant community. Cover is a commonly measured attribute of plant community composition because small, abundant species and large, rare species have comparable cover values. In FIREMON we record foliar cover as the vertical projection of the foliage and supporting parts onto the ground. Therefore, total cover on a plot can exceed 100 percent due to overlapping layers in the canopy. When cover is summed by size class, the total cover can equal more than 100 percent for a plant species due to overlap in the canopy between the different size classes.

Ocular estimates of cover are usually based on cover classes. The two most common are Daubenmire (1959) and Braun-Blanquet (1965). The range of cover values, 0 to 100 percent, are divided into classes, and each class is assigned a rating or number. In broadly defined cover classes, there is little chance for consistent human error in assigning the cover class (Daubenmire 1959). The lowest cover classes are sometimes split into finer units (Bailey and Poulton 1968; Jensen and others 1994), since many species fall into the lowest cover classes. These systems are more sensitive to species with low cover. A finer breakdown of scale toward the lower scale values allows better estimation of less abundant species. In FIREMON we use a cover class system that splits the lowest classes into finer units. The midpoint of each class can be used for numerical computations. The use of midpoints for actual values is based on the assumption that actual cover values are distributed symmetrically about the midpoint.

Plant height measurements are used to estimate the average height of individual species or species by size class. Plant heights give detailed information about the vertical distribution of plant species cover on the plot. In addition, height measurements allow the examiner to calculate plant species volume (cover x height) and to estimate biomass using the appropriate biomass equations based on cover and height. Plant height is measured with a yard stick (or meter stick) for small plants (<10 ft or 3 m) and with a clinometer for larger plants (>10 ft or 3 m).

There are many ways to streamline or customize the SC sampling method. The FIREMON three-tier sampling design can be employed to optimize sampling efficiency. See the sections on **User Specific SC Sampling Design** and **Sampling Design Customization** below.

## SAMPLING PROCEDURE

This method assumes that the sampling strategy has already been selected and the macroplot has already been located. If this is not the case, then refer to the FIREMON **Integrated Sampling Strategy Guide** chapter.

The SC sampling procedure is presented in the order of the fields that need to be completed on the **SC data form**, so it is best to reference the SC data form when reading this section. The sampling procedure described here is the recommended procedure for this method. Later sections will describe how the FIREMON three-tier sampling design can be used to modify the recommended procedure to match resources, funding, and time constraints.

In the **How-To Guide** chapter, see **How To Locate a FIREMON Plot**, **How To Permanently Establish a FIREMON Plot** and **How to Define the Boundaries of a Macroplot** for more information on setting up your macroplot.

### Preliminary Sampling Tasks

Before setting out for your field sampling, layout a practice area with easy access. Try to locate an area with the same species or vegetation life form you plan on sampling. Get familiar with the plot layout and the data that will be collected. This will give you a chance to assess the method and will help you think about problems that might be encountered in the field. For example, use the test plot as an opportunity to check sampler's cover estimates by applying the point intercept method on the same plot and comparing the two. It is better to answer these questions before the sampling begins so that you are not wasting time in the field. This will also let you see if any pieces of equipment will need to be ordered.

Many preparations must be made before proceeding into the field for SC sampling. First, all equipment and supplies in the **SC Equipment List** must be purchased and packed for transport into the field. Because travel to FIREMON plots is usually by foot, it is important that supplies and equipment be placed in a comfortable daypack or backpack. It is also important that there be spares of each piece of equipment so that an entire day of sampling is not lost if something breaks. Spare equipment can be stored in the vehicle rather than the backpack. Be sure all equipment is well maintained and there are plenty of extra supplies such as data forms, map cases, and pencils.

All SC data forms should be copied onto waterproof paper because inclement weather can easily destroy valuable data recorded on standard paper. Data forms should be transported into the field using a plastic, waterproof map protector or plastic bag. The day's sample forms should always be stored in a dry place (office or vehicle) and not be taken back into the field for the next day's sampling.

We recommend that one person on the field crew, preferably the crew boss, have a waterproof, lined field notebook for recording logistic and procedural problems encountered during sampling. This helps with future remeasurements and future field campaigns. All comments and details not documented in the FIREMON sampling methods should be written in this notebook.

It is beneficial to have plot locations for several days of work in advance in case something happens, such as the road to one set of plots is washed out by flooding. Plots should be referenced on maps and aerial photos using pin-pricks or dots to make navigation easy for the crew and to provide a check of the georeferenced coordinates. We found that it is easy to transpose UTM coordinate digits when recording georeferenced positions on the plot sheet, so marked maps can help identify any erroneous plot positions. If possible, the spatial coordinates should be provided if FIREMON plots were randomly located.

A field crew of two people is probably the most efficient for implementation of the SC sampling method. There should never be a one-person field crew for safety reasons, and any more than two people will probably result in some people waiting for critical tasks to be done and unnecessary trampling within the macroplot. The crew boss is responsible for all sampling logistics including the vehicle, plot directions, equipment, supplies, and safety. The crew boss should be the note taker, and the technician should perform most cover and height measurements. However, the SC form can easily be completed by one person, so it may be best for the other crew member to fill out other data forms or perform other FIREMON tasks. The initial sampling tasks of the field crew should be assigned based on field experience, physical capacity, and sampling efficiency, but sampling tasks should be modified as the field crew gains experience. Tasks should also be shared to limit monotony.

## Designing the SC Sampling Method

A set of general criteria recorded on the SC data form makes up the user-specified design of the SC sampling method. Each general SC field must be designed so that the sampling captures the information needed to successfully complete the management objective within time, money, and personnel constraints. These general fields should be determined before the crews go into the field and should reflect a thoughtful analysis of the expected problems and challenges in the fire monitoring project.

### Plot ID construction

A unique plot identifier must be entered on the SC data form. This is the same plot identifier used to describe general plot characteristics in the Plot Description or PD sampling method. See **How to Construct a Unique Plot Identifier** in the **How-To Guide** chapter for details on constructing a unique plot identifier. Enter the plot identifier at the top of the SC data form.

### Macroplot size

The typical macroplot sampled in the SC method is a 0.10 acre (0.04 ha) circular plot having a radius of 37.2 ft (11.28 m). This plot size will be sufficient for most forest ecosystems and should be used if no other information is available. It is more efficient to use the same macroplot shape and size for all the FIREMON sampling methods on the plot, so if you are using other FIREMON sampling methods that require a baseline and transects, then a rectangular plot of 66 x 66 ft (20 x 20 m) should be used. For example, if the FIREMON DE method is being used to estimate density for some species, or if ocular cover is being calibrated using the FIREMON CF or LI methods, then use a rectangular macroplot. The macroplot radius or length and width are recorded in Fields 7 and 8 on the PD data form. If you are sampling with more than one method, see **How To Establish Plots with Multiple Methods,** in the **How-To Guide** chapter.

### Plant species ID level

This field is used to determine the sampling level intensity for the SC method. Enter the percent cover above which all plants are identified in Field 1 on the SC data form. For example, if you were interested in only sampling plants with at least 5 percent cover, then the number 5 would be entered in Field 1. Entering a 0 (zero) in this field indicates that all plant species on the plot will be identified (a full species list will be recorded). Because most changes in plant species cover occur in species that are already present on the plot, full species lists should be collected when feasible. Full species lists are especially useful if your data will be analyzed for biodiversity calculations, community classification, or species inventory. For example, because biodiversity calculations use the number of individual species as part of the calculation, it is important that each species be recorded.

## Conducting the SC Sampling Tasks

### Initial plot survey

Once the plot boundary is delineated (see **How to Define the Boundaries of a Macroplot** in the **How-To Guide** chapter), walk around the plot and become familiar with the plant species and vegetation layers. As you go, use Field 2 to record species and ground cover codes you identify. Only record the items that you are interested in sampling. For example, if you are interested in monitoring only the cover of noxious weeds, you only have to record those species in Field 2.

FIREMON provides plant species codes from the NRCS Plants database. However, local or customized plant species codes are also allowed in FIREMON. See **Populating the Plant Species Codes Lookup Table** in the **FIREMON Database User Manual** for more details. Codes other than plant species codes may also be entered in this field. For example, users may enter codes for estimating cover of snags or downed wood on the plot. See the section on **User-Specific SC sampling design** below for more details.

After examining the macroplot, return to the center and start to record cover and height for all appropriate plant species as described below.

### Sampling cover

Starting with the first species on your list, enter the plant species status in Field 3 on the SC data form. Status describes the general health of the plant species as live or dead using the following codes:

**L – Live:** plant with living tissue

**D – Dead:** plant with no living tissue visible

**NA – Not Applicable**

Plant status is purely qualitative, but it does provide an adequate characteristic for stratification of preburn plant health and determining postburn survival.

### Size class

Plant species size classes represent different layers in the canopy. For example, the upper canopy layer could be defined by large trees, while pole-size trees and large shrubs might dominate the middle layer of the canopy, and the lower canopy layer could include seedlings, saplings, grasses, and forbs. Size class data provide important structural information such as the vertical distribution of plant cover. Size classes for trees are typically defined by height for seedlings and diameter at breast height (DBH) for larger trees. Size classes for shrubs, grasses, and forbs are typically defined by height. If the vegetation being sampled has a layered canopy structure, then cover can be recorded by plant species and by size class. Total size class cover for a plant species can equal more than 100 percent due to overlap between different size classes.

FIREMON uses a size class stratification based on the ECODATA sampling methods (Jensen and others 1994). Group individual plants by species into one or more tree size classes (table SC-1) or shrub, grass, and forb size classes (table SC-2). There can be multiple size classes for each plant species. In the **How-To Guide** chapter, see **How To Measure DBH** for detailed information on measuring DBH to group trees into size classes, and **How to Measure Plant Height** for detailed information on measuring height for grouping shrubs into size classes.

Species Composition (SC) Sampling Method

Table SC-1  Tree size class codes.

| Tree size class | | |
|---|---|---|
| Codes | English | Metric |
| TO | Total cover | Total cover |
| SE | Seedling (<1 inch DBH or <4.5 ft height) | Seedling (<2.5 cm DBH or <1.5 m height) |
| SA | Sapling (1.0 inch <5.0 in. DBH) | Sapling (2.5 <12.5 cm DBH) |
| PT | Pole tree (5.0 inches <9.0 in. DBH) | Pole tree (12.5 <25 cm DBH) |
| MT | Medium tree (9.0 inches <21.0 in. DBH) | Medium tree (25 <50 cm DBH) |
| LT | Large tree (21.0 inches <33.0 in. DBH) | Large tree (50 <80 cm DBH) |
| VT | Very large tree (33.0+ inches DBH) | Very large tree (80+ cm DBH) |
| NA | Not applicable | Not applicable |

Table SC-2  Shrub, grass, and forb size class codes.

| Shrub/herb size class | | |
|---|---|---|
| Codes | English | Metric |
| TO | Total cover | Total cover |
| SM | Small (<0.5 ft height) | Small (<0.15 m height) |
| LW | Low (0.5 <1.5 ft height) | Low (0.15 <0.5m height) |
| MD | Medium (1.5 <4.5 ft height) | Medium (0.5 <1.5 m height) |
| TL | Tall (4.5 <8 ft height) | Tall (1.5 <2.5 m height) |
| VT | Very tall (8+ ft height) | Very tall (2.5+ m height) |
| NA | Not applicable | Not applicable |

If you are recording cover by size class, enter the size class code for each plant species in Field 4 on the SC data form. If size class data are not recorded, then record only the total canopy cover for each plant species. When recording total cover for a species, use the code "TO" for "total" cover.

## Estimating cover

Cover is the vertical projection of the foliage and supporting parts onto the ground (fig. SC-1). See **How to Estimate Cover** in the **How-To Guide** chapter for more details. When estimating total cover for a plant species, do not include overlap between canopy layers of the same plant species (fig. SC-2). When

**Figure SC-1**—Cover is estimated as the vertical projection of vegetation onto the ground.

**Figure SC-2**—Estimating total cover for plant species with overlapping canopies. In this figure the small trees underneath the canopy of the larger trees are the same plant species. Cover is estimated as the projection of the large tree canopy onto the ground, which overlaps the canopy of the smaller trees.

estimating cover by size classes for a plant species, the cover for each size class is recorded and includes canopy overlap between different size classes (fig. SC-3). Select one of the following cover class codes (table SC-3) to describe the cover for the species. Enter the cover class code in Field 5 on the SC data form.

### Measuring average height

Measure the average height for each plant species in feet (meters) within +/– 10 percent of the mean plant height. If plant species are recorded by size class, measure the average height for a plant species by each size class. See **How to Measure Plant Height** in the **How-To Guide** chapter for more details. Enter plant height in Field 6 of the SC data form.

### Using the optional fields

There are two optional fields for user-defined codes or measurements. For example, codes can be entered to record plant species phenology or wildlife utilization of plant species. If standing dead trees (snags) were recorded on the SC data form, one could enter a decay class code in one of the optional fields. Enter the user-defined codes or measurements in Fields 7 and 8 on the SC data form.

## Precision Standards

Use the precision standards listed in table SC-4 for the SC sampling.

# SAMPLING DESIGN CUSTOMIZATION

This section will present several ways that the SC sampling method can be modified to collect more detailed information or streamlined to collect only the most important tree characteristics. First, the suggested or recommended sample design is detailed, then modifications are presented.

**Figure SC-3**—Estimating cover by size class for plant species. In this figure the small trees underneath the canopy of the larger trees are the same plant species but a different size class (seedlings and saplings). Cover is estimated separately for each size class.

**Table SC-3** Cover class codes.

| Code | Cover class |
|---|---|
|  | percent |
| 0 | Zero |
| 0.5 | >0 1 |
| 3 | >1 5 |
| 10 | >5 15 |
| 20 | >15 25 |
| 30 | >25 35 |
| 40 | >35 45 |
| 50 | >45 55 |
| 60 | >55 65 |
| 70 | >65 75 |
| 80 | >75 85 |
| 90 | >85 95 |
| 98 | >95 100 |

**Table SC-4** Precision guidelines for SC sampling.

| Component | Standard |
|---|---|
| Size class | ±1 class |
| Cover | ±1 class |
| Height | ±10 percent average height |

## Recommended SC Sampling Design

The recommended SC sampling design follows the Alternative FIREMON sampling intensity and is listed below:

**Macroplot Size**: 0.10 acre circular plot (0.04 ha circular plot).

**Collect plant species cover.** Species ID level = "0"; record all species present in the plot.

## Streamlined SC Sampling Design

The streamlined SC sampling design follows the Simple FIREMON sampling intensity and is described below:

**Macroplot Size**: 0.10 acre circular plot (0.04 ha circular plot).

**Collect plant species cover.** Species ID level = "5"; record all species with 5 percent cover (cover class 10) or greater in the plot.

## Comprehensive SC Sampling Design

The comprehensive SC sampling design follows the Detailed FIREMON sampling intensity and is detailed below:

**Macroplot Size**: 0.10 acre circular plot (0.04 ha circular plot)

**Collect plant species cover and average height by size class.** Species ID level = "0"; record all species present in the plot.

## User-Specific SC Sampling Design

There are several ways the user can adjust the SC sample fields to make sampling more efficient and meaningful for local situations. Use the species ID level (Field 1) to reduce the number of species recorded. Higher species ID levels yield smaller plant lists. For example, only plant species with 5 percent cover (cover class 10) or greater are recorded with a species ID level of 5, while a species ID level of 15 limits the plant list to species with 15 percent cover (cover class 20) or greater. Sampling a reduced species list can be accomplished in a short time. The SC method can be even more selective by entering 99 in the species ID level, indicating that only specific plant species are being recorded. For example, you might be interested in documenting the presence or absence of rare plants or the invasion of noxious weeds after a fire. In this case only the rare plants or noxious weeds are recorded.

Ocular estimates of cover can be recorded to the nearest 1 percent instead of a cover class. This will allow values to be grouped into different canopy cover classes later when conducting data analysis. Actual cover values are useful when monitoring changes in species with low cover values. If cover changes from 3 to 6 percent, recording 6 percent cover is more accurate than recording 10 percent, the midpoint of the next cover class (5 to 15 percent). However, it is doubtful that cover can be accurately estimated to a 1 percent level using the human eye. If actual cover values are recorded, or if different cover classes are used than the classes listed in the FIREMON SC methods, record the information in the Metadata table.

You can also make ocular estimates of cover for items other than just plant species. In addition, the optional fields give the user flexibility to record their own codes or measurements. Some examples of information recorded in the optional fields might be maturity and vigor for plant species, decay classes for snags, and wildlife utilization of plants.

## Sampling Hints and Techniques

Examiners must be knowledgeable in plant identification.

It is relatively easy to learn to estimate cover to the nearest cover class. Examiners can calibrate their ocular estimates by periodically double sampling with the FIREMON Cover/Frequency (CF) or Point Intercept (PO) method for small shrubs, grasses, and forbs, and the FIREMON Line Intercept (LI) method for large shrubs and trees. Even an experienced investigator may assign an item to the wrong cover. Calibration of ocular estimates should be conducted at the outset of inventory projects and occasionally (usually every five to 10 ocular macroplots) during the project. Variability of cover estimates between trained examiners is usually minimal and is negated by the large number of samples that can be obtained with this method.

Examiners can calibrate their eyes for estimating cover by using the various size cutouts—circular subplots—within the circular macroplot. They should also become familiar with all the subplot sizes and the percent of the entire macroplot each circular subplot represents. Samplers can then mentally group species into a subplot and use the subplot size to estimate percent cover. See **How to Estimate Cover** in the **How-To Guide** chapter for more details.

Height can be difficult to measure for plant species or species by size class since the value must represent an average for all individual plants on the macroplot. One solution is to measure the height of a representative plant. Another solution, which requires more time, is to take additional measurements of individual plants and average the height values.

When entering data on the SC data form, examiners might run out of space on the first page. The form was designed for printing multiple copies so more plant species can be recorded on the additional data forms.

## SPECIES COMPOSITION (SC) FIELD DESCRIPTIONS

Field 1: **Species ID level.** Enter the minimum cover level used to record plant species.

Field 2: **Item Code.** Code of sampled entity. Either the NRCS plants species code, the local code for that species, ground cover code, or other item code. Precision: No error.

Field 3: **Status**: Plant status—Live, Dead, or Not Applicable. (L, D, NA). Precision: No error.

Field 4: **Size Class**. Size of the sampled plant. Valid classes are in tables SC-1 and SC-2 of the sampling method. Precision: $\pm 1$ class.

Field 5: **Cover.** Enter the cover class code for the sampled entity. Precision: $\pm 1$ class.

Field 6: **Height**. Enter the average height for each plant species or life-form on the plot (ft/m). Precision: $\pm 10$ percent mean height.

Field 7: **Local Field 1.** Enter a user-specific code or measurement for the plant species or item being recorded.

Field 8: **Local Field 2.** Enter a user-specific code or measurement for the plant species or item being recorded.

# FIREMON SC Cheat Sheet

### Precision

| Component | Standard |
|---|---|
| Size class | ±1 class |
| Cover | ±1 class |
| Height | ±10 percent average height |

### Status codes

| Code | Description |
|---|---|
| L | Live |
| D | Dead |
| NA | Not applicable |

### Shrub and herbaceous size classes

| | Shrub/herb size class | |
|---|---|---|
| Codes | Description (English) | Description (metric) |
| TO | Total cover | Total cover |
| SM | Small (<0.5 ft height) | Small (<0.15 m height) |
| LW | Low (0.5 <1.5 ft height) | Low (0.15 <0.5m height) |
| MD | Medium (1.5 <4.5 ft height) | Medium (0.5 <1.5 m height) |
| TL | Tall (4.5 <8 ft height) | Tall (1.5 <2.5 m height) |
| VT | Very tall (>8 ft height) | Very tall (>2.5 m height) |
| NA | Not applicable | Not applicable |

### Tree size classes

| | Tree size class | |
|---|---|---|
| Codes | Description (English) | Description (metric) |
| TO | Total cover | Total cover |
| SE | Seedling (<1 inch DBH or <4.5 ft height) | Seedling (<2.5 cm DBH or <1.5 m height) |
| SA | Sapling (1.0 inch < 5.0 in. DBH) | Sapling (2.5 <12.5 cm DBH) |
| PT | Pole tree (5.0 inches <9.0 in. DBH) | Pole tree (12.5 <25 cm DBH) |
| MT | Medium tree (9.0 inches <21.0 in. DBH) | Medium tree (25 <50 cm DBH) |
| LT | Large tree (21.0 inches <33.0 in. DBH) | Large tree (50 <80 cm DBH) |
| VT | Very large tree (33.0+ inches DBH) | Very large tree (80+ cm DBH) |
| NA | Not applicable | Not applicable |

### Canopy cover classes

| Code | Canopy cover |
|---|---|
| | Percent |
| 0 | Zero |
| 0.5 | >0 1 |
| 3 | >1 5 |
| 10 | >5 15 |
| 20 | >15 25 |
| 30 | >25 35 |
| 40 | >35 45 |
| 50 | >45 55 |
| 60 | >55 65 |
| 70 | >65 75 |
| 80 | >75 85 |
| 90 | >85 95 |
| 98 | >95 100 |

### Species composition (SC) equipment list

| | |
|---|---|
| Camera with film | Indelible ink pen (Sharpie, Marker) |
| SC data forms | Lead pencils with lead refills |
| Clinometer | Maps, charts, and directions |
| Clipboard | Map protector or plastic bag |
| Compass | Magnifying glass |
| Diameter tape (inches or cm) (2) | Pocket calculator |
| Field notebook | Plot sheet protector or plastic bag |
| Flagging | Reinforcing bar (to mark plot center) |
| Graph paper | Tape 75 ft (25 m) or longer (2) |
| Hammer | Yard (meter) stick |

# Species Composition (SC) Form

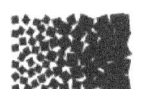

| Field 1 | |
|---|---|
| Species ID Level: | _____ % |

| Field 2 | Field 3 | Field 4 | Field 5 | Field 6 | Field 7 | Field 8 |
|---|---|---|---|---|---|---|
| Item Code | Status | Size Class | Cover | Height (ft/m) | Local Field 1 | Local Field 2 |
| | | | | | | |
| | | | | | | |
| | | | | | | |
| | | | | | | |
| | | | | | | |
| | | | | | | |
| | | | | | | |
| | | | | | | |
| | | | | | | |
| | | | | | | |
| | | | | | | |
| | | | | | | |
| | | | | | | |
| | | | | | | |
| | | | | | | |

Plot Key

SC Page __ __ of __ __

RegistrationID: __ __ __ __ __
ProjectID: __ __ __ __ __
PlotID: __ __ __ __ __
Date: __ __ / __ __ / __ __ __ __

| Field 2 | Field 3 | Field 4 | Field 5 | Field 6 | Field 7 | Field 8 |
|---|---|---|---|---|---|---|
| Item Code | Status | Size Class | Cover | Height (ft/m) | Local Field 1 | Local Field 2 |
| | | | | | | |
| | | | | | | |
| | | | | | | |
| | | | | | | |
| | | | | | | |
| | | | | | | |
| | | | | | | |
| | | | | | | |
| | | | | | | |
| | | | | | | |
| | | | | | | |
| | | | | | | |
| | | | | | | |
| | | | | | | |
| | | | | | | |

# Cover/Frequency (CF)

## Sampling Method

John F. Caratti

## SUMMARY

The FIREMON Cover/Frequency (CF) method is used to assess changes in plant species cover and frequency for a macroplot. This method uses multiple quadrats to sample within-plot variation and quantify statistically valid changes in plant species cover, height, and frequency over time. Because it is difficult to estimate cover in quadrats for larger plants, this method is primarily suited for grasses, forbs, and shrubs less than 3 ft (1 m) in height. Quadrats are placed systematically along randomly located transects. Cover is assessed by visually estimating the percent of a quadrat occupied by the vertical projection of vegetation onto the ground. Plant species frequency is recorded as the number of times a species occurs within a given number of quadrats. Frequency is typically recorded for plant species that are rooted within the quadrat.

## INTRODUCTION

The Cover/Frequency (CF) method is designed to sample within-plot variation and quantify changes in plant species cover, height, and frequency over time. This method uses quadrats that are systematically placed along transects located within the macroplot. First, a baseline is established along the width of the plot. Transects are oriented perpendicular to the baseline and are placed at random starting points along the baseline. Quadrats are then placed systematically along each transect. Characteristics are recorded about the general CF sample design (transect length, number of transects, quadrat size, and number of quadrats per transect) and for individual plant species within each quadrat. Depending on the project objectives, any combination of cover, frequency, and height are recorded for each plant species.

This method is primarily used when the manager wants to monitor statistically significant changes in plant species cover, height, and frequency. The CF sampling method is most appropriate for sampling vascular and nonvascular plants less than 3 ft (1 m) in height. The FIREMON line intercept (LI) method is better suited for estimating cover of shrubs greater than 3 ft (1 m) in height (Western United States shrub communities, mixed plant communities of grasses, trees, and shrubs, and open grown woody vegetation). The CF methods can also be used to estimate ground cover (such as bare soil, gravel, or litter). However, the FIREMON point intercept (PO) methods are better suited for that task. We suggest you use the PO method if you are primarily interested in monitoring changes in ground cover. The PO method may be used in conjunction with the CF method to sample ground cover by using the CF sampling quadrat as a point frame. The PO method is also better suited for sampling fine-textured herbaceous communities (dense grasslands and wet meadows). However, if rare plant species are of interest the CF methods are preferred because it is easier to sample rare species with quadrats than with points or lines. Optionally, use the Rare Species (RS) method.

## Estimating Cover and Height

Cover is an important vegetation attribute that is used to determine the relative influence of each species on a plant community. Cover is a commonly measured attribute of plant community composition because small, abundant species and large, rare species have comparable cover values. In FIREMON we record foliar cover as the vertical projection of the foliage and supporting parts onto the ground. Therefore, total cover on a plot can exceed 100 percent due to overlapping layers in the canopy.

Estimating cover in quadrats is more accurate than estimating cover on a macroplot because samplers record cover in small quadrats more consistently than in large areas. Sampling with quadrats is also more effective than the point intercept (PO) method at locating and recording rare species. Point intercept sampling requires many points to sample rare species (200 points to sample at 0.5 percent cover). Quadrats sample more area and have a greater chance of detecting rare species.

Cover is typically based on a visual estimate of cover classes that range from 0 to 100 percent. These classes are broadly defined, lowering the chance for consistent human error in assigning the cover class. The lowest cover classes are sometimes split into finer units, because many species fall into the lowest cover classes. These systems are more sensitive to species with low cover. A finer breakdown of scale toward the lower scale values allows better estimation of less abundant species. In FIREMON we use a cover class system, which splits the lowest classes into finer units. The midpoint of each class can be used for numerical computations. The use of midpoints for actual values is based on the assumption that actual cover values are distributed symmetrically about the midpoint.

Plant height measurements are used to estimate the average height of individual plant species. Plant heights give detailed information about the vertical distribution of plant species cover on the plot. In addition, height measurements allow the examiner to calculate plant species volume (cover x height) and to estimate biomass using the appropriate bulk density equations. Plant height is measured with a yardstick (meter stick) for plants less than 10 ft tall (3 m) and with a clinometer and tape measure for taller plants.

## Estimating Frequency

Frequency is used to describe the abundance and distribution of species and can be used to detect changes in vegetation over time. It is typically defined as the number of times a species occurs in the total number of quadrats sampled, usually expressed as a percent. Frequency is one of the fastest and easiest methods for monitoring vegetation because it is objective, repeatable, and requires just one decision: whether or not a species is rooted within the quadrat frame. Frequency is a useful tool for comparing two plant communities or to detect change within one plant community over time.

Frequency is most commonly measured with square quadrats. The size and shape of the frequency quadrat influences the results of the frequency recorded. If a plot is too small, rare plants may not be recorded. If you use a large quadrat, you will have individual species in all quadrats and frequency values of 100 percent, which will not allow you to track increases in frequency. If you have small quadrats, you will record low frequency values that are not sensitive to declining frequency values for a species. A reasonable sensitivity to change results from frequency values between 20 to 80 percent. Frequencies less than 5 percent or greater than 95 percent typically result in heavily skewed distributions.

For this reason, nested plots, or subplots, are usually used to sample frequency. Plot sizes are nested in a configuration that gives frequencies between 20 and 80 percent for the majority of species. Nested subplots allow frequency data to be collected in different size subplots of the main quadrat. Because frequency of occurrence can be analyzed for different sized plots, this eliminates the problems of comparing data collected from different size quadrats. In FIREMON, we use a nested plot design of four subplots within one quadrat, and record the smallest subplot number in which the plant is rooted. This frequency measurement is typically referred to as nested rooted frequency (NRF).

Plant species frequency is highly sensitive to the size and shape of quadrats, so changes in frequency may be difficult to interpret, possibly resulting from changes in cover, density, or pattern of distribution. For this reason, if money and time are available, we recommend that you collect cover data along with frequency data. However, if you are only concerned about documenting that a change in vegetation has occurred, then frequency is the most rapid method.

There are many ways to streamline or customize the CF sampling method. The FIREMON three-tier sampling design can be employed to optimize sampling efficiency. See the sections on **User-Specific CF Sampling Design** and **Sampling Design Customization** below.

## SAMPLING PROCEDURE

This method assumes that the sampling strategy has already been selected and the macroplot has already been located. If this is not the case, then refer to the FIREMON **Integrated Sampling Strategy** chapter for further details.

The sampling procedure is described in the order of the fields that need to be completed on the **CF data form**, so it is best to reference the data form when reading this section. The sampling procedure described here is the recommended procedure for this method. Later sections will describe how the FIREMON three-tier sampling design can be used to modify the recommended procedure to match resources, funding, and time constraints.

In the **How-To Guide** chapter, see **How To Locate a FIREMON Plot**, **How To Permanently Establish a FIREMON Plot**, and **How to Define the Boundaries of a Macroplot** for more information on setting up your macroplot.

### Preliminary Sampling Tasks

Before setting out for your field sampling, lay out a practice area with easy access. Try and locate an area with the same species or vegetation lifeform you plan on sampling. Get familiar with the plot layout and the data that will be collected. This will give you a chance to assess the method and will help you think about problems that might be encountered in the field. For example, how will you account for boundary plants? It is better to answer the such questions before the sampling begins so that you are not wasting time in the field. This will also let you see if there are any pieces of equipment that will need to be ordered.

A number of preparations must be made before proceeding into the field for CF sampling. First, all equipment and supplies in the **CF Equipment List** must be purchased and packed for transport into the field. Travel to FIREMON plots is usually by foot, so it is important that supplies and equipment be placed in a comfortable daypack or backpack. It is also important that there be spares of each piece of equipment so that an entire day of sampling is not lost if something breaks. Spare equipment can be stored in the vehicle rather than the backpack. Be sure all equipment is well maintained and there are plenty of extra supplies such as data forms, map cases, and pencils.

All CF data forms should be copied onto waterproof paper because inclement weather can easily destroy valuable data recorded on standard paper. Data forms should be transported into the field using a plastic, waterproof map protector or plastic bag. The day's sample forms should always be stored in a dry place (office or vehicle) and not be taken back into the field for the next day's sampling.

We recommend that one person on the field crew, preferably the crew boss, have a waterproof, lined field notebook for recording logistic and procedural problems encountered during sampling. This helps with future remeasurements and future field campaigns. All comments and details not documented in the

FIREMON sampling methods should be written in this notebook. For example, snow on the plot might be described in the notebook, which would be helpful in plot remeasurement.

Plot locations and/or directions should be readily available and provided to the crews in a timely fashion. It is beneficial to have plot locations for several days of work in advance in case something happens, such as if the road to one set of plots is washed out by flooding. Plots should be referenced on maps and aerial photos using pin-pricks or dots to make navigation easy for the crew and to provide a check of the georeferenced coordinates. If possible, the spatial coordinates should be provided if FIREMON plots were randomly located.

A field crew of two people is probably the most efficient for implementation of the CF sampling method. There should never be a one-person field crew for safety reasons, and any more than two people will probably result in part of the crew waiting for tasks to be completed and unnecessary trampling on the FIREMON macroplot. The crew boss is responsible for all sampling logistics including the vehicle, plot directions, equipment, supplies, and safety. The crew boss should be the note taker, and the technician should perform most quadrat measurements. The initial sampling tasks of the field crew should be assigned based on field experience, physical capacity, and sampling efficiency. As the field crew gains experience, switch tasks so that the entire crew is familiar with the different sampling responsibilities and to limit monotony.

## Designing the CF Sampling Method

There is a set of general criteria recorded on the CF data form that forms the user-specified design of the CF sampling method. Each general CF field must be designed so that the sampling captures the information needed to successfully complete the management objective within time, money, and personnel constraints. These general fields should be determined before the crews go into the field and should reflect a thoughtful analysis of the expected problems and challenges in the fire monitoring project.

### Plot ID construction

A unique plot identifier must be entered on the CF data form. This is the same plot identifier used to describe general plot characteristics in the Plot Description or PD sampling method. Details on constructing a unique plot identifier are discussed in the **How to Construct a Unique Plot Identifier** section in the **How-To Guide** chapter. Enter the plot identifier at the top of the CF data form.

### Determining the sample size

The size of the macroplot ultimately determines the length of the transects and the length of the baseline along which the transects are placed. The amount of variability in plant species composition and distribution determines the number and length of transects and the number of quadrats required for sampling. The typical macroplot sampled in the CF method is a 0.10 acre (0.04 ha) square measuring 66 x 66 ft (20 x 20 m), which is sufficient for most forest understory and grassland monitoring applications. Shrub-dominated ecosystems will generally require larger macroplots when sampling with the CF method. Dr. Rick Miller (Rangeland Ecologist, Oregon State University) has sampled extensively in shrub-dominated systems, and we have included a write-up of his method in **Appendix D: Rick Miller Method for Sampling Shrub-Dominated Systems**. If you are not sure of the plot size to use, contact someone who has sampled the same vegetation that you will be sampling. The size of the macroplot may be adjusted to accommodate different numbers and lengths of transects. In general it is more efficient if you use the same plot size for all FIREMON sampling methods on the plot. However, we recognize that this is not always feasible.

We recommend sampling five transects within the macroplot, and this should be sufficient for most studies. In the **How-To Guide** chapter, see **How To Determine Sample Size** for more details. Enter the number of transects in Field 1 on the CF data form. The recommended transect length is 66 ft (20 m) for a 66 x 66 ft (20 x 20 m) macroplot. However, the macroplot size may be adjusted to accommodate longer or shorter transects based on the variability in plant species composition and distribution. For example, transects may be lengthened to accommodate more quadrats per transect or to allow more distance between quadrats. Enter the transect length in Field 2 of the CF data form. The FIREMON CF data form and data entry screen allow an unlimited number of transects. Enter the number of quadrats per transect in Field 3 of the CF data form. The FIREMON CF data form and data entry screen allow up to 20 quadrats per transect.

## Determining the quadrat size

Frequency is typically recorded in square quadrats. The standard quadrat for measuring nested rooted frequency in a 20 x 20 inch (50 x 50 cm) square with four nested subplot sizes. A nested frame allows frequency data to be collected in different size subplots of the main quadrat. Measuring frequency this way is commonly referred to as nested rooted frequency (NRF). Plot sizes are nested in a configuration. Statistical tests will use the nested plot size that gives frequencies between 20 to 80 percent for the majority of species. Tables CF-1 and CF-2 list common quadrat and subplot sizes in English and metric dimensions for recording nested rooted frequency. See **How To Construct a Quadrat Frame** in the **How-To Guide** chapter for instructions on building and using quadrat frames. Cover can be estimated using the same quadrat frames and recorded at the same time frequency is recorded.

Enter the quadrat length (Field 4) and quadrat width (Field 5) in inches (cm) on the CF data form.

## Recording the subplot size ratio and NRF numbers

If nested rooted frequency is being recorded, then enter the percent area of the quadrat contained by each subplot in Field 6 on the CF data form. Start with the smallest subplot and end with the largest subplot. For example, the subplot ratio for the standard 20 x 20 inch (50 x 50 cm) quadrat would be 1:25:50:100. Subplot 1 is 2 x 2 inches (5 x 5 cm) and is 1 percent of the quadrat. Subplot 2 is 10 x 10 inches (25 x 25 cm) and is 25 percent of the quadrat. Subplots 3 and 4 are 10 x 20 inches (25 x 50 cm) and 20 x 20 inches (50 x 50 cm), which correspond to 50 and 100 percent of the quadrat, respectively. See **How To Construct a Quadrat Frame** in the **How-To Guide** chapter for more details about subplot sizes.

If nested rooted frequency is being recorded, then enter the corresponding frequency numbers for each subplot in Field 7 of the CF data form. Start numbering with the smallest subplot and end with the largest subplot. For example, 1:2:3:4 would correspond with the 1:25:50:100 percentages of total plot when using the standard 20 x 20 inches (50 x 50 cm) quadrat.

Table CF-1 Commonly used quadrat sizes for recording nested rooted frequency (English dimensions).

| NRF numbers | Standard | Grassland communities | Sagebrush communities | Pinyon-juniper |
|---|---|---|---|---|
| | | inches | | |
| Subplot 1 | 2 x 2 | | 2 x 2 | |
| Subplot 2 | 10 x 10 | 2 x 2 | 4 x 4 | 8 x 8 |
| Subplot 3 | 10 x 20 | 4 x 4 | 8 x 8 in. | 20 x 20 |

Table CF-2 Commonly used quadrat sizes for recording nested rooted frequency (metric dimensions).

| NRF numbers | Standard | Grassland communities | Sagebrush communities | Pinyon-juniper |
|---|---|---|---|---|
| | | cm | | |
| Subplot 1 | 5 x 5 | | 5 x 5 | |
| Subplot 2 | 25 x 25 | 5 x 5 | 10 x 10 | 20 x 20 |
| Subplot 3 | 25 x 50 | 10 x 10 | 20 x 20 | 50 x 50 |
| Subplot 4 | 50 x 50 | 20 x 20 | 50 x 50 | 100 x 100 |

## Conducting CF Sampling Tasks

### Establishing the baseline for transects

Once the plot has been monumented, a permanent baseline is set up as a reference from which you will orient all transects. The baseline should be established so that the sampling plots for all of the methods overlap as much as possible. See **How To Establish Plots with Multiple Methods** in the **How-To Guide** chapter. The recommended baseline is 66 ft (20 m) long and is oriented upslope with the 0-ft (0-m) mark at the lower permanent marker and the 66-ft (20-m) mark at the upper marker. On flat areas, the baseline runs from south to north with the 0-ft (0-m) mark on the south end and the 66-ft (20-m) mark on the north end. Transects are placed perpendicular to the baseline and are sampled starting at the baseline. On flat areas, transects are located from the baseline to the east. See **How To Establish a Baseline for Transects** in the **How-To Guide** chapter for more details.

### Locating the transects

Locate transects within the macroplot perpendicular to the baseline and parallel with the slope. For permanent plots, determine the compass bearing of each transect and record these on the plot layout map or the comment section of the PD form. Permanently mark the beginning and ending of each transect (for example, using concrete reinforcing bar). Starting locations for each transect can be determined randomly on every plot or systematically with the same start locations used on every plot in the project. In successive remeasurement years, it is essential that transects be placed in the same locations as in previous visits. If the CF method is used in conjunction with other replicated sampling methods (LI, PO, RS or DE), use the same transects for all methods, whenever possible. See **How To Locate Transects and Quadrats** in the **How-To Guide** chapter for more details.

### Locating the quadrats

We recommend sampling five quadrats located at 12-ft (4-m) intervals along a transect, with the first quadrat placed 12 ft (4 m) from the baseline. See **How To Locate Transects and Quadrats** in the **How-To Guide** chapter for more details. If macroplots are being sampled for permanent remeasurement, quadrats must be placed at specified intervals along a measuring tape, which is placed along each transect. In successive years for remeasurement, quadrats *must* be placed in the same location. When sampling macroplots that are not scheduled for permanent remeasurement, the distance between quadrats may be estimated by pacing after the examiner measures the distance between quadrats.

Each quadrat is placed on the uphill side of the transect line with the quadrat frame placed parallel to the transect. The lower left corner of the quadrat frame will be placed at the foot (meter) mark for the quadrat location. Figure CF-1 displays the proper placement of a quadrat frame.

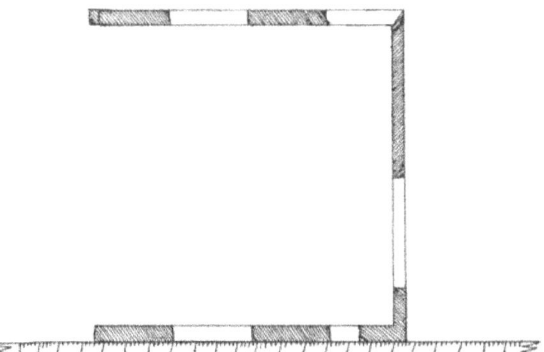

**Figure CF-1**—An example of quadrat placement along a transect.

Cover/Frequency (CF)

## Quadrat Sampling

First, enter the number of the transect that is being sampled in Field 8 of the CF data form.

Next, enter the plant species or item code in Field 9. FIREMON uses the NRCS plants species codes. However, you may use your own species codes. See **Populating the Plant Species Codes Lookup Table** in the **FIREMON Database User Manual** for more details. If ground cover is being sampled, we recommend using the ground cover codes listed in table CF-3.

Next, enter the plant species status in Field 10 on the CF data form. Status describes the general health of the plant species as live or dead using the following codes:

**L—Live:** plant with living tissue.
**D—Dead:** plant with no living tissue visible.
**NA—Not Applicable**.

Although plant status is purely qualitative, it does provide an adequate characteristic for stratification of preburn plant health and in determining postburn survival. Be careful when making this assessment on plants in their dormant season.

### Cover

Cover is the vertical projection of the vegetation foliage and supporting parts onto the ground (fig. CF-2). Estimating cover within quadrats is made easier by using subplot sizes and the percent of quadrat area they represent (fig. CF-3). Subplots are used to estimate cover for a plant species by mentally grouping cover for all individuals of a plant species into one of the subplots. The percent size of that subplot, in relation to the size of the quadrat being sampled, is used to make a cover class estimate for the species. See **How to Estimate Cover** in the **How-To Guide** chapter for more details.

For each plant species or ground cover class in the quadrat, estimate its percent cover within the quadrat and enter a cover class code (table CF-4) in the Cover Class field. Enter the cover class for each 20 x 20 inch (50 x 50 cm) quadrat.

Table CF-3  FIREMON ground cover codes.

| Ground cover | | | |
|---|---|---|---|
| Code | Description | Code | Description |
| ASH | Ash (organic, from fire) | LICH | Lichen |
| BAFO | Basal forb | LITT | Litter and duff |
| BAGR | Basal graminoid | MEGR | Medium gravel (5 20 mm) |
| BARE | Bare soil (soil particles <2 mm) | MOSS | Moss |
| BARR | Barren | PAVE | Pavement |
| BASH | Basal shrub | PEIC | Permanent ice |
| BATR | Basal tree | PEIS | Permanent ice and snow |
| BAVE | Basal vegetation | PESN | Permanent snow |
| BEDR | Bedrock | ROAD | Road |
| BOUL | Boulders (round and flat) | ROBO | Round boulder (>600 mm) |
| CHAN | Channers (2 150 mm long) | ROCK | Rock |
| CHAR | Char | ROST | Round stone (250 600 mm) |
| CML | Cryptogams, mosses and lichens | STON | Stones (round and flat) |
| COBB | Cobbles (75 250 mm) | TEPH | Tephra volcanic |
| COGR | Coarse gravel (20 75 mm) | TRIC | Transient ice |
| CRYP | Cryptogamic crust | TRIS | Transient ice and snow |
| DEVP | Developed land | TRSN | Transient snow |
| FIGR | Fine gravel (2 5 mm) | UNKN | Unknown |
| FLAG | Flag stones (150 380 mm long) | WATE | Water |
| FLBO | Flat boulders (>600 mm long) | WOOD | Wood |
| FLST | Flat stone (380 600 mm long) | X | Not assessed |
| GRAV | Gravel (2 75 mm) | | |

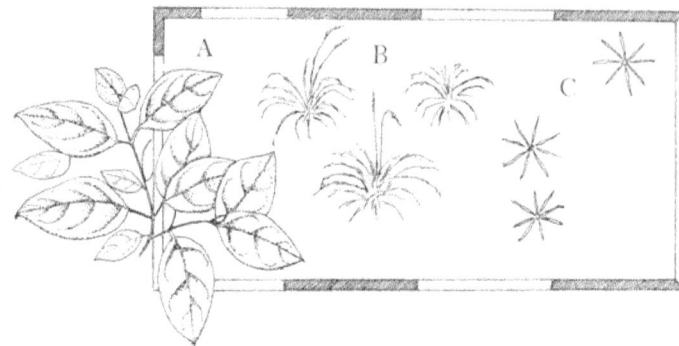

**Figure CF-2**—Cover from species A is estimated even though this species is not actually rooted within the quadrat.

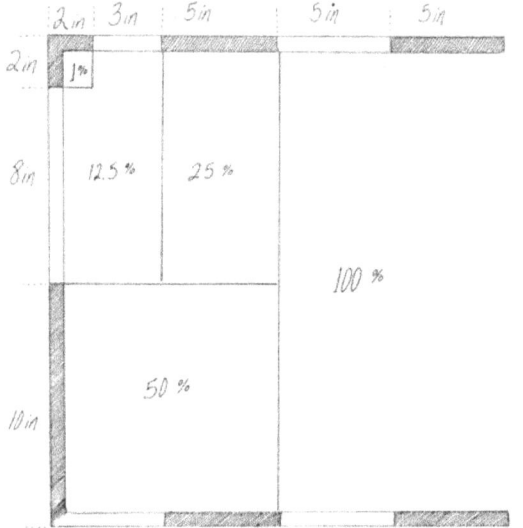

**Figure CF-3**—Subplot dimensions and respective percent of the total plot. Subplots aid the sampler in estimating cover by mentally grouping cover for all individuals of a plant species into one of the subplots.

**Table CF-4** FIREMON cover class codes.

| Code | Cover class |
|------|-------------|
|      | *Percent*   |
| 0    | Zero        |
| 0.5  | >0  1       |
| 3    | >1  5       |
| 10   | >5  15      |
| 20   | >15  25     |
| 30   | >25  35     |
| 40   | >35  45     |
| 50   | >45  55     |
| 60   | >55  65     |
| 70   | >65  75     |
| 80   | >75  85     |
| 90   | >85  95     |
| 98   | >95  100    |

## *Nested rooted frequency*

The standard 20 x 20 inches (50 x 50 cm) quadrat is partitioned into four subplots for recording nested rooted frequency (fig. CF-4 and table CF-5). Species located in the smallest subplot are given the frequency value of 1. Plants in successively larger subplots have frequency values of 2, 3, and 4. Decisions about counting boundary plants—plants that have a portion of basal vegetation intersecting the quadrat—need to be applied systematically to each quadrat. See **How to Count Boundary Plants** in the **How-To Guide** chapter for more details.

Record the smallest size subplot in which each plant species is rooted (fig. CF-5 and table CF-6). Begin with subplot 1, the smallest subplot. If the basal portion of a plant species is rooted in that subplot, record 1 for the species. Next find all plant species rooted in subplot 2, which were not previously recorded for subplot 1, and record a 2 for these plant species. Then identify all plant species, which are rooted in subplot 3, which were not previously recorded for subplots 2 and 1, and record a 3 for these species. Finally, record a 4 for each species rooted in subplot 4, the remaining half of the quadrat, which were not previously recorded in subplots 3, 2, and 1. Enter the subplot number in the NRF field for each species on the CF data form.

Cover/Frequency (CF)

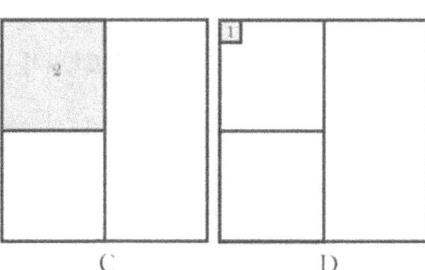

**Figure CF-4**—The numbers inside the plot frame denote the value recorded if a plant is present in that area of the frame. The number 4 corresponds to the entire quadrat (A). The sampling area for number 3 is the entire top half of the quadrat (B). The sampling areas for the numbers 2 and 1 are the upper left quarter and the upper left corner (1 percent) of the quadrat, respectively (C and D). Each larger subplot contains all smaller subplots. Subplots aid the sampler in estimating cover by mentally grouping cover for all individuals of a plant species into one of the subplots.

**Table CF-5** Percent of quadrat represented by the four subplots used to record nested rooted frequency within the standard 20 x 20 inches (50 x 50 cm) quadrat.

| Subplot number for rooted frequency | Size of subplot | Percent area of a 20 x 20 inches (50 x 50 cm) quadrat |
|---|---|---|
| 1 | 2 x 2 inches (5 x 5 cm) | 1 percent |
| 2 | 10 x 10 inches (25 x 25 cm) | 25 percent |
| 3 | 10 x 20 inches (25 x 50 cm) | 50 percent |
| 4 | 20 x 20 inches (50 x 50 cm) | 100 percent |

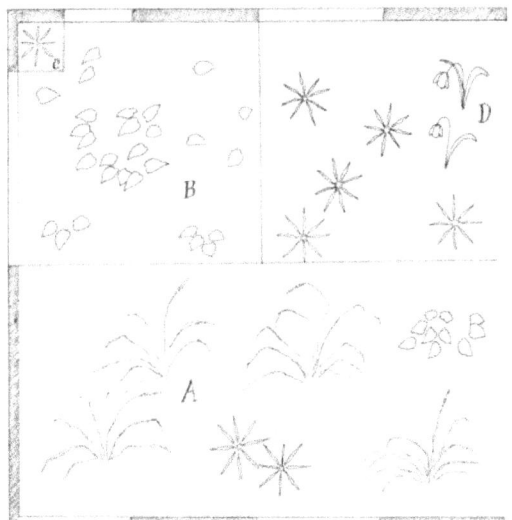

**Figure CF-5**—Example of recording nested rooted frequency values for plant species in a 20 x 20 inches (50 x 50 cm) quadrat frame. Table CF-6 lists the nested rooted frequency value for each plant species displayed in this figure.

**Table CF-6** Standard NRF frame subplot sizes and NRF numbers for the plants illustrated in figure CF 5.

| Species symbol | Smallest subplot size in which species is rooted | NRF value |
|---|---|---|
| C | Smallest, 2 x 2 inches (5 x 5 cm) | 1 |
| B | Next largest, 10 x 10 inches (25 x 25 cm) | 2 |
| D | Next largest, 10 x 20 inches (25 x 50 cm) | 3 |
| A | Largest, 20 x 20 inches (50 x 50 cm) | 4 |

## Estimating Height

Estimate the average height across all the plants of each species in feet (meters) within +/– 10 percent of the mean plant height. See **How to Measure Plant Height** in the **How-To Guide** chapter for more details. Enter plant height in the Height field for each quadrat.

## Precision Standards

Use the precision standards listed in table CF-7 for the CF sampling.

# SAMPLING DESIGN CUSTOMIZATION

This section will present several ways that the CF sampling method can be modified to collect more detailed information or streamlined to collect basic information. First, the suggested or recommended sample design is detailed, then modifications are presented.

## Recommended CF Sampling Design

The recommended CF sampling design follows the Alternative FIREMON Sampling intensity and is listed below:

**Measure only plant species cover and nested rooted frequency within each quadrat.**

**Macroplot Size**: 0.1 acre, 66 x 66 ft (400 m$^2$, 20 x 20 m).

**Quadrat Size**: 20 x 20 inches (50 x 50 cm).

**Number of Transects**:5.

**Number of Quadrats/Transect**: 5.

The quadrat size should be adjusted according to the plant community being sampled.

The number of transects and quadrats sampled should be adjusted according to the applicable sections in the **How To Guide** chapter of the FIREMON manual.

## Streamlined CF Sampling Design

The streamlined CF sampling design follows the Simple FIREMON sample intensity and is designed below:

Measure only nested rooted frequency within each quadrat.

**Macroplot Size**: 0.1 acre, 66 x 66 ft (400 m$^2$, 20 x 20 m).

**Quadrat Size**: 20 x 20 inches (50 x 50 cm).

**Number of Transects**: 5.

**Number of Quadrats/Transect**: 5.

The quadrat size should be adjusted according to the plant community being sampled.

The number of transects and quadrats sampled should be adjusted according to the applicable sections in the **How To Guide** chapter of the FIREMON manual.

Table CF-7   Precision guidelines for CF sampling.

| Component | Standard |
|---|---|
| Cover | ± 1 class |
| NRF | No error |
| Height | ± 10 percent |

## Comprehensive CF Sampling Design

The comprehensive CF sampling design follows the Detailed FIREMON sampling intensity and is detailed below:

**Measure plant species cover, nested rooted frequency, and average plant species height within each quadrat.**

**Macroplot Size**: 0.1 acre, 66 x 66 ft (400 m$^2$, 20 x 20 m).

**Quadrat Size**: 20 x 20 inches (50 x 50 cm).

**Number of Transects:** 5.

**Number of Quadrats/Transect:** 5.

The quadrat size should be adjusted according to the plant community being sampled.

The number of transects and quadrats sampled should be adjusted according to the applicable sections in the **How To Guide** chapter of the FIREMON manual.

## User-Specific CF Sampling Design

There are many ways the user can modify the CF sample fields to make sampling more efficient and meaningful for local situations. Examiners may adjust the number of transects, transect length, and number of quadrats as needed for the specific task.

Quadrat sizes other than the standard 20 x 20 inch (50 x 50 cm) frame can be used for sampling. Small, 4 x 4 inch (10 x 10 cm), quadrats can be used in dense wet meadow communities and large, 40 x 40 inch (1 x 1 m), quadrats can be used in sparse or large vegetation (shrub communities). Nested subplots are not needed when sampling rooted frequency when the plant species being sampled have similar distribution and abundance within the macroplot. Plant species frequency may simply be recorded as presence within the quadrat. The FIREMON sampling forms and databases will accommodate most sampling variations of the recommended procedure.

Ocular estimates of cover can be recorded to the nearest 1 percent instead of a cover class. This will allow values to be grouped into different cover classes later when conducting data analysis. However, it is doubtful that cover can be accurately estimated to a 1 percent level using the human eye. If actual cover values are recorded, or if different cover classes are used than the classes listed in the FIREMON CF methods, record this information in the Metadata table.

## Sampling Hints and Techniques

Examiners must be able to identify many plant species and be able to determine whether a plant species occurs within a quadrat. Examiners must also be familiar with the cover classes used to estimate cover. When collecting rooted frequency data, herbaceous plants (grasses and forbs) must be rooted within the quadrat. However, on many occasions trees and shrubs rooted within the quadrat will provide an inadequate sample size. Counting plants whose cover overhangs the quadrat may increase tree and shrub sample size.

Examiners can calibrate their eyes for estimating cover by using the various subplots within a quadrat frame. Examiners should become familiar with all the subplot sizes and the percent of the entire quadrat each subplot represents. Species are mentally grouped into a subplot. and the subplot size is used to estimate percent cover. See **How to Estimate Cover** in the **How-To Guide** chapter for more details.

Measuring tapes are made from a variety of materials and are available in varying lengths and increments. Examiners should choose tapes with the appropriate units and at least as long, or a little longer, than the transect length being sampled. Because, steel tapes do not stretch, they are the most accurate over long remeasurement intervals. Steel is probably the best choice for permanent transects where remeasurement in exactly the same place each time is important. Cloth and fiberglass tapes will stretch over the life of the tape but are easier to use than steel tapes because they are lighter and do not tend to kink.

The sampling crew may encounter an obstacle, such as a large rock or tree, along one of the transect lines that interferes with the quadrat sampling. If that happens offset using the directions described in **How To Offset a Transect** in the **How-To Guide** chapter.

When entering data on the CF data forms, examiners may run out of space on the first page or sample more than five quadrats per transect. The first page allows only five quadrats per transect. If more quadrats per transect are sampled, use the CF continuation form. The form was designed to print one copy of the first page, and several copies of the second page. The second page of the data form allows the examiner to write the quadrat number on the form. This allows the examiner to design the form to accommodate the number of transects sampled. Print out enough pages to record all species on all transects for the required number of transects.

## COVER/FREQUENCY (CF) FIELD DESCRIPTIONS

Field 1: **Number of Transects**. Total number of transects on the plot.

Field 2: **Transect Length**. Length of transect (ft/m).

Field 3: **Number of Quadrats per Transect**. Number of quadrats sampled per transect.

Field 4: **Quadrat Length**. Length of the quadrat (inches/cm).

Field 5: **Quadrat Width**. Width of the quadrat (inches/cm).

Field 6: **NRF Subplot Ratio**. Percent of the quadrat covered by each subplot starting from the largest subplot to the smallest subplot. For example, the subplot ratio for a typical 20 x 20 inches (50 x 50 cm) quadrat would be 100:50:25:1.

Field 7: **NRF Numbers**. Frequency numbers for the subplots starting from largest to smallest. For example, the frequency numbers for a typical 20 x 20 inches (50 x 50 cm) quadrat would be 4:3:2:1.

Field 8: **Transect Number.** Sequential number of the sample transect.

Field 9: **Item Code.** Code of sampled entity. Either the NRCS plants species code, the local code for that species, ground cover code, or other item code. Or, ground cover code. Precision: No error.

Field 10: **Status**: Plant status—Live, Dead, or Not Applicable. (L, D, NA). Precision: No error.

**Cover Class.** Enter the cover class of sampled entity. Valid classes are in table CF-4 of the sampling methods. Precision: $\pm 1$ class.

**Nested Rooted Frequency.** Enter the NRF number of sampled entity. Precision: No error.

**Height.** Enter the average height for each plant species or life-form in the quadrat. Precision: $\pm 10$ percent of average height.

# FIREMON CF Cheat Sheet

## Ground cover codes

### Ground cover

| Code | Description | Code | Description |
|---|---|---|---|
| ASH | Ash (organic, from fire) | LICH | Lichen |
| BAFO | Basal forb | LITT | Litter and duff |
| BAGR | Basal graminoid | MEGR | Medium gravel (5 20 mm) |
| BARE | Bare soil (soil particles <2 mm) | MOSS | Moss |
| BARR | Barren | PAVE | Pavement |
| BASH | Basal shrub | PEIC | Permanent ice |
| BATR | Basal tree | PEIS | Permanent ice and snow |
| BAVE | Basal vegetation | PESN | Permanent snow |
| BEDR | Bedrock | ROAD | Road |
| BOUL | Boulders (round and flat) | ROBO | Round boulder (>600 mm) |
| CHAN | Channers (2 150 mm long) | ROCK | Rock |
| CHAR | Char | ROST | Round stone (250 600 mm) |
| CML | Cryptogams, mosses and lichens | STON | Stones (round and flat) |
| COBB | Cobbles (75 250 mm) | TEPH | Tephra volcanic |
| COGR | Coarse gravel (20 75 mm) | TRIC | Transient ice |
| CRYP | Cryptogamic crust | TRIS | Transient ice and snow |
| DEVP | Developed land | TRSN | Transient snow |
| FIGR | Fine gravel (2 5 mm) | UNKN | Unknown |
| FLAG | Flag stones (150 380 mm long) | WATE | Water |
| FLBO | Flat boulders (>600 mm long) | WOOD | Wood |
| FLST | Flat stone (380 600 mm long) | X | Not assessed |
| GRAV | Gravel (2 75 mm) | | |

## Cover class codes

| Code | Cover class |
|---|---|
| | *Percent* |
| 0 | Zero |
| 0.5 | >0 1 |
| 3 | >1 5 |
| 10 | >5 15 |
| 20 | >15 25 |
| 30 | >25 35 |
| 40 | >35 45 |
| 50 | >45 55 |
| 60 | >55 65 |
| 70 | >65 75 |
| 80 | >75 85 |
| 90 | >85 95 |
| 98 | >95 100 |

## Status codes

| Code | Description |
|---|---|
| L | Live |
| D | Dead |
| NA | Not applicable |

## Precision

| Component | Standard |
|---|---|
| Cover | ±1 class |
| NRF | No error |
| Height | ±10 percent |

## Cover/frequency (CF) equipment list

| | |
|---|---|
| Camera with film | Maps, charts, and directions |
| CF data forms | Map protector or plastic bag |
| Clipboard | Magnifying glass |
| Compass | Pocket calculator |
| File | Plot sheet protector or plastic bag |
| Field notebook | Quadrat frame |
| Graph paper | Reinforcing bar (2) for baseline plus 2 for each transect |
| Hammer | Steel fence posts (2) and driver (to mark endpoints of baseline) |
| Indelible ink pen (Sharpie, Marker) | Tape 75 ft (25 m) or longer (2) |
| Lead pencils with lead refills | |

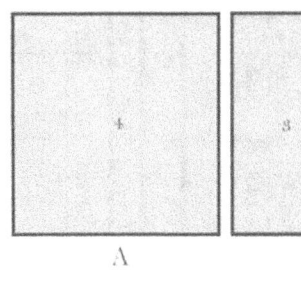
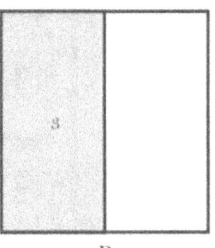
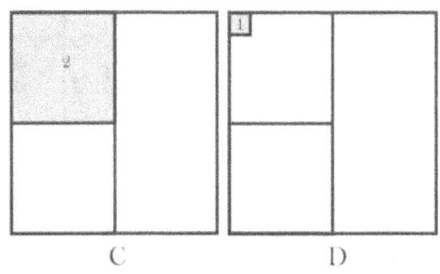

# Cover/Frequency (CF) Form

CF Page ___ of ___

RegistrationID: _ _ _ _ _
ProjectID: _ _ _ _ _
PlotID: _ _ _ _
Date: __/__/____

Plot Key

| Field 1 | Field 2 | Field 3 | Field 4 | Field 5 | Field 6 | Field 7 |
|---|---|---|---|---|---|---|
| Number of Transects | Transect Length | Number of Quad./Tran. | Quadrat Length (in/cm) | Quadrat Width (in/cm) | NRF Subplot Ratio | NRF Numbers |
| | | | | | | |

| Field 8 | Field 9 | Field 10 | Quadrat 1 | | | Quadrat 2 | | | Quadrat 3 | | | Quadrat 4 | | | Quadrat 5 | | |
|---|---|---|---|---|---|---|---|---|---|---|---|---|---|---|---|---|---|
| Transect Number | Item Code | Status | Cover Class | Nested Rooted Frequency | Height (ft/m) | Cover Class | Nested Rooted Frequency | Height (ft/m) | Cover Class | Nested Rooted Frequency | Height (ft/m) | Cover Class | Nested Rooted Frequency | Height (ft/m) | Cover Class | Nested Rooted Frequency | Height (ft/m) |
| | | | | | | | | | | | | | | | | | |

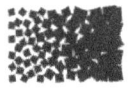

# Cover/Frequency (CF) Form
## Continuation

CF Page __ __ of __ __

| Field 1 | Field 2 | Field 3 | Field 4 | Field 5 | Field 6 | Field 7 |
|---|---|---|---|---|---|---|
| Number of Transects | Transect Length | Number of Quad/Tran. | Quadrat Length (in/cm) | Quadrat Width (in/cm) | NRF Subplot Ratio | NRF Numbers |
|  |  |  |  |  |  |  |

RegistrationID: _ _ _ _ _
ProjectID: _ _ _ _ _
PlotID: _ _ _ _
Date: _ _ / _ _ / _ _ _ _

Plot Key

| Field 8 | Field 9 | Field 10 | Quadrat ___ | | | Quadrat ___ | | | Quadrat ___ | | | Quadrat ___ | | | Quadrat ___ | | |
|---|---|---|---|---|---|---|---|---|---|---|---|---|---|---|---|---|---|
| Transect Number | Item Code | Status | Cover Class | Nested Rooted Frequency | Height (ft/m) | Cover Class | Nested Rooted Frequency | Height (ft/m) | Cover Class | Nested Rooted Frequency | Height (ft/m) | Cover Class | Nested Rooted Frequency | Height (ft/m) | Cover Class | Nested Rooted Frequency | Height (ft/m) |
|  |  |  |  |  |  |  |  |  |  |  |  |  |  |  |  |  |  |
|  |  |  |  |  |  |  |  |  |  |  |  |  |  |  |  |  |  |
|  |  |  |  |  |  |  |  |  |  |  |  |  |  |  |  |  |  |
|  |  |  |  |  |  |  |  |  |  |  |  |  |  |  |  |  |  |
|  |  |  |  |  |  |  |  |  |  |  |  |  |  |  |  |  |  |
|  |  |  |  |  |  |  |  |  |  |  |  |  |  |  |  |  |  |
|  |  |  |  |  |  |  |  |  |  |  |  |  |  |  |  |  |  |
|  |  |  |  |  |  |  |  |  |  |  |  |  |  |  |  |  |  |
|  |  |  |  |  |  |  |  |  |  |  |  |  |  |  |  |  |  |
|  |  |  |  |  |  |  |  |  |  |  |  |  |  |  |  |  |  |
|  |  |  |  |  |  |  |  |  |  |  |  |  |  |  |  |  |  |
|  |  |  |  |  |  |  |  |  |  |  |  |  |  |  |  |  |  |
|  |  |  |  |  |  |  |  |  |  |  |  |  |  |  |  |  |  |
|  |  |  |  |  |  |  |  |  |  |  |  |  |  |  |  |  |  |
|  |  |  |  |  |  |  |  |  |  |  |  |  |  |  |  |  |  |
|  |  |  |  |  |  |  |  |  |  |  |  |  |  |  |  |  |  |

# Point Intercept (PO)

## Sampling Method

John F. Caratti

## SUMMARY

The FIREMON Point Intercept (PO) method is used to assess changes in plant species cover or ground cover for a macroplot. This method uses a narrow diameter sampling pole or sampling pins, placed at systematic intervals along line transects to sample within plot variation and quantify statistically valid changes in plant species cover and height over time. Plant species or ground cover classes that touch the pin are recorded as "hits" along a transect. Percent cover is calculated by dividing the number of hits for each plant species or ground cover class by the total number of points along a transect. This method is primarily suited for vegetation types less than 3 ft (1 m) in height and is particularly useful for recording ground cover.

## INTRODUCTION

The Point Intercept (PO) method is designed to sample within-plot variation and quantify changes in plant species cover and height, and/or ground cover over time. This method uses transects located within the macroplot. First, a baseline is established from which to orient the transects, then transects are placed randomly along the baseline. Characteristics—such as transect length, number of transects, and number of points per transect—are recorded about the general sample design. A sampling pole or sampling pins are systematically lowered along each transect and "hits" are tallied when contact is made with a plant species or ground cover class. Percent cover is calculated as the number of hits for each plant species or ground cover class divided by the total number of points per transect. Height is also recorded for each plant species along the transect.

This method is primarily used when managers want to monitor changes in plant species cover and height or ground cover and is best suited for sampling ground cover and grasses, forbs, and shrubs less than 3 ft (1 m) in height. The Point Intercept method works well for fine textured herbaceous communities, fine leaved plant species, and species with open canopies (pastures, dense grasslands, and wet meadows), which can be more difficult to estimate with the Line Intercept (LI) method. It provides a more objective estimate of cover than the ocular estimates used in the Cover/Frequency (CF) sampling method. It can be difficult to detect rare plants with the PO method unless many points are used for sampling. Point intercept sampling requires many points to sample rare species (200 points to sample at 0.5 percent cover). Quadrats sample more area and have a greater chance of detecting rare species. If rare plant species are of interest the CF or RS methods are preferred because it is more effective to sample rare species using quadrats or by marking individual plants, than with points or lines. We suggest you use the PO method if you are primarily interested in monitoring changes in ground cover. The PO method may be used in conjunction with the CF method to sample ground cover by using the CF sampling quadrat as a point frame.

The Point Intercept method is considered one of the most objective ways to sample cover (Bonham 1989). The observer needs to decide only whether a point intercepts a plant species or the ground. No cover estimates are required. Points offer quick and efficient data collection and can be used to estimate cover values with minimal bias and error. However, errors can be caused by plants moving in the wind or sampling poles lowered incorrectly. The points themselves have dimensions and can be considered small quadrats. In theory, if you sampled an infinite number of points in an area, you could measure the exact cover for each plant species. Points are either the end of the sampling pole or the intersection of crosshairs in a sampling frame.

Cover or ground cover is estimated using individual points or collections of points. Collections of points are sampled either with sampling pins, grouped into a pin frame (typically 10 pins) or cross-hairs grouped into a rectangular sighting frame. When using pin frames, the sampling pole is replaced with a pin. Pins are generally smaller in diameter than a sampling pole so are less prone to sampling error (see below). The pin frame itself helps protect sampling pins from damage. Pin spacing should be determined according to plant species and vegetation patterns. For instance, pins in a collection should not be placed so close together that all pins hit bare ground between clumps of grasses or all fall on one clump of grass. The number of points used determines the percent cover values that can be estimated. For example, if 50 points are sampled along a transect, then cover can be estimated in 2 percent intervals (1/50, 2/50, and so forth) for that transect. Cover is estimated by counting the number of "hits" per species or ground cover category divided by total number of points measured. More than one species may be tallied for each pin location depending on project objectives.

Sampling pole and pin diameter can influence the accuracy of cover estimates. This is mostly an issue with large diameter sampling poles, which overestimate cover, especially for narrow or small-leafed species. Pins less than 0.1 inch (2.5 mm) are impractical in the field because they move in the wind and are easily damaged (Bonham 1989). Overestimation of cover is not a problem if the monitoring objective is to note relative cover changes rather than absolute change in cover. Because of the effect of sampling pole and pin diameter on cover estimations, it is necessary to always use the same diameter poles or pins when remeasuring cover.

## Point Sampling Techniques

### Single points

Each sample point is defined by a sampling pole guided vertically to the ground. We recommend using a sturdy 0.25 inch (0.635 cm) diameter sampling pole when sampling with the FIREMON PO method. Smaller diameter poles (0.125 inch, 0.3175 cm) may be used for more precise measurements and less observer decisions. However, thin poles are more flexible, require more finesse to place in a straight line, and are easily bent in the field. A fiberglass tent pole, wooden dowel, or aluminum rod could all be used as a sampling pole. It should be longer than the vegetation that will be sampled is tall and long enough that field crews can sample without leaning over (40 inches, 100 cm), sharpened on one end with a loop or bend on the other.

Individual points placed at systematic intervals along a transect can give a more precise cover estimate than points grouped into point frames or grid frames, given the same number of points are sampled (Blackman 1935; Goodall 1952; Greig-Smith 1983). Using individual points requires approximately one-third the number of points than using points in groups (Bonham 1989). The distance between systematically located pins along a transect depends on plant size, plant distribution, and the distance between plants. The recommended FIREMON PO sampling method uses the single point sampling approach.

## Point frame

Point frames are more practical and more commonly used than grid frames. Point frames are built using wood or metal and consist of two legs and two cross arms, typically containing 10 pins (fig. PO-1). The pins can be made of any material as long as they are relatively small diameter (0.25 inch, 0.635 cm), rigid enough not to bend or break, and long enough to touch the ground.

When sampling, the pins are lowered to the ground cover through the holes and the interceptions are tallied. The size of the frame needs to be designed to suit local vegetation conditions because plant height and distribution patterns affect the spacing of pins and height of the frame. In vegetation types with large plants or clumped distributions, groups of points may intercept plants more frequently with more hits per frame, resulting in an overestimation of cover. Some point frames are built to allow the pins to be slid at an angle into the vegetation, and this can have some sampling benefits in certain types of vegetation. FIREMON provides an alternative data entry form and data entry software to accommodate data gathered with the point frame method. See the section on **User-Specific PO Sampling Design** for more details. See **How to Construct Point Frames and Grid Frames** in the **How-To Guide** chapter for more information.

## Grid quadrat frame

A grid frame is made from metal or wood. Rows of thin wire or light string are attached to the inside vertical and horizontal pieces of the frame resulting in a number of intersections or "cross-hairs." Cross-hairs of a grid quadrat are considered point quadrats, and the vertical interception of cross-hairs with plant parts are considered hits. A double grid of cross-hairs prevents error due to observers viewing the cross-hairs at different angles (fig. PO-2).

Stanton (1960) designed a grid frame for estimating cover shrub in communities. This grid of cross-hairs consisted of 25 points spaced 2.95 inches (7.5 cm) apart and was supported with metal legs. This type of frame is good for measuring cover up to 4.5 ft (1.5 m) tall in sparse vegetation. FIREMON provides a data entry form and data entry software to accommodate data gathered with the grid quadrat frame method. See the section on **User-Specific PO Sampling Design** for more details.

There are a number of ways to streamline or customize the PO sampling method. The FIREMON three-tier sampling design can be employed to optimize sampling efficiency. See the sections on **User-Specific PO Sampling Design** and **Sampling Design Customization** below.

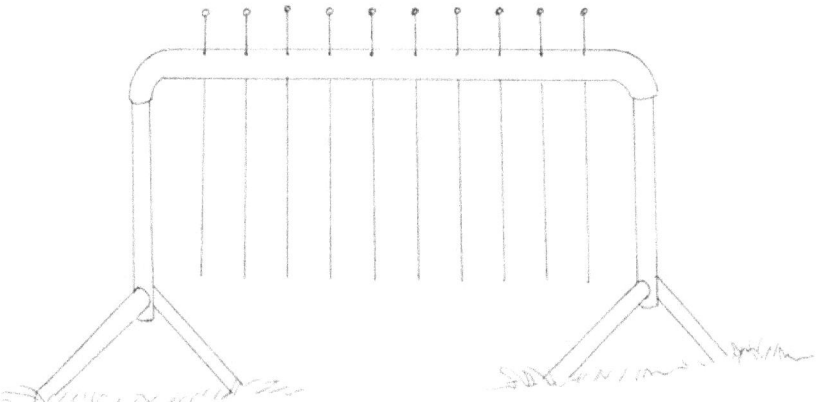

**Figure PO-1**—Example of a point frame with 10 pins.

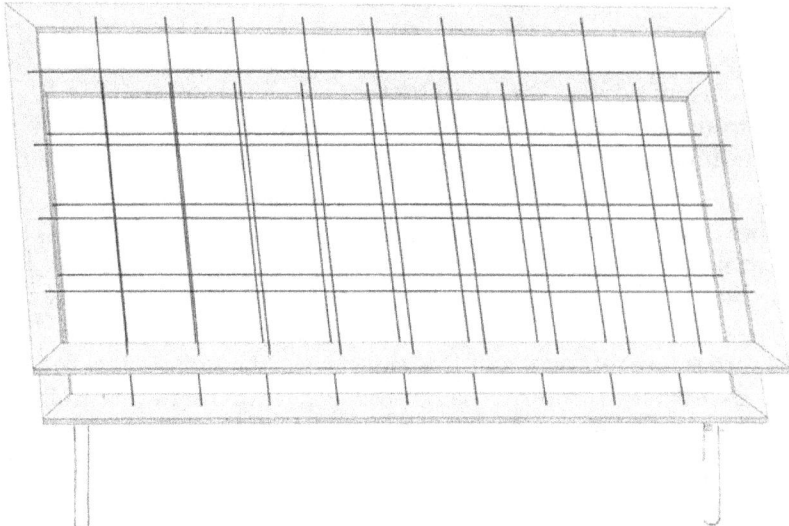

**Figure PO-2**—Example of a grid frame with 36 points (4 x 9).

## SAMPLING PROCEDURE

This method assumes that the sampling strategy has already been selected and the macroplot has already been located. If this is not the case, then refer to the FIREMON **Integrated Sampling Strategy** and for further details.

The sampling procedure is described in the order of the fields that need to be completed on the **PO Transect data form**, so it is best to reference the data form when reading this section. The sampling procedure described here is the recommended procedure for this method. Later sections will describe how the FIREMON three-tier sampling design can be used to modify the recommended procedure to match resources, funding, and time constraints.

In the **How-To Guide** chapter, see **How To Locate a FIREMON Plot, How To Permanently Establish a FIREMON Plot,** and **How to Define the Boundaries of a Macroplot** for more information on setting up your macroplot.

### Preliminary Sampling Tasks

Before setting out for your field sampling, lay out a practice area with easy access. Try to locate an area with the same species or vegetation life form you plan on sampling. Get familiar with the plot layout and the data that will be collected. This will give you a chance to assess the method and will help you think about problems that might be encountered in the field. For example, will you be recording the plant status—dead or alive—for the part of the plant hit by the sampling pin or the entire plant? It is better to answer these questions before the sampling begins so that you are not wasting time in the field. This will also let you see if there are any pieces of equipment that will need to be ordered.

Many preparations must be made before proceeding into the field for PO sampling. First, all equipment and supplies in the **PO Equipment List** must be purchased and packed for transport into the field. Travel to FIREMON plots is usually by foot, so it is important that supplies and equipment be placed in a comfortable daypack or backpack. It is also important that there be spares of each piece of equipment so that an entire day of sampling is not lost if something breaks. Spare equipment can be stored in the vehicle rather than the backpack. Be sure that all equipment is well maintained and there are plenty of extra supplies, such as data forms, map cases, and pencils.

All PO data forms should be copied onto waterproof paper because inclement weather can easily destroy valuable data recorded on standard paper. Data forms should be transported into the field using a plastic, waterproof map protector or plastic bag. The day's sample forms should always be stored in a dry place (office or vehicle) and not be taken back into the field for the next day's sampling.

We recommend that one person on the field crew, preferably the crew boss, have a waterproof, lined field notebook for recording logistic and procedural problems encountered during sampling. This helps with future remeasurements and future field campaigns. All comments and details not documented in the FIREMON sampling methods should be written in this notebook.

It is beneficial to have plot locations for several days of work in advance in case something happens, such as if the road to one set of plots is washed out by flooding. Plots should be referenced on maps and aerial photos using pin-pricks or dots to make navigation easy for the crew and to provide a check of the georeferenced coordinates. We found that it is easy to transpose UTM coordinate digits when recording georeferenced positions on the plot sheet, so marked maps can help identify any erroneous plot positions. If possible, the spatial coordinates should be provided if FIREMON plots were randomly located.

A field crew of two people is probably the most efficient for implementation of the PO sampling method. There should never be a one-person field crew for safety reasons, and any more than two people will probably result in part of the crew waiting for tasks to be completed and unnecessary trampling on the FIREMON macroplot. The crew boss is responsible for all sampling logistics including the vehicle, plot directions, equipment, supplies, and safety. The crew boss should be the note taker and the technician should perform most quadrat measurements. The initial sampling tasks of the field crew should be assigned based on field experience, physical capacity, and sampling efficiency. As the field crew gains experience, switch tasks so that the entire crew is familiar with the different sampling responsibilities and to limit monotony.

## Designing the PO Sampling Method

A set of general criteria recorded on the PO data form allows the user to customize the design of the PO sampling method so that the sampling captures the information needed to successfully complete the management objective within time, money, and personnel constraints. These general fields should be decided before the crews go into the field and should reflect a thoughtful analysis of the expected problems and challenges in the fire monitoring project. However, some of these fields, in particular the number of points per transect and number of transects, might be adjusted after preliminary sampling is conducted in the field to determine a sufficient sample size.

### Plot ID construction

A unique plot identifier must be entered on the PO sampling form. This is the same plot identifier used to describe general plot characteristics in the Plot Description or PD sampling method. Details on constructing a unique plot identifier are discussed in the **How to Construct a Unique Plot Identifier** section in the **How-To Guide** chapter. Enter the plot identifier at the top of the PO data form.

### Determining sample size

The size of the macroplot ultimately determines the length of the transects and the length of the baseline along which the transects are placed. The amount of variation in plant species composition and distribution determines the number and length of transects and the number of quadrats required for sampling. The typical macroplot sampled in the PO method is a 0.10 acre (0.04 ha) square measuring 66 x 66 ft (20 x 20 m), which is sufficient for most monitoring applications. If you are not sure of the plot size to use, contact someone wo has sampled the same vegetation that you will be sampling. The size of the macroplot may be adjusted to accommodate different numbers and lengths of transects, and number of points per transect. It is more efficient if you use the same plot size for all FIREMON sampling methods on the plot.

The sampling unit for Point Intercept is the transect. We recommend sampling five transects within the macroplot. However, there are situations when more transects should be sampled. See **How To Determine Sample Size** in the **How-To Guide** chapter for more details. Enter the number of transects in Field 1 on the PO Transect data form. The recommended transect length is 66 ft (20 m) for a 66 x 66 ft (20 x 20 m) macroplot. However, the macroplot size may be adjusted to accommodate longer or shorter transects based on the variability in plant species composition and distribution. For example, transects may be lengthened to accommodate more points per transect or more widely spaced points. Enter the transect length in Field 2 of the PO Transect data form. The FIREMON PO data form and data entry screen allow a maximum of 10 transects.

We recommend that 66 points be placed 1 ft (0.3 m) apart along each 66-ft (20-m) transect. However, when sampling with a metric tape, we recommend sampling at every 0.25 m for a total of 80 points per transect. The number of points and spacing should be adjusted based on plant species size and spacing. For example, points should not be placed so close together that all sample points hit bare ground between clumps of grasses or all sample points fall on grass clumps. The number of points along a transect determines the resolution of cover recorded. If 50 points are recorded along a transect, cover values can be recorded in increments of 2 percent (1/50, 2/50, and so forth). At a minimum, you want enough points to sample at least some of the species of interest along each transect. Enter the number of points per transect in Field 3 of the PO data form.

## Conducting PO Sampling Tasks

### Establish the baseline for transects

Once the plot has been monumented, a permanent baseline is set up as a reference from which you will orient all transects. The baseline should be established so that the sampling plots for all of the methods overlap as much as possible. See **How To Establish Plots with Multiple Methods** in the **How-To Guide** chapter. The recommended baseline is 66 ft (20 m) long and is oriented upslope with the 0-ft (0-m) mark at the lower permanent marker and the 66-ft (20-m) mark at the upper marker. On flat areas, the baseline runs from south to north with the 0-ft (0-m) mark on the south end and the 66-ft (20-m) mark on the north end. See **How To Establish a Baseline for Transects** in the **How-To Guide** chapter for more details.

### Locating the transects

Transects are placed perpendicular to the baseline and are sampled starting at the baseline. On flat areas, transects are laid out east starting at the baseline. For permanent plots, determine the compass bearing of each transect, record these on the plot layout map and permanently mark each end of the transect. Starting locations for each transect are determined by selecting a sampling scheme using the FIREMON random transect locator or from supplied tables. If the PO method is used in conjunction with other replicated sampling methods (CF, LI, RS, or DE), use the same transects for all methods. In successive remeasurement years, it is essential that transects be placed in the same location. See **How To Locate Transects and Quadrats** in the **How-To Guide** chapter for more details.

### Sampling points

Points are sampled at equal intervals along the length of a transect by lowering the sampling pole vertically to the ground, not perpendicular to the slope (fig. PO-3). If 66 points are sampled along a 66-ft transect, the first point is recorded at 1 ft, then every foot to the end of the tape. If 80 points are sampled at 0.25 m along a 20-m transect, the first point is sampled at 0.25 m and the last point at 20 m.

## Point Intercept Sampling

### Recording hits

The FIREMON PO method may be used to sample just species cover, just ground cover, or species and ground cover together. If the sampling crew is collecting species cover data, record only the plant species that are "hit" by the sampling pole. FIREMON provides plant species codes from the NRCS Plants database. However, local or customized plant species codes are allowed in FIREMON. See **Populating**

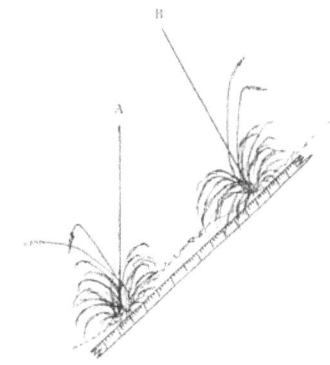

**Figure PO-3**—Points are sampled by lowering the pin vertically to the ground (point A) and not perpendicular to the slope (point B).

the **Plant Species Codes Lookup Table** in the **FIREMON Database User Manual** for more details. When using the PO method to sample species cover in clumped or sparse vegetation, you may find that data are only recorded at a subset of the sampling locations because there may not be vegetation at every point that is sampled. If ground cover is being sampled, use the cover codes in table PO-1 to record sampling hits. Unlike species sampling, every point should have a ground cover code recorded. If species and ground cover are being sampled at the same time, record the species name for the plant that the pole hits and the appropriate ground cover code from table PO-1 (fig. PO-4). For instance, if you are lowering a sampling pole and it first contacts a blade of blue grama grass and then, as you continue to lower, it hits the basal portion of the plant, record the NRCS species code, (BOGR2) and the ground cover code (BAGR or BAVE). Enter the plant species and ground cover code in Field 4 on the PO Transect data form.

The number of hits that are recorded for each sampling point is dependent on the project objectives. If the objective is just to monitor ground cover, samplers need to record only the ground cover hits. To develop a complete species list, samplers should record all unique species hits at each sampling point. Multiple hits for each species can be recorded if measuring biomass, volume, or species composition. Again, this is important information to be recorded in the Metadata table.

Table PO-1 FIREMON ground cover codes

| Ground cover | | | |
|---|---|---|---|
| Code | Description | Code | Description |
| ASH | Ash (organic, from fire) | LICH | Lichen |
| BAFO | Basal forb | LITT | Litter and duff |
| BAGR | Basal graminoid | MEGR | Medium gravel (5 20 mm) |
| BARE | Bare soil (soil particles <2 mm) | MOSS | Moss |
| BARR | Barren | PAVE | Pavement |
| BASH | Basal shrub | PEIC | Permanent ie |
| BATR | Basal tree | PEIS | Permanent ice and snow |
| BAVE | Basal vegetation | PESN | Permanent snow |
| BEDR | Bedrock | ROAD | Road |
| BOUL | Boulders (round and flat) | ROBO | Round boulder (>600 mm) |
| CHAN | Channers (2 150 mm long) | ROCK | Rock |
| CHAR | Char | ROST | Round stone (250 600 mm) |
| CML | Cryptogams, mosses and lichens | STON | Stones (round and flat) |
| COBB | Cobbles (75 250 mm) | TEPH | Tephra volcanic |
| COGR | Coarse gravel (20 75 mm) | TRIC | Transient ice |
| CRYP | Cryptogamic crust | TRIS | Transient ice and snow |
| DEVP | Developed Land | TRSN | Transient snow |
| FIGR | Fine gravel (2 5 mm) | UNKN | Unknown |
| FLAG | Flag stones (150 380 mm long) | WATE | Water |
| FLBO | Flat boulders (>600 mm long) | WOOD | Wood |
| FLST | Flat stone (380 600 mm long) | X | Did not assess |
| GRAV | Gravel (2 75 mm) | | |

Point Intercept (PO) Sampling Method

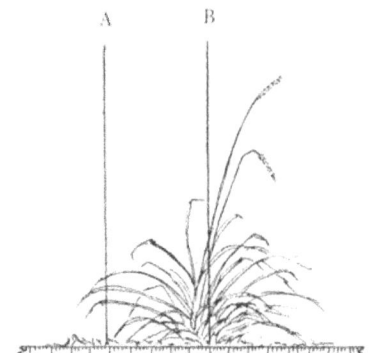

**Figure PO-4**—If the sampling pole eventually hits ground instead of the basal portion of the plant, then the appropriate ground cover code is recorded, even if it intercepts the aerial portion of the plant (A). Basal vegetation is recorded for ground cover when the pin hits the basal portion of the plant (B).

The angle of the sampling pole has an effect on cover estimates. Vertically lowered sampling poles hit flat bladed species (forbs) more often than grasses. A pole lowered at an angle tends to favor grasses (Winkworth 1955). Most cover measurements use vertical placement of poles but will underestimate narrow leafed species (such as grasses). Other angles are used to increase the number of hits. However, angled sampling eliminates the intuitive visualization of vegetation on to the ground. In FIREMON we recommend using the vertical orientation of the sampling pole. The angle used in sampling should be entered in the FIREMON Metadata table so that data collected in subsequent visits is compatible—especially if some orientation other than vertical pole placement is used for point sampling.

At each interval, lower the sampling pin to the ground and record one hit for each plant species that touches the pole. Record only one hit for each plant species, even if the pole touches the same plant or plant species more than once (fig. PO-5). When measuring ground cover you will generally record only the first or upper most hit. For instance, if the pin passes through an ash layer to bare soil, record only the ash layer.

Tally the "hits" for each species using a dot tally in the workspace column for each transect on the PO Transect data form. See **How to Dot Tally** in the **How-To Guide** chapter for more details. Enter the total number of hits for each species or ground cover class, by transect, in the HITS field on the PO Transect data form.

### Recording plant status

Next enter the plant species status in Field 5 on the PO Transect data form. Status describes the general health of the plant species as live or dead using the following codes:

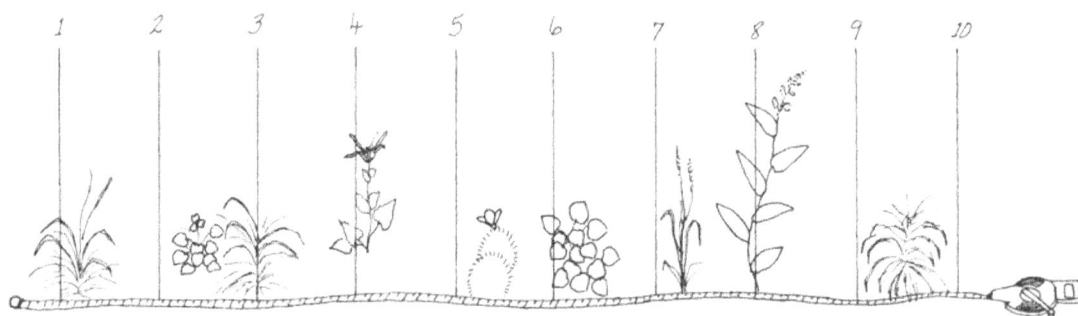

**Figure PO-5**—In this illustration points 1, 3, 4, 6, and 8 intersect plants and are recorded as "hits" for each species. Points 2, 5, 9, and 10 "miss" plants and are only recorded if ground cover is being sampled. Samplers will usually record just the first ground cover hit at each sampling point.

Point Intercept (PO) Sampling Method

**L—Live:** plant with living tissue.

**D—Dead:** plant with no living tissue visible.

**NA—Not Applicable**.

This may be an evaluation of the entire plant or just the part of the plant that comes in contact with the sampling pole depending on the project objectives. In FIREMON we recommend recording the status of the plant part that touches the sampling pole. Recognize that an accurate assessment of plant status may be difficult during the dormant season.

Plant status is purely qualitative, but it provides an adequate characteristic for stratification of preburn plant health and determining postburn survival.

### Estimating average height

At the end of each transect, estimate the average height of each plant species you tallied, in feet (m) within +/– 10 percent of the mean plant height. This is not the height above the ground where the sampling pole touches the vegetation, but rather the average total height of the plants that are tallied. See **How to Measure Plant Height** in the **How-To Guide** chapter for more details. Enter plant height in the Height field for each transect.

## Precision Standards

Use the precision standards listed in table PO-2 for the PO sampling.

# SAMPLING DESIGN CUSTOMIZATION

This section presents several ways that the PO sampling method can be modified to collect more detailed information or streamlined to collect only the most important tree characteristics. First, the suggested or recommended sample design is detailed, then modifications are presented.

## Recommended PO Sampling Design

The recommended PO sampling design follows the Alternative FIREMON sampling intensity and is listed below:

**Macroplot Size**: 0.1 acre, 66 x 66 ft (400 m$^2$, 20 x 20 m).

**Number of Transects:** 5.

**Number of Points per Transect:** 66 per 66 ft transect or 80 per 20 m transect, vertically oriented.

The number of transects sampled should be adjusted according to the appropriate methods in the **How-To Guide** chapter of the FIREMON manual. The number of points per transect should be adjusted based on plant species size and spacing.

Collect plant species cover data and ground cover data.

Table PO-2  Precision guidelines for PO sampling.

| Component | Standard |
|---|---|
| Hits | ±3 percent total hits |
| Height | ±10 percent average height |

## Streamlined PO Sampling Design

The streamlined PO sampling design follows the Simple FIREMON sampling intensity and is designed below:

**Macroplot Size**: 0.1 acre, 66 x 66 ft (400 m$^2$, 20 x 20 m).

**Number of Transects:** 3.

**Number of Points per Transect:** 66 per 66-ft transect or 80 per 20-m transect, vertically oriented.

The number of transects sampled should be adjusted according to the appropriate methods in the **How-To Guide** chapter of the FIREMON manual. The number of points per transect should be adjusted based on plant species size and spacing.

Collect plant species cover and average plant height data.

## Comprehensive PO Sampling Design

The comprehensive PO sampling design follows the Detailed FIREMON sampling intensity and is detailed below:

**Macroplot Size**: 0.1 acre, 66 x 66 ft (400 m$^2$, 20 x 20 m).

**Number of Transects:** 5.

**Number of Points per Transect:** 100 placed at 8-inch intervals on a 66-ft transect or placed at 0.20-m intervals on a 20-m transect, vertically oriented.

The number of transects sampled should be adjusted according to the appropriate methods in the **How-To Guide** chapter of the FIREMON manual. The number of points per transect should be adjusted based on plant species size and spacing.

Collect plant species cover, ground cover, and measure average plant height.

## User-Specific PO Sampling Design

There are a number of ways the user can adjust the PO sample fields to make sampling more efficient and meaningful for local situations. Examiners may adjust the number of transects and points per transect based on plant species size and distribution.

Points could be sampled in quadrats or frames placed along a transect rather than using individual points. Collections of points are pins grouped into a pin frame (usually 10 pins) or cross-hairs grouped into a sighting frame. See **How To Construct Point Frames and Grid Frames** in the **How-To Guide** chapter for more details. If 10 pins are used in a frame, percent cover can be estimated within 10 percent intervals (1/10, 2/10 and so forth) for each frame. If the point frames or grid frames are placed far enough apart (they are independent samples), the frames can be the sample units rather than the transects.

The PO method may be used in conjunction with the CF method to sample ground cover by using the CF sampling quadrat as a point frame. A pencil or pen is used to record ground cover "hits" at the four corners and the four midpoints on each side of the quadrat. A total of eight points are recorded for each quadrat.

If grid frames or point frames are being sampled along transects rather than individual points, then determine the number of quadrats or point frames sampled per transect using the methods in **How To Determine Sample Size** in the **How-To Guide** chapter. Record the number of frames per transect in Field 3, the number of points per frame in Field 4, and the transect number in Field 5 on the PO Point Frames data form. The FIREMON data entry screen and database accommodates 20 frames per transect and an unlimited number of transects.

Point Intercept (PO) Sampling Method

The PO method is typically used for grasses, forbs, and small shrubs less than 3 ft (1 m) in height. However, this method can be modified to sample large shrubs and trees as well. Instead of using pins that drop to the ground or a grid frame that the observer looks down on, you could use a sighting device, such as a sighting tube or "moosehorn," which allows you to look down for small plants and look up into the canopy for larger species.

The type of cover typically estimated by points is total cover. Cover can also be measured within defined vegetation layers, or by different species. If measuring cover, record only the first hit of each pin. Multiple hits for each species can be recorded if measuring biomass, volume, or species composition.

## Sampling Hints and Techniques

Examiners must be able to identify plant species, be familiar with ground cover category codes, and know how to collect cover data using a pin. If only ground cover is being sampled, plant identification skills are not required.

Measuring tapes are made from a variety of materials and are available in varying lengths and increments. Examiners should choose English (metric) tapes for this method and select a tape that is at least as long, or a little longer, than the transect length being sampled. Steel tapes do not stretch and are the most accurate over the life of the tape. Steel is probably the best choice for permanent transects where remeasurement in exactly the same place each time is important. Cloth and fiberglass tapes will stretch over the life of the tape but are easier to use than steel tapes because they are lighter and do not tend to kink.

Point intercept cover is easily calculated if 100 points are sampled per transect. Cover values are equal to the number of "hits" for each item on the transect. Metric tapes are easily divided into 100 intervals, and sampling a 20-m tape at 20-cm intervals is relatively simple. However, if sampling with English tapes marked in inches and feet, sampling 100 points is impractical unless the transect length is a multiple of 100 (50, 100, 200). On a 66-ft transect, 100 points must be placed at 8-inch intervals. It is time consuming to place the point at the right mark on the tape. If a high resolution of cover is desired (for example, 1 percent), one solution is to sample along a 66-ft transect at every 0.5 ft for 132 points. Another solution is to increase the transect length to 100 ft, and place a point at every foot. When deciding how many points to sample per transect, it is better to sample more transects than place more points per transect. Sampling with fewer points and more transects will often sample more variability within the plot at a slightly lower resolution of cover (for example, 1/66 = 1.5 percent versus 1/100 = 1 percent).

The sampling crew may encounter an obstacle, such as a large rock or tree, along one of the transect lines that interferes with the quadrat sampling. If that happens, offset using the directions described in **How To Offset a Transect** in the **How-To Guide** chapter.

When entering data on the PO Transect data forms, examiners will most likely run out of space on the first page. The form was designed to print one copy of the first page, and several copies of the second page. The second page can be used to record more plant species on the first three transects or to record data for additional transects. The second page of the data form allows the examiner to write the transect number on the form. This allows the examiner to design the form to accommodate the number of transects sampled. Print out enough pages to record all species on all transects for the required number of intercepts. The FIREMON data entry screens and database allow a maximum of 10 transects.

When entering data on the PO Point Frames data forms, examiners will most likely run out of space on the first page. As with the PO Transect form, the Frame form was designed to print one copy of the first page and several copies of the second page. The second page can be used to record more plant species on the first five point frames or to record data for additional frames. The second page of the data form allows the examiner to write the frame number on the form. This allows the examiner to design the form to accommodate the number of frames sampled per transect. Print out enough pages to record all species on all transects for the required number of intercepts. The FIREMON PO data entry screens and database allow a maximum of 20 transects per plot.

# POINT INTERCEPT (PO) FIELD DESCRIPTIONS

**Transect Form**

Field 1: **Number of Transects**. Total number of transects on the plot.

Field 2: **Transect Length.** Length of transect (ft/m).

Field 3: **Number of Points per Transect**. Number of points sampled per transect.

Field 4: **Item Code**. Code of sampled entity. Either the NRCS plants species code, local code for that species, ground cover code, or other item code.

Field 5: **Status**: Plant status—Live, Dead, or Not Applicable. (L, D, NA). Precision: No error.

**Hits.** Enter the total number of hits for the item on the transect. Precision: $\pm 3$ percent of total

**Average Height**. Enter the average height for each plant species or life-form on the transect (ft/m). Precision: $\pm 10$ percent mean height.

**Point Frame Form**

Field 1: **Number of Transects**. Total number of transects on the plot.

Field 2: **Transect Length**. Length of transect (ft/m).

Field 3: **Number of Frames per Transect**. Number of frames sampled per transect.

Field 4: **Number of Points per Frame**. Number of points sampled per frame.

Field 5: **Transect Number**. Sequential number of the sample transect.

Field 6: **Item Code**. Code of sampled entity. Either the NRCS plants species code, local code for that species, ground cover code, or other item code.

Field 7: **Status**: Plant status—Live, Dead, or Not Applicable. (L, D, NA). Precision: No error.

**Hits**. Enter the total number of hits for the item in the frame. Precision: $\pm 3$ percent of total.

**Average Height**. Enter the average height for each plant species or life-form in the frame (ft/m). Precision: $\pm 10$ percent mean height.

# FIREMON PO Cheat Sheet

## Ground cover codes

| Code | Description | Code | Description |
|------|-------------|------|-------------|
| ASH  | Ash (organic, from fire) | LICH | Lichen |
| BAFO | Basal forb | LITT | Litter and duff |
| BAGR | Basal graminoid | MEGR | Medium gravel (5 20 mm) |
| BARE | Bare soil (soil particles <2 mm) | MOSS | Moss |
| BARR | Barren | PAVE | Pavement |
| BASH | Basal shrub | PEIC | Permanent ie |
| BATR | Basal tree | PEIS | Permanent ice and snow |
| BAVE | Basal vegetation | PESN | Permanent snow |
| BEDR | Bedrock | ROAD | Road |
| BOUL | Boulders (round and flat) | ROBO | Round boulder (>600 mm) |
| CHAN | Channers (2 150 mm long) | ROCK | Rock |
| CHAR | Char | ROST | Round stone (250 600 mm) |
| CML  | Cryptogams, mosses and lichens | STON | Stones (round and flat) |
| COBB | Cobbles (75 250 mm) | TEPH | Tephra volcanic |
| COGR | Coarse gravel (20 75 mm) | TRIC | Transient ice |
| CRYP | Cryptogamic crust | TRIS | Transient ice and snow |
| DEVP | Developed Land | TRSN | Transient snow |
| FIGR | Fine gravel (2 5 mm) | UNKN | Unknown |
| FLAG | Flag stones (150 380 mm long) | WATE | Water |
| FLBO | Flat boulders (>600 mm long) | WOOD | Wood |
| FLST | Flat stone (380 600 mm long) | X | Did not assess |
| GRAV | Gravel (2 75 mm) | | |

## Point cover (PO) equipment list

| | |
|---|---|
| Camera with film | Maps, charts and directions |
| PO data forms | Map protector or plastic bag |
| Clipboard | Magnifying glass |
| Compass | Pocket calculator |
| File | Point frame (pn or grid) (optional) |
| Field notebook | Pole 0.25 inch diameter |
| Graph paper | Plot sheet protector or plastic bag |
| Hammer | Rebar stakes (2) for baseline plus 2 for each transect |
| Indelible ink pen (Sharpie, Marker) | Steel fence post (2) and driver (to mark endpoints of baseline) |
| Lead pencils with lead refills | Tape 75 ft (25 m) or longer (2) |

## Precision

| Component | Standard |
|-----------|----------|
| Cover | ±1 class |
| NRF | No error |
| Height | ± 10 percent |

## Status codes

| Code | Description |
|------|-------------|
| L | Live |
| D | Dead |
| NA | Not applicable |

## Canopy cover classes

| Code | Cover |
|------|-------|
|      | Percent |
| 0    | Zero |
| 0.5  | >0 1 |
| 3    | >1 5 |
| 10   | >5 15 |
| 20   | >15 25 |
| 30   | >25 35 |
| 40   | >35 45 |
| 50   | >45 55 |
| 60   | >55 65 |
| 70   | >65 75 |
| 80   | >75 85 |
| 90   | >85 95 |
| 98   | >95 100 |

# Point Intercept (PO) Form
## Transects

RegistrationID: _ _ _ _ _
ProjectID: _ _ _ _ _
PlotID: _ _ _ _
Date: _ _ / _ _ / _ _ _ _

PO Page _ _ _ of _ _ _

| Field 1 | Field 2 | Field 3 |
|---|---|---|
| Number of Transects | Transect Length | Number of Pts./Tran. |
| | | |

| Field 4 | Field 5 | Transect 1 | | Transect 2 | | Transect 3 | | Transect 4 | | Transect 5 | |
|---|---|---|---|---|---|---|---|---|---|---|---|
| Item Code | Status | Hits | Height (ft/m) | Hits | Height (ft/m) | Hits | Height (ft/m) | Hits | Height (ft/m) | Hits | Height (ft/m) |
| Dot Tally Space → | | | | | | | | | | | |
| | | | | | | | | | | | |
| | | | | | | | | | | | |
| | | | | | | | | | | | |
| | | | | | | | | | | | |
| | | | | | | | | | | | |
| | | | | | | | | | | | |
| | | | | | | | | | | | |

Notes:

Crew:

# Point Intercept (PO) Form
## Transect continuation

RegistrationID: _ _ _ _ _ _
ProjectID: _ _ _ _ _ _
PlotID: _ _ _ _
Date: _ _ / _ _ / _ _ _ _

PO Page _ _ _ of _ _ _

Plot Key

| Field 1 | Field 2 | Field 3 |
|---|---|---|
| Number of Transects | Transect Length | Number of Pts./Tran. |
| | | |

| Field 4 | Field 5 | Transect _ | | Transect _ | | Transect _ | | Transect _ | | Transect _ | | Transect _ | |
|---|---|---|---|---|---|---|---|---|---|---|---|---|---|
| Item Code | Status | Hits | Height (ft/m) | Hits | Height (ft/m) | Hits | Height (ft/m) | Hits | Height (ft/m) | Hits | Height (ft/m) | Hits | Height (ft/m) |
| Dot Tally Space ⟶ | | | | | | | | | | | | | |
| | | | | | | | | | | | | | |
| | | | | | | | | | | | | | |
| | | | | | | | | | | | | | |
| | | | | | | | | | | | | | |
| | | | | | | | | | | | | | |
| | | | | | | | | | | | | | |
| | | | | | | | | | | | | | |
| | | | | | | | | | | | | | |

Notes:

Crew:

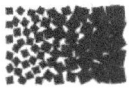

# Point Intercept (PO) Form
## Point Frames

PO Page _ _ of _ _ _

| Field 1 | Field 2 | Field 3 | Field 4 |
|---|---|---|---|
| Number of Transects | Transect Length | Number of Frames/Trans | Number of Pts./Frame |
|  |  |  |  |

RegistrationID: _ _ _ _ _
ProjectID: _ _ _ _ _
PlotID: _ _ _ _
Date: _ _ / _ _ / _ _ _ _

Plot Key

| Field 5 | Field 6 | Field 7 | Frame 1 | | Frame 2 | | Frame 3 | | Frame 4 | | Frame 5 | |
|---|---|---|---|---|---|---|---|---|---|---|---|---|
| Transect Number | Item Code | Status | Hits | Height (ft/m) | Hits | Height (ft/m) | Hits | Height (ft/m) | Hits | Height (ft/m) | Hits | Height (ft/m) |
|  | Dot Tally Space → |  |  |  |  |  |  |  |  |  |  |  |
|  |  |  |  |  |  |  |  |  |  |  |  |  |
|  |  |  |  |  |  |  |  |  |  |  |  |  |
|  |  |  |  |  |  |  |  |  |  |  |  |  |
|  |  |  |  |  |  |  |  |  |  |  |  |  |
|  |  |  |  |  |  |  |  |  |  |  |  |  |
|  |  |  |  |  |  |  |  |  |  |  |  |  |
|  |  |  |  |  |  |  |  |  |  |  |  |  |
|  |  |  |  |  |  |  |  |  |  |  |  |  |
|  |  |  |  |  |  |  |  |  |  |  |  |  |
|  |  |  |  |  |  |  |  |  |  |  |  |  |

Notes:

Crew:

# Point Intercept (PO) Form
## Point Frames Continuation

RegistrationID: _ _ _ _ _ _
ProjectID: _ _ _ _
PlotID: _ _ _
Date: _ _ / _ _ / _ _ _ _

Plot Key

PO Page _ _ _ of _ _ _

| Field 1 | Field 2 | Field 3 | Field 4 |
|---|---|---|---|
| Number of Transects | Transect Length | Number of Frames/Trans | Number of Pts./Frame |
|  |  |  |  |

| Field 5 | Field 6 | Field 7 | Frame_ | | Frame_ | | Frame_ | | Frame_ | | Frame_ | | Frame_ | |
|---|---|---|---|---|---|---|---|---|---|---|---|---|---|---|
| Transect Number | Item Code | Status | Hits | Height (ft/m) | Hits | Height (ft/m) | Hits | Height (ft/m) | Hits | Height (ft/m) | Hits | Height (ft/m) | Hits | Height (ft/m) |
| | Dot Tally Space → | | | | | | | | | | | | | |
| | | | | | | | | | | | | | | |
| | | | | | | | | | | | | | | |
| | | | | | | | | | | | | | | |
| | | | | | | | | | | | | | | |
| | | | | | | | | | | | | | | |
| | | | | | | | | | | | | | | |
| | | | | | | | | | | | | | | |
| | | | | | | | | | | | | | | |
| | | | | | | | | | | | | | | |
| | | | | | | | | | | | | | | |

Notes:

Crew:

# Density (DE)

## Sampling Method

John F. Caratti

## SUMMARY

The FIREMON Density (DE) method is used to assess changes in plant species density and height for a macroplot. This method uses multiple quadrats and belt transects (transects having a width) to sample within plot variation and quantify statistically valid changes in plant species density and height over time. Herbaceous plant species are sampled with quadrats while shrubs and trees are sampled with belt transects. Quadrats for sampling herbaceous plants are placed systematically along randomly located transects. Belt transects for sampling shrub and tree density use the same randomly located transects. The number of individuals for each plant species in a quadrat or belt transect are calculated. Density is calculated as the number of individuals per unit area using the area of the sampling unit, quadrat, or belt transect. This method is primarily suited for grasses, forbs, shrubs, and small trees in which individual plants or stems can be distinguished. However, we recommend using the FIREMON TD sampling methods for estimating tree density.

## INTRODUCTION

The Density (DE) method was designed to sample within-plot variation and quantify changes in plant species density and height over time. This method uses multiple quadrats to sample herbaceous plant density and belt transects to sample shrub and tree density. First, a baseline is established from which to run the transects from. Transects are placed randomly along the baseline. Quadrats for sampling herbaceous plants are then placed systematically along each transect. Belt transects for sampling shrub and tree density are placed along the length of each transect. Characteristics are recorded about the general sample design and for individual plant species. First the transect length, number of transects, and number of quadrats per transect are recorded. Within each quadrat or belt transect, density and average height are recorded for each plant species. The quadrat length and width or belt transect width are also recorded for each species. Different size quadrats and belts can be used for different plant species.

This method is primarily used when the fire manager wants to monitor changes in plant species numbers. This method is primarily suited for grasses, forbs, shrubs, and small trees that are easily separated into individual plants or counting units, such as stems. However, we recommend using the FIREMON TD sampling method for estimating tree density. The DE sampling method uses density to assess changes in plant species numbers over time. The quadrat size and belt width varies with plant species or item and size class, allowing different size sampling units for different size plants or items. Quadrat size and belt width should be adjusted according to plant size and distribution.

Density is defined as the number of items per unit area. When sampling density, the examiner must be able to recognize and define individual plants. This may be relatively easy for single-stem plants but is more difficult for plants that reproduce vegetatively, such as rhizomatous plants (western wheatgrass) or clonal species (aspen). It is critical to define exactly what item will be counted before sampling. The item counted may be individual plants for single-stem plants or individual stems for a clonal species, such as quaking aspen.

Density is used for monitoring an increase or decrease in the number of individuals or counting units. Density is more effective for detecting changes in recruitment or mortality than changes in vigor. It is not a practical monitoring method when plants respond to management treatments or disturbance with decreased cover rather than mortality. In such cases, density may not change much although cover and biomass may change considerably.

The accuracy of the density estimate depends largely on the size and shape of the quadrat or belt transect. Note that a belt transect is essentially a long, narrow quadrat. Pound and Clements (1898) considered plant dispersion, quadrat size and shape, and number of observations required as important characteristics to sample design. Van Dyne and others (1963) reviewed results from studies on quadrat sizes and shapes used to sample grassland communities of the Western United States. Long narrow quadrats tend to include more species because vegetation tends to occur in clumps rather than be randomly distributed. The desired size and shape for a quadrat depends largely on the distribution of plant species being sampled. In general, sampling in sparse vegetation requires the use of larger quadrats. However, quadrats should not be too large, because counting large numbers of plants in a quadrat can be overwhelming and can lead to errors. In order to increase the sampling area, it is better to sample more quadrats than to use overly large quadrats.

Small quadrats pose a greater chance of boundary error because of the greater perimeter to area ratio. Boundary problems are due to erroneously including or excluding plants near the quadrat perimeter. Some portion of basal vegetation must intersect the quadrat boundary for a plant to be considered a boundary plant. Boundary rules must be established before sampling. We recommend counting boundary plants "in" on two adjacent sides of a quadrat and "out" on the other adjacent sides. See **How to Count Boundary Plants** in the **How-To Guide** chapter.

There are many ways to streamline or customize the DE sampling method. The FIREMON three-tier sampling design can be employed to optimize sampling efficiency. See the sections on **User-Specific DE Sampling Design** and **Sampling Design Customization** below.

## SAMPLING PROCEDURE

This method assumes that the sampling strategy has already been selected and the macroplot has already been located. If this is not the case, then refer to the FIREMON **Integrated Sampling Strategy** for further details.

The sampling procedure is described in the order of the fields that need to be completed on the **DE data form**, so it is best to reference the DE data form when reading this section. The sampling procedure described here is the recommended procedure for this method. Later sections will describe how the FIREMON three-tier sampling design can be used to modify the recommended procedure to match resources, funding, and time constraints.

In the **How-To Guide** chapter, see **How To Locate a FIREMON Plot**, **How To Permanently Establish a FIREMON Plot** and **How to Define the Boundaries of a Macroplot** for more information on setting up your macroplot.

## Preliminary Sampling Tasks

Before setting out for your field sampling, lay out a practice area with easy access. Try to locate an area with the same species or vegetation life form you plan on sampling. Get familiar with the plot layout and the data that will be collected. This will give you a chance to assess the method and will help you think about problems that might be encountered in the field. For example, how close do two bunchgrasses have to be before they are counted as one? It is better to answer these questions before the sampling begins so that you are not wasting time in the field. This will also let you see if there are any pieces of equipment that will need to be ordered.

Many preparations must be made before proceeding into the field for DE sampling. First,

all equipment and supplies in the **DE Equipment List** must be purchased and packed for transport into the field. Travel to FIREMON plots is usually by foot, so it is important that supplies and equipment be placed in a comfortable daypack or backpack. It is also important that there be spares of each piece of equipment so that an entire day of sampling is not lost if something breaks. Spare equipment can be stored in the vehicle rather than the backpack. Be sure that all equipment is well maintained and there are plenty of extra supplies such as data forms, map cases, and pencils.

All DE data forms should be copied onto waterproof paper because inclement weather can easily destroy valuable data recorded on standard paper. Data forms should be transported into the field using a plastic, waterproof map protector or plastic bag. The day's sample forms should always be stored in a dry place (office or vehicle) and not be taken back into the field for the next day's sampling.

We recommend that one person on the field crew, preferably the crew boss, have a waterproof, lined field notebook for recording logistic and procedural problems encountered during sampling. This helps with future remeasurements and future field campaigns. All comments and details not documented in the FIREMON sampling methods should be written in this notebook. For example, snow on the plot might be described in the notebook, which would be helpful in plot remeasurement.

It is beneficial to have plot locations for several days of work in advance in case something happens, such as if the road to one set of plots is washed out by flooding. Locations and/or directions to the plots you will be sampling should be readily available to reduce travel times. If the FIREMON plots were randomly located within the sampling unit, it is critical that the crew is provided plot coordinates before going into the field. Plots should be referenced on maps and aerial photos using pin-pricks or dots to make navigation easy for the crew and to provide a check of the georeferenced coordinates. We found that it is easy to transpose UTM coordinate digits when recording georeferenced positions on the plot sheet, so marked maps can help identify any erroneous plot positions. If possible, the spatial coordinates should be provided if FIREMON plots were randomly located.

A field crew of two people is probably the most efficient for implementation of the DE sampling method. There should never be a one-person field crew for safety reasons, and any more than two people will probably result in some people waiting for critical tasks to be done. The crew boss is responsible for all sampling logistics including the vehicle, plot directions, equipment, supplies, and safety. The crew boss should be the note taker, and the technician should perform most quadrat measurements. The initial sampling tasks of the field crew should be assigned based on field experience, physical capacity, and sampling efficiency, but sampling tasks should be modified as the field crew gains experience. Tasks should also be shared to limit monotony.

## Designing the DE Sampling Design

A set of general criteria recorded on the DE data form allows the user to customize the design of the DE sampling method so that the sampling captures the information needed to successfully complete the management objective within time, money, and personnel constraints. These general fields should be determined before the crews go into the field and should reflect a thoughtful analysis of the expected problems and challenges in the fire monitoring project.

## Plot ID construction

A unique plot identifier must be entered on the DE sampling form. This is the same plot identifier used to describe general plot characteristics in the Plot Description or PD sampling method. Details on constructing a unique plot identifier are discussed in the **How to Construct a Unique Plot Identifier** section in the **How-To Guide** chapter. Enter the plot identifier at the top of the DE data form.

## Determining sample size

The size of the macroplot ultimately determines the length of the transects and the length of the baseline along which the transects are placed. The amount of variability in plant species composition and distribution determines the number and length of transects and the number of quadrats required for sampling. The typical macroplot sampled in the DE method is a 0.10 acre (0.04 ha) square measuring 66 x 66 ft (20 x 20 m), which is sufficient for most monitoring applications. The size of the macroplot may be adjusted to accommodate different numbers and lengths of transects. However, it is more efficient if you use the same plot size for all FIREMON sampling methods on the plot.

If you are sampling shrubs and trees, we recommend sampling five belt transects within the macroplot. This should be sufficient for most studies. However, there are situations when more transects should be sampled. See **How To Determine Sample Size** in the **How-To Guide** chapter for more details. Enter the number of transects in Field 1 on the DE Belt Transect data form. The recommended transect length is 66 ft (20 m) for a 66 x 66 ft (20 x 20 m) macroplot. However, the macroplot size may be adjusted to accommodate longer or shorter transects based on the variability in plant species composition and distribution. The FIREMON DE data form and data entry screen allow a maximum of 10 transects per macroplot.

We recommend sampling at least five quadrats per transect for herbaceous plants, and this should be sufficient for most studies. There are situations when more quadrats should be sampled. Additional quadrats may be sampled by placing more quadrats along a transect or by sampling more transects within the macroplot. Enter the number of quadrats per transect in Field 2 of the DE Quadrat data form. The FIREMON DE Quadrat data form and data entry screen allow a maximum of 25 quadrats per transect.

## Determining the belt transect size and quadrat size

Density is typically recorded in quadrats for herbaceous species and in belt transects for shrubs. We recommend using a 66 x 3 ft (20 x 1 m) belt width for sampling smaller shrubs (< 3 ft, 1 m average diameter) and a 66 x 6 ft (20 x 2 m) belt width for larger shrubs (> 3 ft, 1 m average diameter). Belt length and width may be adjusted to accommodate different sizes and densities of shrubs and trees. Longer, wider transects can be sampled for larger or sparsely distributed shrubs and trees, and shorter, narrower transects can be sampled for smaller, and more dense shrubs and trees. Enter the belt transect length in Field 5 and width in Field 6 of the DE Transect data form. You may vary the belt length and width for different plant species you encounter on the macroplot, but be sure to enter the appropriate length and width for each species on the data form.

We recommend using 3 x 3 ft (1 x 1 m) quadrats for sampling herbaceous vegetation. However, quadrat size and transect length may be adjusted to accommodate the size and spacing of the plants being sampled. Larger plants can be counted in larger quadrats on longer transects and smaller plants in smaller quadrats on shorter transects. See **How To Construct a Quadrat Frame** in the **How-To Guide** chapter for instructions on building and using quadrat frames. Enter the quadrat length and width in feet (m) in Fields 7 and 8 on the DE Quadrat data form. Quadrat length and width is entered by item and size class and may vary for different items and size classes. Be sure to record all of the length and width measurements on the data form.

In the Metadata table record the criteria used to determine the DE transect, quadrat, and belt sizes, if they are different than recommended here.

## Conducting DE Sampling Tasks

### Establish the baseline for transects

Once the plot has been monumented, a permanent baseline is set up as a reference from which you will orient all transects. The baseline should be established so that the sampling plots for all of the methods overlap as much as possible. See **How To Establish Plots with Multiple Methods** in the **How-To Guide** chapter. The recommended baseline is 66 ft (20 m) long and is oriented upslope with the 0-ft (0-m) mark at the lower permanent marker and the 66-ft (20-m) mark at the upper marker. On flat areas, the baseline runs from south to north with the 0-ft (0-m) mark on the south end and the 66-ft (20-m) mark on the north end. See **How To Establish a Baseline for Transects** in the **How-To Guide** chapter for more details.

### Locating the transects

Transects are placed perpendicular to the baseline and are sampled starting at the baseline. On flat areas, transects are located from the baseline to the east. Starting locations for each transect are determined using the FIREMON random transect locator or from supplied tables. If the CF method is used in conjunction with other replicated sampling methods (LI, PO, RS, or CF), use the same transects for all methods, if possible. In successive remeasurement years, it is essential that transects be placed in the same location. See **How To Locate Transects and Quadrats** in the **How-To Guide** chapter for more details.

### Locating the quadrats

There will typically be five 3 x 3-ft (1 x 1-m) quadrats located at 12-ft (4-m) intervals along a transect, with the first quadrat placed 12 ft (4 m) from the baseline. However, the spacing of the quadrats will depend on the size of the quadrat and the length of the transect. See **How To Locate Transects and Quadrats** in the **How-To Guide** chapter for more details. If macroplots are being sampled for permanent remeasurement, quadrats must be placed in the same location in successive sampling. When sampling macroplots that are not scheduled for permanent remeasurement, the distance between quadrats may be estimated by pacing.

## Density Sampling

When sampling herbaceous species, enter the transect number in Field 3 on the DE Quadrat data form. The transect number is not entered on the DE Transect data form for sampling shrub or tree species using belt transects. Enter the species code or item code (moose pellets) in Field 4 on the DE Quadrat data form or Field 2 on the DE Transect data form. FIREMON uses the NRCS plants species codes. However, you may use your own species codes. See **Populating the Plant Species Codes Lookup Table** in the **FIREMON Database User Manual** for more details.

Next enter the plant species status in Field 5 on the DE Quadrat data form and Field 3 on the DE Transect data form. Status describes the general health of the plant species as live or dead using the following codes:

**L—Live:** plant with living tissue.

**D—Dead:** plant with no living tissue visible.

**NA—Not Applicable**.

Plant status is purely qualitative, but it does provide an adequate characteristic for stratification of preburn plant health and in determining postburn survival. Use care in determining plant status during the dormant season.

## Size class

Plant species size classes represent different layers in the canopy. For example, the upper canopy layer could be defined by large trees, while pole-size trees and large shrubs might dominate the middle layer of the canopy, and the lower canopy layer could include seedlings, saplings, grasses, and forbs. Size class data provide important structural information such as the vertical distribution of plant cover. Size classes for trees are typically defined by height for seedlings and diameter at breast height (DBH) for larger trees. Size classes for shrubs, grasses, and forbs are typically defined by height. If the vegetation being sampled has a layered canopy structure, then density can be recorded by plant species and by size class.

FIREMON uses a size class stratification based on the ECODATA sampling methods (Jensen and others 1994). Group individual plants by species into one or more trees size classes (table DE-1) or shrub, grass, and forb size classes (table DE-2). There can be multiple size classes for each species. In the **How-To Guide** chapter, see **How To Measure DBH** for detailed information on measuring DBH to group trees into size classes and see **How to Measure Plant Heights** for detailed information on measuring height for grouping shrubs into size classes.

If recording density by size class, enter the size class code for each plant species in Field 6 on the DE Quadrat data form and Field 4 on the DE Transect data form. If size class data are not recorded, indicate that by entering the code "TO."

## Density

Record the number of individual plants for each plant species or individual plants for each plant species by size class within the quadrat (fig. DE-1). Decisions about counting boundary plants (plants that have a portion of basal vegetation intersecting the quadrat) need to be applied systematically to each quadrat. See **How to Count Boundary Plants** in the **How-To Guide** chapter for more details. Enter the count for each plant species or plant species by size class in the Count field by quadrat on the DE Quadrat data form or by transect on the DE Transect data form. Use the workspace below each Count field for a dot tally. See **How to Dot Tally** in the **How-To Guide** chapter for more details.

Table DE-1 Tree size class codes.

| Codes | Description (English) | Description (metric) |
|---|---|---|
| | **Tree size class** | |
| TO | Total count | Total count |
| SE | Seedling (<1 inch DBH or <4.5 ft height) | Seedling (<2.5 cm DBH or <1.5 m height) |
| SA | Sapling (1.0 inch <5.0 inches DBH) | Sapling (2.5 <12.5 cm DBH) |
| PT | Pole tree (5.0 inches <9.0 inches DBH) | Pole tree (12.5 <25 cm DBH) |
| MT | Medium tree (9.0 inches <21.0 inches DBH) | Medium tree (25 <50 cm DBH) |
| LT | Large tree (21.0 inches <33.0 inches DBH) | Large tree (50 <80 cm DBH) |
| VT | Very large tree (>33.0 inches DBH) | Very large tree (>80 cm DBH) |
| NA | Not applicable | Not applicable |

Table DE-2 Shrub, grass, and forb size class codes.

| Codes | Description (English) | Description (metric) |
|---|---|---|
| | **Shrub/herb size class** | |
| TO | Total count | Total count |
| SM | Small (<0.5 ft height) | Small (<0.15 m height) |
| LW | Low (0.5 <1.5 ft height) | Low (0.15 <0.5 m height) |
| MD | Medium (1.5 <4.5 ft height) | Medium (0.5 <1.5 m height) |
| TL | Tall (4.5 <8 ft height) | Tall (1.5 <2.5 m height) |
| VT | Very tall (>8 ft height) | Very tall (>2.5 m height) |
| NA | Not applicable | Not applicable |

# Density (DE) Sampling Method

**Figure DE-1**—Count individual plants by plant species or plant species by size class. In this example, there are five individuals of a grass species, three individuals of a tree species, and two individuals of a forb species.

## Estimating average height

Measure the average height for each plant species in feet (m) within +/– 10 percent of the mean plant height. If plant species are recorded by size class, measure the average height for individual species by each size class recorded. See **How to Measure Plant Heights** for more details. Enter plant height in the Height field for each transect on the DE Transect data form or for each quadrat on the DE quadrat form.

## Precision Standards

Use the precision standards listed in table DE-3 for the DE sampling.

# SAMPLING DESIGN CUSTOMIZATION

This section will present several ways that the DE sampling method can be modified to collect more detailed information or streamlined to collect basic information.

## Recommended DE Sampling Design

The recommended DE sampling design follows the Alternative FIREMON sampling intensity and is listed below:

**Macroplot Size:** 0.1 acre, 66 x 66 ft (400 $m^2$, 20 x 20 m).

**Quadrat Size (herbaceous plant sampling):** 3 x 3 ft (1 x 1 m).

Table DE-3 Precision guidelines for DE sampling.

| Component | Standard |
|---|---|
| Count | ±10 percent total count |
| Average height | ±10 percent average height |
| Size class | ±1 class |

**Belt Transect Width (shrubs and tree sampling):** 3 ft (1 m) or 6 ft (2 m).

**Number of Transects:** 5.

**Number of Quadrats/Transect:** 5.

**Count plant species within each quadrat**.

The quadrat size and shape and belt transect width should be adjusted according to the plant community being sampled.

The number of transects and quadrats sampled should be adjusted according to the appropriate methods in the **How-To Guide** chapter of the FIREMON manual.

## Streamlined DE Sampling Design

The streamlined DE sampling design follows the Simple FIREMON sampling inensity and is designed below:

**Macroplot Size:** 0.1 acre, 66 x 66 ft (400 m$^2$, 20 x 20 m).

**Quadrat Size (herbaceous plant sampling):** 3 x 3 ft (1 x 1 m).

**Belt Transect Width (shrubs and tree sampling):** 3 ft (1 m) or 6 ft (2 m).

**Number of Transects:** 3.

**Number of Quadrats/Transect:** 5.

**Count plant species within each quadrat**.

The quadrat size and shape and belt transect width should be adjusted according to the plant community being sampled.

The number of transects and quadrats sampled should be adjusted according to the appropriate methods in the **How-To Guide** chapter of the FIREMON manual.

## Comprehensive DE Sampling Design

The comprehensive DE sampling design follows the Detailed FIREMON sampling intensity and is detailed below:

**Macroplot Size:** 0.1 acre, 66 x 66 ft (400 m$^2$, 20 x 20 m).

**Quadrat Size (herbaceous plant sampling):** 3 x 3 ft (1 x 1 m).

**Belt Transect Width (shrubs and tree sampling):** 3 ft (1 m) or 6 ft (2 m).

**Number of Transects:** 5.

**Number of Quadrats/Transect:** 5.

**Count plant species by size class and measure average plant species height by size class within each quadrat**.

The quadrat size and shape and belt transect width should be adjusted according to the plant community being sampled.

The number of transects and quadrats sampled should be adjusted according to the appropriate methods in the **How-To Guide** chapter of the FIREMON manual.

## User-Specific DE Sampling Design

The user can modify the DE sample fields a number of ways in order to make sampling more efficient and meaningful for local situations. This will usually mean adjusting the number of transects, transect length, transect width, quadrat size, or number of quadrats as needed for the specific task. Use the MD form to record any changes in sampling methods that are modified from the standard or to remark on any other DE matter that needs to be explained or defined for subsequent sampling and data use.

Many different sizes and shapes of quadrats and belt transects can be used to sample density. It is probably most efficient to conduct a pilot study to determine the size and shape of quadrat or belt transect that allows number of plants to be easily counted and also minimizes the variance between sampling units. See Elzinga and others (1998) for a detailed discussion on this topic. Plant species density may be sampled using quadrats of various sizes, belt transects of various sizes, or a combination of both belts and quadrats. If circular plots are being used to count plants, then enter the radius of the plot in the quadrat length field and leave the quadrat width field blank on the DE data forms. The FIREMON sampling forms and databases will accommodate most sampling variations of the DE method.

## Sampling Hints and Techniques

Examiners must be able to identify plant species, identify individual plants, and be able to determine whether a plant species occurs within a quadrat. It can be difficult to distinguish individual plants for some species such as sod-forming grasses. If individual plants are difficult to identify, guidelines should be determined before sampling as to what constitutes the individual counting unit. Some examples include counting individual stems in aspen communities, culm groups in rhizomatous grasses, and flowering stems for mat-forming forbs. However, the counting unit chosen to monitor should reflect a real change in the plant community.

Measuring tapes are made from a variety of materials and are available in varying lengths and increments. Examiners should choose English (metric) tapes for this method and select a tape that is at least as long, or a little longer, than the transect length being sampled. Steel tapes do not stretch and are the most accurate over the life of the tape. Steel is probably the best choice for permanent transects where remeasurement in exactly the same place each time is important. Cloth and fiberglass tapes will stretch over the life of the tape but are easier to use than steel tapes because they are lighter and do not tend to kink.

The sampling crew may encounter an obstacle, such as a large rock or tree, along one of the transect lines that interferes with the quadrat sampling. If that happens, offset using the directions described in **How To Offset a Transect** in the **How-To Guide** chapter.

When entering data on the DE Transect data forms, examiners will likely run out of space on the first page. The form was designed to print one copy of the first page and several copies of the second page. The second page can be used to record more plant species on the first three transects or to record data for additional transects. The second page of the data form allows the examiner to write the transect number on the form. This allows the examiner to design the form to accommodate the number of transects sampled. Print out enough pages to record all species on all transects for the required number of intercepts. The FIREMON data entry screens and database allow a maximum of 10 transects.

When entering data on the DE Quadrat data forms, examiners will likely run out of space on the first page. If so, use the DE continuation form. The continuation page can be used to record more plant species on the first five quadrats or to record data for additional frames. The continuation page allows the examiner to write the quadrat number on the form, thus allowing the examiner to design the form to accommodate the number of frames sampled per transect. Print out enough pages to record all species on all transects for the required number of counts. The FIREMON data entry screens and database allow a maximum of 20 frames per transect and an unlimited number of transects.

# DENSITY (DE) FIELD DESCRIPTIONS

## Quadrat Form

Field 1: **Number of Transects**. Total number of transects on the plot.

Field 2: **Number of Quadrats per Transect**. Number of quadrats sampled per transect.

Field 3: **Transect Number**. Sequential number of the sample transect.

Field 4: **Item Code**. Code of sampled entity. Either the NRCS plants species code or the local code for that species. Precision: No error.

Field 5: **Status**: Plant status—Live, Dead, or Not Applicable. (L, D, NA). Precision: No error.

Field 6: **Size Class**. Size of the sampled plant. Valid classes are in tables DE-1 and DE-2 of the sampling method. Precision: $\pm 1$ class.

Field 7: **Quadrat Length**. Length of the quadrat. May be different for different species/life forms (ft/m).

Field 8: **Quadrat Width**. Width of the quadrat. May be different for different species/life forms (ft/m).

**Count**. Total number of individuals for the plant species or lifeform inside the transect. Precision: $\pm 10$ percent of total count.

**Average Height**. Average height for each plant species or lifeform in transect (ft/m). Precision: $\pm 10$ percent mean height.

## Belt Transect Form

Field 1: **Transect Number**. Sequential number of the sample transect.

Field 2: **Item Code**. Code of sampled entity. Either the NRCS plants species code or the local code for that species. Precision: No error.

Field 3: **Status**: Plant status—Live or Dead. Precision: No error.

Field 4: **Size Class**. Size of the sampled plant. Valid classes are in tables DE-1 and DE-2 of the sampling method. Precision: $\pm 1$ class.

Field 5: **Transect Length**. Length of transect. May be different for different species/lifeforms (ft/m).

Field 6: **Transect Width**. Width of transect. May be different for different species/lifeforms (ft/m).

**Count**. Enter the total number of individuals for the plant species or life-form inside the transect. Precision: $\pm 10$ percent total count.

**Average Height**. Enter the average height for each plant species or lifeform in transect (ft/m). Precision: $\pm 10$ percent mean height.

#  FIREMON DE Cheat Sheet

### Status codes

| Code | Description |
|------|-------------|
| L    | Live |
| D    | Dead |
| NA   | Not Applicable |

### Precision

| Component | Standard |
|-----------|----------|
| Count | ±10 percent total count |
| Average height | ±10 percent average height |
| Size class | ±1 class |

### Tree size classes

| | Tree size class | |
|---|---|---|
| Codes | Description (English) | Description (metric) |
| TO | Total count | Total count |
| SE | Seedling (<1 inches DBH or <4.5 ft height) | Seedling (<2.5 cm DBH or <1.5 m height) |
| SA | Sapling (1.0 inches < 5.0 inches DBH) | Sapling (2.5 <12.5 cm DBH) |
| PT | Pole tree (5.0 inches <9.0 inches DBH) | Pole tree (12.5 <25 cm DBH) |
| MT | Medium tree (9.0 inches <21.0 inches DBH) | Medium tree (25 <50 cm DBH) |
| LT | Large tree (21.0 inches <33.0 inches DBH) | Large tree (50 <80 cm DBH) |
| VT | Very large Tree (>33.0 inches DBH) | Very large Tree (>80 cm DBH) |
| NA | Not applicable | Not applicable |

### Shrub and herbaceous size classes

| | Shrub/herb size class | |
|---|---|---|
| Codes | Description (English) | Description (metric) |
| TO | Total count | Total count |
| SM | Small (<0.5 ft height) | Small (<0.15 m height) |
| LW | Low (0.5 <1.5 ft height) | Low (0.15 <0.5 m height) |
| MD | Medium (1.5 <4.5 ft height) | Medium (0.5 <1.5 m height) |
| TL | Tall (4.5 <8 ft height) | Tall (1.5 <2.5 m height) |
| VT | Very tall (>8 ft height) | Very tall (>2.5 m height) |
| NA | Not applicable | Not applicable |

### Density (DE) equipment list

| | |
|---|---|
| Camera, film and flash | Maps, charts and directions |
| DE data forms | Map protector or plastic bag |
| Clipboard | Magnifying glass |
| Compass | Pocket calculator |
| Field notebook | Plot sheet protector or plastic bag |
| File | Rebar (2) for baseline plus 2 for each transect |
| Graph paper | Steel fence posts (2) and driver (to mark endpoints of baseline) |
| Hammer | Tape 75 ft (25 m) or longer (3) |
| Indelible ink pen (e.g., Sharpie, Marker) | Folding rulers 6 ft (2 m) (2 3) |
| Lead pencils with lead refills | |

# Density (DE) Form
## Quadrats

DE Page _ _ of _ _ _

RegistrationID: _ _ _ _ _
ProjectID: _ _ _ _ _ _
PlotID: _ _ _ _
Date: _ _ / _ _ / _ _ _ _

Plot Key

| Field 1 | Field 2 |
|---|---|
| Number of Transects | Number of Quad/Tran |
|  |  |

| Field 3 | Field 4 | Field 5 | Field 6 | Field 7 | Field 8 | Quadrat 1 | | Quadrat 2 | | Quadrat 3 | | Quadrat 4 | | Quadrat 5 | |
|---|---|---|---|---|---|---|---|---|---|---|---|---|---|---|---|
| Transect Number | Item Code | Status | Size Class | Quadrat Length (ft/m) | Quadrat Width (ft/m) | Count | Height (ft/m) | Count | Height (ft/m) | Count | Height (ft/m) | Count | Height (ft/m) | Count | Height (ft/m) |

Dot Tally Workspace →

Notes:

Crew:

# Density (DE) Form
## Quadrats - continuation

DE Page _ _ of _ _

**Plot Key**
- RegistrationID: _ _ _ _ _ _
- ProjectID: _ _ _ _ _
- PlotID: _ _ _ _
- Date: _ _ / _ _ / _ _

| Field 1 | Field 2 |
|---|---|
| Number of Transects | Number of Quad/Tran |
| | |

Notes:

Crew:

| Field 3 | Field 4 | Field 5 | Field 6 | Field 7 | Field 8 | Quadrat 1 | | Quadrat 2 | | Quadrat 3 | | Quadrat 4 | | Quadrat 5 | |
|---|---|---|---|---|---|---|---|---|---|---|---|---|---|---|---|
| Transect Number | Item Code | Status | Size Class | Quadrat Length (ft/m) | Quadrat Width (ft/m) | Count | Height (ft/m) | Count | Height (ft/m) | Count | Height (ft/m) | Count | Height (ft/m) | Count | Height (ft/m) |
| | | | | | Dot Tally Workspace ↑ | | | | | | | | | | |
| | | | | | | | | | | | | | | | |
| | | | | | | | | | | | | | | | |
| | | | | | | | | | | | | | | | |
| | | | | | | | | | | | | | | | |
| | | | | | | | | | | | | | | | |
| | | | | | | | | | | | | | | | |
| | | | | | | | | | | | | | | | |
| | | | | | | | | | | | | | | | |

# Density (DE) Form
## Belt Transects

DE Page _ _ of _ _

RegistrationID: _ _ _ _ _
ProjectID: _ _ _ _ _
PlotID: _ _ _ _
Date: _ _ / _ _ / _ _ _ _

Plot Key

| Field 1 |
|---|
| Number of Transects |
| |

| Field 2 | Field 3 | Field 4 | Field 5 | Field 6 | Transect 1 | | Transect 2 | | Transect 3 | | Transect 4 | | Transect 5 | |
|---|---|---|---|---|---|---|---|---|---|---|---|---|---|---|
| Item Code | Status | Size Class | Transect Length (ft/m) | Transect Width (ft/m) | Count | Height (ft/m) | Count | Height (ft/m) | Count | Height (ft/m) | Count | Height (ft/m) | Count | Height (ft/m) |
| | | | | ← Dot Tally Workspace | | | | | | | | | | |
| | | | | | | | | | | | | | | |
| | | | | | | | | | | | | | | |
| | | | | | | | | | | | | | | |
| | | | | | | | | | | | | | | |
| | | | | | | | | | | | | | | |
| | | | | | | | | | | | | | | |
| | | | | | | | | | | | | | | |

Notes:

Crew:

# Density (DE) Form
## Belt Transects Continuation

DE Page _ _ _ of _ _ _

**Plot Key**

| Field 1 |
|---|
| Number of Transects |

RegistrationID: _ _ _ _ _ _
ProjectID: _ _ _ _ _ _
PlotID: _ _ _ _
Date: _ _ / _ _ / _ _ _ _

Notes:

Crew:

| Field 2 | Field 3 | Field 4 | Field 5 | Field 6 | Transect_ | | Transect_ | | Transect_ | | Transect_ | | Transect_ | | Transect_ | |
|---|---|---|---|---|---|---|---|---|---|---|---|---|---|---|---|---|
| Item Code | Status | Size Class | Transect Length (ft/m) | Transect Width (ft/m) | Count | Height (ft/m) | Count | Height (ft/m) | Count | Height (ft/m) | Count | Height (ft/m) | Count | Height (ft/m) | Count | Height (ft/m) |
| | | | | | | | | | | | | | | | | |
| | | | | | | | | | | | | | | | | |
| | | | | | | | | | | | | | | | | |
| | | | | | | | | | | | | | | | | |
| | | | | | | | | | | | | | | | | |
| | | | | | | | | | | | | | | | | |
| | | | | | | | | | | | | | | | | |

←— Dot Tally Workspace —→

# Line Intercept (LI)

## Sampling Method

John F. Caratti

## SUMMARY

The FIREMON Line Intercept (LI) method is used to assess changes in plant species cover for a macroplot. This method uses multiple line transects to sample within plot variation and quantify statistically valid changes in plant species cover and height over time. This method is suited for most forest and rangeland communities, but is especially useful for sampling shrub cover greater than 3 ft (1 m) tall, because it is difficult to ocularly estimate the cover of tall shrubs. The LI method can be used in conjunction with cover-frequency (CF) transects when vegetation over 3 ft (1 m) exists. Line intercept can also be used to calibrate ocular estimates of shrub cover when the Species Composition (SC) method is used. Cover is recorded as the number of ft (m) intercepted by each species along a transect. Percent cover is calculated by dividing the number of ft (m) intercepted by each species by the total length of the transect.

## INTRODUCTION

The Line Intercept (LI) method is designed to sample within-plot variation and quantify changes in plant species cover and height over time using transects located within the macroplot. Transects have random starting points and are oriented perpendicular to the baseline. First, samplers record the transect length and number of transects. Then along each transect, cover intercept and average height are recorded for each plant species.

This method is primarily used when the fire manager wants to monitor changes in plant species cover and height. This method is primarily designed to sample plant species with dense crowns or large basal areas. The LI method works best in open grown woody vegetation (Western United States shrub communities), especially shrubs greater than 3 ft (1 m) in height. The CF method is generally preferred for sampling herbaceous plant communities with vegetation less than 3 ft (1 m) in height. However, the LI method can be used in junction with the CF method if shrubs greater than 3 ft (1 m) exist on the plot (CF quadrats can be used to sample herbaceous vegetation, then the transect used to locate the quadrats can be used to sample shrubs using the LI methods). This is probably the best method of sampling cover in mixed plant communities with grasses, shrubs, and trees. This method is not well suited for sampling single-stemmed plants or dense grasslands. The PO method is better suited for sampling fine textured herbaceous communities such as dense grasslands and wet meadows. Cover measured with line intercept is less prone to observer bias than ocular estimates of cover in quadrats (CF method). However, if rare plant species are of interest, the CF methods are preferred because rare species are easier to sample with quadrats than with points or lines.

Tansley and Chipp (1926) introduced the line intercept method. A line transect—typically a measuring tape stretched taut on the ground or at a height that just contacts the vegetation canopy—is used to make observations of plant cover. The method consists of measuring the length of intercept for each plant that occurs over or under the tape. If basal cover is of interest, then the tape is placed at ground level. Percent cover is sampled by recording the length of intercept for each plant species measured along a tape by noting the point on the tape where the plant canopy or basal portion begins and the plant canopy or basal portion ends. When these intercept lengths are summed and divided by the total tape length, the result is a percent cover for the plant species along the transect.

The line transect can be any length and, if modified, is usually done so based on the type of vegetation being sampled (Bonham 1989). In general, cover in herbaceous communities can be estimated with short lines (typically less than 50 m), while longer lines (50 m or greater) should be used in some shrub and tree communities. Canfield (1941) recommended using a 15 m transect for areas with 5 to 15 percent cover and using a 30 m line when cover is less than 5 percent. The amount of time needed to measure a transect can also be used to determine the length of the transect (Bonham 1989). Canfield recommended a transect length in which canopy intercepts can be measured by two people in approximately 15 minutes.

The line intercept method is most efficient for plant species that have a dense crown cover (shrubs or matted plants) or have a relatively large basal area (bunch grasses), and is best suited where the boundaries of individual plants are easily determined. Line intercept is not an effective method for estimating the cover of single-stemmed plant species or dense grasslands (rhizomatous species).

Most plant species have some gaps in their canopies, such as bunchgrasses with dead centers or shrubs with large spaces between branches. Because observers treat gaps differently, rules for dealing with gaps must be clearly defined. One solution is for the observer to assume a plant has a closed canopy unless a gap is greater than some predetermined width. We recommend that gaps less than 2 inches (5 cm) be considered part of the canopy.

There are many ways to streamline or customize the LI sampling method. The FIREMON three-tier sampling design can be employed to optimize sampling efficiency. See the sections on **User-Specific LI Sampling Design** and **Sampling Design Customization** below.

## SAMPLING PROCEDURE

This method assumes that the sampling strategy has already been selected and the macroplot has already been located. If this is not the case, then refer to the FIREMON **Integrated Sampling Strategy** and for further details.

The sampling procedure is described in the order of the fields that need to be completed on the **LI data form**, so it is best to reference the form when reading this section. The sampling procedure described here is the recommended procedure for this method. Later sections will describe how the FIREMON three-tier sampling design can be used to modify the recommended procedure to match resources, funding, and time constraints.

In the **How-To Guide** chapter, see **How To Locate a FIREMON Plot, How To Permanently Establish a FIREMON Plot**, and **How to Define the Boundaries of a Macroplot** for more information on setting up your macroplot.

### Preliminary Sampling Tasks

Before setting out for your field sampling, lay out a practice area with easy access. Try to locate an area with the same species or vegetation life form you plan on sampling. Get familiar with the plot layout and the data that will be collected. This will give you a chance to assess the method and will help you think about problems that might be encountered in the field. For example, how will you take into account gaps in the foliage of the same plant? It is better to answer these questions before the sampling begins so that you are not wasting time in the field. This will also let you see if there are any pieces of equipment that will need to be ordered.

A number of preparations must be made before proceeding into the field for LI sampling. First, all equipment and supplies in the **LI Equipment List** must be purchased and packed for transport into the field. Travel to FIREMON plots is usually by foot, so it is important that supplies and equipment be placed in a comfortable daypack or backpack. It is also important that there be spares of each piece of equipment so that an entire day of sampling is not lost if something breaks. Spare equipment can be stored in the vehicle rather than the backpack. Be sure that all equipment is well maintained and there are plenty of extra supplies such as data forms, map cases, and pencils.

All LI data forms should be copied onto waterproof paper because inclement weather can easily destroy valuable data recorded on standard paper. Data forms should be transported into the field using a plastic, waterproof map protector or plastic bag. The day's sample forms should always be stored in a dry place (office or vehicle) and not be taken back into the field for the next day's sampling.

We recommend that one person on the field crew, preferably the crew boss, have a waterproof, lined field notebook for recording logistic and procedural problems encountered during sampling. This helps with future remeasurements and future field campaigns. All comments and details not documented in the FIREMON sampling methods should be written in this notebook. For example, snow on the plot might be described in the notebook, which would be helpful in plot remeasurement.

It is beneficial to have plot locations for several days of work in advance in case something happens, such as if the road to one set of plots is closed. Plots should be referenced on maps and aerial photos using pin-pricks or dots to make navigation easy for the crew and to provide a check of the georeferenced coordinates. We found that it is easy to transpose UTM coordinate digits when recording georeferenced positions on the plot sheet, so marked maps can help identify any erroneous plot positions. If possible, the spatial coordinates should be provided if FIREMON plots were randomly located.

A field crew of two people is probably the most efficient for implementation of the LI sampling method. There should never be a one-person field crew for safety reasons, and any more than two people will probably result in some people waiting for critical tasks to be done and unnecessary trampling on the plot. The crew boss is responsible for all sampling logistics including the vehicle, plot directions, equipment, supplies, and safety. The crew boss should be the note taker, and the technician should perform most point intercept measurements. The initial sampling tasks of the field crew should be assigned based on field experience, physical capacity, and sampling efficiency. Sampling tasks can be modified and shared, to limit monotony, as the field crew gains experience.

## Designing the LI Sampling Method

A set of general criteria recorded on the LI data form allows the user to customize the design of the LI sampling method so that the sampling captures the information needed to successfully complete the management objective within time, money, and personnel constraints. These general fields should be decided before the crews go into the field and should reflect a thoughtful analysis of the expected problems and challenges in the fire monitoring project. However, some of these fields, in particular the number and length of transects, might be adjusted after a pilot study is conducted in the field to determine a sufficient sample size.

### *Plot ID construction*

A unique plot identifier must be entered on the LI sampling form. This is the same plot identifier used to describe general plot characteristics in the Plot Description or PD sampling method. Details on constructing a unique plot identifier are discussed in the **How to Construct a Unique Plot Identifier** section of the **How-To Guide** chapter. Enter the plot identifier at the top of the LI data form.

## Line Intercept (LI) Sampling Method

### Determining the sample size

The size of the macroplot ultimately determines the length of the transects and the length of the baseline along which the transects are placed. The amount of variation in plant species composition and distribution determines the number and length of transects required for sampling. The typical macroplot sampled in the LI method is a 0.10 acre (0.04 ha) square measuring 66 x 66 ft (20 x 20 m), which is sufficient for most monitoring applications. Shrub-dominated ecosystems will generally require larger macroplots when sampling with the LI method. Dr. Rick Miller (Rangeland Ecologist, Oregon State University) has sampled extensively in shrub-dominated systems, and we have included a write-up of his method in **Appendix C: Rick Miller Method for Sampling Shrub Dominated Systems**. If you are not sure of the plot size to use, contact someone who has sampled the same vegetation that you will be sampling. The size of the macroplot should be adjusted to accommodate the size of the vegetation. However, it is more efficient if you use the same plot size for all FIREMON sampling methods on the plot.

The recommended transect length is 66 ft (20 m) for a 66 x 66 ft (20 x 20 m) macroplot. However, the macroplot size may be adjusted to accommodate longer or shorter transects based on the variability in plant species composition and distribution.

We recommend sampling five transects within the macroplot. However, there are situations when more transects should be sampled. See **How To Determine Sample Size** in the **How-To Guide** chapter for more details. Enter the number of transects in Field 1 on the LI data form.

The following section is designed to help the sampling crew lay out the FIREMON LI sampling plot. For simplicity these directions assume that the crew has decided to use the recommended macroplot size. However, the size of the LI plot may need to be modified based on resource constraints and vegetation attributes. Permanent sampling plots will need to be laid out only once. On subsequent sampling visits, field crews will simply need to locate each end of the line intercept transects, stretch the measuring tape between them, and resample the vegetation.

Once the permanent FIREMON plot has been monumented (see **How To Permanently Establish a FIREMON Plot** in the **How-To Guide** chapter) the sampling crew can begin laying out the LI sampling plot, which is accomplished in two steps: 1) locate the baseline and 2) locate the transects where vegetation will be sampled.

## Conducting LI Sampling Tasks

### Locating the baseline for transects

Once the plot has been monumented, a permanent baseline is set up as a reference from which you will orient all transects. The baseline should be established so that the sampling plots for all of the methods overlap as much as possible. See **How To Establish Plots with Multiple Methods** in the **How-To Guide** chapter. The recommended baseline is 66 ft (20 m) long and is oriented upslope with the 0-ft (0-m) mark at the lower permanent marker and the 66-ft (20-m) mark at the upper marker. On flat areas, the baseline runs from south to north with the 0-ft (0-m) mark on the south end and the 66-ft (20-m) mark on the north end. See **How To Establish a Baseline for Transects** in the **How-To Guide** chapter for more details.

### Locating the transects

Transects are located within the macroplot, perpendicular to the baseline and across the slope. For permanent plots, determine the compass bearing of the transects and record it on the plot layout map. All transects should be at the same azimuth. Starting locations for each transect are determined using the FIREMON random transect locator or from supplied tables. If the LI method is used in conjunction with other replicated sampling methods (CF, PO, RS or DE), use the same transects for all methods. In successive remeasurement years, it is essential that transects be placed in the same location. See **How To Locate Transects and Quadrats** in the **How-To Guide** chapter for more details.

Carefully stretch a measuring tape, which represents the transect, from the starting point on the baseline out 66 ft (20 m) at an azimuth perpendicular to the baseline. The measuring tape will be stretched taut and straight on the ground or at a height above the vegetation canopy (fig. LI-1) if measuring crown cover. If basal cover is recorded, then the tape is always placed at ground level.

There are two reasons that the tape must be as straight as possible and not zigzagging around the vegetation. First, a tape stretched straight between the permanently marked transect ends will ensure that the same vegetation sampled during the initial visit will be resampled on subsequent visits. Second, the crown cover estimate could be biased if the tape is bent around the stems because more of the tape lies under the plant canopy.

If the tape is to be stretched above vegetation, the crew will need some way to hold it taught and above the canopy. One method is to drive a rebar at each end of the transect then slip a piece of metal electrical conduit over the bar and attach the tape ends to the conduit with wire, hooks, or tape. Rebar is an excellent way to permanently mark the ends of the transects. However, it should not be left in place if there are horses, other livestock, or people that frequent the study site because the rebar can injure feet and legs. Also, in areas where people are recreating, any visible rebar may be objectionable because it is incongruous with natural surroundings.

## Line Intercept Sampling

First, enter the transect number in Field 2 on the LI data form. Next, enter the plant species or item code for each item recorded in Field 3. FIREMON provides plant species codes from the NRCS Plants database. However, local or customized plant species codes are allowed in FIREMON. See **Populating the Plant Species Codes Lookup Table** in the **FIREMON Database User Manual** for more details.

Next enter the plant species status in Field 4 on the LI data form. Status describes the general health of the plant species as live or dead using the following codes:

**Figure LI-1** — The measuring tape is stretched taught below, in, or above the canopy vegetation, whichever position allows the easiest estimation of cover without the tape zigzagging around plants.

**L—Live:** plant with living tissue.
**D—Dead:** plant with no living tissue visible.
**NA—Not Applicable.**

Plant status is purely qualitative, but it does provide an adequate characteristic for stratification of preburn plant health and for determining postburn survival.

## Size class

Plant species size classes represent different layers in the canopy. For example, the upper canopy layer could be defined by large trees, while pole-size trees and large shrubs might dominate the middle layer of the canopy, and the lower canopy layer could include seedlings, saplings, grasses, and forbs. Size class data provide important structural information such as the vertical distribution of plant cover. Size classes for trees are typically defined by height for trees less than 4.5 ft (1.37 m) tall and diameter at breast height (DBH) for larger trees. Size classes for shrubs, grasses, and forbs are typically defined by height. If the vegetation being sampled has a layered canopy structure, then cover can be recorded by plant species and by size class. Total size class cover for a plant species could equal more than 100 percent for each plant species due to overlap between different size classes.

FIREMON uses a size class stratification based on the ECODATA sampling methods (Jensen and others 1994). Field crews can group individual plants by species into one or more trees size classes (table LI-1) or shrub, grass, and forb size classes (table LI-2). There can be multiple size classes for each species. In the **How-To Guide** chapter, see **How To Measure DBH** for detailed information on measuring DBH to group trees into size classes and see **How to Measure Plant Height** for detailed information on measuring height for grouping shrubs into size classes.

If cover is being recorded by size class, enter the size class code for each plant species in Field 5 on the LI data form. If size class data is not recorded, then record only the total cover for each plant species. When recording total cover for a species, enter the TO code in Field 5 to indicate that the cover estimate is for all of the size classes.

Table LI-1  Tree size class codes.

| | Tree size class | |
|---|---|---|
| Codes | English | Metric |
| TO | Total cover | Total cover |
| SE | Seedling (<1 inch DBH or <4.5 ft height) | Seedling (<2.5 cm DBH or <1.5 m height) |
| SA | Sapling (1.0 inch <5.0 inches DBH) | Sapling (2.5 <12.5 cm DBH) |
| PT | Pole tree (5.0 inches <9.0 inches DBH) | Pole tree (12.5 <25 cm DBH) |
| MT | Medium tree (9.0 inches <21.0 inches DBH) | Medium tree (25 <50 cm DBH) |
| LT | Large tree (21.0 inches <33.0 inches DBH) | Large tree (50 <80 cm DBH) |
| VT | Very large tree (>33.0 inches DBH) | Very large tree (>80 cm DBH) |
| NA | Not applicable | Not applicable |

Table LI-2  Shrub, grass, and forb size class codes.

| | Shrub/herb size class | |
|---|---|---|
| Codes | English | Metric |
| TO | Total cover | Total cover |
| SM | Small (<0.5 ft height) | Small (<0.15 m height) |
| LW | Low (0.5 <1.5 ft height) | Low (0.15 <0.5m height) |
| MD | Medium (1.5 <4.5 ft height) | Medium (0.5 <1.5 m height) |
| TL | Tall (4.5 <8 ft height) | Tall (1.5 <2.5 m height) |
| VT | Very tall (>8 ft height) | Very tall (>2.5 m height) |
| NA | Not applicable | Not applicable |

Enter the transect length in Field 6 of the LI data form. Transect length is entered by item and size class allowing transect length to vary by species and size class.

## Estimating cover (intercept)

The procedure for measuring the live crown intercept bisected by the transect line is illustrated in figure LI-2. Proceed from the baseline toward the opposite end of the tape and measure the horizontal linear length of each plant that intercepts the line. The start and stop point for each intercept are recorded in feet (m). When measuring intercepts in feet, use a tape that is marked in 10ths and 100ths of feet. Measure the intercept of grasses and grass-like plants, along with rosette-forming plants, at ground level. For forbs, shrubs, and trees measure the vertical projection of the vegetation intercepting one side of the tape. Be sure not to inadvertently move the tape and carefully look under tall dense crowns to be sure you are sampling all species and size classes. The measurements are recorded by plant species or item in Start and Stop fields on the LI data form to the nearest 0.1 ft (0.03 m). The FIREMON data entry screens populate the Intercept field automatically when the start and stop points are entered.

Canopy overlap within a species is not distinguished but canopy overlap between different species *is* recorded (fig. LI-3).

Percent cover is calculated by totaling the intercept measurements for all individuals of that species (in the Intercept field) along the transect and dividing by the total length of the transect. Most plant species have some gaps in their canopies, such as bunchgrasses with dead centers or shrubs with large spaces between branches. Examiners must determine how to deal with gaps in the canopy. One solution is for the observer to assume a closed canopy unless the gap is greater than some predetermined length. We recommend that gaps less than 2 inches (5 cm) be considered part of the canopy (fig. LI-4).

The FIREMON data entry screens and database allow an unlimited number of intercepts for each plant species along a transect.

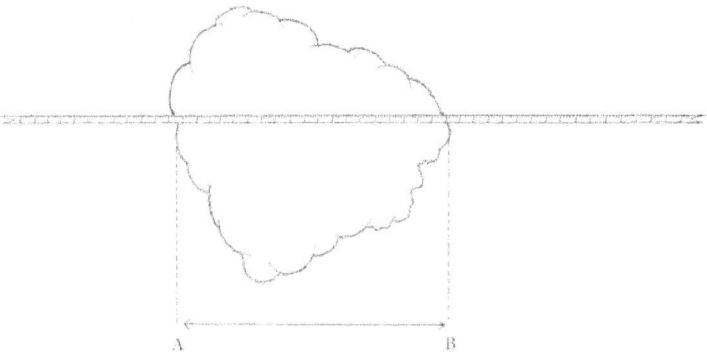

**Figure LI-2**—Measure cover intercept in feet (m) along the measuring tape. Since canopy intercept can vary on each side of the measuring tape, measure intercept on one side of the measuring tape only. We suggest using the right side as you move along the tape. Record the start of the plant intercept (A) in the Start field and the end intercept (B) in the Stop field.

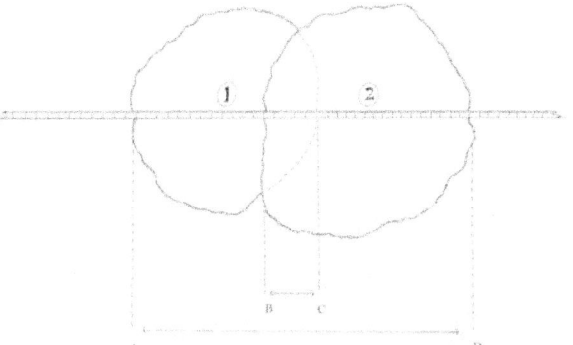

**Figure LI-3**—Canopy overlap (points B to C) is not measured if the canopy of two or more plants of the same species overlap. For example, if shrubs 1 and 2 are the same species, then the canopy intercept is measured from points A to D. If shrubs 1 and 2 are different species, then canopy intercept is measured from points A to C for shrub species 1 and from points B to D for shrub species 2.

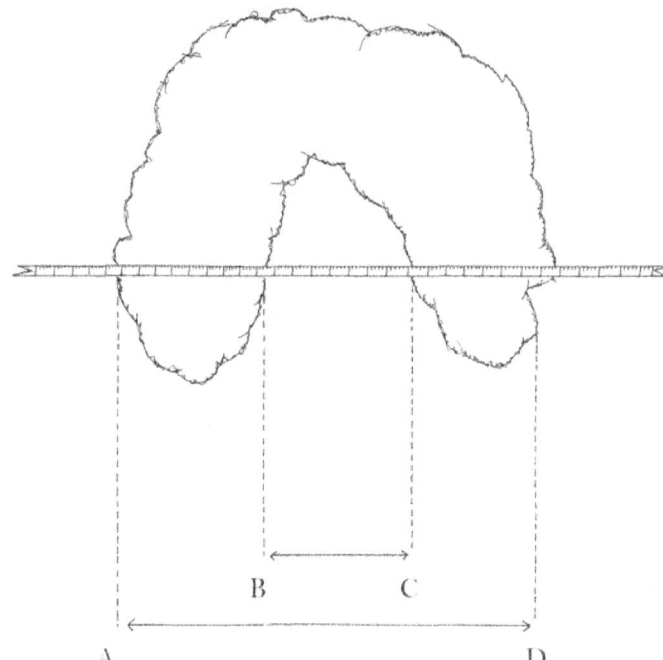

**Figure LI-4**—Gaps in the canopy (points B to C) greater than 2 inches (5 cm) are not measured. The canopy intercept for this shrub is measured from point A to D if the distance from B to C is less than or equal to 2 inches (5 cm) or measured from points A to B and points C to D if the gap is greater than 2 inches (5 cm).

## Estimating average height

Estimate the average height in feet (meters), within +/– 10 percent, for each plant species (fig. LI-5). The estimation should be for only the part of the plant that is intercepted by the tape, not the entire plant.

Enter plant height in the Height field on the LI data form for each item or species intercept. If plant species are recorded by size class, measure the average height for the plant species by each size class

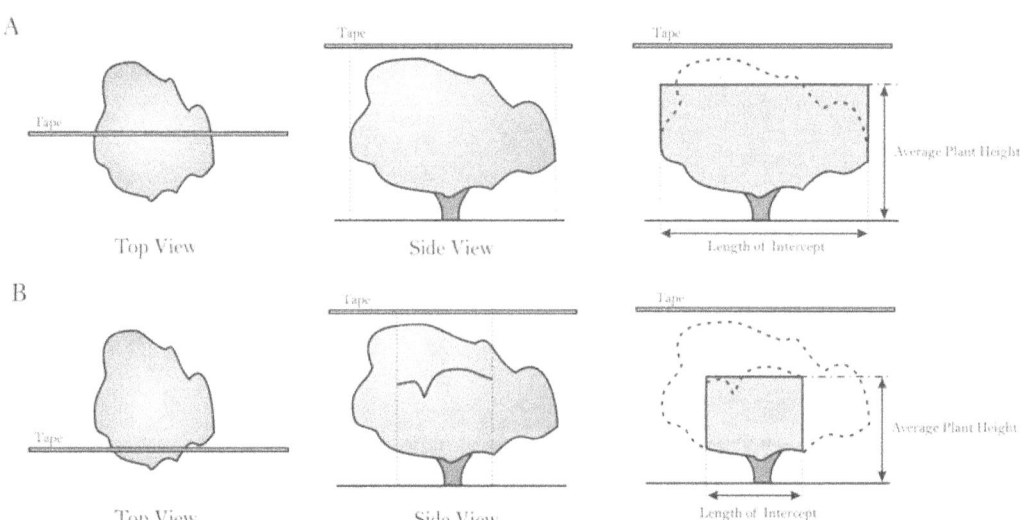

**Figure LI-5**—Estimate the average plant height only for the portion of the plant intercepted by the tape. If the tape crosses the entire plant then average the height for the entire plant (A). If the tape only crosses a portion of the plant, estimate the average height for only the part that is intercepted (B).

recorded. See **How to Measure Plant Height** in the **How-To Guide** chapter for more details. Plant height may be recorded at one intercept representing an average for the entire transect, at a few intercepts, or at every intercept. Be sure to record this information in the FIREMON Metadata table.

## Precision Standards

Use the precision standards listed in table LI-3 for the LI sampling.

# SAMPLING DESIGN CUSTOMIZATION

This section will present several ways that the LI sampling method can be modified to collect more detailed information or streamlined to collect only the most important tree characteristics. First, the suggested or recommended sample design is detailed, then modifications are presented.

## Recommended LI Sampling Design

The recommended LI sampling design follows the Alternative FIREMON sampling intensity and is listed below:

Collect plant species data. Make one estimate of height for each species that is representative of the entire transect.

**Macroplot Size**: 0.1 acre, 66 x 66 ft (400 m$^2$, 20 x 20 m).

**Number of Transects:** 5.

The number of transects sampled should be adjusted according to the sample size determination described in the **How-To Guide** chapter of the FIREMON manual.

## Streamlined LI Sampling Design

The streamlined LI sampling design follows the Simple FIREMON sampling intensity and is designed below:

The number of transects sampled should be adjusted according to the sample size determination described in the **How-To Guide** chapter of the FIREMON manual.

Collect plant species data.

**Macroplot Size**: 0.1 acre, 66 x 66 ft (400 sm$^2$, 20 x 20 m).

**Number of Transects:** 3.

## Comprehensive LI Sampling Design

The comprehensive LI sampling design follows the Detailed FIREMON sampling intensity and is detailed below:

Collect plant species data by size class and measure average plant height at each intercept.

**Macroplot Size**: 0.1 acre, 66 x 66 ft (400 m$^2$, 20 x 20 m).

**Number of Transects:** 5.

Table LI-3 Precision guidelines for LI sampling.

| Component | Standard |
|---|---|
| Size class | ±1 class |
| Start | ±0.1 ft/0.03 m |
| Stop | ±0.1 ft/0.03 m |
| Height | ±10 percent average height |

The number of transects sampled should be adjusted according to the sample size determination described in the **How-To Guide** chapter of the FIREMON manual.

## User-Specific LI Sampling Design

There are many ways the user can adjust the LI sample fields to make sampling more efficient and meaningful for local situations. Adjust the number and length of transects based on plant species size and distribution. Longer transects capture more variability in plant species cover, reducing the number of transects required to accurately estimate cover (Elzinga and others 1998).

The LI method is generally used for sampling shrub communities with vegetation greater than 3 ft (1 m) in height. However, this method can be used to sample taller vegetation if sighting devices are used to record the start and stop points for each intercept along a transect. Sighting devices may be mounted to a tripod and pointed downward to sample shorter vegetation (grasses, forbs, and small shrubs) and pointed upward to sample taller vegetation (tall shrubs and trees).

## Sampling Hints and Techniques

Examiners must be able to identify plant species and know how to collect cover data by measuring canopy intercepts along a measuring tape. Line intercepts should be recorded along only one edge of the measuring tape—the right side as you proceed to the end of the tape. It is important to prevent the tape from moving so that certain plants are not inadvertently included or excluded. For instance it could be difficult to sample using the LI method when it is windy because the tape will not be stationary.

The accuracy of this method depends on how well the FIREMON crew can estimate the vertical projection of vegetation along the tape. Observer bias occurs because the sighting line used to determine canopy starts and stops is not perpendicular to the tape. This bias can be minimized by using two measuring tapes (one above and one below) and sighting along the right side of the top tape to the right side of the bottom tape. Another solution is to suspend the measuring tape above the vegetation and use a plumb bob to record intercepts. For overhead vegetation, a pole with a level can be used. When measuring low and high vegetation, the most accurate method is to use a type of optical sighting device.

Measuring tapes are made from a variety of materials and are available in varying lengths and increments. Examiners should choose English (metric) tapes for this method and select a tape that is at least as long, or a little longer, than the transect length being sampled. When measuring plant species intercepts in feet, use a tape that is marked in 10ths and 100ths of feet. Steel tapes do not stretch and are the most accurate over the life of the tape. Steel is probably the best choice for permanent transects where remeasurement in exactly the same place each time is important. Cloth and fiberglass tapes will stretch over the life of the tape but are easier to use than steel tapes because they are lighter and do not tend to kink.

The sampling crew may encounter an obstacle, such as a large rock or tree, along one of the transect lines that interferes with the quadrat sampling. If that happens, offset using the directions described in **How To Offset a Transect** in the **How-To Guide** chapter.

When entering data on the LI data forms, examiners will most likely run out of space on the first page. The form was designed to print one copy of the first page, and several copies of the second page. The second page can be used to record more plant species intercepts for the plant species or items recorded on the first page or for additional plant species and items. The second page of the data form allows the examiner to write the intercept number on the form. This allows the examiner to design the form to accommodate the number of intercepts sampled. Print out enough pages to record all species on all transects for the required number of intercepts.

# LINE INTERCEPT (LI) FIELD DESCRIPTIONS

Field 1: **Number of Transects**. Total number of transects on the plot.

Field 2: **Transect Number**. Sequential number of the sample transect.

Field 3: **Item Code.** Code of sampled entity. Either the NRCS plants species code or the local code for that species, or ground cover code. Precision: No error.

Field 4: **Status**: Plant status—Live, Dead. or Not Applicable. (L, D, NA). Precision: No error.

Field 5: **Size Class**. Size of the sampled plant. Valid classes are in tables LI-1 and LI-2 of the sampling method. Precision: $\pm 1$ class

Field 6: **Transect Length**. Length of transect. May be different for different species/life forms (ft/m).

**Start**. Enter the starting point of each intercept for the plant species or life-form along the transect (ft/m). Precision: $\pm 0.1$ ft/0.03 m

**Stop**. Enter the stopping point of each intercept for the plant species or life-form along the transect (ft/m). Precision: $\pm 0.1$ ft/0.03 m

**Height**. Enter the average height for each plant species or life-form at one or more intercepts along the transect. Precision: $\pm 10$ percent of average height

#  FIREMON LI Cheat Sheet

**Status codes**

| Code | Description |
|------|-------------|
| L    | Live |
| D    | Dead |
| NA   | Not applicable |

**Precision**

| Component | Standard |
|-----------|----------|
| Size class | ± 1 class |
| Start | ±0.1 ft/0.03 m |
| Stop | ±0.1 ft/0.03 m |
| Height | ±10 percent average height |

**Tree size class**

| | Tree size class | |
|---|---|---|
| Codes | Description (English) | Description (Metric) |
| TO | Total cover | Total cover |
| SE | Seedling (<1 inches DBH or <4.5 ft height) | Seedling (<2.5 cm DBH or <1.5 m height) |
| SA | Sapling (1.0 inches < 5.0 inches DBH) | Sapling (2.5 <12.5 cm DBH) |
| PT | Pole tree (5.0 inches <9.0 inches DBH) | Pole tree (12.5 <25 cm DBH) |
| MT | Medium tree (9.0 inches <21.0 inches DBH) | Medium tree (25 <50 cm DBH) |
| LT | Large tree (21.0 inches <33.0 inches DBH) | Large tree (50 <80 cm DBH) |
| VT | Very large tree (>33.0 inches DBH) | Very large tree (>80 cm DBH) |
| NA | Not applicable | Not applicable |

**Shrub and herbaceous size classes**

| | Shrub/herb size class | |
|---|---|---|
| Codes | Description (English) | Description (Metric) |
| TO | Total cover | Total cover |
| SM | Small (<0.5 ft height) | Small (<0.15 m height) |
| LW | Low (0.5 <1.5 ft height) | Low (0.15 <0.5m height) |
| MD | Medium (1.5 <4.5 ft height) | Medium (0.5 <1.5 m height) |
| TL | Tall (4.5 <8 ft height) | Tall (1.5 <2.5 m height) |
| VT | Very tall (>8 ft height) | Very tall (>2.5 m height) |
| NA | Not applicable | Not applicable |

**Line intercept (LI) equipment list**

| | |
|---|---|
| Camera with film | Map protector or plastic bag |
| LI data forms | Magnifying glass |
| Clipboard | Pocket calculator |
| Compass | Plot sheet protector or plastic bag |
| File | Field notebook |
| Graph paper | Reinforcing bar (2) to mark baseline |
| Hammer | Reinforcing bars or bridge spikes to mark transects |
| Indelible ink pen (Sharpie, Marker) | Metal electrical conduit (lengths) to attach tape |
| Lead pencils with lead refills | Tape 75 ft (25 m) or longer, marked in 0.01 units (2) |
| Maps, charts and directions | Yardstick or meter stick |

# Line Intercept (LI) Form

RegistrationID: _ _ _ _ _
ProjectID: _ _ _ _
PlotID: _ _ _ _
Date: _ _ / _ _ / _ _ _ _

LI Page _ _ of _ _

Plot Key

Field 1: _____
Number of Transects: _____

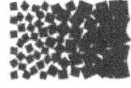

| Field 2 | Transect Number | | Field 2 | Transect Number | | Field 2 | Transect Number | | Field 2 | Transect Number | |
|---|---|---|---|---|---|---|---|---|---|---|---|
| Field 3 | Item Code | | Field 3 | Item Code | | Field 3 | Item Code | | Field 3 | Item Code | |
| Field 4 | Status | | Field 4 | Status | | Field 4 | Status | | Field 4 | Status | |
| Field 5 | Size Class | | Field 5 | Size Class | | Field 5 | Size Class | | Field 5 | Size Class | |
| Field 6 | Transect Length (ft/m) | | Field 6 | Transect Length (ft/m) | | Field 6 | Transect Length (ft/m) | | Field 6 | Transect Length (ft/m) | |
| Start | Stop | Height | Start | Stop | Height | Start | Stop | Height | Start | Stop | Height |
|  |  |  |  |  |  |  |  |  |  |  |  |
|  |  |  |  |  |  |  |  |  |  |  |  |
|  |  |  |  |  |  |  |  |  |  |  |  |
|  |  |  |  |  |  |  |  |  |  |  |  |
|  |  |  |  |  |  |  |  |  |  |  |  |
|  |  |  |  |  |  |  |  |  |  |  |  |
|  |  |  |  |  |  |  |  |  |  |  |  |
|  |  |  |  |  |  |  |  |  |  |  |  |
|  |  |  |  |  |  |  |  |  |  |  |  |
|  |  |  |  |  |  |  |  |  |  |  |  |
|  |  |  |  |  |  |  |  |  |  |  |  |
|  |  |  |  |  |  |  |  |  |  |  |  |
|  |  |  |  |  |  |  |  |  |  |  |  |
|  |  |  |  |  |  |  |  |  |  |  |  |
|  |  |  |  |  |  |  |  |  |  |  |  |
|  |  |  |  |  |  |  |  |  |  |  |  |

# Rare Species (RS)

## Sampling Method

Steve Sutherland

## SUMMARY

The FIREMON Rare Species (RS) method is used to assess changes in uncommon, perennial plant species when other monitoring methods are not effective. This method monitors individual plants and statistically quantifies changes in plant survivorship, growth, and reproduction over time. Plants are spatially located using distance along and from a permanent baseline, and individual plants are marked using a permanent tag. Data are collected for status (living or dead), stage (seedling, nonreproductive, or reproductive), size (height and diameter), and reproductive effort (number of flowers and fruits). This method is primarily used for Threatened and Endangered species and uncommon grass, forb, shrub, and tree species of special interest.

## INTRODUCTION

When plants become rare, most sampling methods are not effective because the species will not likely occur in the sampling unit. Therefore, it becomes necessary to follow individual plants in order to determine fire effects for that species. The Rare Species (RS) method was designed to quantify temporal changes in plant survivorship, growth, and reproduction for uncommon, perennial plant species. This method is not effective for rare annual species. First, a permanent baseline is established, and baseline length and characteristics about the general sample design are recorded. For each plant, a unique ID number, distance along the baseline, and perpendicular distance from the baseline is recorded. Data are collected for status (living or dead), stage (seedling, nonreproductive, or reproductive), size (height and diameter), and reproductive effort (number of flowers and fruits) for each individual plant.

This method is primarily used when the fire manager wants to monitor changes in Threatened and Endangered perennial plant species and uncommon species of special interest. This method is suited for rare grass, forb, shrub, and tree species that are not effectively monitored by other methods. This sampling method uses attributes of individual plants to assess changes in survivorship, growth, and reproduction over time.

There are many ways to streamline or customize the RS sampling method. The FIREMON three-tier sampling design can be employed to optimize sampling efficiency.

See the sections on **User-Specific RS Sampling Design** and **Sampling Design Customization** below.

# SAMPLING PROCEDURE

This method assumes that the sampling strategy has already been selected and the macroplot has already been located. If this is not the case, then refer to the FIREMON **Integrated Sampling Strategy** and for further details.

The sampling procedure is described in the order of the fields that need to be completed on the **RS data form**, so it is best to reference the form when reading this section. The sampling procedure described here is the recommended procedure for this method. Later sections will describe how the FIREMON three-tier sampling design can be used to modify the recommended procedure to match resources, funding, and time constraints. A field-by-field description of the sampled elements is provided in the **RS Field Descriptions**.

More than any of the other FIREMON methods, the RS method is especially sensitive to the timing of sampling in relation to the phenological stage of the plants being sampled. Plant size and stage, and stem, flower, and fruit counts change as the growing season progresses. A monitoring plan that includes rare species sampling but does not take phenological stage into consideration will probably not result in useful data. This topic is covered more thoroughly in Chapter 12 of *Measuring and Monitoring Plant Populations* (Elzinga and others 1998). The publication is free and is available for download from the Bureau of Land Management library Web site: http://www.blm.gov/nstc/library/techref.htm. Select T.R. number 1730-1.

In the **How-To Guide** chapter, see **How To Locate a FIREMON Plot**, **How To Permanently Establish a FIREMON Plot,** and **How to Define the Boundaries of a Macroplot** for more information on setting up your macroplot.

## Preliminary Sampling Tasks

Before setting out for your field sampling, lay out a practice area with easy access. Try and locate an area with the same species or vegetation life form you plan on sampling. Get familiar with the plot layout and the data that will be collected. This will give you a chance to assess the method and will help you think about problems that might be encountered in the field. For example, will you really be able to identify the species of interest once you are in the field? It is better to answer these questions before the sampling begins so that you are not wasting time in the field. This will also let you see if there are any pieces of equipment that will need to be ordered.

Many preparations must be made before proceeding into the field for RS sampling. First, all equipment and supplies in the **RS Equipment List** must be purchased and packed for transport into the field. Travel to FIREMON plots is usually by foot, so it is important that supplies and equipment be placed in a comfortable daypack or backpack. It is also important that there be spares of each piece of equipment so that an entire day of sampling is not lost if something breaks. Spare equipment can be stored in the vehicle rather than the backpack. Be sure that all equipment is well maintained and there are plenty of extra supplies such as data forms, map cases, and pencils.

All RS data forms should be copied onto waterproof paper because inclement weather can easily destroy valuable data recorded on standard paper. Data forms should be transported into the field using a plastic, waterproof map protector or plastic bag. The day's sample forms should always be stored in a dry place (office or vehicle) and not be taken back into the field for the next day's sampling.

We recommend that one person on the field crew, preferably the crew boss, have a waterproof, lined field notebook for recording logistic and procedural problems encountered during sampling. This helps with future remeasurements and future field campaigns. All comments and details not documented in the FIREMON sampling methods should be written in this notebook. For example, snow on the plot might be described in the notebook, which would be helpful in plot remeasurement.

It is beneficial to have plot locations for several days of work in advance in case something happens, such as if the road to one set of plots is washed out by flooding. Plots should be referenced on maps and aerial photos using pin-pricks or dots to make navigation easy for the crew and to provide a check of the georeferenced coordinates. We found that it is easy to transpose UTM coordinate digits when recording georeferenced positions on the plot sheet, so marked maps can help identify any erroneous plot positions. If possible, the spatial coordinates should be provided if FIREMON plots were randomly located.

A field crew of two people is probably the most efficient for implementation of the RS sampling method. There should never be a one-person field crew for safety reasons, and any more than two people will probably result in some people waiting for critical tasks to be done. The crew boss is responsible for all sampling logistics including the vehicle, plot directions, equipment, supplies, and safety. The crew boss should be the note taker and the technician should perform most quadrat measurements. The initial sampling tasks of the field crew should be assigned based on field experience, physical capacity, and sampling efficiency, but sampling tasks should be modified as the field crew gains experience. Tasks should also be shared to limit monotony.

## Designing the RS Sampling Design

A set of general criteria recorded on the RS data form allows the user to customize the design of the RS sampling method so that the sampling captures the information needed to successfully complete the management objective within time, money, and personnel constraints These general fields should be determined before the crews go into the field and should reflect a thoughtful analysis of the expected problems and challenges in the fire monitoring project.

### Plot ID construction

A unique plot identifier must be entered on the RS sampling form. This is the same plot identifier used to describe general plot characteristics in the Plot Description (PD) sampling method. Details on constructing a unique plot identifier are discussed in the **How to Construct a Unique Plot Identifier** section of the **How-To Guide** chapter. Enter the plot identifier at the top of the RS data form.

### Determining the sample size

The size of the rare species population ultimately determines the length of the baseline from which the individual plants are located. The baseline length is recorded in Field 1 on the RS data form. If the population is divided along several patches, several baselines can be established. The size of the rare species population also determines the number of individuals sampled. We recommend sampling 25 individuals within the population; this should be sufficient for most studies. However, there are situations when more individuals should be sampled, and in these cases the project objectives should identify the sampling intensity. The FIREMON RS data form and data entry screen allow an unlimited number of individuals to be measured per baseline. If the population is smaller than 25 individuals, measure all individuals. This will then be a census rather than a sample, and a statistical analysis will not be necessary.

## Conducting RS Sampling Tasks

### Establish the baseline for locating individual plants

The baseline serves as a georeferenced starting point for relocating individual plants. If there are other FIREMON methods implemented at the same sample site, then, as much as possible, the plots should be set up so that they correspond with one another. See **How To Establish Plots with Multiple Methods** in the **How-To Guide** chapter. In most cases, this will not be advisable when monitoring rare plants as the additional sampling will negatively impact the individuals being monitoring (trampling). Once the rare species population is located, a permanent baseline is set up as a reference from which you will locate individual plants. On flat areas, the baseline runs from south to north with the 0-ft (0-m) mark on the south end. On slopes, the baseline runs upslope with the 0-ft (0-m) mark on the bottom (down slope) end. The length of the baseline will be determined by the size of the rare species population.

## Locating individual plants

Locate individual plants within the rare species population by running a tape perpendicular to the baseline to the plant. If the rare species population being sampled is to the right of the baseline when looking up hill (or north on flat ground) then distances to plants are recorded as positive (+) numbers and distances to plants to the left are recorded as negative (-) numbers. Because the distance along the baseline and the distance from the baseline will be used to relocate plants in successive years, it is essential that these measurements are exact and the lines are perpendicular (fig. RS-1). Measure the distances to the nearest 0.1 ft (0.02 m)

## Tagging the individual plants

Individual plants need to be permanently tagged with a unique plant number. If the plant is woody (shrub or tree), the tag can be attached to the plant. If the plant is herbaceous (grass or forb), the tag should be anchored in the soil adjacent to the plant in a standard location (for example, directly down slope of the plant). Because these tags will uniquely identify the individual plant, it is essential that they be located in a manner that will eliminate any confusion between individuals.

# Rare Species Sampling

## Plant identity

First enter the species code in Field 2 on the RS data form. FIREMON uses the NRCS plants species codes. However, you may use your own species codes. See **Populating the Plant Species Codes Lookup Table** in the **FIREMON Database User Manual** for more details.

To uniquely identify each individual of the rare species, enter the plant number in Field 3, distance along baseline in Field 4, and distance from the baseline in Field 5 on the RS data form.

## Status

Next enter the plant status in Field 6 on the RS data form. Status describes the individual plant as live or dead using the following codes:

**L—Live:** plant with living tissue.

**D—Dead:** plant with no living tissue visible.

**NA—Not Applicable**.

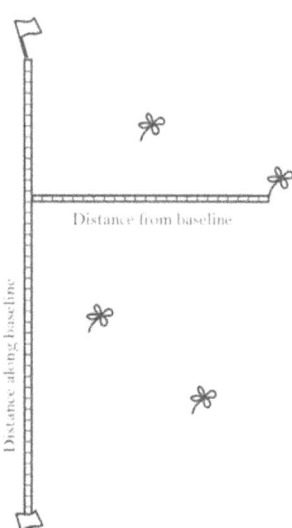

**Figure RS-1**—Measuring the location of rare plants. It is critical that the distance locating each plant—along and from the baseline—be accurately measured and that the measuring tapes are perpendicular.

Plant status is purely qualitative but determines postburn survivorship and population health. Use care in determining plant status during the dormant season.

## Stage

Enter the plant species stage in Field 7 on the RS data form. Stage describes the individual plant as a seedling, nonreproductive adult, or reproductive adult.

**S—Seedling**: plant less than 1 year old.

**NR—Non-Reproductive adult**: plant 1 year old or older without flowers or fruits.

**R—Reproductive adult**: plant 1 year old or older with flowers or fruits.

Plant stage is also qualitative but provides information on plant growth and reproduction and population health. Use care in determining plant reproductive status during the dormant season.

## Size

Size information provides data on growth rates and population vigor. There are four measures of plant size that will be entered on the RS field data form: two diameter measures, maximum height, and number of stems.

For grasses, forbs, shrubs, and tress less than 1 inch (2 cm) DBH, measure the diameter of the plant canopy at two places. First, measure the widest part and record the diameter in Field 8. Make a second measurement at a right angle to the first and record in Field 9. Make both measurements in inches (cm) to the nearest 0.1 inch (0.2 cm).

For trees at least 1 inch (2 cm) DBH, record DBH in Field 8. Measure the maximum height of the plant in feet (m) and record in Field 10 to the nearest 0.1 ft (0.03 m). For more information on measuring plant heights see **How to Measure Plant Height** in the **How-To Guide** chapter.

Regardless of the plant size, in Field 11 record the number of plant stems for the individual you are measuring. Use care in determining individual plants when measuring size.

## Reproduction

Flower and fruit counts provide data on reproductive rates and population viability. Count the number of flowers and fruits on the plant and record in Fields 12 and 13, respectively, of the RS field data form.

## Precision Standards

Use the precision standards in table RS-1 for the RS sampling.

# SAMPLING DESIGN CUSTOMIZATION

This section presents several ways that the RS sampling method can be modified to collect more detailed information or streamlined to collect basic information.

**Table RS-1** Precision guidelines for RS sampling.

| Component | Standard |
|---|---|
| Distance along baseline | ±0.1 ft/0.03 m |
| Distance from baseline | ±0.1 ft/0.03 m |
| Maximum diameter | ±0.1 inch/0.2 cm |
| Diameter 2 | ±0.1 inch/0.2 cm |
| Height | ±0.1 ft/0.03 m |
| Stems | ±3 percent total count |
| Flowers | ±3 percent total count |
| Fruits | ±3 percent total count |

## Recommended RS Sampling Design

The recommended RS sampling design follows the Alternative FIREMON sampling intensity and is listed below:

**Number of Individual Plants Sampled:** 25.
**Record:** Status and Stage.
**Measure:** Maximum Diameter, Diameter at right angles, Height.
**Count:** Stems, Flowers, Fruits.

The baseline length should be adjusted according to the size of the rare species population being sampled.

The number of individual plants sampled should be adjusted according to the appropriate methods in the **How-To Guide** chapter of the FIREMON manual.

## Streamlined RS Sampling Design

The streamlined RS sampling design follows the Simple FIREMON sampling intensity and is designed below:

**Number of Individual Plants Sampled:** 15.
**Record:** Status and Stage.
**Measure:** Maximum Diameter, Height.

The baseline length should be adjusted according to the size of the rare species population being sampled.

The number of individual plants sampled should be adjusted according to the appropriate methods in the **How-To Guide** chapter of the FIREMON manual.

## Comprehensive RS Sampling Design

The comprehensive RS sampling design follows the Detailed FIREMON sampling intensity and is detailed below:

**Number of Individual Plants Sampled**: 25 per stage (25 seedlings, 25 nonreproductive adults, and 25 reproductive adults) adding additional plants to the stage categories as individuals die or move into another stage (for example, seedling to nonreproductive adult).

**Record:** Status and Stage.
**Measure:** Maximum Diameter, Diameter at right angles, Height.
**Count:** Stems, Flowers, Fruits, Seeds, Vegetative Reproduction.

These data could be used to conduct a population viability analysis; see Elzinga and others (1998) for a detailed discussion on this topic.

The baseline length should be adjusted according to the size of the rare species population being sampled.

## User-Specific RS Sampling Design

The user can modify the RS sample fields a number of ways in order to make sampling more efficient and meaningful for local situations. This will usually mean adjusting the number of individuals sampled as needed for the specific task. Use the Metadata form to record any changes in sampling methods that are modified from the standard, or to remark on any other RS matter that needs to be explained or defined for subsequent sampling and data use.

## Sampling Hints and Techniques

Examiners must be able to identify the target plant species and identify individual plants. It can be difficult to distinguish individual plants for some species such as sod-forming grasses. If individual plants are difficult to identify, guidelines should be determined before sampling as to what constitutes the individual counting unit. Some examples include counting individual stems in aspen communities, culm groups in rhizomatous grasses, and flowering stems for mat-forming forbs. However, the counting unit chosen to monitor should reflect a real change in the plant community.

Measuring tapes come in a variety of lengths, increments, and materials. Examiners should choose English (metric) tapes for this method and select a tape that is at least as long, or a little longer, than the baseline length being sampled. Steel tapes do not stretch and are the most accurate over the life of the tape. Steel is probably the best choice where remeasurement in exactly the same place each time is important. Cloth and fiberglass tapes will stretch over the life of the tape but are easier to use than steel tapes because they are lighter and do not tend to kink.

The sampling crew may encounter an obstacle, such as a large rock or tree, along one of the transect lines that interferes with the quadrat sampling. If that happens, offset using the directions described in **How To Offset a Transect** in the **How-To Guide** chapter.

Because the purpose of resampling is to determine change over time, it is essential that the plots are resampled when plants are in the same phonologic condition as when they were originally sampled. Often this means resampling on the same date as when you originally sampled. If you do not resample when the plants are in the same phonologic condition, then you may be documenting annual growth cycles rather than fire effects.

When entering data on the RS data forms, examiners will most likely run out of space on the first page. Print out enough pages to record all individuals that you will be sampling.

## RARE SPECIES (RS) FIELD DESCRIPTIONS

Field 1: **Baseline Length**. Length of baseline (ft/m).
Field 2: **Species Code**. Code of sampled species. Either the NRCS plants species code or the local code for that species.
Field 3: **Plant Number**. Sequential number for individual plants.
Field 4: **Distance Along Baseline**. Measured distance from 0-ft (0-m) mark on baseline to point at which a perpendicular line intersects the individual plant location (ft/m). Precision: ±0.1 ft/0.02 m.
Field 5: **Distance From Baseline**. Measured perpendicular distance from baseline to individual plant location (ft/m). Precision: ±0.1 ft/0.02 m.
Field 6: **Status**: Plant status—Live, Dead, or Not Applicable. (L, D or NA). Precision: No error.
Field 7: **Stage**. Plant stage–Seedling, Non-Reproductive adult, Reproductive adult (S, NR, or R). Precision: No error.
Field 8: **Maximum Diameter**. For plants less than 1 inch DBH, maximum plant canopy diameter. Or record DBH for plants 1 inch (2 cm) and greater DBH (inch/cm). Precision: 0.1 inch/0.2 cm.
Field 9: **Diameter 2**. Plant canopy diameter measured at right angles to maximum canopy diameter (inch/cm). Precision: 0.1 inch/0.2 cm.
Field 10: **Height**. Height for each plant marked on the plot (ft/m). Precision: 0.1 ft/0.03 m.
Field 11: **Stems**. Number of stems for each plant marked on the plot. Precision: ±3 percent.
Field 12: **Flowers**. Number of flowers for each plant marked on the plot. Precision: ±3 percent.
Field 13: **Fruits**. Number of fruits for each plant marked on the plot. Precision: ±3 percent.
Field 14: **Local Field 1**. User-specific code or measurement for the individual plant being recorded.
Field 15: **Local Field 2**. User-specific code or measurement for the individual plant being recorded. Do not embed blanks in your codes to avoid confusion and database problems. Document your coding method in the Metadata table.
Field 16: **Local Field 3**. User-specific code or measurement for the individual plant being recorded. Do not embed blanks in your codes to avoid confusion and database problems. Document your coding method in the Metadata table.

**Rare species (RS) equipment list**

| | |
|---|---|
| Camera, film and flash | Maps, charts and directions |
| RS data forms | Map protector or plastic bag |
| Clipboard | Magnifying glass |
| Compass | Markers to mark small plants (at least 25 per plot) |
| Field notebook | Pocket calculator |
| File | Plot sheet protector or plastic bag |
| Folding rulers, 6 ft (2 m) (2 3) | Rebar (2) for baseline plus 2 for each transect |
| Graph paper | Steel fence posts (2) and driver (to mark endpoints of baseline) |
| Hammer | Tags to mark large plants (at least 25 per plot) |
| Indelible ink pen (Sharpie, Marker) | Tape, 75 ft (25 m) or longer (3) |
| Lead pencils with lead refills | |

# Rare Species (RS) Form

| Field 1 |
|---|
| Baseline Length (ft/m) |

RegistrationID: _ _ _ _
ProjectID: _ _ _ _
PlotID: _ _ _ _
Date: _ _ / _ _ / _ _ _ _

Plot Key

RS Page _ _ of _ _

Notes

| Field 2 | Field 3 | Field 4 | Field 5 | Field 6 | Field 7 | Field 8 | Field 9 | Field 10 | Field 11 | Field 12 | Field 13 | Field 14 | Field 15 | Field 16 |
|---|---|---|---|---|---|---|---|---|---|---|---|---|---|---|
| Species | Plant Number | Dist Along Baseline (ft/m) | Dist From Baseline (ft/m) | Status | Stage | Maximum Diameter (in./cm) | Diameter 2 (in./cm) | Maximum Height (ft/m) | Stems | Flowers | Fruits | Local Code 1 | Local Code 2 | Local Code 3 |
| | | | | | | | | | | | | | | |
| | | | | | | | | | | | | | | |
| | | | | | | | | | | | | | | |
| | | | | | | | | | | | | | | |
| | | | | | | | | | | | | | | |
| | | | | | | | | | | | | | | |
| | | | | | | | | | | | | | | |
| | | | | | | | | | | | | | | |
| | | | | | | | | | | | | | | |
| | | | | | | | | | | | | | | |
| | | | | | | | | | | | | | | |
| | | | | | | | | | | | | | | |
| | | | | | | | | | | | | | | |
| | | | | | | | | | | | | | | |
| | | | | | | | | | | | | | | |
| | | | | | | | | | | | | | | |

Crew

# Landscape Assessment (LA)

## Sampling and Analysis Methods

**Carl H. Key**
**Nathan C. Benson**

## SUMMARY

Landscape Assessment primarily addresses the need to identify and quantify fire effects over large areas, at times involving many burns. In contrast to individual case studies, the ability to compare results is emphasized along with the capacity to aggregate information across broad regions and over time. Results show the spatial heterogeneity of burns and how fire interacts with vegetation and topography. The quantity measured and mapped is "burn severity," defined here as a scaled index gauging the magnitude of ecological change caused by fire. In the process, two methodologies are integrated. Burn Remote Sensing (BR) involves remote sensing with Landsat 30-meter data and a derived radiometric value called the Normalized Burn Ratio (NBR). The NBR is temporally differenced between pre- and postfire datasets to determine the extent and degree of change detected from burning (fig. LA-1). Two timeframes of acquisition identify effects soon after fire and during the next growing season for Initial and Extended Assessments, respectively. The latter includes vegetative recovery potential and delayed mortality. The Burn Index (BI) adds a complementary field sampling approach, called the Composite Burn Index (CBI). It entails a relatively large plot, independent severity ratings for individual strata, and a synoptic rating for the whole plot area. Plot sampling may be used to

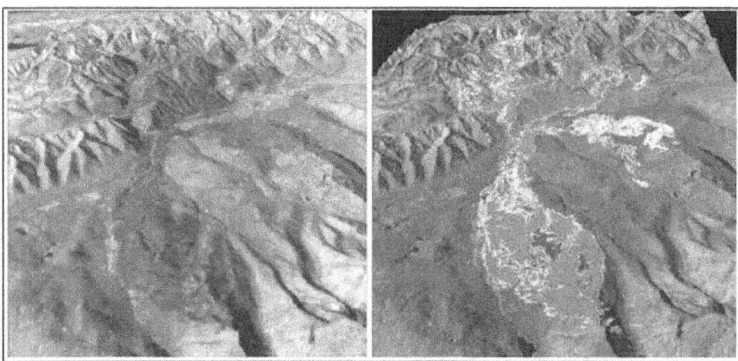

**Figure LA-1**—A three-D view of the Moose fire, northwestern Montana, taken by Landsat ETM+ on 9 September 2001. On the left, spectral Band 4 and Band 7 are displayed as a composite of green and red, respectively. On the right, differencing the NBR before and after fire has derived an initial assessment of burn severity. The gradient of differenced NBR has been stratified to identify burn severity levels, including: unburned, low (green), moderate-low (yellow), moderate-high (orange), and high (red).

calibrate and validate remote sensing results, to relate detected radiometric change to actual fire effects on the ground. Alternatively, plot sampling may be implemented in stand-alone field surveys for individual site assessment.

## INTRODUCTION

Methods in this chapter are designed to provide a landscape perspective on fire effects. That is, spatial data on burn severity throughout a whole burn. They show the results of fire in context of regional biophysical characteristics, such as topography, climate, vegetation, hydrography, fuels, and soil. At this level, one can isolate burned from unburned surroundings, measure the amount burned at various levels of effect, and gauge the spatial heterogeneity of the burn (fig. LA-2). Such methods provide a quantitative picture of the whole burn as if viewed from the air. They are adapted to remote sensing and GIS technologies, which in turn produce a variety of derived products such as maps, images, and statistical summaries.

The authors first developed the methods in 1996, following analysis of wildfires that occurred in Glacier National Park during 1994. Since that time, efforts were made to bring the remote sensing and field approaches of dNBR and CBI into an operational setting for national burn area mapping. As a component of that, the first version of documentation appeared in FIREMON in 2001, and was made available on the FIREMON Web site. Subsequently, minor revisions were made in 2002 and 2003. Documentation has been updated in Version 4 to reflect experience gained over recent field seasons. Changes were made to clarify some issues with the timing of Landsat data acquisitions and how that relates to burn severity, and also to refine severity rating-factor definitions to make ground sampling more broadly applicable across ecosystems of the United States. To implement landscape assessment of burns, several factors must be considered; among them scale, resolution, standardization, and cost effectiveness. Methods herein furnish information covering potentially several tens of thousands of square kilometers at a time, with capability to monitor large or inaccessible burns (fig. LA-3). Small burns of a few hectares can be monitored as well, but that may not be cost effective unless those are

**Figure LA-2**—The Kootenai Complex from the interior of Glacier National Park captured by Landsat 12 September 1999, a year after the fire. The range of colors within the perimeter shows patterns of burning, gradations of reddish-brown (no burn to low severity) through dark blue (high severity). Information about how the fire interacted with the landscape is also evident. Red areas are perennial snowfields and glaciers nestled among the rocky peaks (light blue to white).

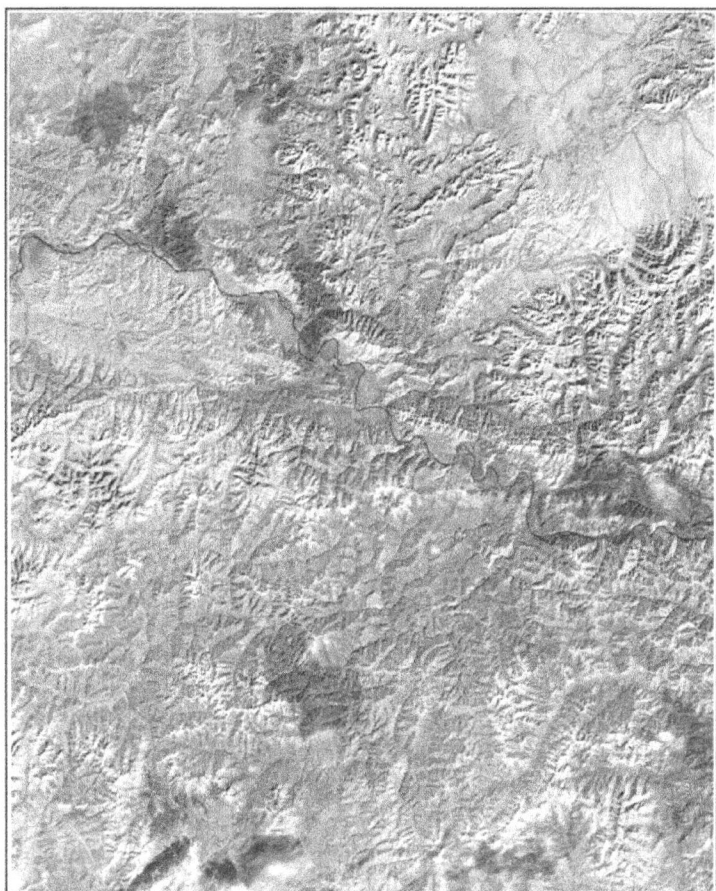

**Figure LA-3**—This mosaic of two Landsat 7 ETM+ scenes from 12 Sept 1999, shows the Yukon Charlie region of Alaska. Nine large burns are clearly visible (dark purple) from fires earlier in the summer. This type of remote sensing data allows managers to discover details of burns, which because of daunting logistics, may never be visited on the ground. The challenge has been to find the best measure of severity from such data, and to develop protocols that offer standardization, so burns can be compared spatially and temporally.

covered in conjunction with other large burns. Cost per unit area diminishes as burn area increases over a region. Products also are useful to the manager dealing with local burn issues. Important to note is that objectives are for standard approaches that can be applied uniformly over multiple concurrent burns and that yield comparable metrics from region to region over time. This section identifies data sources and methods that can be broadly implemented on a national level for repeatable and routine assessment at an affordable cost—that is, in terms of the Federal and State land-based agencies accountable for wildland burn programs.

The Landsat satellite program has been well suited for burn area assessment. Landsat archives contain near global repeat coverage of multispectral data acquired since 1982 at 30-meter spatial resolution. With two operational satellites as of 2003 (Landsat 5 and Landsat 7), data acquisition is possible every 8 days. Unfortunately in spring of 2003, Landsat 7 developed the now well-known scan line corrector problem, which results in missing lines of data through portions of each scene. This data can still be used in some burn assessments, however, where missing data does not impact the burned area, or where multiple scenes can be patched together to fill the missing lines. Most important, Landsat is the only source for temporally and spatially consistent information on a continual basis nationwide. It allows one to compare both prefire and postfire conditions when evaluating the magnitude of fire-caused change (fig. LA-4). Moreover, resolution is efficient for broad-area coverage, in terms of computer resources and funds available to most land managers today. Such characteristics are key to methodologies presented here for whole-burn monitoring. For more information, contact the USGS Landsat 7 Web site at: http://landsat7.usgs.gov/.

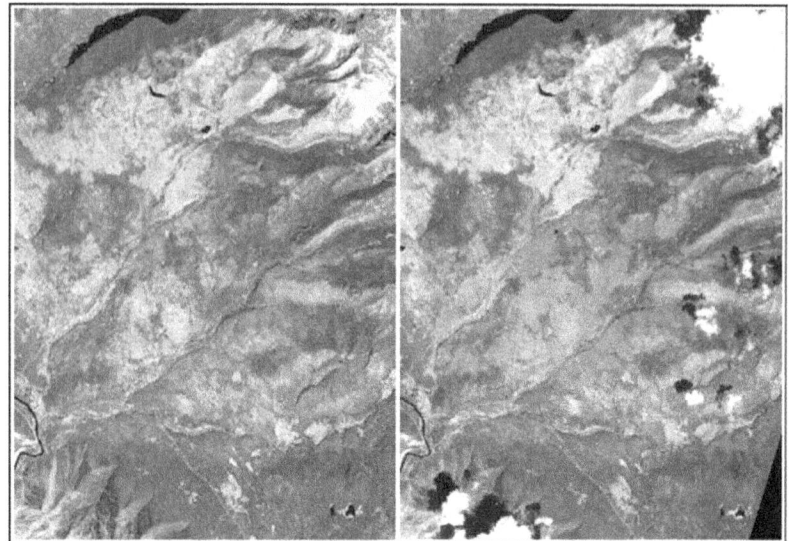

**Figure LA-4**—These images show the Anaconda burn area, on the west side of Glacier National Park, on 10 July 1999 (left) soon before the fire, and on 25 June 2000 (right) about 9 months after the burn. The light area above is a 1994 burn still evident on the landscape. Clouds appear as bright white in the 2000 image. Knowledge of prefire conditions allows one to gauge severity through change detection techniques and explore relationships between burn patterns and vegetation structure and composition.

To be applied, Landsat data must be statistically related to particular features of interest on the ground. One must determine target characteristics that are important and find ways to measure those that are complementary with the sensor (fig. LA-5). Ground measures provide the basic way to gauge usefulness and to understand the meaning of results. Thus, to assess burned areas, a field-based sampling strategy has been developed to be compatible with the resolution and spectral characteristics of the Landsat TM/ETM+ data. Though relevant to signals relayed from satellites, field information also can be used independently, where applications on the ground call for broad-area coverage or synoptic levels of detail.

The FIREMON Landscape Assessment methods were developed along the lines of: 1) optimizing satellite-derived information; 2) matching ground-based methods to the constraints of remote sensing; and 3) standardizing procedures to meet the needs for comparable results and implementation. The following sections in this chapter cover the three interrelated elements of the approach.

***Definition of burn severity.*** Adapted to moderate resolution, mesoscale perspectives, the definition influences how we interpret severity on the ground. It is the basis for understanding fire effects at the landscape level, encompassing perhaps many types of communities over large areas. Definition is critical to correctly apply methods, use information appropriately, communicate results, and avoid misconceptions. To some extent, this may differ from concepts of severity based on individual trees, small-area microplots, or subsurface evidence of heating.

***Ground measure of severity (Burn Index, BI).*** The protocol is designed to match field sampling with the definition of severity and the characteristics of TM/ETM+ data. The measure is called the Composite Burn Index (CBI). It also can be used for a variety of applications to estimate the general, average burn conditions of stands or communities.

***Remote sensing measure of severity (Burn Remote Sensing, BR).*** This section shows how to process and map burn severity using Landsat TM/ETM+ data. A particular algorithm is used, called the Normalized Burn Ratio (NBR). Pre- and postfire NBR datasets are differenced to isolate the burn from surroundings and provide a scale of change caused by fire. In most cases, the approach reliably separates burned from unburned surfaces, and optimally identifies a broad gradient of fire-effect levels within the burn.

The LA Cheat Sheets follow the BI and BR sections, and a field form is provided at the end of the BI section. Additional techniques specific to the LA methods are described in the **LA How-To** section in this chapter. The **LA Glossary** follows the **How-To** document.

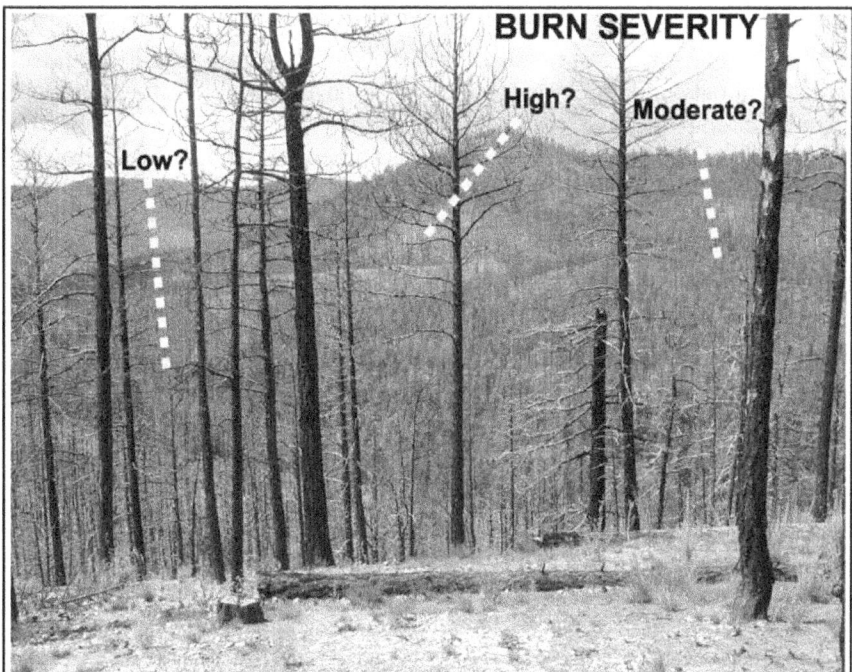

**Figure LA-5**—Fire creates heterogeneity on the landscape. How does one define severity so this mosaic of effects can be mapped across broad regions? Viewed from space, features on the ground become aggregated across multiple levels to make up the spectral signals received by satellite. Upper canopies are important, but so too are lower strata that may show through the canopy or become more prominent after fire. To assess severity, key features within each stratum are evaluated that have relevance to ecological effects on the site as well as potential for contributing to the satellite signal.

## DEFINITION OF BURN SEVERITY FOR LANDSCAPE ASSESSMENT

Admittedly, there is still some discrepancy in the way researchers and managers use the term "burn severity." In this section we clarify how we intend to use it so at least one might better understand our discussion regarding Landscape Assessment of burns. Whether or not these concepts become standard practice depends on repeated trial and acceptance, but we hope this chapter contributes to more discussion and common understanding of the issues involved.

Some of the discrepancy arises from inconsistency in the combination of the relevant terms: *fire, burn, severity, and intensity*. It is useful, therefore, to first define these for LA methods. The meanings we aim to convey are brief excerpts taken from the dictionary, followed by nuances imparted in the context of wildland fire.

**FIRE (n).** The phenomenon of combustion manifested in light, flame, and heat. The period of active flaming and smoldering.

**BURN (n).** Injury, damage, or effect produced by heating. The result(s) of fire, also an area where fire has occurred in the past.

**INTENSITY (n).** The strength of a force, or the amount of energy expended. The level of heat produced by fire.

**SEVERITY (n).** The quality or state of distress inflicted by a force. The magnitude of environmental change caused by fire, or the resulting level of cost in socioeconomic terms.

Based on the definitions it seems reasonable to apply the following two terms:

**FIRE INTENSITY:** The magnitude of heat produced by fire is an empirical measure that gauges the fire's status during combustion. This is commonly defined in reference to fire line intensity, which equals energy output per length of fire front per unit time. It may be measured by thermocouple readings in time series, as in experimental situations, or more commonly on wildfires, in proportion to observed flame length and rate of spread. Fire intensity may be divided into two heat components: downward penetration into soil, and upward transfer to vegetation and the atmosphere. These depend on residual flame time and are a function of fuel and weather characteristics. An analogy to fire intensity is storm intensity, which uses such parameters as wind speed and precipitation rate to describe the strength of a storm.

**BURN SEVERITY:** Socio-economic impacts associated with fire can be measured directly in terms such as cost of suppression, cost of rehabilitation, property loss, or human causality. For this discussion, however, we focus on the degree of environmental change caused by fire. This result of fire is the cumulative after-the-fact effect of fire on ecological communities that compose the landscape. An analogy to burn severity would be storm severity, which refers to the damage or outcome left in the wake of the storm. For example, you might say an intense storm resulted in severe consequences. The ecological criteria to judge burn severity differ, naturally, from those of storms. Here we are talking about physical and chemical changes to the soil, conversion of vegetation and fuels to inorganic carbon, and structural or compositional transformations that bring about new microclimates and species assemblages. The scope includes all degrees of effect, ending with the most extreme where essentially all aboveground organisms are eliminated, and the community must regenerate from "ground zero."

Of the remaining two terms, "burn intensity" seems least sensible and should be avoided. "Fire severity," though, does make sense, so long as one clearly understands it references conditions left after fire. We have simply chosen to use the term "burn" with severity, mainly to reinforce the notion of an area where fire occurred some time in the past.

## Discussion of Ecological Burn Severity

No common standard has emerged to measure burn severity ecologically. There may be, in fact, many valid ways to view burn severity, depending on the scale and the particular means available to measure it. On the other hand, a fundamental concept of burn severity—as we suggest here as a magnitude of change—may lead to designing measures of severity that are at least compatible over multiple scales.

How investigators choose to measure ecological burn severity is closely linked to the objectives of burn evaluation. In most cases, it is scale dependent, so definitions reflect the detail and complexity of systems described. You may be interested, for example, primarily in only one factor, such as potential for herbaceous recovery. In that sense, severity may be understood and scaled directly by a single measure, such as depth of charring or scorching into soil, and this measure may be well suited for evaluating small areas, but the method would be difficult to implement over large areas. There are literally thousands of individual ecological components that might be used to indicate severity. To some extent, each species potentially responds in a unique way to fire, and depending on objectives, change in abundance of just one species may be most relevant to describing severity.

In landscape ecology, however, we tend to look at burn severity holistically, such that it represents an aggregate of effects over large areas. This enables you to map and compare whole burns composed of many communities that occupy various topographic, climatic, and edaphic situations. Here, severity is three dimensional, spread over multiple components and strata of the community and across units of area that almost always display considerable heterogeneity. The overall severity of the site, then, can be viewed as the average of all that variability. Besides specific ecological consequences such as tree mortality, burn heterogeneity or patchiness is also a primary variable of interest. It reveals large-scale interactions of fire behavior with the environment (useful for fire modeling) and influences the kind and rate of recovery (useful for ecological projections). At the same time, it is advantageous for assessment of burns to retain some level of information about individual components so you can break those out to evaluate specific conditions.

In a broad sense, the consequences of fire in a particular area are governed by short- and long-term processes, so overall severity is an amalgamation of factors. The most immediate effects are on the biophysical components that existed on a site before the fire. Downward and upward heat transfer generated from fire intensity directly causes those effects. The amount of downed woody fuel consumed, or the biomass of living canopy that was killed, are examples of this, and we refer to it as the short-term severity or first-order fire effects (fig. LA-6). Those effects, though, are dependent on sensitivities of the components where fire occurs, which are far from equal across the landscape. For instance, it is well known that some species have adaptations that make them more resistant to fire than others. The implication is that the same fire intensity can produce different degrees of initial burn severity, depending on the community's prefire composition and structure. Thus, severity likely does not vary in parallel with intensity, especially through low-to-moderate ranges, though the two variables are obviously related. At highest ranges of fire intensity, however, even fire-adapted species are likely to be severely impacted.

Beyond that, the longevity of impacts and the nature of postfire responses are influenced by a number of locally unique conditions, including:

- The kind of seed bank species present, and whether or not they are able to mature under fire-altered microclimate and soil.
- Proximity to adaptive seed sources from unburned areas.
- Localized site properties, such as slope, aspect, and soil moisture-holding capacity.
- Successional pathways and the successional stage when the community burned.
- Subsequent climate, which may differ from historic climate existing when the prefire community became established and matured.
- Secondary ecological effects initiated by fire, such as erosion and mass wasting.

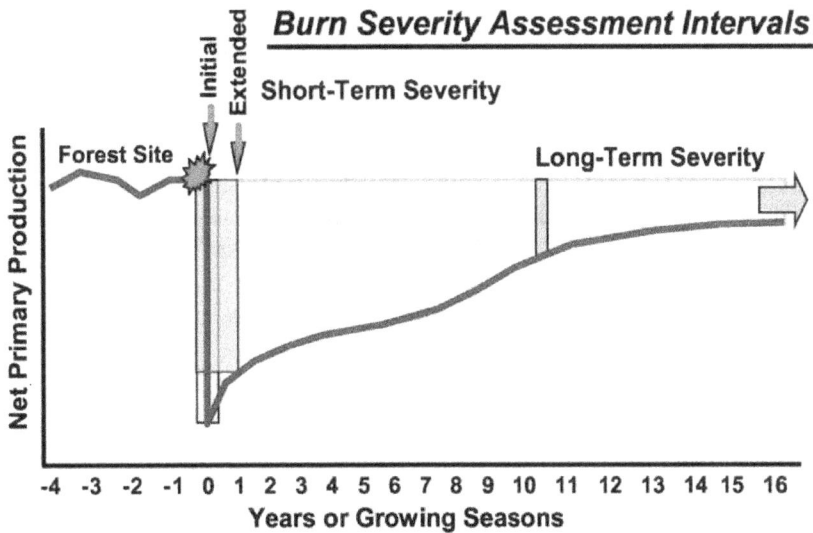

**Figure LA-6**—A hypothetical 30 x 30 m forest site experiences a fire. Short-term severity reflects change to prefire community components (vertical bars). Long-term, it reflects both that change, plus unique site conditions that prevail into the future. In reality, many response variables are aggregated over many strata of the forest to determine the overall severity on the site.

These combine with initial effects on established components to shape *long-term severity* (fig. LA-6), the magnitude of long-term change brought by fire. In most cases those local circumstances can be estimated, or at least inferred for an area, so it is really the spatial variation in short-term severity that must be determined. If that is known, then projections about long-term severity can be worked out. Thus, in the LA methods, we focus on first-order effects and attempt to define and map severity as it relates to the *magnitude of change to components existing at the time of fire*. To an extent, that includes near-term vegetative survivorship of the next growing season, which incorporates recovery and delayed mortality of burned vegetation as major expressions of short-term severity.

The measure of severity across landscapes, we propose, is first a combination of factor effects within strata, and then a combination of strata effects within communities or subregions of the burn. Such may be difficult to conceive, but if the degree of change is the focus, one can envision a numeric scale with zero (no change) being the starting point, and some positive number as the highest possible amount of change. We can apply the same scale to each stratum of the landscape, and combine those to derive an overall measure for an area. At the lowest level, factors within strata are rated independently. Then, factor values are averaged per stratum, and likewise, strata are integrated into higher levels to ultimately derive the severity of the whole community. The criteria may differ by stratum, but the scale applied to all is the same. It is the full range of change between no effect and greatest possible effect (due to fire), which forms a common denominator. The measure of severity, then, is a consistent numeric scale gauging the amount of change. It may represent a single factor or a composite of multiple factors, depending on intent.

To successfully assess burn effects across landscapes, two methods are required, one for remote sensing and one for field validation and calibration. If comparable remote sensing data are available from before fire and after fire, magnitude of change can be determined empirically, as described in the **Remote Sensing Measure of Severity: The Normalized Burn Ratio** section later in this chapter. Thus, the proposed definition of severity can fit relatively easily with available remote sensing technology, so long as guidelines on timing are followed. For field estimation, however, you must judge how much change occurred relative to prefire conditions for the individual rating factors and that can be difficult, given a typical lack of prefire data for most burns. Also, it is not ordinarily the case that significant portions of many large burns can be visited within 1 year after fire, and alternative information, such as aerial photography, is rarely available both from before and after fire. Consequently, you must rely heavily on expert knowledge and judgment when gathering field data.

Ways to sample for ecological severity in Landscape Assessment are presented in the **Ground Measure of Fire Severity: The Composite Burn Index** section in this chapter. The breakout of strata and the rating factors for the CBI are discussed in detail in the **Field Documentation** section. Basically, they boil down to phenomena we can observe. Some pertain to the amount of organic material consumed and characteristics of residual inorganic carbon and ash, while others address short-term potential for vegetative regeneration and mortality. The amount of heating is also inferred by estimates of scorching, or changes in amount and color of exposed mineral soil. The selected factors are only a manageable subset of all the possibilities for judging severity. They were the ones that collectively seemed most recognizable and significant, while being most relevant to remote sensing and radiometric response.

## GROUND MEASURE OF SEVERITY: THE COMPOSITE BURN INDEX

These methods are used to derive index values that summarize general fire effects within an area, that is, the average burn condition on a plot. They are designed for moderate-resolution remote sensing applications, assuming a landscape perspective of entire burned regions. As such, plots are fairly big and widely spaced (>90 meters apart). Field data are relatively quick to collect (about 30 minutes per plot), relying mostly on ocular estimation and judgment. This allows a representative number of plots to be sampled effectively over large areas. The primary task is to encompass the range of variation found within burns, covering as many fire effects and biophysical settings as possible.

A characteristic of sampling is that average conditions of many factors are considered across multiple strata to derive the severity value for a plot. As such, the approach has been named the Composite Burn Index, or CBI. It logically parallels the way Landsat satellite sensors average all features within a pixel to record the multispectral brightness values used to model burn severity.

CBI information is not solely limited to remote sensing, however. Field data can stand alone for general burn assessment, as a way to summarize conditions exceeding a few hectares. Methods work at stand or community levels to estimate the combined severity of individual factors. Data may be useful for reconnaissance, rapid assessment after burning, planning rehabilitation, documenting results of prescribed burns, or any activity where burn information needs to be gathered relatively quickly over large areas. Other methods within FIREMON address smaller scale sampling for detailed fire effects on individual components of a community. Those can complement the CBI when more site-specific information is needed.

The landscape sampling design is hierarchical and multilayered. Each stratum of a vegetative community is evaluated independently by several criteria and given a rating. Scores are decimal values between 0.0 and 3.0, spanning the possible range of severity between unburned and highest burn effect. Scores may be combined (averaged) to yield aggregate CBI ratings for the understory, the overstory, and the total plot. Table LA-1 shows the three composite levels (lettered) and five strata (numbered) currently used. The total plot CBI comprises all five strata, when all strata are present. When plots do not contain all strata, those missing strata are simply not counted. Ratings may be reported separately by strata or in their composite forms, depending on objectives.

## What Do CBI Values Mean?

The CBI provides an index to represent the magnitude of fire effects combined across all strata per sample area of a community. Ratings incorporate such factors as condition and color of the soil, amount of vegetation or fuel consumed, resprouting from burned plants, establishment of new colonizing species, and blackening or scorching of trees. As a continuous numeric value, the CBI is useful for correlation with environmental variables, such as plant productivity or fuel loading, and is well suited to communicate to managers and researchers the salient attributes of burns. For example, you might calculate a CBI score of 1.4 for one area and over time compare vegetative recovery there to another area with a CBI of 2.3. In addition, the CBI may be stratified into ordinal levels as a basis for tabulating area statistics or aggregating effects. You then may report impacts in terms of low, medium, and high severity merged over multiple burns, for example.

The CBI attempts to answer how ecologically significant the consequences of a given fire are, or how much fire has altered the biophysical conditions of a site, by providing the numeric scale for gauging such changes. CBI is not uniquely weighted for different community types. Rather, by defining severity, it attempts to gauge the magnitude of change from prefire conditions; thus, it should provide comparable values regardless of community type, location, or time. It is inherently related to prefire conditions and not an absolute value, like weight of fuel per unit area. A distinction is made that a given fire intensity can produce variable degrees of burn severity, depending on site or vegetation characteristics. For instance, low to moderate fire intensity will likely generate more severe consequences in stands of thin-barked tree species such as spruce, than on thick barked species such as ponderosa pine. The CBI should register such differences appropriately, yielding higher values in the spruce stand than in the pine

**Table LA-1** Three composite levels (A C) encompass the five strata level (1 5). CBI scoring is completed for each strata and averaged to the desired composite level.

| A. Total plot | B. Understory | 1. Substrates |
| | | 2. Herbs, low shrubs and trees less than 3 ft (1 m) |
| | | 3. Tall shrubs and trees 3 to 16 ft (1 to 5 m) |
| | C. Overstory | 4. Intermediate trees (pole sized trees, subcanopy) |
| | | 5. Big trees (upper canopy, dominant/co dominant trees) |

stand, even though the fire intensities for both were about the same. Thus, CBI does tend to reflect a community's sensitivity to fire when fire intensity is constant.

A second example is worth considering. In herbaceous communities, fire intensity is normally lower than in forested areas, due to less fuel loading and stratification. Consequently, burn severity is usually less and more ephemeral than in a forest. In most cases, fire actually enhances productivity of herbs through the first or second growing seasons after fire. This is partly a response to nutrient cycling and reduction of aboveground competition, while belowground heating is low and minimally damages roots or rhizomes. In such situations, CBI is predominantly based on the lower two strata of the understory, yielding low but positive overall severity scores. It rarely attains the levels observed in forest burns, and it reflects the true nature of low fire impacts to herb communities. It is not a relative measure considering herb communities alone. Enhanced productivity, above 100 percent of what it was before fire, is presently being evaluated for CBI. Currently, it is captured by variables in the herb and shrub strata, but is not yet averaged into CBI ratings.

Keep in mind, CBI is geared to correlation and validation of 30-meter Landsat data. It helps answer the question of satellite mapping performance for large burned areas. Hence, emphasis is on a large plot size, multistrata average conditions, covering the range of effects with sufficient replicates, and sampling broad areas efficiently.

## Sampling for the Composite Burn Index

Time since fire is relevant to how factors appear when a plot is sampled. Therefore, it is essential to plan field work for specific objectives, and to enter the "fire date" on the field form so you can track the timing of field data. If you are interested in short-term severity (first-order fire effects), the optimum time for fieldwork is during the first postfire growing season. This would correspond with the timing for *extended assessment*, which is the primary reference point for change from prefire conditions, as it reveals survivorship potential and delayed mortality. That timing naturally varies by ecosystem, however. It can be as long as 9 to 11 months in relatively cold climates, or as short as a few weeks in sub-tropical regions.

If plots are visited soon after fire, as in *initial assessment*, many effects will be evident, but ability to estimate survivorship and delayed mortality will be diminished. New seed germination and resprouting likely will be missed, as will effects on trees stressed but not immediately killed by fire. Soil properties may also be obscured by ash that has not had time to wash off. A solution for initial assessment may be to simply omit some factors from consideration, but that could weaken the validity of CBI. At the least, detective work may be needed to make reasonable ratings of such questionable factors.

If plots are visited beyond the first growth period after fire, short-term effects become increasingly obscure. Data then reflects elements of recovery that may largely depend on postfire climate, soil, or other factors, and only partly on the first-order fire effects establishing a site's new ecological starting point. Intervening litter fall and prolonged growth, for example, may lessen the apparent magnitude of short-term severity. Under these circumstances, assessment may require calibration of observed responses back to what they were like one growing season after fire. Counting back the annual nodes of growth on shrubs, for example, can help determine the resprout status in previous years.

On the other hand, revisiting plots several years after fire provides useful information about long-term severity and recovery rates. Imagery and plots can be compared if they have similar sampling intervals since fire. A time-series of CBI data can be used for multiple change detection experiments in remote sensing and to help verify predictions made about severity from initial and extended assessments. The timeframe may be compressed in semitropical regions to perhaps as short as 1 year, and still be practical for monitoring prescribed fire effects. For such objectives, a plot can be read straight up as it appears at the time of sampling. Some factors may be relatively static, such as percent of black tree canopy, while others may change markedly over time, such as percent of trees felled or shrub and sapling regrowth. Record "fire date" on the CBI form and do not lose sight of time as an important variable in analysis.

We recognize that ground data might never be available on many burns, given the constraints of time, funding, and logistics. Therefore, we expect some remote sensing results to be calibrated or validated with field data collected from different areas. We presently see no problem with this, so long as the burns are similar and remote sensing data are acquired with similar timing. Once a number of burns have been sampled in a region, statistical confidence in the remote sensing results should increase to a point where the need for new ground data should diminish. It can continue to be collected, but only to spot check results. That is, in fact, one goal of the whole process, so field time and expense can be minimized, without sacrificing reliability and availability of burn information. In all cases, the level of validation should be documented in the FIREMON Metadata table.

## Sample Design and Site Selection

For remote sensing applications, we prefer a stratified sampling design that attempts to represent the full range of severity with equal sampling effort across it. The main objective is to analyze statistical association between burn severity observed on the ground and the variation derived from satellite data. There is no need for CBI samples to estimate the spatial composition of the burn itself because the remote sensing product will eventually provide a complete tally of the entire "population" of burned pixels. Thus, the sample of CBI plots only calibrates and/or validates remote sensing with less necessity for strict spatial randomization.

In most cases, depending on burn size and complexity, 50 to 100 plots per burn are adequate. This should yield suitable sample size to determine how well remotely sensed severity correlates with field measures. If several burns from the same year are being assessed in one area, plots can be distributed over multiple burns. If total burn area is larger than about 2,500 acres (1,000 ha), the number of plots is somewhat independent of burn size. Large burns tend to exhibit an adequate range of severity in proportional area sufficient for sampling. On the other hands, smaller burns are typically not as diverse and may be covered by as few as 20 to 40 plots. To spatially stratify the sample, limit plot selection to suitable locations defined by accessibility and data-content factors. Ownership, topography, and distance to roads and trails are key elements when overlaid in a GIS to mask out unsuitable areas.

If the delta NBR (dNBR) is available (see the **Remote Sensing Measure of Severity: The Normalized Burn Ratio** section), additional stratification based on the severity level and the amount of localized heterogeneity may be warranted. You would attempt to draw locations from small areas that show minimal spatial variation in dNBR. Such stratification can be based on the 3 x 3 pixel matrix of dNBR for each pixel being within a range of about 0.15 dNBR of its neighbors (maximum–minimum– $\leq 0.15$), or 150 if dNBR is scaled by $10^3$. Some experimentation may be necessary to arrive at minimum variation that offers sufficient areas for sampling, with the cutoff being as low as practical. One may find enough sample pixels, for example, with the cutoff for local variation being set as low as 0.10 dNBR (100 if dNBR is scaled by $10^3$). Pixels satisfying that requirement would tend to be located in areas of fairly homogeneous severity, and would generally represent surrounding pixels. Sampling those pixels would lessen problems associated with georectification accuracy of satellite imagery and locational error in the field. It would help ensure that plots were sampled from the intended areas. It would also help associate a given CBI value with the appropriate dNBR value in analysis. Moreover, it would reduce the chance of vastly dissimilar or nonrepresentative nearby pixels from affecting the site through autocorrelation of reflectance from adjacent pixel values. Likewise, such pixels would have less chance of corrupting a local dNBR average, if a pixel-neighborhood approach were used to compare dNBR to the CBI.

After suitable sample areas are delineated, attempt to randomly draw roughly equal numbers of target locations from each of the general ranges of severity: unburned, low, moderate-low, moderate-high, and high. Plan for about 10 to 20 plots in each of these levels. Unburned sites will be visited only to verify that none of the plot burned, so effort per plot should be minimal in those cases. A histogram of pixel dNBR frequencies within the burn can be used to check whether or not selected locations adequately sample the range of variability within the burn. At this stage, the severity levels are just based on best available knowledge. They are likely to change slightly during the course of the survey, so anticipate some adjustment in sample target areas as work progresses.

The way locations are selected within the stratified sampling area can be either random or nonrandom. Stratified random locations raise fewer doubts from independent reviewers and may be more appropriate for some statistical analysis. There is often a trade-off in effort and doable sample size with spatial randomization, however. Objective but nonrandom selection gains substantial field efficiency by the ability to group plots near one another or along routes traversable in a day. More plots can be done that way, and sampling can often achieve adequate representation with fewer plots across the total range of severity than can random locations. Furthermore, judgment in the field can be exercised to pick up additional plots while walking through large areas that seem to represent certain burn conditions but were overlooked in preselection.

Spatially random design without stratification is not advised in large or remote burns because access to many sites will be prohibitively time consuming, and many will end up being excluded because they fall in highly heterogeneous or otherwise unsuitable areas. Moreover, randomization leads to sampling the levels of severity in proportion to the area covered within each burn, not in equal numbers per severity level, and each burn is unique in regards to its spatial composition of effects. Total randomization generally requires more plots than stratified random or nonrandom sampling in order to collect enough plots from severity levels of limited distribution within a burn. Thus, random sampling would only be called for if one hopes to represent the entire population of burned pixels from the sample or attempts to estimate the spatial composition of the burn. By contrast, the objective here for CBI plots is to represent all levels of severity more or less equally for the limited purpose of ascertaining the nature of the relationship between severity on the ground and the magnitude of change detected by dNBR.

Keep in mind that the greatest expense of time is usually getting to sites in the field. It is best to err on the side of collecting more sample plots than to get fewer plots while trying to reach remote and hard-to-access areas of the burn.

## Plot Layout

In the field, navigate to preselected target areas by GPS and locate the plot center. In most cases that will be at the preselected coordinates. However, at times that location may not be suitable for one reason or another. Try to select locations that 1) represent the range of variability found at the site and 2) fall within relatively large homogeneous areas, preferably 200 x 200 ft (60 x 60 m) of basically the same fire effects. This allows a plot to be placed somewhat centrally in the larger area, to be representative yet not too close to adjacent areas exhibiting different fire effects. "Too close" depends on remote sensing resolution. Here we are considering 30-m Landsat data, so try to stay at least 150 ft (45 m) from the edges. Plots should be spaced at least 300 ft (90 m) apart. If more than one plot can fit within the target area, attempt to sample that as well, depending on time constraints. Though an area may look patchy or mottled with burn, it may still represent a level of severity that is characteristic of the burn, especially where impacts are light to moderate. In those cases, look for areas with the same degree of small-scale patchiness throughout. Remember, the plot and the Landsat data are integrating surface characteristics over about 0.25 acre (900 $m^2$), so try to envision the land from that perspective when in the field.

In some validation studies, it may be necessary to have a subset of preselected plot locations that are *never* deviated from in the field. In those cases, proceed to locate the plot center as precisely as possible at the specified coordinates. Attempt to read the plot even if it seems unacceptable due of edge effects, for example, and note such problems on the CBI form. A decision to keep or reject the plot can be made later. If a more representative site exists nearby, it may be worthwhile to add another plot there since the crew has already made the effort to reach the area. That plot could be used for calibration even if it may not serve in validation.

At plot center, set the GPS to coordinate averaging mode, and let it acquire data over about 10 minutes or so. Record the GPS location in UTM coordinates to the nearest meter, noting the zone number, geodetic datum, and the amount of error. Mark the plot center so it can be identified temporarily while taking plot data. From plot center, stretch two tapes out to locate the plot perimeter with a 49-ft (15-m) radius or 98-ft (30-m) diameter. In open plant communities, it may be sufficient to just lay two tapes on the ground crossing at 90 degrees at plot center. Mark the plot perimeter at the end of each tape with flagging so that the boundary is discernable. At times, if understory effects are relatively complex or the plot is too difficult to walk through, one can use a nested 66-ft (20 m) diameter plot for the understory, while keeping the 98-ft (30-m) diameter plot for the overstory. If 66-ft (20-m) understory and 98-ft (30-m) overstory plots are used, the two different plot sizes need to be marked. If you do not plan to return to the plot within a short time, remove all flagging before leaving. Recorded GPS coordinates will serve to relocate the plot if revisited in the future.

## Plot Sampling, Using the Field Form

The **LA data form** (provided in the **Field Documentation** section of this chapter) cues all information needed to calculate CBI values for the plot. Each stratum present on the plot is evaluated independently by a number of factors. The basic data are severity scores in decimal increments (0.0 to 3.0) that express the magnitude of fire's impact on the individual rating factors, such as litter and duff consumption. The criteria used to score each factor are given in more detail in the **Field Documentation** section, but they generally correspond to break points along an escalating scale of effects. For example, if the proportion of tall shrub resprouting is around 90 percent, you would assign that factor a score of 1.0; for 30 percent regrowth the score would be 2.0.

It is beneficial to work in teams of two or three. The crew can either evaluate the plot independently or together, but at some point before recording final ratings, try to reach consensus on each factor. It is beneficial to compare impressions and discuss why each score was given. This process aids consistency, adds confidence in the ratings, and generally leads to better understanding of fire ecology on the site. The team may consider understory criteria first and then the overstory. With a day or two of experience, the team can complete a plot in about 30 minutes (not including travel time to the plot).

Take a while to become familiar with the strata on the plot and the factors used to rate each stratum on the **LA data form**. Refer to the **LA cheat sheet** at chapter's end, and especially to the definitions of strata and rating factors provided in the **Field Documentation** section. Take time to walk over the whole plot, looking for clues that point to particular levels of burning. For example, dig through litter and duff and examine stems focusing on amount consumed, depth of charring, and survivorship. Consider the condition of tufts of grasses. Charred bark or buds may conceal living tissue with potential for regeneration, even though elapsed time may not be sufficient for that to occur. Also, remnant woody material on burn plots usually indicates previous vegetation structure; look for the snags or stubs of former trees and shrubs. Finally, it helps to compare nearby unburned areas to gauge what might have been present on the plot before burning; base similarity for comparison on topographic factors, soil type, and the density and size (age) of trees. Keep this in mind when traversing unburned terrain on the way to burn plots.

Once comfortable with general understanding of the plot, enter the plot descriptive information at the top of the CBI form. Then determine and enter the amount of burned area within the 98-ft (30-m) plot (the percent of plot area showing *any* impact from fire). Make that determination also for the nested 66-ft (20-m) plot if it is used. Area estimators at the bottom of the form can be used throughout the exercise to help resolve percent cover in a plot. Also, for each stratum there are a few variables to describe prefire conditions. They are entered early on to serve as benchmarks for estimating the amount of change called for later in some severity factors. These preliminary entries are important, so please make every effort to record them.

Proceed down the data form, stratum by stratum, scoring each factor between 0.0 (no burn effect) and 3.0 (highest burn effect). Factors are designed to take decimal scores, so you may score a 2.8, for example, if you feel the condition is not quite a 3.0 but definitely more than 2.5. It is important to decide where a rating falls in reference to average conditions over the whole plot. In other words, if there is patchy distribution of both moderate and low burn effects, average those aerially together and mentally determine an overall score. It likely will be somewhere in the middle, depending on proportions of each level observed across the plot. Scores generally reflect the degree of change from the preburn state, for example, the proportion of fuel consumed. If herbaceous cover was sparse before fire, for example, you would not necessarily give a high score to sparse herb resprouting observed after fire. It would depend on how the herbs are doing relative to how dominant they were before fire.

If some factors (or strata) do not apply to the plot, then do not count them. On the data form, however, you should note why the factor was not assessed so people do not think you simply forgot to record it. The entry may either be "not applicable" (N/A) when the rating factor is insignificant or not present, or "uncertain" (UC) when the factor may be present, but one cannot make a reasonable determination about it. Such fields are ignored during analysis. For example, the plot may have big trees but no intermediate trees. Then the intermediate trees simply would be N/A and not be rated, so the overstory CBI would be based on just the big trees. Do not score nonapplicable factors as zero. A zero score means that a factor was present and it was ratable at a level determined to be unaffected by fire. Zero ratings *do* get averaged into CBI scores. The N/A and UC codes cannot be entered in the FIREMON database. When entering data in FIREMON leave blank any fields where you recorded N/A or UC. An explanation of why certain fields were not assessed can be entered in the comments section on the data entry screen.

When all factor scores are entered, calculate an average rating for each stratum and the CBI for understory, overstory, and total plot. That is done by adding up scores within each hierarchical level and dividing by the number of *rated* factors. To get overstory CBI, for example, total intermediate tree and big tree scores and divide by the number of factors rated for both intermediate plus big trees. For total plot, add up all factor scores and divide by total number of scores rated in all strata. Note, when entering data using the FIREMON software, CBI values are calculated automatically by the database, which helps to double check scores entered on the field form.

Before leaving a plot, review the CBI ratings. See that they make sense and adequately correspond to interpretations of the plot. If not, examine strata ratings; consider and discuss which one(s) may be questionable and why. You will often find that severe impacts in one stratum are mitigated by lesser impacts in other strata. This is an intended result when aggregate CBI ratings are calculated. Differences of a point or less between observers on individual factors usually make little difference in overall CBI ratings. Thus, it generally is not worth adjusting the disparity in single factor scores unless they are quite different after complete CBI ratings have been calculated. Remember, you are interested in cumulative impacts of fire over the vertical structure in a community and over the whole area of a plot. The first impression of a plot may be biased toward obvious conditions in one stratum, but when all strata are considered independently and averaged, one often agrees that the "real" composite effect is different from that first impression.

Finally, enter community notes or comments about burn patterns within the plot. These are important attributes to know in subsequent analysis. Things to consider here include: height and density of the various strata, dominant species present per strata, general fuel characteristics, general microclimate and moisture, topography, evidence of insects or disease, and any descriptors about the burn mosaic. Other comments may refer to the suitability of the plot, for example, if the plot straddles an edge or has signs of postfire disturbance like salvage logging.

There are two reasons to note observations that may not directly relate to the CBI. First, fieldwork is expensive and time consuming, so it is most efficient to gather all potentially useful information the first time and avoid a site revisit by you or others. Second, fire affects many things that may be of interest to others in a variety of disciplines, and it is always beneficial to demonstrate a service to the areas one visits; thus, recording any information that may potentially be useful is encouraged. That may include observations of rare plants or weeds, cultural resources exposed or affected by fire, interesting wildlife sightings (including carcasses), as well as erosion or water quality evidence. It is always good to communicate with local resource and cultural specialists before fieldwork to be better informed on what to look for and to find out what may be of interest.

Field interpretations for the CBI are forced to be a little fuzzy and based on best judgment. They must assess change to a site, usually without quantifiable data on what was there before fire. In addition, some estimated effects are time dependent and may not become manifest for a while after fire. These are just inherent circumstances of burn evaluation. It is much like forensic ecology. Do the best you can, and with experience you will get increasingly comfortable untangling the sometimes puzzling evidence of burn severity.

## Plot Photos

It is a good idea to take photos after completing the rating exercise, when you are most familiar with plot burn conditions. There are many approaches to this, but some recommended procedures include the following. Use a high-resolution digital camera, or a 35 mm camera with color slide film, ASA of about 125. Take at least two photos approximately 180 degrees opposite one another, showing the plot center and about half of the plot in each. Avoid taking photos directly toward the sun, especially under darkened forest canopy. Include a signboard for scale and to identify the date and plot number in the picture. As time or objectives allow, take other photos targeting features of interest, such as typical charring patterns on substrates or trees, and regrowth on perennial herbs and shrubs. Try to capture tree canopy effects as well as ground effects. It is also useful for training or presentations to take photos specifically to represent different severity characteristics or fire scenarios, and problems encountered in the field, whether or not they fall on a plot.

# FIELD DOCUMENTATION

## Strata Definitions

With the CBI, fire effects are assessed somewhat independently within strata because vertical levels in a community have different biophysical components, and multiple levels impart structural complexities that profoundly influence fire behavior. Structure affects how wind shapes fire and the nature of available fuels. Though strata are spatially contiguous, each level may have unique combustible properties and effects may differ strongly between them.

At times in the field, one may find vague or discontinuous boundaries between strata where distinctions are not always clear cut. In such cases, try to reach consensus on what is included or excluded from each stratum and be consistent from plot to plot. If distinctions between some strata cannot be resolved, it may help to simplify the overall structure; try combining strata, or drop one out. Even if only one factor in a stratum appears to be applicable, continue to score that stratum. The main objective is to make reasonable interpretations on one to five factors within each stratum, and then to combine those to derive composite ratings that summarize severity over the overstory, understory, and the total plot. The goal is not so much a high degree of precision in rating a specific factor, as it is a consistent summary rating of severity that aggregates a variety of burn effects over multiple levels of the plot.

The strata listed below are commonly identified in a complex forest community, segregated principally by the vertical space they occupy. Their arrangement and component parts determine, to some extent, the character and connectivity of fuel, variation in flammability, and timing of seasonal drying. When evaluating burn severity with the CBI, fuel and fire behavior relationships are emphasized while strata species composition is less important. Because species may exhibit multiple life forms and occupy several strata, consider primarily where individual plants or other materials fit within prefire community structure.

*Strata hierarchical structure*

    A. Total plot (overall)

        **B. *Understory***

            1) Substrates

2) Herbs, low shrubs and trees less than 3 ft (1 m) tall

3) Tall shrubs and trees 3 to 16 ft (1 to 5 m) tall

### C. Overstory

4) Intermediate trees (pole-sized trees, subcanopy)

5) Big trees (dominant/codominant trees, upper canopy)

**Substrates**—Inert surface materials of rock, soil, duff, litter, and downed woody fuels. We include just the surface characteristics of soil, even though soil in general could be broken into substrata on its own. This is an artifact of remote sensing objectives with emphasis on understanding changes to surface reflectance. Exposed soil is considered soil or rock surface that is visible from eye level and not covered by litter, duff, or low herbaceous cover less than about 12 inches (30 cm) high. Such surfaces that are likewise visible but under taller shrubs and trees are considered exposed soil.

**Herbs, Low Shrubs and Trees (less than 3 ft (1 m) tall)**—All grasses and forbs, plus shrubs and small trees less than 3 ft (1 m) tall. Herbs are plants that die back to ground level each year. Shrubs retain persistent aboveground woody stems, from which subsequent years' growth develops. Small trees, including tree seedlings, are like shrubs, but typically have only one central stalk and eventually grow to heights far exceeding this 3-ft (1 m) size class.

**Tall Shrubs and Trees (3 to 16 ft (1 to 5 m) tall)**—Shrubs and trees generally greater than 3 ft (1 m) and less than 16 ft (5 m) tall. If trees or shrubs are between 16 ft (5 m) and 25 ft (8 m) tall, decide which stratum the life form fits best. They could be scored with intermediate trees, but only if they are distinctly treelike and have characteristics of other intermediate trees. When there is question between this stratum and intermediate trees, look at the community at large, beyond the plot if necessary, and consider whether there really are two strata with one being distinctly taller and over the other. If not, then in most situations you would elect to score only one stratum. Also, consider the life form, and whether there is dense branching that extends nearly to the ground such that fire behavior may be influenced by that in particular ways. This occurs on both counts for many pinyon/juniper communities, for example, where they would be scored for tall shrubs and trees 3 to 16 ft (1 to 5 m), but not for intermediate trees.

**Intermediate Trees (pole-sized trees, subcanopy)**—Trees occupying space between the tall-shrub/tree layer (3 to 16 ft) and the uppermost canopy; generally 4 to 10 inches (10 to 25 cm) DBH, and 25 to 65 ft (8 to 20 m) tall. If trees of this size are the uppermost canopy, then consider them as intermediate trees while not counting a big tree stratum. This stratum may itself be of stratified heights, with crown tops extending into the upper canopy. Still consider, however, that they are intermediate trees if they receive little direct sunlight from above. Actual size of the intermediate trees is relative to height of upper canopy and may vary from community to community.

**Big Trees (dominant, and codominant trees, upper canopy)**—Dominant and codominant trees that are larger than intermediate trees. They occupy the uppermost canopy and usually receive direct sunlight from above. These tree crowns form the general or average level of the upper canopy, while some individuals may extend above that.

**Understory**—This region comprises substrates, herbs/low shrubs/trees less than 3 ft (1 m) tall, and tall shrubs/trees 3 to 16 (1 to 5 m) tall.

**Overstory**—The region above the understory, consisting of intermediate and big trees.

**Total Plot, or Overall**—All strata of the plot combined.

Note: Composite scores reflect only those strata that existed before fire. As a rule, strata that are not applicable on a plot cannot be rated and are not considered in the composite ratings. If a plot contains no trees, for example, then only the first three strata would be rated. In that case, we would consider the assemblage an "understory" (even though an implied "overstory" is missing), and use only the combined understory factor scores for the overall rating. The hierarchy intends to accommodate structurally complex communities, while not requiring the presence of all strata.

Landscape Assessment (LA) Sampling and Analysis Methods

## Initial Summary of Area Burned

**Percent Plot Area Burned**: Before examining the individual severity factors within strata, record the percent surface area showing *any* impact from fire for the 98-ft (30-m) diameter plot, and for the nested 66-ft (20-m) plot, if that is used for the understory. This always reflects the area of burned substrates and low-growing plants. If there is a rare case with area of burned overstory but unburned understory, count that overstory burn as well, as if viewed from the air. Do not subtract, however, unburned overstory from the burned area of the understory. The percentages are entered early on to set upper limits for rating factors that reference burn effects by proportional area of the plot. These preliminary entries are important, so please make every effort to record them. Note that the 98-ft (30-m) plot covers about two and a quarter times the area of the 66-ft (20-m) plot. Rectangular dimensions are provided at the bottom of the CBI form to help equate plot percentages with ground area as a means of visualizing these quantities.

**Prefire Conditions**: The CBI form contains a few fields in each stratum for estimating prefire variables such as cover, depth, and density. Complete these as possible, trying to most reasonably represent those specific conditions as they would have appeared before fire. Make sure prefire nonburnable areas within the plot (soil, rock, and so forth) are estimated, so they can be used to calibrate scores later on. Reference within-plot or nearby unburned areas, or evidence such as char heights, amount of charcoal, or number of standing snags. Report cover by percent of plot area, depth in inches, and density of trees as an estimated number of individuals on a plot. The intent is to get general approximate information about prefire conditions. The values will be useful later to possibly weight strata, or to categorize plots and group them in analysis across the potential range of prefire starting points. Remember to take time to compare burned areas with unburned areas to get as good an understanding of prefire conditions as possible. If any prefire information is not feasible to estimate, then enter "not applicable" (N/A) or "uncertain" (UC) on the data form. Do not leave the fields blank on the CBI data form.

**Enhanced Growth Factors**: Fields for Enhanced Growth are provided on the CBI data form under Herb/Low Shrub/Tree and Tall Shrub/Tree strata. They are used to record whether or not fire has actually enhanced the productivity of herbs, shrubs, or trees above and beyond the level that was on the plot before fire. Productivity can be regarded as amount of green living biomass, in terms of cover, volume, and density. If plots show about the same or less estimated productivity than before fire, then these variables should be entered as not applicable (N/A). If there is cause to believe that a plot shows enhanced growth, then enter the percent productivity that is judged to be augmented by fire, with 100 percent being the same postfire productivity as prefire. An entry of 200 percent, for example, would represent double the estimated productivity that was present before fire, and 150 percent would constitute one and a half times more green vegetation than prefire. Reference similar but unburned areas in or near the plot to gauge the possible effect of fire on enhancing growth.

## Rating Factor Definitions

Calculated CBI scores for understory, overstory, and total plot depend mainly on a variety of factors being independently examined and comparably scored. Factors are grouped by strata so field samplers can focus on a particular group of potentially related effects. Strata can then be evaluated in sequence to reflect on whether or not each unit is being scored appropriately. A goal is transferability across regions, and as generically as possible, factors are designed as a framework of components that respond to fire. In addition to the CBI scores, strata-level information can be retained and used in other applications.

It may be tempting to adjust rating factors and criteria to a particular area, but that is not encouraged without carefully considering the entire framework and strategy of the approach. Factors and criteria are designed to balance each other such that ratings do not double-count the same effect or exaggerate one effect over other mitigating effects. In the end, modifications may only add unnecessary detail and diminish the applicability to other areas.

Within strata, the factors that are rated are generally common and relatively easy to observe after fire. Those selected are also ones that may influence surface reflectance directly and be likely elements of a collective signal detected by satellite. As such, rating factors may not include all the effects that may be of interest to fire ecologists and managers (for example, subsurface soil properties). Those other conditions of interest, such as exotic species, can be documented in the Community Notes section of the CBI form.

At present, severity factors are considered equal when averaged into composite levels. In reality, that may not be the case. If one could judge the overall ecological significance of a factor relative to another, factors could be weighted before averaging to improve the measure. For example, it is likely that removal of the tree canopy has longer lasting consequences than removal of litter or fuel from the substrate. However, those relationships are not easily quantified, so, at present, effects are simply considered equal contributors when averaged for CBI severity.

Most Strata Rating Factors are interpreted relative to conditions that existed before fire and not in absolute quantities. This responds to the definition of severity as a magnitude of ecological change, such that the amount of change depends on the state of the community before fire. It is particularly true for all understory ratings, and why the prefire estimates on the CBI form are so important. In addition, all factors are considered in terms of the area of the whole plot. Thus, all areas of a plot are averaged together to derive each rating, adding in unburned spots and mottled burn patterns of varying severity.

Like strata, factors that are not applicable or cannot be resolved in a plot are not rated; they are omitted from that plot's composite ratings. Moreover, if there is much uncertainty about how a specific factor should be rated, or whether it is even relevant to the plot, then that factor should be left unranked. Only the number of rated factors is used to compute averages. If a factor is not rated, enter not applicable (N/A) or uncertain (UC) on the CBI data form. Do not just leave the field blank. As stated previously, such factors are not part of the CBI average, but one wants to know whether these factors were actually assessed and it was decided not to rate them, or just accidentally overlooked and skipped.

Zeros, on the other hand, are valid entries and do get averaged into composite scores. Zeros should be used when a rating factor is applicable and exhibits an unburned condition. A zero represents no detected change in an observable factor.

Field personnel need to use judgment as to whether a factor to be rated has some minimal level of significance as a reference to burn severity on a plot. That pertains to whether or not the factor had enough presence on the plot before fire so as to show representative effects after the fire, or whether it contributed some influence on fire behavior. If, for example, there is only one large fuel item, and it covered an insignificant portion of a plot, then it may not be worth rating. That one piece of wood is not likely to provide much information about severity realized across the plot. Other examples are provided under the specific rating factors below.

If an area has burned more than once in recent years, try to find places where you can compare sites that burned only once in the most recent fire and once in the older fire(s). Look for clues that might identify the age of the evidence, so that when you are on a plot that burned twice, one can separate out the most current fire effects from older ones. For example, the sheen on charcoal dulls with age, and annual nodes on burned shrubs can indicate the years of regrowth since fire. Older burn indicators are not to be used in the evaluation of current fire effects. If reliable indicators to distinguish multiple burns on a site cannot be found, it may be best to reject the plot or skip those rating factors that are not definitive.

If a site has been rehabbed after fire, the added mulch, straw, or woody barriers should not be counted; rather, substrate estimates should be made as if that new material were not present. Any planted and growing vegetation, however, can be tallied where appropriate, such as change to species composition/relative abundance. But rehabbed vegetation should not be included as new colonizers because its response was through cultural activities rather than because of fire. The extent of rehabilitation should be recorded on the data form under Community Notes/Comments.

The primary objective for rating factors is to reach a reasonable cumulative score for each stratum based on field personnel expert knowledge of fire effects. The CBI simply helps to focus that knowledge in discrete directions and provides a standard structure to quantify severity. It is the combination of many ratings, not the precision of any one specific factor score, that gives strength to the CBI.

## Substrate rating factors

These factors are rated in relation to the substrate components that existed before fire. Recent postfire additions to substrates, *excluding soil*, are generally not considered in these ratings. That means that you should not count litter, duff, or woody fuels that accumulated after fire. Rather, you should identify and mentally remove that newly fallen material to rate what is underneath. Interpretations should be based on those substrates that were in place at the time of fire.

Make sure to include substrate areas that did not burn as unchanged (such as unburned patches and prefire areas of exposed mineral soil or rock) when estimating average plotwide changes to duff, litter, and soil. Prefire nonburnable area within the plot (soil and rock) must be estimated and entered in the prefire field on the data form so that it can be referred to later on. It is potentially important as a weighting factor for calibration of the dNBR. Exposed soil is considered soil or rock surface that is visible from eye level and not covered by litter, duff, or low herbaceous cover less than about 12 inches (30 cm) tall. Such surfaces that are likewise visible, but under taller shrubs and trees, count as exposed soil.

If there are areas of a plot that cannot possibly burn, such as large prefire exposed rock, then those areas should be treated as unburned. Some caution needs to be exercised, however, because burnable materials, such as litter or vegetation, can cover prefire rock. In these cases, the rock surfaces could have "burned" and they could be treated as burned area within the plot. Charring of rock surfaces and pockets of charcoal can provide clues to that effect. Cobbles and stones less than a 1.5 ft (0.5 m) in diameter may or may not fall in this category, depending on prefire overstory characteristics and to what degree they were covered by litter, duff, or vegetation before fire. Sites generally should be avoided if they contained more than 50 percent exposed rock before fire. Such plots would not reveal much about fire effects and would appear as essentially unburned. There may be occasion, however, to include a few such areas to confirm remote sensing results.

**Litter and Light Fuels**—Relative amount consumed of small organic materials lying on the surface of the ground, including leaves, needles, and woody material less than 3 inches (7.6 cm) in diameter. All litter is counted even though some may occur under living vegetation. Incorporate nonliving attached basal material, such as dead grass or rosette leaves. In deciduous forest, late season burns are often followed by significant leaf fall, so rate the litter as if the freshly fallen leaves were not present. The same applies to freshly fallen conifer needles. If no light fuels are present, the maximum score, based solely on litter, is a 2.0. Scores above 2.0 need to include some significant portion of light woody fuel consumption. This rating relates to percent change in the litter and light fuels cover estimated at the time of fire, not in relation to total plot area. For example, if litter and light fuels covered 70 percent of the plot and the fire consumed all litter and light fuels, there would be 100 percent change in cover, even though the change only amounted to 70 percent of the plot area. The rating in that case would be a 3.0. Prefire estimates of litter and light fuel cover can be used to weight this factor to determine percent of plot wide change. Litter should probably be assessed as "not applicable" if cover was less than 20 to 25 percent of a plot before fire.

**Duff Condition**—Relative amount consumed and charring of organic materials that lie beneath the litter and above the soil. Duff is organic material that has undergone considerable decomposition prior to the fire. All duff is counted even though some may occur under living vegetation. Do not consider fine root mass left after duff is consumed to be part of the duff rating. If there was a deep prefire duff layer, then one could use the absence of fine root mass as an indicator of intense fire, which could affect soil structure or chemistry, and influence the soil severity rating, below. Like litter, this rating relates to percent change in postfire duff cover compared to prefire cover; not in relation to total plot area. For example, if duff covered 70 percent of the prefire plot and the fire consumed all duff, the change in cover would be 100 percent resulting in a CBI score of 3.0. Duff should probably be rated as "not applicable" if it is less than 0.25 inch (0.6 cm) deep and covers less than 20 to 25 percent of a plot before fire. At such low occurrence, duff can be treated as part of the Litter and Light Fuels rating factor.

**Medium Fuels 3 to 8 inch**—This factor gauges primarily consumption of downed woody fuels between about 3 inches (7.6 cm) to 8 inches (20.3 cm) in diameter. Base consumption on the percent of volume or weight lost in relation to estimated plotwide prefire fuel load for this class. Consumption includes conversion of woody material to inorganic carbon (charcoal), as well as the complete loss of woody fuel. Generally, this factor should not be used when such fuels cover less than about 5 percent of the 65-ft (20 m) diameter understory plot (an area about 3 x 5 ft or 10 x 25 m) or when they are distributed in only one localized area of the plot. If there is not enough fuel to separately score medium and large fuels, but the fuel could be scored if size classes were combined, go ahead and score one of the two factors on the form, using the one that seems most common, and make a note to that effect in the comments field. Stumps that existed before fire can be included in this fuel size category or ignored all together, as deemed appropriate.

**Large Fuels >8 inches**—Includes consumption and charring of downed woody fuels greater than 8 inches (20.3 cm) in diameter. Base consumption on the percent of volume or weight lost in relation to plotwide prefire fuel load for this class. Consumption includes conversion of woody material to inorganic carbon (charcoal), as well as the complete loss woody fuel. This factor should not be used when such fuels cover less than about 5 percent of the 65-ft (20-m) diameter understory plot or when they are distributed in only one localized area of the plot. See note above under **Medium Fuels**, to determine when to combine medium and large fuels. Stumps that existed before fire can be included in this fuel size category or ignored all together, as deemed appropriate.

**Change in Soil Percent Cover and Color**—Increase in percent cover of newly exposed mineral soil and rock, over and above estimated prefire levels plotwide. Exposed soil is considered soil or rock surface that is visible from eye level and not covered by litter, duff, or low herbaceous cover less than about 12 inches (30 cm) high. Such surfaces that are likewise visible, but under taller shrubs and trees, count as exposed soil. Exposed rock that "burned" due to prefire covering of litter, duff or vegetation should be treated as newly exposed soil. The key interpretation is change in the percent cover. Ash and charcoal from consumed woody fuel, as well as newly exposed fine root mass within consumed duff layers, are overlooked when estimating exposed soil (that is, all the new soil below those components is considered). Change in soil color may also provide clues to severity. Base ratings on the proportion of exposed soil changing from native color to a general lightening with loss of organics at moderate to moderate-high severity, and up to 10 percent soil cover changing to a reddish color from oxidation at high severity. The amount of reddish soil varies by soil type, thus adaptation to particular ecosystems is warranted.

The following five examples are provided to help you sample the substrates:

1. If a plot had 20 percent exposed prefire rock (meaning no litter, duff, down woody fuels, or smaller herbs or shrubs covering the rock before fire), then only up to 80 percent of the plot could have burned. If fire consumed all litter, light fuels, and duff that covered the remaining 80 percent of the plot, then the estimated litter and duff consumption on the plot would be 100 percent resulting in ratings of 3.0 for both litter/light fuels and duff, and an entry of 80 percent for the percent of plot burned.

2. If 70 percent of a plot appears to be exposed rock or soil that existed before fire, you should not sample that plot in most cases.

3. A plot had essentially continuous cover of litter, light fuels, and duff before fire, and 80 percent of the plot burned (20 percent was unburned). If fire consumed all litter and light fuels and 50 percent of the duff within the area that burned, then plotwide litter and light fuel consumption would be 80 percent, and plotwide duff consumption would be 40 percent (half of the 80 percent of the plot that burned).

4. If fire consumed all litter and duff, and a fine layer of ash covers the soil, ignore the ash and treat the area as newly exposed mineral soil when rating that factor. You should scrape through the ash in places to examine the soil.

5. If a burn occurs in deciduous forest and it consumes all litter and duff before the current leaves have fallen, you should consider litter and duff consumption and newly exposed mineral soil to be 100 percent. Newly fallen leaves are not included in any of the substrate ratings.

### Herbs, low shrubs and trees less than 3 ft (1m) rating factors

As with substrates, field personnel must determine initially whether herb, low shrub, and/or small tree rating factors are sufficiently represented on a plot to justify scoring them. The stratum should have been sufficiently present to indicate severity after fire. In general, suspected prefire coverage of less than about 5 percent of the plot, or limited distribution throughout might not be enough and may lead you not to count at least some of the factors. Such cases may occur under dense conifer canopies where the prefire understory consisted solely of needle-cast litter and duff, or in other cases, where vegetation was sparse and exposed soil was relatively high. At times, however, even small cover of herbs and low shrubs can be diagnostic, so take time before concluding not to score them.

**Percent Foliage Altered**—Percent of prefire woody-species cover that was impacted by fire as estimated by change in cover from green to brown or black. This only concerns the prefire low shrubs and small trees, not grasses and herbs. It includes girdle, scorch, and torch of needles, leaves, and stems. Resprouting from the base of shrubs or trees is not considered in this estimate of altered foliage, only the prefire foliage is. In other words, the entire blackened crown of a low shrub counts as prefire foliage altered, even though it may be resprouting. The amount of resprouting does not lessen the percent of prefire foliage altered. At high levels of severity, consumption of outer fine branching on low shrubs and small trees has occurred.

**Frequency Percent Living**—Percent of prefire vegetation that is still alive after fire. This is a measure of survivorship based on numbers of individuals and not necessarily on change in cover. Include unburned as well as burned, resprouting perennial herbs, low shrubs, and small trees plotwide. Resprouting plants are ones that burned but survive from living roots and stems. Include all green vegetation as well as burned plants that have not had enough time to resprout but remain viable. Burned plants may need to be examined for viable cambium or succulent buds near growth points. Dead stems will be brittle when bent; living ones will be supple. Do not include new colonizers or other plants newly germinating from seed. Make sure to take in the whole plot in the average score, including unburned areas.

**New Colonizers**—Potential dominance within 2 to 3 years postfire of plants newly generating from seed (native or exotic) averaged over the plot. The basis for this rating is the proliferation of such species due to fire, that is, above and beyond what might be expected had fire not occurred. Relative frequency of colonizers compared to established plants may be more recognizable at first, with relative cover increasing over time. This includes herbs such as like fireweed, thistle and pokeweed, as well as new tree or shrub seedlings. It also includes increased dominance of nonvascular plants that proliferate after fire in some areas, such as fungi, bryophytes, lycopodium, and small fine-leaved moss. Such plants should be rated with an understanding of how such species respond to fire. If you are not familiar enough with the species in a particular area to be confident in the appropriateness of rating these plant populations, then you should consult local botanists for assistance.

New colonizers also include aspen suckers that generate from former trees as well as similar tree-to-shrub responses from other species. Suckers are defined as stem growth originating from underground roots or rhizomes, as opposed to originating from branches or central trunks. All suckers are counted even though some may exceed 3 ft (1 m) in height, because 1) they represent a change in life form from top-killed trees; 2) they often disperse widely from spreading root masses; and 3) they are functionally equivalent to colonizers occupying new ground recently prepared by fire. Such effects are not counted in the tall shrub stratum, because 1) other colonizers in that stratum do not seem to exist, so the factor was omitted from that stratum; 2) it seems most efficient and representative to rate suckers within only one stratum and not two; 3) tree seedlings, which suckers functionally resemble, are counted only within the herb and low shrub stratum; and 4) no matter where they are counted, they will contribute the same to the understory CBI score.

For trees, include newly seed-established conifers, such as lodgepole pine, ponderosa pine, table mountain pine, long leaf pine, slash pine, or other coniferous or deciduous trees that colonize after fire. Tree seedling response after fire can be site specific, but in general, certain tree species are adapted to fire, taking advantage of fire-prepared soil and openings.

**Species Composition and/or Relative Abundance**—Change in species composition, and/or relative abundance of species anticipated within 2 to 3 years postfire. This is a community-based assessment that gauges the ecological resemblance of the postfire community compared to the community that existed before fire. It represents alterations in dominance among species (biomass or cover) as well as potential change in the species present, such as absence of prefire species and/or presence of new postfire species. Consider the distribution of abundance or dominance among the species present after fire, compared to before fire. Such factors qualitatively determine the similarity or dissimilarity of the site from before to after fire. Increases or decreases in certain species abundance and dominance, or changes in the species present after fire, raise the score for this rating factor.

These three examples are provided to help you sample the Herb, Low Shrub, and Trees less than 3 ft (1 m) tall.

1. A plot had understory cover distributed throughout all sections that seemed to consist only of perennial herbs. If 50 percent of the plot did not burn and fire appeared to have killed 90 percent of perennial herbs on the remaining 50 percent of the plot, then "frequency percent living" would be about 55 percent plotwide (50 percent unburned plus 5 percent resprouting). Foliage altered, on the other hand, would not be rated because the site did not appear to contain low shrubs or small trees before fire.

2. By observing nearby unburned areas, it was reasonable to assume that a burned plot still contained most, if not all, of the eight to 12 herb and low shrub species present before fire. It was also apparent, however, that a few small shrub species, such as *Vaccinium* and *Ribes,* were knocked back by the fire, while dominance of two herbs, *Calamagrostis* and *Epilobium,* was enhanced. Because original community composition was largely intact, and it was mainly just two species increasing with two decreasing, change in species composition/relative abundance could fall in a range of low to moderate on the CBI form. This would depend on the magnitude of change in dominance exhibited by the species diminished and enhanced by the fire.

3. If "frequency percent living" is low, and there are large numbers of new colonizers, then most likely change in species composition/relative abundance would be fairly high.

## Tall shrub and trees 3 to 16 ft (1 to 5 m) rating factors

**Percent Foliage Altered**—Percent of prefire foliage for tall shrubs and trees 3 to 16 ft (1 to 3 m) that was impacted by fire as estimated by change in crown volume from green to brown or black. This includes girdle, scorch, and torch of needles, leaves, and stems. Resprouting from the base of shrubs or trees is not considered in this estimate of altered, only the prefire foliage is. In other words, the entire blackened top-killed crown of a tall shrub counts as prefire foliage altered, even though there may be a portion that is resprouting. The volume of the resprouting is ignored; it does not lessen the amount of prefire foliage altered. At high levels of severity, consumption of outer fine branching on shrubs and trees is evident. In fall burns and leaf-off conditions, base the score on effects to remaining boles and branches, the degree of outer branch consumption, and whether or not fire top-killed plants.

**Frequency Percent Living**—Percent of prefire tall shrubs and trees (3 to 16 ft) that are still alive after fire. This is a measure of survivorship based on numbers of individuals, not on change in cover or crown volume. Include unburned area as well as burned but resprouting tall shrubs and trees 3 to 16 ft (1 to 5 m) tall plotwide. Resprouting plants are ones that burned and survive from living roots and stems. Include all green plants, plus burned plants that have not had enough time to resprout but remain viable. Burned plants may need to be examined for viable cambium or succulent buds near growth points. Dead stems will be brittle when bent; living ones will be supple. Do not include new colonizers, such as aspen suckers or other plants newly germinating from seed. Account for potential mortality that could occur up to 2 years postfire (for example, conifer saplings that are 70 percent brown will likely die in 2 years), and make sure to average plotwide, including unburned areas.

**Percent Change in Cover**—Overall *decrease* in cover of shrubs and trees between 3 and 16 ft (1 and 5 m) tall, relative to the area occupied by those plants before fire. Count resprouting from plants that burned, plus the unburned plants, as cover that mitigates against or lessens the amount of decrease in cover. Do not include new colonizers or other plants newly germinating from seeds, including suckers that represent tree-to-shrub responses. Suckers from aspen and other species are counted as new colonizers that generate from underground roots or rhizomes, as opposed to coming from branches or central trunks. Make sure to average plotwide, including unburned areas. Account for potential mortality that could occur up to 2 years post fire. For example, conifer saplings that are 70 percent brown will likely die in 2 years.

Here is one example of how to score the Percent Change in Cover. A plot had 20 percent estimated cover for tall shrub before fire. An estimated half of that cover *did not* burn, and a tenth of prefire cover, though burned, was still green from resprouting. The remainder was burned and completely killed (no evidence of resprouting potential). The overall decrease in cover, would be estimated at 40 percent (100 percent minus 50 percent minus 10 percent), and given a factor rating of about 1.3. The prefire estimated cover of 20 percent would also be entered on the form.

**Species Composition and/or Relative Abundance**—Change in species composition, and/or relative abundance of species anticipated within 2 to 3 years postfire. Include prefire tall shrubs and trees 3 to 16 ft (1 to 5 m) tall as well as big and intermediate trees resprouting from the base. Basal sprouting from larger trees is included here because 1) the growth form changes from treelike to shrubby and then to multistemmed saplings; 2) basal large tree resprout does not seem to represent a fire effect that would be counted under tall-shrub/tree Frequency Percent Living; 3) growth tends to be restricted around root crowns and not spreading widely from the burned trunk, as in suckering; and 4) this needs to be counted somewhere, but only once, and these individuals will contribute to the large shrub/sapling community for some extended period of years. This is a community-wide assessment that gauges the ecological resemblance of the postfire community compared to the community that existed before fire. It represents alterations in dominance among species (biomass or cover), as well as potential change in the species present, such as absence of prefire species and/or presence of new postfire species. Consider the distribution of abundance or dominance among the species present after fire, compared to before fire. Such factors qualitatively determine the similarity or dissimilarity of the burned site to the prefire site. Increases or decreases in certain species abundance and dominance, or changes in the species present after fire raise the score for this rating factor.

### Intermediate and big tree rating factors (combined)

Generally for northern and western conifer forests in the United States, the sum of the first three factors—Percent Unaltered, Percent Black, and Percent Brown—will be 100 percent. That may not be the case, however, in some deciduous forests or southeastern pine forests, where crowns may have been blackened or torched but not killed and subsequently resprout. In such cases, continue to score the unaltered, black and brown factors as they appear on the site, even though they may add up to more than 100 percent. The balance of the three factors should still maintain appropriate overall ratings for severity in the overstory.

Often insects or disease can affect fire-stressed trees soon after burning. If such effects are suspected and observed within a year or two of the fire, our tendency is to include them as fire-caused change. They would be relevant to the percent brown and canopy mortality rating factors below, as well as to supplemental factors, such as percent girdled and tree mortality.

**Percent Unaltered (Green)**—Percent of prefire crown foliage volume (living or dead) unaltered by fire, relative to estimated prefire crown volume of the plot. Include resprouting from burned crowns, but not from tree bases, as unaltered/green.

**Percent Black (torch)**—Percent black is prefire crown foliage (living or dead) that actually caught fire, stems and leaves included, relative to estimated prefire crown volume plotwide; may or may not be viable crown foliage after fire. At high severity, consumption of fine branching is evident. Do not consider resprouting from black branches as lessening the percent black. In many cases, deciduous trees will not torch especially when leaves are off; yet high flame lengths from the ground may blacken virtually the entire tree. Due to the aerial intensity of such fire and its similarity to crown fire, this type of burn is also included in the percent black rating.

**Percent Brown (scorch)**—Percent of tree canopy affected by scorch or killed by girdling, in relation to the estimated prefire crown volume over the whole plot. This is foliage killed by proximal heating without direct flame contact (such as brown foliage that did not actually catch fire). It includes scorching effects at the time of fire as well as delayed mortality, often from heat impacts around tree boles and roots. Suspected insect and disease effects also may be included, if that is manifested in the crowns relatively soon after fire (within about 2 years). This avoids a need to separate burn impacts from similar-appearing and related foliage conditions caused by fire-induced pathogens. Include crowns obviously impacted by these effects, even though brown foliage may have fallen to the ground. Include deciduous trees burned in leaf-off condition that are not resprouting from crowns. In those cases, look for dead crowns or portions of crowns that do not contain any black but may show severe charring around lower trunks or at ground level.

**Percent Canopy Mortality**—Of trees killed by fire or expected to die within 2 years, this should represent the proportion of crown volume now contributed by fire-killed trees (the proportion of once-living crown volume that is now dead). Consider in relation to crown volume still contributed by surviving trees. Only count trees that are completely dead, not the fire-killed portions of crowns that may still exist on living trees. One can count completely top-killed crowns on trees that show shrubby basal resprout. The factor is viewed as the proportion of a plot's total once-living canopy now lost because of recently dead trees. Suspected insect and disease effects also may be included, if that has contributed to killing whole trees relatively soon after fire (within about 2 years). This avoids a need to separate burn impacts from similar appearing and related foliage conditions caused by fire-induced insects or disease.

**Char Height**—The average height of char on tree trunks resulting from ground flames. This is the mean height on individual trees averaged over all trees in the plot. The mean height on a given tree is determined as halfway between upper char height and lower char height. Trees on slopes typically have char running up higher on the up-slope side, and wind-driven flames usually result in char running up higher on the leeward side of trees. Include unburned trees (char height = 0) and burned trees only where demarcation of ground char height is discernable. This rating does not include the black on upper boles resulting from crown fire. Trees that do not clearly show where ground flames ended and crown fire began are not included in the score. Thus, char height may not be applicable where crown fire predominates.

Three additional overstory factors are estimated but not averaged into CBI scores. These are entered as plotwide average conditions. They provide useful information to help interpret the rating factors for the overstory. Include recent postfire insect and disease effects where appropriate.

**Percent Girdled (at root or lower bole)**—Percent of trees effectively killed by heat through the lower bark, affecting the cambium around the circumference of lower boles or buttress roots. Girdling may or may not actually char through the bark and into the wood. It is often indicated a year or more after fire by sheets of bark loosely attached or sloughing off the lower bole. Include trees either dead or likely to die within 1 to 2 years. Do not include trees killed by crown fire or other scorching to crown.

**Percent Felled (downed)**—Percent of trees, whether dead or alive, that were standing before fire but now are lying on the ground. Such trees usually result from wind throw after fire. They typically exhibit fresh upturned root masses and charring patterns different from trees that were down when the fire occurred.

**Percent Tree Mortality**—Estimated percent of trees in each of the two size classes that died on the plot from the most recent fire or are expected to die within 1 to 2 years. The percent is based on the number of dead trees postfire compared to the estimated number of living trees before fire. Mortality should be judged on compelling evidence that, with 90 percent certainty, the trees are dead or will die.

## Community Notes and Comments

Community notes or comments about burn patterns within the plot are optional but helpful in subsequent analysis. Attributes to consider here include 1) height and density of the various strata; 2) dominant species present per strata; 3) general fuel characteristics; 4) general microclimate, moisture, and topography; 5) evidence of insects or disease; and 6) any descriptors about the burn mosaic. Other comments may refer to the suitability of the plot, for example, when the plot straddles an edge or has signs of disturbance other than fire, such as postfire salvage logging or other rehab. Record any information that may potentially be useful to others, such as observations of rare plants, cultural resources exposed or affected by fire, interesting wildlife (including carcasses) as well as erosion or water quality evidence.

## Data Sheets

The **BI Field Form** is for use in the field when direct digital entry is not possible. It includes the scaled criteria for rating severity factors, which helps calibrate field interpretations among observers. The **Cheat Sheets** and **Field Documentation**, in this chapter should also accompany field crews to aid in standardizing definitions of strata and rating factors. Once back in the office, recorded data can be entered using the FIREMON database software for PD (Plot Description) and BI (Burn Index), or the National Park Service FEAT database software. Note, strata mean scores and the CBI ratings will be computed on the fly as data are entered in BI or FEAT. Check these values against those calculated in the field, and correct the data sheets if necessary. At the time plot photos are digitized and given filenames, that information is entered in the FIREMON PD table. Entry of digital plot photo filenames can also be made on the appropriate data sheet.

# REMOTE SENSING MEASURE OF SEVERITY: THE NORMALIZED BURN RATIO

## If Unfamiliar With the Science of Remote Sensing

Many excellent references to general principles, methods, and applications of remote sensing can be found on the Internet. A recommended subject matter text is *Remote Sensing and Image Interpretation*, by T.M. Lillesand and R.W. Kiefer (1994). Most fires over 500 acres that occur on public lands are now routinely processed for perimeters and severity using the dNBR. Data are continually made available by USGS and USFS over the Internet; therefore, there is less cause for individual managers or scientists to produce these results, except perhaps under special circumstances or for research. Still, the discussion below serves to document the methodology, and provide the necessary background for both producers and users of results.

## Introduction to the Normalized Burn Ratio (NBR)

Raw Landsat multispectral data contain a wealth of information about earth's features. Each spectral band responds in unique ways to surficial characteristics such as water content, vegetation structure, productivity, and mineral composition. When brightness values of multiple bands are combined in mathematical algorithms, information about targeted features can be enhanced, isolated, and analyzed. From available raw data, the challenge is to develop a specific index providing an optimum measure and useful signal for fire-effects, given the available bandwidths to work with. The index we developed is called the Normalized Burn Ratio (NBR). It is similar in construct to another standard index, called the Normalized Difference Vegetation Index (NDVI). The primary difference is that NBR integrates the two Landsat bands that respond most, but in opposite ways to burning (fig. LA-7). Those were determined to be TM/ETM+ Band 4 and Band 7. The NBR is calculated as follows:

Equation LA-1 $\qquad NBR = (R4\text{-}R7) / (R4+R7)$

where $R$ values are the calculated per pixel "at satellite" reflectance quantities per band, which have been corrected for atmospheric transmittance.

Based on experience in generally forested ecosystems of the Western United States, $R4$ decreases while $R7$ increases from prefire to postfire TM/ETM+ acquisitions. The change is greatest in magnitude compared to other bands, and the variance within burns is greatest for $R7$. The combination of those traits appears to provide the best distinction between burned and unburned areas. It also provides an optimum signal, over other Landsat band combinations, for information about variation of burn severity found within the burn. The $(R4\text{-}R7)$ difference is scaled by the sum of the two bands to normalize for overall brightness that is consistent across the bands. It helps remove within-scene topographic effects and between-scene solar illumination effects. This effectively isolates the real reflective differences between the bands, which enables spatial and multitemporal comparison of the derived NBR values.

To isolate burned from unburned areas and to provide a quantitative measure of change, the NBR dataset derived after burning is subtracted from the NBR dataset obtained from before burning, such that:

Equation LA-2 $\qquad dNBR = NBR_{prefire} - NBR_{postfire}$

This measured change in NBR, delta NBR, or dNBR, is hypothesized to be correlative in magnitude to the environmental change caused by fire (the burn severity as it relates to fire effects on previously existing vegetative communities). Assuming unburned terrain is relatively similar in phenology and moisture between the two sample dates, and the two datasets are adequately coregistered, background areas take on values near zero in dNBR. Likewise, burned areas assume strongly positive or negative values, depending on whether the fire distresses or actually enhances productivity on the site. The latter

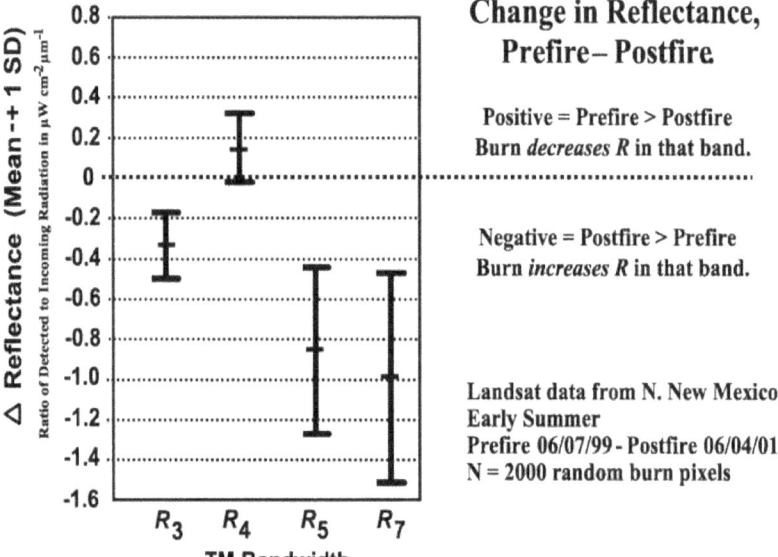

**Figure LA-7**—Change in band reflectance and the variation from before to after fire is indicative of each band's information content for severity. This example is typical for most burns. R4 shows the greatest decrease while R7 has the greatest increase after fire. Their difference yields the greatest range of change across all bandwidths.

can occur in herbaceous communities where severity is light and ephemeral, and burned vegetation responds quickly with renewed vigor from the release of nutrients or other factors after fire. Strongly positive dNBR is more typical, however, in forested and shrub-dominated areas where fire generally creates longer lasting conversions of biomass to less productive or earlier successional states.

In either case, burned areas can be suitably distinguished from unburned, and potential exists for a wide range of dNBR within the burn (depending on actual characteristics of the subject fire). This range appears to resolve the breadth of fire effects, revealing the complexity and spatial heterogeneity of the burn. It also appears to provide a broader range than other radiometric indices tested, such as the differenced NDVI. Results are constrained by the 30-meter Landsat resolution, however, which makes the index appropriate for landscape perspectives that yield whole-burn spatial data on severity.

## Timing of Landsat Acquisitions

The dates from which pre- and postfire data are acquired by the TM/ETM+ are extremely important. If not carefully considered, they may be the primary pitfall leading to unsatisfactory results.

One contributing reason is that our approach involves change detection specifically targeting burned areas. Ideally, results show only change caused by fire, with all other surface features remaining neutral so as to elucidate the unburned areas that did not change. Unfortunately, unburned features in the landscape do not remain static over time, because they are naturally altered by wetting and drying, and cycles of productivity (fig. LA-8). As a result, the NBR difference, though apparently very good at resolving burn variations, still may be affected by such factors. (Other band combinations tested were influenced by seasonal acquisition timing to a greater degree than NBR.)

Therefore, to better isolate and enhance the burn signal, the pre- and postfire Landsat datasets should be chosen to represent moisture content and phenology as similarly as possible. This timing is relative to localized growing seasons, which may vary by date and location from year to year. Landsat scenes should be compared in false color for indications of seasonal differences (Bands 4, 5, 7 or 7, 4, 3 for red, green, blue, respectively). Two helpful characteristics to key on are 1) the productivity indicated by Band 4 in herbaceous and shrub communities, which typically shows strong seasonal patterns of regrowth, decline and curing out, and 2) the pattern of seasonal snowmelt or other regular change in surface moisture. Available Landsat scenes can be viewed on line to appraise these trends:

**Figure LA-8**—Seasonal difference in phenology is evident in false-color Landsat images from the Lamar Valley of Yellowstone National Park, Wyoming. The unburned landscape is not static as communities progress from productive (left) to unproductive states (right).

http://edcsns17.cr.usgs.gov/EarthExplorer/

http://edclxs2.cr.usgs.gov/

(See the links for ordering Landsat data provided in the **Recent Landsat Satellites** section of this chapter).

Of annual periods, early to middle growing season dates seem to yield best results. That is when unburned vegetation is green and lush (orange-red in fig. LA-8), showing peak contrast with areas affected by fire. Results remain good through late in the growing season, but as areas dry and deciduous plants cure (blue-gray in fig. LA-8), some resolution of the burn may be lost. In tests, NBR tended to minimize this issue, unlike NDVI, which degraded more strongly late in the season. Especially, distinction between unburned and low burn severity tends to diminish late in the growing season. By then, cured-out vegetation can mimic fire effects, and burn effects show less contrast against the background of unchanged but dry vegetation.

With these issues in mind, optimal timing of TM/ETM+ acquisitions may be difficult to achieve given cloudiness and the 16-day return interval of the satellite. If so, consider increasing options by reviewing even predominantly cloudy scenes. Cloud-free areas need only extend enough to encompass the burn(s). In addition for the prefire scene, data acquisition can safely occur within 2 or perhaps 3 years before the burn as long as other landscape disturbances (including interim fires) are accounted for and do not interfere with the subject burn. See discussion below for options on the postfire scene.

## Two Strategies, Initial Versus Extended Assessment

With some exceptions, many severity indicators are apparent soon after fire is out. These have to do with scorching, charring, and consumption of living vegetation and dead fuel, and with changes in the nature of exposed mineral soil and ash. The exceptions, though, are important. They concern initial recovery of vegetation and delayed mortality, which also contribute to near-term severity. Often those factors may not become evident until at least the following growing season, so some passage of time may be required to get the fullest assessment of the burn. Given these circumstances, we recommend two scenarios for processing and applying dNBR severity measures (fig. LA-9).

### Initial assessment

The Initial Assessment bears upon the most immediate fire effects to biophysical components that existed at the time of fire. It uses a postfire TM/ETM+ scene acquired when 1) the fire is as completely out as possible, and 2) when good remote sensing conditions exist. Note, these conditions cannot always be met for rapid response when rehabilitation plans need to be developed within about 2 weeks of significant burning. In order to match phenology (see discussion on Timing), the prefire scene generally

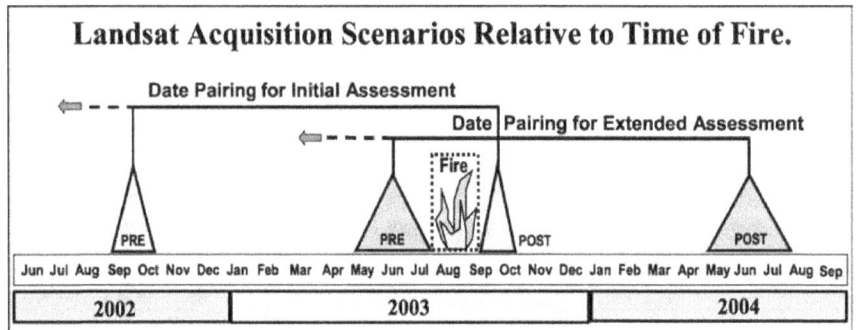

**Figure LA-9**—Initial and extended assessments require imagery from different periods before and after fire. The timing significantly affects what the NBR is measuring.

comes from a similar period of the previous year, or if necessary, the year before that. The only exception is if the fire was short-lived, and both scenes could be acquired within 8 to 24 days of one another. The latter also assumes there were no marked changes in moisture or productivity over the period. Initial assessment can provide good delineation of burned area and preliminary estimates of severity. It may not be optimum, though, for the following reasons: 1) vegetative regeneration will likely be missing, which may lead to overestimating severity; 2) unburned vegetation may be naturally cured out by the end of the fire season, diminishing burn contrast (see Timing); and 3) when the fire season extends from late summer into fall, sun angles may be low and there may be limited time before bad weather and/ or snow obscures the burn. Late-season initial assessment, though perhaps the only timing available for some emergency response applications, may show less definition of the perimeter, less data range and contrast within the burn, and poorer correlation with field data. When possible however, severity will certainly be suggested because this timing integrates at least part of the suite of factors responding to fire. Importantly, we found that dNBR was less affected by late season effects than other indices, such as $\Delta$NDVI, and concluded that dNBR held up better and was still useful in late-season. As such, initial assessment can yield figures on burn size, composition and complexity within a month or two of the fire. Products may suffice for many needs concerning public information, planning, and rehabilitation.

### Extended assessment

The second scenario provides an extended assessment, which we believe is more representative of the actual severity. It postpones acquiring the postfire TM/ETM+ scene until the next growing season, which may be as soon as a few weeks or as long as 11 months after fire, depending on the ecosystem and climate. The prefire scene is then usually taken from that same seasonal period but during the year of the fire, since that period typically falls before a given fire season. If necessary, a prefire scene can come from a year or two before fire, so long as conditions are comparable and no interim disturbances overlap the burn. By waiting until the following season, burned vegetation has had a short time to recover and demonstrate additional responses indicative of initial severity. Delayed mortality may also be evident, revealing that plants green right after fire had died by the next growth period. Extended assessments would be most useful for final portrayal and summary of short-term severity or first-order fire effects. They would be suited for projects that depend on more accurate delineation of burn heterogeneity than initial assessment, like those comparing multiple burns over space and time, testing methods, or modeling. They also might better address long-term ecological consequences, such as impacts to sensitive communities or species, or risk factors like erosion and future fire potential.

## Prefire Considerations

### General software requirements

Several image processing systems can fulfill the needs of these procedures, such as ERDAS Imagine or GRASS, which analyze geographically referenced image data. Whatever system is used, there are two capabilities that, while not unusual, may be unavailable in some systems. First, one will have to be able to write and execute algorithms that incorporate raster datasets as variables, and output raster datasets as results. Second, one will need to be able to do mathematical operations in floating-point math, which requires manipulating and storing rasters in signed 16-bit or 32-bit data formats, that is, two or four bytes per pixel, because pixels can take on positive or negative values with at least four significant digits. The software also should be able to generate topographic slope and aspect data from a Digital Elevation Model (DEM). If not part of the image processing system, a statistics package is recommended to perform regression analysis for normalization of atmospheric effects.

### General hardware requirements

Anticipate working with a large number of big data files in excess of 40 megabytes each. Steps can be taken to subset large scenes into smaller working regions, but some processes are best done on full scenes, especially if there is any possibility the data might be used for future fires or other applications.

Minimum starting disk space should be in the 10-gigabyte range; minimum RAM should be at least 256 megabytes. You will find a need to eventually manage around these limitations if work expands much beyond a couple of fire years in any one-scene area. A 21-inch graphics monitor is also recommended for on-screen interpretations and digitizing.

### Digital geographic data needs

Besides two Landsat scenes (prefire and postfire), other types of thematic GIS data are highly useful for checking registration, orienteering, or supplementing maps. A digital elevation model (DEM) of the burn surroundings is recommended. DEMs are available at no cost from the EROS Data Center if "terrain" or "precision" corrected Landsat data are ordered. Those DEMs encompass the entire area of the scene, so the file is quite large. Some basic vector datasets for ownership, roads, trails, watersheds, lakes, and streams are also helpful, as well as digital raster graphics (DRG) and digital orthophotoquads (DOQ). Finally, there may be digitized fire perimeters from the fire incident teams that can help locate the general area of the burn. Keep in mind, these perimeters may not include all burned area and may vary in the quality of reference data available, even over the time of one fire.

## Ordering the Data

Once the timing and assessment types have been decided, and the seasonal requirements for pre- and postfire Landsat datasets are understood (see above), search the Landsat archives for available scenes that best meet requirements. Both Landsat 7 and Landsat 5 data may be available, so both archives should be searched. You will need to obtain the chosen scene identifiers prior to ordering. These relate to the satellite path/row (or the geographic area covered) and the date. For more information on how to preview and order data, see the **Recent Landsat Satellites** section of this chapter.

When ordering the data, there are a number of format and processing options to choose from. For complete listings with definitions see:

http://edcdaac.usgs.gov/tutorial/daacdef.htm

For description of processing levels see:

http://landsat7.usgs.gov/l7_processlevels.html.

A sample specification is shown below for one Landsat 7 scene. Options in bold are recommended for all orders. The rest depend on your specific area, date, and working map base. If the datum is other than the default NAD83, it must be clearly specified. The highest standard processing level for ETM+ is L1G, which is geometrically corrected without ground control or relief models. We recommend, however, special-order level L1T, terrain corrected from ground control and relief models. It must be unambiguously specified, however, or processors will assume one of the lower levels. It costs extra, but it is well worth it, because terrain registrations are time consuming and may not be possible to do "in house." One additional option is to order the DEM for the scene area. This can be included at no cost, but will probably only be needed once per path/row, or not at all if a DEM is already available. As with Landsat data, map projection and datum should match what is in current use by the end user.

### Item 001

Data granule: E1SC:L70RWRS.002:2001192063
Data set: LANDSAT-7 LEVEL-1 WRS-SCENE V002
Path/Row: 43/34
Acquisition Date: 02 July 2000
Ordering Option: E1SC:L70RWRS.002:2001192063:
L1T Product - **TERRAIN CORRECTED** UTM Projection
Cost: US $800.00
Format/Media: FASTL7A: CD-ROM: ISO 9660
Additional Information: WITH TERRAIN CORRECTION USING NAD27 DATUM

ORDER Options:
  Product: **L1T**
  Projection: UTM, ZONE 11, NAD27 DATUM
  Radiometric Correction Method: **CPF**
  Band Combination: **1; 2; 3; 4; 5; 6L; 6H; 7; 8**
  Image Orientation: **NUP**
  Resampling Method: **CC**
  Grid cell size for the pan band (8): **15.0**
  Grid cell size for the reflective bands (1-5, 7): **30.0**
  Grid cell size for the thermal bands (6L, 6H): **60.0**
  Zone Number: 11

These are the recommended options, given the following factors. Further processing involves reflectance-based calculations, so terrain registration is important. Multitemporal comparisons will be made, and pixel boundaries will shift scene to scene, so the resampling method should provide the best estimate of reflectance for the geographically rectified space.

## Steps to Process NBR and dNBR

The steps outlined below are intended to be somewhat generic, recognizing that differences exist between various processing systems available to users. In general, steps identify what is needed along the way and not so much how to get there. One will need to find the proper procedures and syntax available within one's specific system. Systems usually provide analogous functions, so one should be able to adapt steps easily enough. We assume those undertaking these procedures are well grounded in remote sensing principles, and the functionality of the processing system in use. If not, one may want to consult a local expert.

**I. Initial setup for operations for each Landsat scene**:

1. Review the known facts about the burn, location, start and end dates, approximate size, and geographic distribution.

2. Plan a naming convention for the large number of files to be generated. A recommended sequence is provided in the **How-To** section of this chapter.

3. Extract the Landsat scene header file and print it out for reference.

4. Import Raw Bands 4 and 7 into image processing system (or GIS) for analysis.

5. Recommended, but optional, import other bands to explore data in false color.

6. Review the burn area displayed on the pre- and postfire imagery. Become familiar with its distribution within the surrounding landscape and juxtaposition to geographic features and named places.

7. Ensure compatible map projections exist between all data layers, and check registration of Landsat scenes, one to another and in conjunction with basic reference data (such as lake boundaries). Misregistration of more than one-half pixel should be corrected.

Note: Some image processors save rasters only as integer data, while the following calculations are done in floating point math and generate datasets of real numbers. If that is a problem, reflectance and subsequent NBR calculations can be scaled by 1,000 to retain positive or negative integer values with four significant digits.

**II. Radiance and reflectance transformations**: The procedure is used to derive "at satellite" reflectance for Band 4 and Band 7 of each Landsat scene. To expedite calculations, an executable script can be written and calculations can be combined in one algorithm. The script can then be modified for subsequent analyses by simply replacing the scene-specific parameters.

Note: these transformations are not specific to remote sensing of burns. Rather, they are recommended for any analysis that involves quantitative comparison of different Landsat scenes. They standardize the bands, account for drift in the multispectral scanner, normalize daily variation in sunlight, and optionally, correct illumination differences caused by topography. Standard order for processing mathematical operators is used.

The radiance per pixel per band is calculated as:

Equation LA-3 $$L_i = DN_i * G_b + B_b$$

where $i$ is a particular pixel, $DN_i$ is the per-pixel raw Landsat band brightness value (density number, or digital number), $G_b$ is the gain and $B_b$ is the bias for a particular band, $b$ (in this case, Band 4 and Band 7). The $G_b$ and $B_b$ are reported per band in the scene header file.

The reflectance per pixel per band is calculated as:

Equation LA-4 $$R_i = (L_i * \pi * d^2) / (\mathrm{Esi}_b * \cos(z_s))$$

where $d^2$ is an eccentricity factor for Earth-to-Sun distance, $\mathrm{Esi}_b$ is the per-band exoatmospheric solar irradiance constant, and $z_s$ is the per-scene Sun zenith angle. Here, the underlying assumption is that the Earth surface is flat, and only the sun zenith angle is used to calibrate $\mathrm{Esi}_b$ in the denominator. While topographic variables (pixel slope and aspect) have a bearing on surface reflectance, subsequent ratioing of NBR mathematically cancels those factors out, so it makes no difference whether those factors are included or not. Thus, in the case of calculating NBR, we use the simplified at-satellite reflectance algorithm. For more information on variable terms, refer to the **Glossary** section of this chapter; also, refer to the sections on **Reflectance Terms and Reflectance Incorporating Topography** in the **How-To** section of this chapter.

The reflectance algorithm yields values with a theoretical valid range of 0.0 to +1.0, or if scaled by $10^3$, that is 0 to +1,000. They are a per-band ratio of detected surface brightness to the incoming solar radiation available at the top of the atmosphere, which allows the bands to be compared directly, as in the NBR.

Note: Landsat scenes acquired through the Multi-Resolution Land Characteristics Program (MRLC) from the USGS EROS Data Center may already be in reflectance units, and would not need the processing outlined above. Those data are rescaled to an 8-bit byte range of 0 to 255, however, and can be used without modification. The data will have a slight impact by reducing the spectral resolution of input bands, but the range and other statistical qualities of NBR will be approximately the same as using 16-bit or 32-bit data.

**III. Transmittance normalization, for Band 4 and Band 7 of one Landsat scene**: This addresses the fact that atmospheric clarity varies spatially and temporally, and the ability of light to penetrate the atmosphere (transmittance) varies per bandwidth as light is scattered by particulates and moisture. If one compares multiple dates of Landsat data, such effects should be minimized to avoid influences on surface reflectance. In some cases, multitemporal datasets are so similar in atmospheric clarity that this normalization is not necessary and may be skipped. Steps should still be taken, however, to first determine if that is the case. For those wishing more rational and detail on this method, see **Atmospheric Normalization** in the **How-To** section of this chapter.

Note: Atmospheric normalization is not currently done by the USGS EROS Data Center for scenes used in national programs, such as MRLC or the NPS-USGS National Burn Severity Mapping Project. This is due to a judgment that the effort would be too complex and costly for the improvement realized, given the large number of scenes involved in such national repeat coverage.

**IV. Computing NBR for each Landsat scene, prefire, and postfire**: At this point one has pre- and postfire reflectance datasets for TM/ETM+ Bands 4 and 7. If need be, one of the datasets has been transformed by regression to normalize for atmospheric effects. The calculation of pre- and postfire NBR is then straightforward:

Equation LA-5 $$NBR_s = (R4_s - R7_s) / (R4_s + R7_s)$$

Where, $NBR_s$ is the per-pixel normalized burn ratio for scene $s$, and $R4$ and $R7$ are the calculated reflectance quantities, as described above, for the respective bands per scene. Note, the NBR has a theoretical range of –1.0 to +1.0, or if scaled by $10^3$, –1,000 to +1,000 (fig. LA-10).

**V. Computing dNBR from the pair of NBR datasets**: The NBR difference is then computed:

Equation LA-6
$$dNBR = NBR_{prefire} - NBR_{postfire.}$$

This integrates multitemporal NBR datasets into a single gradient, or a one-dimensional scale. The difference measures change in NBR that occurred from time before fire to time after fire. The dNBR has a theoretical range of –2.0 to +2.0, or if scaled by $10^3$, –2,000 to +2,000 (fig. LA-10).

## NBR Responses

By understanding how individual TM/ETM + Bands respond, one can grasp relationships of the NBR to burn characteristics. The NBR incorporates Band 4 reflectance ($R4$), which naturally reacts positively to leaf area and plant productivity, and Band 7 reflectance ($R7$), which positively responds to drying and some nonvegetated surface characteristics. Band 7 has low reflectance (it is absorbed) over green vegetation and moist surfaces, including wet soil and snow—just the opposite from Band 4.

Because NBR measures the difference of $R4$ minus $R7$, it is positive when $R4$ is greater than $R7$ (fig. LA-10). This is the case over most vegetated areas that are productive. When it is near zero, $R4$ and $R7$ are about equal, as occurs with clouds, nonproductive vegetation (cured grasses), and drier soils or rock. When NBR is negative, $R7$ is greater than $R4$. This suggests severe water stress in plants and the nonvegetative traits created within burns. There one finds, for example, decreased vegetation density and vigor that $R4$ responds negatively to, coupled with increased exposed substrates and charred fuels, which $R7$ registers positively. Charring of living and inert components, drying, and soil exposure

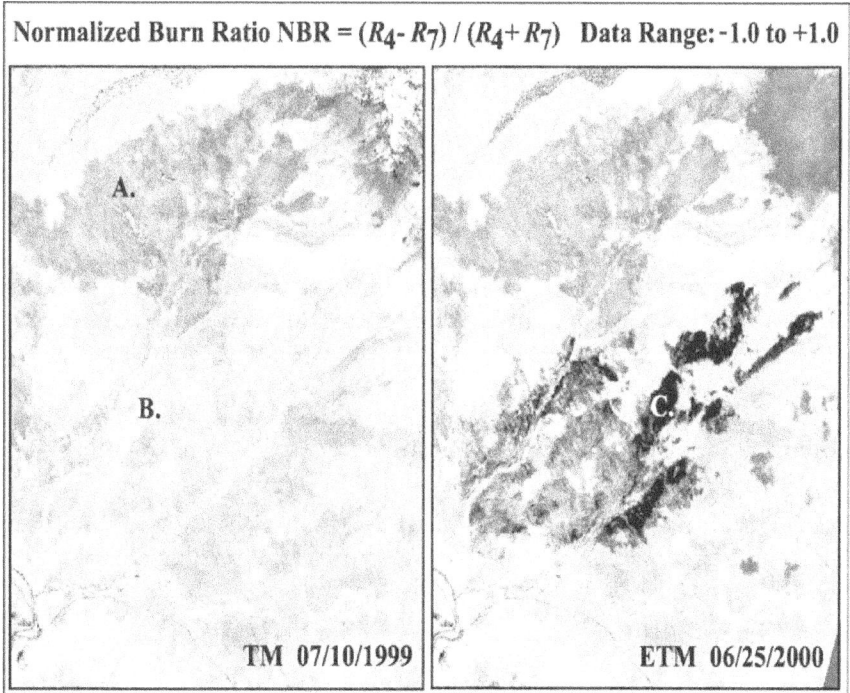

**Figure LA-10**—Pre- and postfire images of NBR illustrate A) values near zero from old burn; B) highly positive values of living vegetation; and C) strongly negative values of recent burn. Compare with figure LA-4 showing false-color composite images of this area.

enhance the signal registered by *R7* in comparison to *R4*. Results over recent burns, then, typically show near-zero to strongly negative NBR.

## Interpreting Results of dNBR

### Continuous data

Results are a model of severity. They measure the change that Landsat TM/ETM+ has been able to detect in NBR, a normalized difference of bands known to be highly sensitive to fire effects. The initial dNBR product is a continuous range of values that can be used directly for mapping and analysis (fig. LA-10). Assign a *linear* grayscale to the range of the data—one that provides good contrast—from unburned to highest burn conditions. We use a range of –800 to +1,100 dNBR that is assigned to the gray-level range of 0 to 255 (black to white). Unburned areas generally fall out as medium gray, and burns display a gradient of lighter grays with white at the high end. As hypothesized, the sequence of brightness corresponds to the gradient of severity.

Because they are a difference of normalized ratios, the units of dNBR are dimensionless. We tend to speak in terms of "points" that gauge magnitude of positive or negative change in NBR. From that we infer how strongly fire has affected a site. Individual values reference conditions averaged over the whole area of a pixel. Thus, a given value may represent either uniform distribution of one severity within the pixel, or small-scale patchy distribution of multiple levels of severity. Overall though, brightness for dNBR generally corresponds to a steady progression of effects relative to the prefire community:

1) Increasing char and consumption of downed fuels.

2) Increasing exposure of mineral soil and ash.

3) Change to lighter colored soil and ash.

4) Decreasing moisture content.

5) Increasing scorched-then-blackened vegetation.

6) Decreasing aboveground green biomass and vegetative cover.

Theoretically, dNBR (scaled by 1,000) can range between –2,000 and +2,000, but in reality it is rare for valid data to vary much beyond –550 to +1,350 (based on the scope of disturbance factors potentially affecting natural landscapes so far encountered).

Negative values result from postfire NBR being greater than prefire NBR. This may be due to clouds in the prefire image, or increased plant productivity in the postfire image. Enhanced vegetative regrowth is detected in approximately the -500 to -100 range of dNBR. A recent burn may exhibit this after one growing season if severity is light and the burn is in mostly herbaceous communities that recover quickly to exceed the productivity existing before fire. Also, older burns exhibit this as they recover vegetatively from the first year postfire into subsequent years. Pixels below about –550 are likely cloud effects, or noise caused by misregistration or anomalies in original Landsat data. Extreme negative (or positive) values also appear where data from one scene overlaps missing data in the other scene, as occurs near scene edges.

Ignoring the extremes, typical unburned signals fall approximately in the range near zero, –100 to +100. This indicates relatively little or no change over the time interval. Phenological differences between pre- and postfire scenes can shift the distribution of unburned values, sometimes as much as 50 to100 points.

Positive values occur when postfire NBR is less than prefire NBR (fig. LA-11). These may result from clouds in the post-fire scene, or from fire effects within a burn. The latter typically occupy a range between about +100 to +1,300. Values above about +1,350 are likely cloud effects. Cloud shadows do not have a pronounced affect on dNBR because NBR is a normalized ratio and is not influenced as much by brightness variation that is consistent across all bands—like that caused by shadow—as it is by inconsistent spectral differences between the bands. Cloud shadows, though, tend to boost dNBR slightly when in the prefire image, and decrease it slightly when in the postfire scene.

Landscape Assessment (LA) Sampling and Analysis Methods

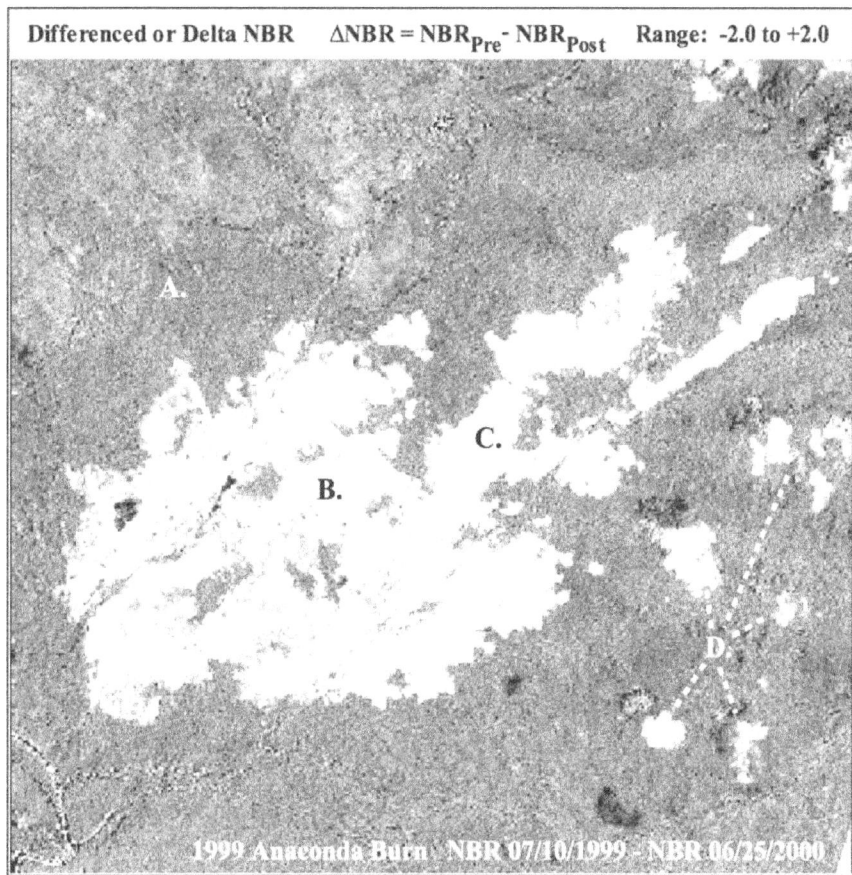

**Figure LA-11**—The difference of NBR images from figure LA-10 yields a gradient of detected change. A) medium-gray areas are near zero, indicating no change; B) lighter gray is moderate change; C) near-white is highly positive, indicating a large magnitude of change; D) is cloud in the postfire dataset.

## Burn perimeter

One of the first things to do with the continuous data is to interpret the burn perimeter. Software may be available to automate this, but even so, those results should be reviewed, and then manually edited if necessary. If one does not have much knowledge of the burn, it is highly recommended to consult with others who do. Discrimination of burned area is enhanced when guided by direct field observation as much as possible. The objective of the perimeter, as we see it, is to delineate polygons that minimally encompass all burn areas. They can be used for graphic purposes, to calculate area statistics, or as a mask for isolating burn areas in GIS overlay processes (fig. LA-12). A manually digitized perimeter is quick and suitably accurate for 1:24,000 mapping. It provides a good way to plan sampling or to communicate information about burn size and distribution promptly. With a linear grayscale image of dNBR displayed on the computer monitor, digitize on screen following the boundary of the burned area. Zoom up to be able to faithfully follow the edge. You will need to decide on a level of generalization for the perimeter because the actual boundary can be quite complex and convoluted. The amount of detail is a matter of scale, limited by data resolution and intended use. It is also useful to have the pre- and postfire false-color composite images to refer to on screen.

As a rule of thumb, try to retain the obvious character of the shape of the burn when displayed at a scale where individual pixels are not so obvious. (For example, 1:24,000). Do not attempt so much detail as to be outlining each individual pixel. By and large, err on the liberal side; try to stay one-half to a whole

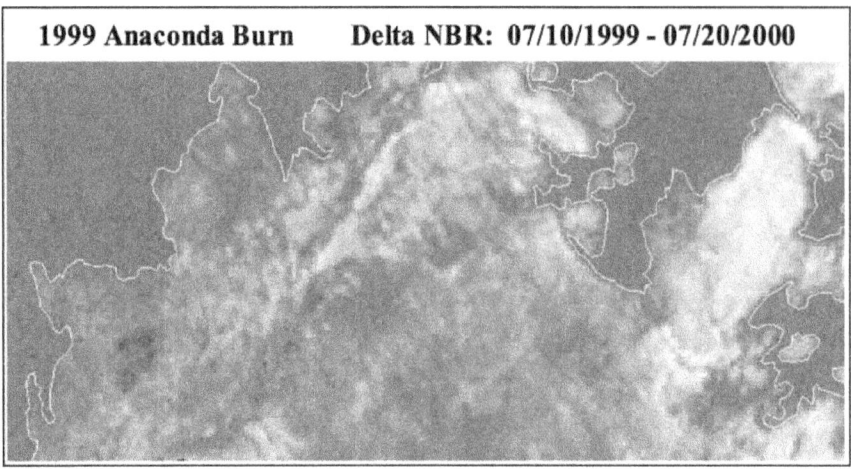

**Figure LA-12**—Enlargement of figure LA-11 showing detail of the digitized perimeter.

pixel outside the burn, and pass across small peninsulas of unburned areas that project into the burn, if less than about 3 pixels wide. Also, do not digitize around interior unburned islands, unless some specific objective requires it. In all these cases, the burned and unburned pixels included within the digitized perimeter should eventually be adequately identified by their dNBR values (for example, when it comes time for statistical summaries of the burn). Continue digitizing all disjunct patches of the burn created by spot fires, and label all polygons with a unique identifier.

**Subsequent Procedures**—Using the digitized perimeters, one can extract a histogram of all dNBR pixel values occurring within the burn(s). This is a basic reference for comparison to other burns (fig. LA-13). It supplies information on mean and variance of severity, and the frequency (or area) of values

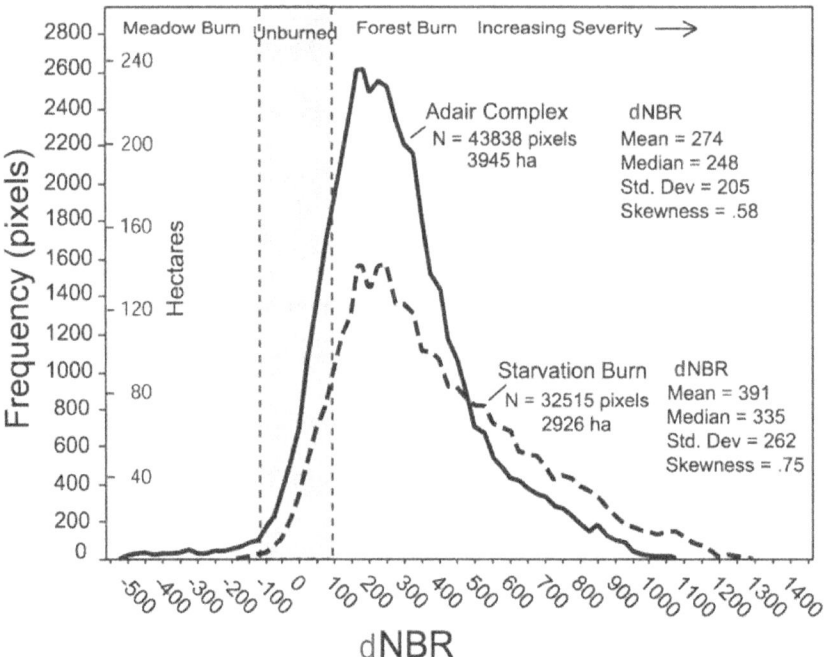

**Figure LA-13**—Frequency distribution of pixel values within two burns from 1994 in Glacier National Park, Montana. Starvation had proportionately more area burned at higher severity than Adair.

occurring across the dNBR gradient. Histograms also may be used to break down the composition and aerial extent of burned patches, such as watersheds or forest stands.

Next, you will be able to identify target areas for field sampling with the perimeter, trails, roads, and other references (lakes and streams) overlaid on the grayscale dNBR. This allows you to find accessible areas large enough to represent the range of variability within the burn, as described in **Ground Measure of Fire Severity: The Composite Burn Index** section in this chapter.

Once field plots are located and sampled, plot locations can be mapped onto the continuous dNBR dataset using GIS overlay functions, and the pixel values from those locations extracted. There are many creative ways of doing this, including multipoint averages or weighted averages within a local neighborhood of the plot. Due to burn heterogeneity and improved GPS locational accuracy, we looked for alternatives to the commonly used 3 x 3 pixel average. We found that a straight average of 9 pixels too often interjects values in the average that are greatly dissimilar, and not representative of the plot. One option is to weight the center pixel by two or three times, and throw out one pixel that has the most different value from the center pixel. Instead, however, we tend to use a five-point pixel average, where the points for sampling dNBR are the plot center, and theoretically plus or minus 50 ft (15 m) from plot center. This results in 1 to 4 pixels being sampled per plot, depending on the juxtaposition of the plot center to the pixel center. The center pixel—the one containing the plot center—is always counted at least twice, providing extra weight to the center pixel in the average. If the plot lays dead center within a pixel, then this five-point sampling yields the value of only that 1 pixel.

The dNBR values extracted for all plots can then be imported into a database that contains corresponding plot CBI ratings, or other measures of severity determined in the field. At that time, analysis of the association between dNBR and observed severity can be undertaken. Field work may also detect where revisions to the perimeter are needed. Each subsequent step adds a level of verification that should be specified in the metadata.

## Discrete ordinal data

Continuous dNBR datasets can be stratified into ordinal classes or severity levels to simplify description and comparison of burns (fig. LA-14).

The breadth and number of levels is entirely up to the user, based on requirements of the application. However, we commonly employ a seven-tiered configuration proven useful in a variety of ways (table LA-2). Value ranges of dNBR may vary between paired scenes. Values less than about –550, or greater than about +1,350 may also occur. If they do, they are not considered burned. Rather, they are masked out as anomalies caused by misregistration, clouds, or other factors not related to real land cover differences.

The first two severity levels (table LA-2) reflect areas where productivity increased after burning. They occur almost exclusively in herb communities where dNBR can be strongly negative from enhanced productivity after fire (the postfire NBR is much greater than the prefire). Typical unburned pixels occupy the range near zero. The last four levels include all other burned areas where dNBR is distinctly positive (the postfire NBR is much less than the prefire). They cover what is normally recognized as recently burned, including forest, shrub, and some herb communities.

Ordinal or nominal classes such as these are useful for a wide array of purposes like reporting aerial statistics, aggregating the statistics of many burns, stratifying for study of ecological consequences or treatment, quantifying burn heterogeneity, and mapping. Ordinal or nominal classes, however, can be quite variable case to case, depending on each project's objectives and individual perceptions of burn severity.

## Severity thresholds

Unfortunately, the threshold levels reported above are not hard and fast for all dNBR scenarios; they are somewhat flexible. Recent experience shows shifts for some burns in the range of about $\pm$ 10 to 100 points for a given severity level. At this time, we believe the primary causes for this variation are

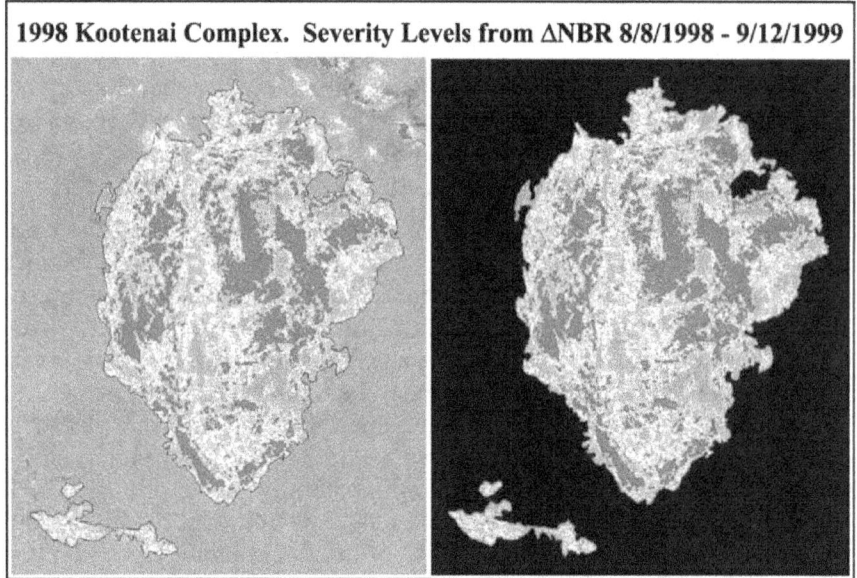

**Figure LA-14**—Stratified delta NBR image (left) and the data masked by the burn perimeter (right). Severity levels were derived from thresholds in table LA-2, showing unburned, low, moderate-low, moderate-high, and high as gray, green, yellow, orange, and red, respectively. Areas outside perimeter with elevated values are relatively easy to discern as snow, clouds, or dry patches away from the burn; refer to figure LA-2.

**Table LA-2** Ordinal severity levels and example range of dNBR (scaled by $10^3$), to the right.

| Severity level | dNBR range |
| --- | --- |
| Enhanced regrowth, high | −500 to −251 |
| Enhanced regrowth, lo | −250 to −101 |
| Unburned | −100 to +99 |
| Low severity | +100 to +269 |
| Moderate low severity | +270 to +439 |
| Moderate high severity | +440 to +659 |
| High severity | +660 to +1300 |

(dNBR value ranges are flexible; scene pair dependent; shifts in thresholds ±100 points are possible. dNBR less than about −550, or greater than about +1,350 may also occur, but are *not* considered burned. Rather, they likely are anomalies caused by misregistration, clouds, or other factors not related to real land cover differences.)

1) seasonality of the images, and 2) whether the timing is for initial assessment or extended assessment. Under extended assessment, thresholds tend to elevate for early-to-middle season dNBR as that assessment exhibits greater range overall compared to late-season extended assessment. On the other hand, initial assessment may indicate considerably higher severity—requiring higher thresholds—when the postfire scene comes soon after burning, as opposed to the following growing season. When the postfire scene is drier overall than the prefire scene, the burn-unburned threshold tends to elevate somewhat, and there is a greater chance of confusion between the driest unburned pixels and lowest severity burned pixels.

If image timing is indeed a controlling factor—relative to time since fire and time of year—then scenario-specific scales for severity may be possible to achieve in the future. In the mean time, fine calibration of thresholds can only be done individually for each dNBR model using a combination of expert knowledge and correlation to ground data. We intend to add further guidance along these lines as the number of burns analyzed in different ecosystems expands.

Before ground data are analyzed, one can interactively color up ranges of the continuous dNBR data to determine preliminary severity level thresholds by computer (fig. LA-13). We strongly advise that this procedure be done by, or in consultation with, someone who has direct knowledge of at least portions of the burn on the ground. That will greatly facilitate and improve the initial classification of severity levels by introducing an ability to recognize spatial patterns as they were observed in the field.

Display the dNBR linear grayscale, and from the low end at about –100, progressively "color up" increasing values with one color. You will find the distribution of colored pixels being randomly scattered at first around the burn, then incrementally becoming gradually more localized, until mainly pixels near the edge of the burn are being colored. When the burn area is clearly delimited, yet not excessively to crop out potentially low severity pixels, the end value marks the approximate upper limit of unburned in terms of dNBR. The same procedure should be done in reverse, from the top down, to find the bottom threshold of low severity level. Then compare the burned-unburned endpoints, and revise the threshold as needed.

Do not be alarmed if some spurious "low severity" pixels show up well outside the perimeter where it did not appear to burn. That is to be expected when setting discrete boundaries for categories based on continuous data. On the other hand, you want to reach a threshold where those spurious pixels are at minimum, and at the same time have a "burned" pixel distribution that most faithfully defines the actual burn area. The quality of severity-level discrimination can be determined later statistically, when final judgments about reliability can be weighed with field data. At this time, however, some idea of how well dNBR is working can be inferred by examining how well the burn-unburned threshold either includes or excludes the burn, based on the distribution of seemingly errant pixels relative to correct ones.

Most errant pixels should be scattered randomly and relatively few in number. If there are some well-defined patches of pixels that seem to be wrong, the most likely cause is difference in moisture or phenology between the two Landsat scenes. For example, unburned meadow patches may be lighter than surroundings if the postfire scene is particularly drier than the prefire. In a case study of Yellowstone National Park burns at elevations above 6,000 ft, where early September and mid-October NBR values were differenced, the October scene was notably drier. Deciduous plants had become dormant in meadows and along riparian corridors, and those appeared faintly with elevated dNBR (fig. LA-7). If location of the burn is generally known, these cases should not be cause for concern. However, thresholds may need to be shifted up or down, depending on relative scene conditions indicated by unburned background.

Note there is always overlap in dNBR values between severity levels. To explore statistical properties of a particular dNBR, sample the population of unburned pixels. Large samples of unburned pixels (5,000 or more) typically display normal distributions with a mean of ±50 and a standard deviation less than 50. Include only areas that are not affected by extreme phenological or atmospheric differences, such as where a meadow was green in one scene and completely brown in the other, or where there was snow in only one scene. If such differences predominate over the unburned area, then the scene pair is probably not appropriate to use in the first place. The sample mean should indicate the bias from zero dNBR, which marks the theoretical value for no detectable change. If the unburned mean is shifted from zero, that indicates some nonfire related difference between the scenes, and the whole dNBR histogram could be shifted by that amount to standardize multiple dNBR scenarios (shifts up to $\pm$ 50 points can occur). Furthermore, two standard deviations should mark an approximate 95 percent confidence interval for the unburned class. From our experience, a standard deviation of about 50 points or less indicates a good pairing of the NBR scenes, with little nonfire related difference between them.

Next, it is useful to establish a lower threshold for high severity. Change to a different color and start at the high end of dNBR to progressively color in values decreasing from the highest. You will notice a point (value) where the first few pixels within the burn take on the color. This is the upper realistic limit of dNBR for that burn. If the burn is one that seems to include some extreme high severity, that value should be within a range of about +1,000 to +1,300. If that value is lower, it may be that severity did not reach maximum levels within the burn. If it is below about +600, it is possible that only moderately high levels were reached, and the burn contains no high severity. If it is above about +1,300, bear in mind it may result from clouds in the postfire scene or some other data anomaly. Examine the raw Landsat scenes to confirm that high values are not where clouds interfere with dNBR in the burn.

From that highest realistic value, continue coloring values incrementally down the scale. You will notice colored pixels appearing in a number of new locations, and then progressively fewer new locations. You will see increased clumping of newly colored pixels around previously colored patches, gradually appearing to fill out and expand existing patches. Isolated uncolored pixels within colored patches may indicate that you are missing some high severity, and the threshold should go lower still. At some point after patches are well formed, but before most of these patches coalesce and before lone pixels start to frequently show up in dispersed new areas of the burn, note the dNBR value. If it is within a range of about +600 to +750, you can assume you are within the range of a preliminary lower threshold for high severity. Based on knowledge of the burn—aerial photos or some other source—you may want to go further down or back up the scale to try to settle on the most reasonable breakpoint. This may be influenced in part by the size of the area identified as high severity. Remember that there may be a tendency for initial assessments to have higher thresholds for high severity, depending on the ecosystem.

If there are a few isolated pixels that are not colored within larger surrounding patches of "high severity," that may indicate the threshold for high severity is too high and needs to be lowered. In fire ecology terms, one might question the validity of a few isolated pixels remaining within large patches of high severity. Query the values of those pixels to determine if they should be included with high severity, as would be likely if those values are only a few points below the current high threshold. If they are quite a bit lower than the current high threshold, then those truly may be isolated areas of a lower level severity class. In the end, one should see a distribution of high severity that is not excessively fragmented or "speckled," while at the same time, not too broadly contiguous over an unreasonably large portion of the burn.

Apply the range between unburned and high severity just determined to partition the remaining positive severity levels. Refer to table LA-2 for proportions of those levels relative to a comparative span of about 560 points between unburned and high severity. If reliable information on the burn is available, the levels can be adjusted to fit what is known. For example, one may know that certain areas burned with low severity, so thresholds can be adjusted accordingly to correctly identify those areas.

The Enhanced Regrowth levels (strongly negative values) can be determined much as described above by reversing the progression of coloring the negative values. Start by going down from unburned into the Enhanced Regrowth Low level. Focus on areas within the burn where dNBR appears darker than the unburned medium-gray levels outside the burn. These should correspond to meadow or grassland habitats, if present. It helps to key on the distribution and shape of familiar meadow patches evident in false-color images or aerial photographs. The range of valid negative values is basically divided by one-third and two-thirds to split out low and high enhanced levels, respectively, assuming the total is a span of about 420 points or more. If much less than that, retain a span of about 100 points for the Enhanced Regrowth, Low level.

**Subsequent procedures**—The initial stratification of the burn can be analyzed for preliminary assessments, or taken into the field where adjustments to the model thresholds may become evident. Walk- or drive-through surveys are recommended with severity maps in hand to initially spot check for obvious agreements or discrepancies. More in-depth field work is usually required for statistical validation and calibration, and the map of burn levels is useful for locating target sample areas. Refer to the **Ground Measure of Fire Severity: The Composite Burn Index** section in this chapter for discussion of field protocols. The dNBR thresholds can then be revised as soon as consistent field observations are made. Once sufficient ground data have been analyzed, that should ultimately guide where thresholds most appropriately fall, based on statistically determined intervals.

Recognizing that ground data may never be available on some burns, thresholds for severity levels (classes) may need to be based on field results from other burns within similar ecotypes and similar timing of dNBR. After a number of burns have been sampled in a region, statistical confidence in thresholds for dNBR should increase to a point where subsequent ground data are less essential. At that time, plots can be sampled less frequently and used mainly to spot check results. That is, in fact, one goal of the whole process, so field time and expense can be minimized, without impacting availability and reliability of burn information.

Each subsequent procedure adds a level of validation that should be documented in the metadata. Once results are improved and verified as much as possible, the burn severity model can be used to compile final reports and statistics and to address a variety of issues.

# LANDSCAPE ASSESSMENT HOW TO

## How To Record a GPS Location

Basically, there are a number of options today for GPS receivers and ways to acquire locational data. Here are recommend protocols on only a few issues.

### Acceptable accuracy

The two-dimensional accuracy of X (Easting), Y (Northing) coordinates reported by the GPS should be less than 33 ft (10 m), preferably less than 23 ft (7 m). There is no procedural requirement on elevation, or the Z coordinate.

### Geodetic datum

The more accurate and more recent NAD83 should be selected, unless there is a strong need to use the data predominantly within a local GIS and the local standard is for some other datum. For example, many National Park Service sites still use NAD27 in order to reference data taken from their older base maps. These were some of the first areas mapped with USGS 1:24,000 quadrangles. In any event, it is important to note the datum used for plot location so conversions can be made if necessary.

### When digitizing a single point

Y-Code receivers such as Rockwell's PLGR, should be set to an averaging mode and allowed to log a number of points until the coordinates stabilize, at which time the plot center is either jotted down or saved in memory. These receivers are only available to approved Federal government employees.

P-Code receivers such as Trimble's Pathfinder and GeoExplorer, or various Garmin models should be treated similarly only if "selective availability" is off. If "selective availability" is on, a differential correction should be used. Set the receiver to log and save a 100 or so points for each plot center. Options are to do differential correction "on the fly," or later by receiving suitable reference control data from a surveyed base station. You will have to check with GPS-knowledgeable people in your area to learn how to access local base station data.

## How To Take Plot Photos

It is a good idea to take photos *after* completing the rating exercise, when one is most familiar with plot burn conditions. There are many approaches to this, but some recommended procedures include the following. Use a high-resolution digital camera, or a 35 mm camera with color slide film, ASA of about 125. Take at least two photos approximately 180 degrees opposite one another, showing the plot center and about half of the plot in each. Avoid taking photos directly toward the Sun. Include a signboard for scale and to identify the date and plot number in the picture. As time or objectives allow, take any number of photos targeting features of interest; for example, typical charring patterns on substrates or trees, and regrowth of perennial herbs and shrubs. Try to capture tree canopy effects as well as ground effects.

## How To Name Files for Burn Remote Sensing

Generally, it is beneficial to keep filenames as concise as possible and be consistent from analysis to analysis. Filenames can get quite involved, but try to include some indication of data content and level of processing, to simplify retracing steps and file lineage. These data are a time series, so it is helpful to use names that sort into a date sequence. Here are some suggestions:

**YYDDD.bB**—Raw Landsat bands can be named by date and band number, most simply with a two-digit year (four digits if necessary), three-digit Julian day (two-digit month and two-digit day, if necessary), "b" for Band, and the Band number, B.

Example: 01179.b4 designates Band 4 of a scene from 2001 Julian day 179, or 28 June.

Scenes of the same path have the same date sequence, so if scenes of the same path and date are kept separate and not seamlessly patched together (as could be), names may become ambiguous when used in the same analysis. In that case, a designator for row number can be added. Scenes of the same path that will always be used in separate analyses (such as one from California and one from Montana) are not a problem. They can be separated within directories of the file system. Landsat TM and ETM+ data will never be acquired from the same path/row on the same date, so redundancies should not occur between the two satellites. If it is found useful to distinguish them, a prefix of "t" or "e" may be used (for example, t01171.b4 and e01179.b4).

**YYDDD.rB and YYDDD.trB**—Band reflectance and transmittance-corrected reflectance data can use the same designation above, with substitution of "r" or "tr" in the extension.

Example: 01179.r4, 01179.tr4

**YYDDD.n or nYYDDD**—The normalized burn ratio for scene YYDDD carries the same form as above but with an "n" as an extension or as a prefix.

**dYYDDDYYDDD.n or dnYYDDDYYDDD**—The name identifies the operation ("d" for delta or difference), the two scene dates, and the index used in the difference ("n" for NBR). Differencing presents many possible scene combinations, so it helps to keep scenes explicitly identified in the name.

Example: d0018201179.n = NBR difference between 1 July 2000 and 28 June 2001—Julian days 182 and 179, respectively.

Variations might include different indices such as "v" or "vi" for the NDVI, or subsets of scenes that may require added codes for individual burns. A classification or rescaling of continuous data might add a "c" or "r" to the extension, such as dnYYDDDYYDDD.c to represent a file containing the ordinal severity classes derived from the delta NBR.

## How To Handle Reflectance

This section is additional comment on the reflectance algorithm described in the **Steps to Process NBR and dNBR** section of this chapter.

Equation LA-7
$$L_i = DN_i * G_b + B_b$$

The radiance term ($L_i$) would be straight forward, except for the fact that gains and biases ($G_b$ and $B_b$) have been reported differently over time, depending on the satellite generation, postprocessing software, and the vendor.

It is important to read the technical documentation accompanying the data to understand what units are being used and whether maximum and minimum radiance is being reported instead of true gain and bias. Generally TM data from EOSAT/Space Imaging Corp., a private corporation, are processed in "Fast Format," and reports gain and bias in milliwatts per square centimeter per steradian per micrometer. For a period, though, those values were actually maximum and minimum radiance, so a conversion is required to generate appropriate $G_b$ and $B_b$. Recent procurement of TM and ETM+ data from the USGS EROS Data Center, on the other hand, will most likely be processed with NLAPS or L7 software, which generates gain and bias in watts per square meter per steradian per micron (a micron is equivalent to 1 micrometer).

Equation LA-8
$$R_i = (L_i * \pi * d^2) / (Esi_b * \cos(z_s))$$

In the reflectance term ($R_i$), the exoatmospheric solar irradiance, $Esi_b$ is the mean irradiance striking the uppermost atmosphere per-bandwidth, $b$. It is important for $Esi_b$ to be in the same units as the gain and bias used to calculate radiance. The $Esi_b$ also differs slightly between Landsat satellites, as the bandwidths are slightly different. Generally, for Landsat 5 TM data processed by EOSAT/Space Imaging Corp., $Esi_b$ would be applied in milliwatts per square centimeter per micrometer:

| Band: | 1 | 2 | 3 | 4 | 5 | 7 |
|---|---|---|---|---|---|---|
| L5, TM $Esi_b$: | 195.7 | 182.9 | 155.7 | 104.7 | 21.93 | 7.452 |

For TM and ETM+ processed by the USGS EROS Data Center, $Esi_b$ values are reported in watts per square meter per micron (a micron is equivalent to a micrometer):

| Band: | 1 | 2 | 3 | 4 | 5 | 7 |
|---|---|---|---|---|---|---|
| L5, TM $Esi_b$: | 1957. | 1829. | 1557. | 1047. | 219.3 | 74.52 |
| L7, ETM+ $Esi_b$: | 1970. | 1843. | 1555. | 1047. | 227.1 | 80.53 |

Note: Watts per square meter per micron is a factor of 10 greater than milliwatts per square centimeter per micron.

The $Esi_b$ decreases steadily as wavelength increases to the right. For example, only about 1/13th the amount of incoming irradiance occurs in Band 7 as in Band 4. Thus, it would be difficult to associate the two bands directly, because Band 7 is naturally so much darker by comparison. The ratio calculated by reflectance, though, normalizes for this initial difference in irradiance magnitude. It produces a value between 0.0 and 1.0, which is then comparable between the bands. Reflectance is functionally equivalent to a simple percentage calculation, associating the amount of detected light with the total available.

The factor, $d^2$, makes minor adjustment in exoatmospheric solar irradiance, $Esi_b$, due to orbital eccentricity, or the daily deviation from average distance between Earth and the Sun (see the **Glossary** in this chapter). The $d^2$ ranges between 0.9666 and 1.0350, and it is closest to 1.0 in early October and early April. For a given day, it can be obtained from a look-up table. Refer to a reference dealing with solar radiation (for example, Muhammad Iqbal, *An introduction to solar radiation*).

In its most basic form, if one discounts surface topography and Sun angles, "at satellite" reflectance boils down to a simple ratio of radiance (or detected brightness) over the amount of available incoming solar radiation per bandwidth; in other words:

Equation LA-9
$$R_i = (L_i * \pi * d^2) / Esi_b$$

All other factors in the algorithm simply modify $Esi_b$ in ways to more realistically gauge the amount of incoming radiant energy that has potential for reflection off the Earth's surface. That, of course, initially involves the angle of the Sun in relation to zenith, as is the case assuming a flat surface:

Equation LA-10
$$R_i = (L_i * \pi * d^2) / Esi_b * \cos(z_s)$$

This denominator simplifies the influence of incidence angles on $Esi_b$, where only the Sun zenith angle of a particular scene, $z_s$, is applied (see the **Glossary** in this chapter). When the Sun is directly overhead, $z_s = 0$ degrees and $\cos(z_s) = 1$, so the entire quantity of $Esi_b$ is available for detection. As the Sun angle moves away from zenith, $\cos(z_s)$ decreases, serving to decrease the amount of available light ($Esi_b$) for detection. In a more complex form below, the algorithm incorporates topographic angles, which more accurately model the amount of incoming light hitting a pixel's surface. That, however, is deemed unnecessary for NBR analyses, as NBR per-pixel normalization cancels those factors, so use of either reflectance algorithm is mathematically equivalent.

## Reflectance incorporating topography

The following is provided only for added understanding of reflectance, but is not required to complete the analysis of burns using the NBR. Such an approach may be preferred in other cases when normalized band ratios are *not* used, and comparisons between scenes depend directly on calculated per-band reflectance. The more complete rendering of reflectance has Sun direction and topographic factors built into the derivation:

Equation LA-11  $R_i = (L_i * \pi * d^2) / (Esi_b * (\cos(z_s) * \cos(slope_i) + \sin(z_s) * \sin(slope_i) * \cos(a_s - aspect_i)))$

where, $a_s$ is the per-scene Sun azimuth angle, and *slope$_i$* and *aspect$_i$* are the per-pixel topographic variables derived from a DEM. Note that aspect in degrees from north has "flat" reassigned to equal the Sun azimuth angle; such that, $a_s - aspect_i = 0$ for flat pixels.

When a DEM is available, the added terms in the denominator adjust $z_s$ to approximately the actual incidence angles created by interplay of sunlight on topography. Generally, the terms increase available light as slope angles increase and aspects face toward the Sun. Conversely, reflective light potential decreases as slopes steepen and aspects face away from the Sun. On a flat surface, the topographic terms reduce to 1.0; so again, $Esi_b$ is only modified by $\cos(z_s)$.

Because $Esi_b$ and terms associated with modifying potential irradiance occur in the denominator, the effect of angular adjustments on reflectance, $R_i$, is the inverse. That is, a given radiance value, $L_i$, will yield higher reflectance on a slope facing away from the Sun than the same $L_i$ on a slope facing toward the Sun. This occurs because available incoming radiation is modeled to be less on the slope facing away from the Sun. The given radiance value indicates that proportionately more incoming light is being reflected there than on the slope facing the Sun, and the surface, therefore, is more highly reflective. The net effect of this transformation tends to remove some of the contrast created by topographic shading, so images appear somewhat "flatter." As it were, the distribution of pixel values is now on a spatially "level playing field."

The natural range of $R_i$ should be 0.0 to +1.0, or by some order of magnitude (to scale reflectance from 0 to 1,000). Unfortunately, this reflectance model tends to overcorrect when *all* the following conditions are true: 1) surfaces face nearly directly opposite the Sun ($a_s - spect \approx 180°$); 2) slopes are steeper than the angle of the Sun above the horizon ($slope_i > 90 - z_s$); and 3) detected radiance, $L_i$, is above zero. In these cases, the product of the denominator evaluates to near zero and, with some level of detected $L_i$, the quotient $R_i$ becomes unrealistically high (either positive or negative). This occurs because the algorithm assumes total potential for illumination comes only from $Esi_b$, and thus solar irradiance is essentially absent from areas in shadow. In reality, however, the situation demonstrates that some contribution to detected radiance, $L_i$, may come from reflected light off surrounding surfaces or the atmosphere, and not only from $Esi_b$ directly. Overcorrecting may also arise from misregistration, error in the digital slope, and aspect models, or it may indicate clouds, where pixels obviously do not have appropriate slope and aspect angles associated with them.

Ignoring the known errors, which typically can be corrected or excluded from analysis, solutions to the steep shadowed slope problem are variable and potentially complicated. They generally only pertain to a small portion of a scene, however. If those areas are not widespread and/or do not interfere with a particular analysis, it is best to ignore them and mask out the overcorrected pixels ($R_i < 0.0$ or $R_i > 1.0$) by rescaling them to zero. Exclude them from subsequent calculations, generally, if such pixels are on steep sheltered slopes without sufficient vegetation to warrant concern in burn applications. (Assuming images over mountainous terrain are from mid-latitudes, within 6 weeks of summer solstice.)

If such pixels apparently interfere with acceptable accuracy of objectives, one can reclassify the slope dataset such that slope angles greater than or equal to the angle of the Sun above the horizon are reassigned to equal the Sun elevation angle minus 1:

$$\text{IF } (slope_i \geq 90-z_i) \text{ THEN } (slope_i = 90-z_i-1)$$

This minimally alters the response curve of $\text{Esi}_b$ on steep slopes over the range of aspects but prevents that term from being reduced to zero in the denominator. At the same time, it accommodates a small amount of irradiance (equivalent to 1 degree of slope) contributing to the brightness of those pixels, which may naturally come from indirect reflected light.

A third option is to filter the topographically corrected solar irradiance of the denominator to establish a lower near-zero limit. Values less than approximately 1 percent of $\text{Esi}_b$ should be reassigned to a minimum level (the value of 0.01 is flexible and may be modified with experimentation) equal to about 1 percent of $\text{Esi}_b$:

$$\text{IF } (t\text{Esi}_b < 0.01 * \text{Esi}_b) \text{ THEN } (t\text{Esi}_b = 0.01 * \text{Esi}_b);$$

where $t\text{Esi}_b$ is the modified incoming solar irradiance per band that accounts for incidence angles from the Sun and the Earth's surface. To facilitate this, it may be necessary to create an interim raster dataset of $t\text{Esi}_b$ and then employ available GIS functions to make the reassignment. The resulting dataset is then reinserted in the algorithm to complete the reflectance calculation.

Pixels with zero radiance (no detected brightness) will not be affected by these two adjustments, as should be the case, because both the numerator and the quotient will evaluate to zero overall.

Other solutions to the problem involve expression of more complex geometry within the algorithm. At this time, we are not prepared to recommend any one in particular, but we would be interested in hearing about practical solutions that have been tried and found successful.

## Atmospheric normalization

There are a number of solutions, and good literature exists on performance of different methods. Here, a relative normalization of one scene to another is undertaken, considering that results are based on band ratios, and geographic scope is typically limited to subsets of scenes. A list of procedural steps follows.

1. Determine if transmittance is a factor, and if so, which scene is most affected by atmospheric scattering. The scene less clear—the one with less transmittance—will be the one corrected and will become the independent variable in regression analysis performed later. Usually, this scene will exhibit brighter reflectance over dark targets, such as lakes, compared to values from the same targets in scenes with greater transmittance. Relative difference in transmittance can be determined quantitatively by comparing pixel reflectance sampled over dark targets. Lower transmittance also appears to decrease contrast, compared to scenes where the air is clear. The most affected bandwidths include the visible Bands (1, 2, and 3), so base comparisons on those bands to enhance differences. (Band 1 is most influenced by atmospheric factors, so use it when comparing single bands.) Compare bands between scenes individually as grayscales, in false-color combinations, or by their histograms. The scene with less transmittance should show less variance in the histogram and a shift to the right or brighter region.

2. Using both scenes for reference, digitize small polygons consisting of a few (50 to 200) pixels each, which collectively represent areas of low, mid-range, and high reflectance. Each polygon should be restricted to quasiinvariant targets that appear to have the same relative reflectance in both scenes. Targets should not be subject to seasonal changes or other disturbances that normally affect reflectance. Examples of acceptable targets include: 1) lakes or deep shadow for the dark sample; 2) high density mature conifer stands for the mid-range; and 3) rock, snow, or parking lots for the bright sample. Avoid areas that differ noticeably between scenes, such as where clouds occur, or where snowmelt patterns are not the same. Attempt to obtain sample polygons from within about one-quarter of the scene surrounding burn study area(s). Sample sizes should be in the range of four to eight polygons and 600 to 1200 pixels per level of reflectance, or more.

3. For each Band (4, 7) of each scene, extract pixel reflectance values that occur within the set of digitized polygons. Import them to statistics software capable of performing linear and quadratic regression. The statistics data file should sequentially contain a record for each pixel, with each pixel's reflectance values as the variables (Bands 4 and 7 of one scene, and Bands 4 and 7 of the other scene).

4. Perform curve-fitting operations on each band using both linear and quadratic models to associate the two dates. Assign the evidently less clear scene to the X axis (independent variable) and the scene with greater transmittance to the Y axis (dependent variable), per band. Make sure that scenes are properly assigned to dependent and independent variables by reviewing the regression coefficients. If they are not, reverse the order of the scenes in the regression analysis.

5. Notice which model best fits the data of each band by reviewing the regression statistics, and a scattergram of plots with regression lines overlaid. If possible, include a 98 percent confidence interval on those plots. Select the regression model that most adequately explains the intraband differences observed between scenes. This is a judgment call. Generally, one would not resort to the more complex quadratic model unless it obviously appeared to improve the fit throughout the distribution of points.

6. Remove widely deviant pixels from the sample before further analysis of each band. Such pixels likely differ between scenes due to reasons other than atmospheric clarity, such as phenology or some kind of disturbance. Use a broad confidence interval (CI) on the regression model just selected per band, such as 98 to 99 percent. Determine the breadth of those intervals per band in reflectance units plus-or-minus. Goodness of fit is generally quite high, since sampled pixels have been selected for their comparable reflectance. The typical $r^2$ should be in the range of 0.93 to 1.00, so even the 99 percent CI will usually be quite narrow. Construct a filter that eliminates records from the statistics data file, if a record has either Band 4 or Band 7 differing between scenes by more than the per-band CI. The objective is to define a set of pixels that shows a consistent trend in reflectance difference between the two scenes, caused only by atmospheric effects.

7. Rerun the selected regression model on the remaining pixels, with deviant pixels removed. Evaluate regression statistics and coefficients. If $r^2$ value is high, and coefficients indicate essentially no difference between the two distributions per band (for reflectance that is a *slope* in the range of 0.995 to 1.005, and an *offset* of ±0.010) then atmospheric normalization is probably not necessary, and one can use both original reflectance datasets to calculate NBR. If the $r^2$ value is too low, or elimination of deviant pixels leaves too small a sample size, consider redefining the sample polygons, and rerunning the regression. The initial sample may contain an excess of nonatmospheric effects or perhaps clouds.

8. Use regression coefficients just derived to transform the bands from the scene with greater atmospheric effects (independent variable above) and normalize them to the scene with apparently clearer air (dependent variable above). A linear transformation will have the form:

Equation LA-12 $$R_{Yi} = b_1 * R_{Xi} + b_0 ;$$

where, $R_{Yi}$ is the new normalized per-pixel reflectance per band of the scene with less atmospheric clarity. $R_{Xi}$ is the original per-pixel reflectance for that scene (the original independent variable above), $b_0$ is the regression coefficient of bias or offset, and $b_1$ is second regression coefficient for gain or slope. A quadratic transformation will have the form:

Equation LA-13 $$R_{Yi} = b_2 * R_{Xi}^2 + b_1 * R_{Xi} + b_0 ;$$

where variables are the same as above with the addition of $b_2$, which is the third regression coefficient for the square of the independent variable.

The resulting band reflectance data, $R_{Yi}$, are used in subsequent NBR calculations. Interactively on the graphics monitor, take a while to compare the results per band to the original reflectance data and to the scene not transformed. One should notice only subtle changes in the normalized data, but where visible, they should be in the direction that makes the corrected bands more similar to the scene with high atmospheric clarity, compared to the original bands with low clarity.

Finally, compare results to original band reflectance datasets. The results of regression transformation, though visually subtle, will be used in subsequent NBR calculations.

# LANDSCAPE ASSESSMENT GLOSSARY

**Azimuth angle**—Reported in the Landsat scene header file, it is the angle in degrees from north of the position of the Sun when scene acquisition occurred, relative to scene center.

**Band or Bandwidth**—A discrete region of the electromagnetic spectrum, representing a range of wavelengths; the breadth of the region is the bandwidth. Sensors usually are designed to record the amount of energy detected within specific regions of the spectrum, hence the derivation of bands. In order of increasing wavelength, the spectrum begins with gamma rays and the ultraviolet, and progresses through the blue, green, and red zones of the relatively narrow visible range. The entire visible range expressed as one bandwidth is considered panchromatic and would appear like a black-and-white photograph. The near (NIR), middle (short-wave, SWIR), and thermal (long-wave, LWIR) infrared bandwidths follow in a broad region. Beyond infrared regions encompass much of what is used for communication, including microwaves, TV, and the longest, radio waves.

**Burn severity**—The degree or magnitude of environmental change caused by fire. The change may be represented by single or multiple biophysical variables on a continuous scale from no change to high change. The gradient may be partitioned into nominal levels, such as low, moderate, and high. For mesoscale landscape perspectives, it is the degree that fire affected an area or community, measured as a composite value over the horizontal and vertical dimensions of the area. In socioeconomic terms, burn severity is often measured by cost or human casualty incurred during and after a fire, including loss of resources. There are usually short- and long-term implications of burn severity, see **Fire effect**.

**Community**—For our purposes, an ecological community, consisting of all plants and animals, as well as the various inert materials, both organic and inorganic, which occupy an area. Generally, a community can be divided into multiple zones with different subcomponents (see **Stratum/strata**). Groups of communities often have similar, repeatable characteristics that lead to classification of specific community types, often based on dominant species, microclimate, and soil. Landscapes consist of assemblages of different communities.

**Eccentricity factor**—Variation in the radius of Earth orbit that accounts for daily deviation from an average circular orbit, as applied to **reflectance**.

Equation LA-14 $$d^2 = 1 / E_o ;$$

where $d^2$ is the eccentricity factor, and $E_o$ is the eccentricity correction quotient for day o.

Equation LA-15 $$E_o = (r_o / r )^2 ;$$

where $r_o$ is the Sun-to-Earth orbital radius on day o, and r is the mean orbital radius over a year period.

**Edaphic**—Characteristics resulting from properties of the soil.

**Fire effect**—Any result of fire. It may be related to biological or physical components of ecosystems, or to ecological processes that in turn impact biological or physical components. It may also be related to biophysical systems, such as communities, the atmosphere, or landscapes.

**Fire effects, initial or first order**—Those effects manifested on the biophysical components or systems that existed at the time of fire. First order fire effects are the direct result of combustion processes, including plant injury and death, fuel consumption, and smoke production (Reinhardt and others 2001).

**Fire effects, long term or second order**—Time-dependent responses to fire over the long term, where initial fire effects are influenced by many biophysical factors subsequent to the fire, such as: 1) seed-bank species and proximity to postfire seed sources; 2) localized site characteristics like topography and soils; 3) subsequent climate; and 4) secondary effects from erosion and mass wasting.

**GIS**—Geographic Information System(s). Integrated software, hardware, and data used to store and manipulate information that combines thematic and locational attributes about geographic features.

**Herb**—Annual, biennial, or perennial plants, including grasses, that do not develop persistent woody tissue, but tend to die back and regrow (or reseed) on a seasonal basis.

**Histogram**—A frequency distribution over a range of values. For example, the number of pixels that occur at each numeric value of reflectance. May be presented in tabular or graph form.

**Hydrography**—The geography of hydrologic features, principally surface waters comprising all variations of lakes, rivers, and streams; as well as the drainages within which these features are nested.

**Irradiance**—The quantity of light reaching Earth, as measured in energy units per unit area per bandwidth (watts per square meter per micron). The amount of incoming radiation available for detection.

**Multispectral**—Data received by a sensor, recorded as brightness values that occur within a few (usually four to 20) relatively narrow ranges, or **bandwidth**, of the electromagnetic spectrum. Each band is recorded independently but simultaneously from the same surface area, providing information about surface composition. The number of bands and the bandwidths define the sensor's spectral resolution. Spectral values recorded in many (100 to 300) relatively narrow bands are considered hyperspectral and more closely approximate the continuum of energy observed in the spectrum.

**Normalized Difference Vegetation Index (NDVI)**—A normalized difference of Landsat Band 4 (NIR) and Band 3 (visible red), expressed as:

Equation LA-16 $$NDVI = (R4\text{-}R3)/(R4+R3)$$

The index is directly related to the amount of green biomass, per unit area. It has been used as a measure of leaf area, primary plant production, and when temporally differenced, as an index of dryness or drought.

**Pixel**—Literally, "picture element." The smallest unit area that has a data value assigned to it. Pixels within an image generally are all the same size, and arranged in a contiguous rectangular grid of rows and columns. Spatial orientation of the grid can be registered to a map projection so that individual pixels may be located on the ground.

**Radiance**—The brightness detected by a sensor in a particular bandwidth.

**Radiometric**—Having to do with measurements related to the intensity of radiant energy.

**Raster**—A digital image stored in one of many grid cell formats, where the cells (pixels) are represented as binary numeric values referenced by byte position within the file. Byte position can be translated into pixel row and column such that the grid models some two-dimensional space.

**Reflectance**—For our purposes, a per-pixel ratio of the amount of reflected energy measured per bandwidth by the satellite, to the amount of available incoming solar radiation per bandwidth. The latter quantity is the exo-atmospheric solar **irradiance** per bandwidth, a constant, modified by the daily eccentricity of Earth-Sun distance and the incidence angle of sunlight striking Earth and reflecting back to the satellite. Incidence angles may simply incorporate the solar zenith angle derived from the Landsat scene header file, or may be refined by also including the solar azimuth angle, and the per-pixel slope and aspect angles derived from a Digital Elevation Model (DEM).

**Scattergram**—A two-dimensional plot of points with X and Y axes assigned respectively to an independent and a dependent variable. Point locations indicate the value observed in one variable at a given value of the other variable; for example, the Band 4 reflectance of a pixel before fire versus Band 4 reflectance of that same pixel after fire.

**Spatial heterogeneity**—The mix and diversity of identifiable landscape features, incorporating not only the types of features, but also their size, shape, and location in relation to each other. Useful dimensions or units include: number of different patches; patch size, shape, and diversity; fractal dimension (one of several ratios of patch size to perimeter distance); juxtaposition (a weighted length of edges surrounding a central area); and contagion (the degree of clumping). For burn heterogeneity, such measures indicate how complex a burn was, and the prevalence of particular levels of burning.

**Spatial resolution**—The aerial dimension of the smallest element that can be resolved, or identified on a map, image, or ground surface. For Landsat TM and ETM+ sensors, it is approximately 30 x 30 $m^2$, which constitutes a **pixel**. Smaller features, or parts of multiple features that co-occupy a single pixel, become averaged together to make up the overall spectral signature recorded for that area (see **Multispectral**).

**Spectral signature**—The combined values of one or more **bands** that uniquely define a particular area or feature, such as an individual pixel or a type of vegetation. The bands and the ways they are combined are highly variable and dependent on user objectives. It may be as simple as a range of values from one band, or as complex as involved mathematical algorithms incorporating many bands, as in clustering techniques. Usually statistical reliabilities are associated with signatures to help the analyst determine which ones best identify the features of interest.

**Stratified sampling**—Where the entire population to be sampled is divided into subgroups, and samples are drawn by rules pertaining to each subgroup. For the population of pixels representing a whole burn, one might divide the area by drainages or by perceived severity levels, and choose a number of sample points from each area. The draw might be done either randomly or in equal numbers per subgroup.

**Stratum or strata**—Referring to one or more layers of a community, arranged vertically and having a continuous sequential order from below ground to ground level, and from ground level to the top of the uppermost vegetative canopy. Strata typically are based on within-stratum similarities of physical organization, species composition, and/or microclimate. Heights of strata usually differ, increasing upward. A few too many strata may be used to characterize a given community, depending on recognizable traits and consistency of occurrence, as well as objectives for doing so. For burns, we identify strata that likely influence fire behavior and show potentially unique responses to burning.

**Transmittance**—For our purposes, a ratio of the amount of energy actually passing through the atmosphere and reaching the ground, to the maximum amount that can possibly reach the ground. It designates the clarity of the atmosphere, and is inversely related to the amount of atmospheric scattering. When transmittance equals 1, the air is perfectly clear. Progressively lesser values indicate increased scattering of light, as influenced by clouds, humidity, and particulates, including smoke. Areas of low transmittance generally appear brighter than areas of high transmittance, because the atmospherically scattered portions of light are not diminished by absorption from ground surfaces. Transmittance varies by **bandwidth**; a property of energy, such that capacity for atmospheric penetration increases with wavelength. For Landsat, this means that infrared Bands 4 through 7 are less influenced by scattering factors than visible bands.

**UTM**—The Universal Transverse Mercator is a map projection. Widely used in natural science applications, it is suitable for maps of 1:100,000 and greater scale, (1:24,000). Each hemisphere of the world is divided into 60, 6-degree, zones by longitude. Within each zone, the reference is an X,Y equidistant grid in meters, with origin at the lower left zone corner (western most point on the equator). Coordinate pairs are given in meters northing and easting (for example, 5437689N, 278334E), increasing from the origin to the north and to the east, respectively. For more UTM information, see:

http://erg.usgs.gov/isb/pubs/factsheets/fs07701.pdf

**Vector**—Geographic data represented as numeric X, Y coordinates, and usually some attribute identifier. Vector data define features by point, line, or polygon topology, and are displayed as such on maps or graphics.

**Zenith angle**—The angle of the Sun in degrees from zenith (the position directly overhead at scene center) when scene acquisition occurred. The Sun elevation angle, $e$, is reported in the Landsat scene header file as degrees above the horizon. To obtain the zenith angle, subtract that value from 90 degrees: ($z_s = 90 - e_s$).

## RECENT LANDSAT SATELLITES

### Landsat 5

Launched in 1984, Landsat 5 carries the Thematic Mapper (TM), which records 30-m data in six spectral Bands and 60-meter data in one Band. The Bands include the blue, green, red, and near infrared (NIR) portions of the spectrum (Bands 1 to 4, respectively), and two Bands (5 and 7) in the middle infrared, or short-wave infrared (SWIR), range. These all measure reflected energy. The final Band (6) records emitted thermal infrared or long-wave infrared (LWIR) that registers heat. All bands are spatially coregistered. The orbit is near-polar, with a swath width, or path, recorded across about 180 km. The continuous stream of data for each swath is segmented out into approximately square areas called "scenes" for purchase. A scene covers roughly 32,400 km$^2$ (180 x 180 km) of the Earth's surface. The orbital sequence is continual and iterative, such that repeat coverage for any particular swath is 16 days. Scenes initially are path-oriented—but one can request various levels of geo-rectification—to register a dataset of all bands to a user-specified map projection. Both side-lap and end-lap occurs between adjacent scenes. The latter is typically constant at about 5 percent, while the former increases from the equator toward the poles, within a range of about 5 to over 60 percent. Regions that fall within overlap areas can have multiple scene dates for expanded temporal coverage outside of the 16-day interval.

Recent reduction in Congressional appropriations may necessitate decommissioning Landsat 5. As a consequence, this source may not be available in the near future. For more information visit the Landsat 5 Web site at:

http://www.earth.nasa.gov/history/landsat/landsat5.html

Marketing and pricing of Landsat 5 data used to be fairly complicated. Basically, there were two exclusive sources for the U.S. scenes more that 10 years old, and all scenes previously purchased by the Federal Government were available through the USGS EROS Data Center (EDC) outside Sioux Falls, SD. The cost was fixed and relatively inexpensive, ranging between $600 and $800, depending on the postprocessing options requested by the buyer. Scenes less than 10 years old, and not previously purchased, were sold by a private company, Space Imaging Corp., formerly EOSAT Corp. Pricing was roughly double the above, at about $1,200 to $1,600 per scene. One could also order these scenes through EDC. However, the cost included the normal EDC processing, plus a tape origination fee, plus a fee for every scene that preceded the scene of interest on the tape. Since the latter was highly variable, the least cost alternative had to be determined on a case-by-case basis. Presently, all Landsat data are marketed by EDC, and costs have been fixed at lower prices. Data are distributed on CD and various tape media.

For ordering and viewing available Landsat 1 through 5 scenes, check the following Web site:

http://edcsns17.cr.usgs.gov/EarthExplorer/

Space imaging continues to sell derived Landsat 5 products, such as map-registered photographs from space, and products from other satellites; their Web site is:

http://www.spaceimaging.com/products/25ms.html

## Landsat 7

Launched in 1999, Landsat 7 carries an Enhanced Thematic Mapper Plus (ETM+) sensor. It records essentially the same spectral and spatial characteristics for Bands 1 through 7 as the TM (above). In addition, ETM+ has a 15-meter panchromatic band, Band 8, spanning the whole visible spectrum (blue through red) in one bandwidth, and an additional thermal infrared band, Band 9 at 60-m resolution. Orbital and scene characteristics are similar to Landsat 5, with closely overlapping paths. Overpasses between the two satellites are staggered by 8 days. This provides more frequent coverage for a given area, at least as long as both satellites remain operational. As of May 2003, Landsat 7 developed a scan line corrector problem, which leaves a regular pattern of missing lines within the scenes. The problem has a negative impact on fire-related applications, but some of that data may still be useful.

For more information about Landsat 7, see:

http://landsat7.usgs.gov/

http://landsat.gsfc.nasa.gov/

All Landsat 7 data is available through the USGS EROS Data Center near Sioux Falls, SD. Prices for data collected before May 2003 range between $475 and $800, depending on buyer-specified postprocessing options. Those include four levels of radiometric and geometric correction. The highest level, "precision corrected," is recommended because some of these procedures are difficult to perform "in-house," and the added cost is low to ensure that all data holdings have been treated in standard and well documented ways. Data are distributed on CD, various tape media, or by ftp over the Internet. There are options for data after May 2003, with the scan line corrector problem, at much lower prices. Consult the USGS EROS Data Center for up-to-date information.

For ordering and viewing available Landsat 7 scenes, contact either of the two Web sites:

http://edcdaac.usgs.gov/landsat7/

http://edclxs2.cr.usgs.gov/

## OTHER REMOTE SENSING DATA SOURCES

Several remote sensing technologies besides Landsat may be available to address fire management objectives. Applicability depends on scale, scope, and cost requirements, which may be different from those of FIREMON Landscape Assessment. For example, continental geographic coverage and daily sampling frequency can be obtained at low cost from AVHRR data. The resolution is 1 km, however, and may be too course for local resource managers.

MODIS, a relatively new sensor geared mostly to global dynamics, may be suited to landscape monitoring of burned areas, but has limited resolution of 250 m in two bands. The remaining 34 bands have resolutions of either 500 or 1,000 m. Currently released data products are provisional data sets, primarily of interest to the research community. Standardization of these products may not be optimal as incremental improvements are still occurring. Moreover, data are not continuously archived, and there is limited availability over designated target sites. At this writing, the data are free however, and may be worth considering for burn monitoring in some areas.

Detail finer than Landsat can be achieved with 1- to 20-m resolution from a host of airborne or satellite sensors, including AVIRIS, ASTER, SPOT, and IKONOS. Price is significantly greater than Landsat per unit area. However, these might be appropriate for individual case studies of high ecological or socioeconomic significance, where interregional standardization is not so much a concern. Generally, none provide continual continent-wide coverage, and acquisition is intermittent. Missions are contracted project to project and prescheduled at designated target areas. Acquisition is usually limited to a few attempts, so unsuitable conditions, such as excessive cloudiness, can force cancellation of a mission when adequate data cannot be had in the allotted time. Because future fire locations cannot be known, there also is little chance to acquire preburn information. That notwithstanding, some of these sensors may be added to FIREMON methodology in the future, as they become further developed and evaluated for routine burn monitoring in a variety of ecosystems.

# FIREMON LA BR Cheat Sheet

**The Normalized Burn Ratio (NBR)**—Brief Outline of Processing Steps

Acquire adequate Landsat TM or ETM+ scenes:

- Determine timing requirements: *initial* or *extended* assessment.
- Pre- and postfire scenes should match phenologically as much as possible.
- Search for available scenes using Web browsers
  - Landsat 7 ETM+: http://edclxs2.cr.usgs.gov/L7ImgViewer.shmtl/
  - Landsat 5 TM: http://earthexplorer.usgs.gov/ (follow links to Landsat TM)
- Check availability of already-purchased data before ordering.
- Get Terrain Corrected data.

With data in hand, explore data in false-color composite images, study burn characteristics.
Transform raw data to *Radiance* ($L_i$) and "at-satellite" *Reflectance* ($R_i$) for Bands 4 and 7.

$$L_i = DN_i * G_b + B_b; \qquad R_i = (L_i * \pi * d^2) / (Esi_b * \cos(z_s))$$

$DN_i$ = per-pixel raw brightness value.
$G_b$ **and** $B_b$ = per-band gain and bias from scene header.
$d^2$ = daily earth-sun eccentricity from lookup table.
$Esi_b$ = per-band exoatmospheric solar irradiance from published L5 and L7 tables.
$z_s$ = per-scene solar zenith angle (90-solar elevation angle reported in scene header).

Determine if atmospheric normalization is necessary, and if so, do it if for Bands 4 and 7.
Generate an NBR image for each scene, pre- and postfire:

$$NBR = (R_4 - R_7) / (R_4 + R_7);$$

Generate the differenced (or delta) NBR:

$$dNBR = NBR_{prefire} - NBR_{postfire}$$

This isolates burned from unburned areas, provides a quantitative measure of absolute change in NBR. Practical data range ≈ −500 to +1,300 when scaled by $10^3$.
Apply a linear grayscale to the data range of −800 to 1,100, and study this image carefully.
Define the burn perimeter using combined automated and on-screen digitizing from the *dNBR*.
Make an initial cut at severity thresholds in false color. A seven-tiered configuration may be useful. Ordinal severity levels and example range of
NBR (scaled by $10^3$) are shown:

**Ordinal severity levels and example range of dNBR (scaled by $10^3$), to the right.**

| Severity level | dNBR range |
| --- | --- |
| Enhanced regrowth, high | 500 to 251 |
| Enhanced regrowth, low | 250 to 101 |
| Unburned | 100 to +99 |
| Low severity | +100 to +269 |
| Moderate low severity | +270 to +439 |
| Moderate high severity | +440 to +659 |
| High severity | +660 to +1300 |

(dNBR value ranges are flexible; scene pair dependent; shifts in thresholds ±100 points are possible. dNBR less than about 550, or greater than about +1,350 may also occur, but are *not* considered burned. Rather, they likely are anomalies caused by misregistration, clouds, or other factors not related to real land cover differences.)

 **FIREMON LA CBI Cheat Sheet**

## STRATA

*Substrates*—Inert surface materials of soil, duff, litter, and downed woody fuels.
*Herbs, Low Shrubs and Trees*—All grasses + forbs, and shrubs + small trees <3 ft (<1 m).
*Tall Shrub and Trees*—Shrubs and trees 3–16 ft (1–5 m) tall.
*Intermediate Trees (pole-size, subcanopy)*—Trees between tall shrubs/trees and upper canopy, approximately 4–10 inches (10–25 cm) diameter, and 25–65 ft (8–20 m) tall. May be stratified heights and extend to upper canopy, but crowns receive little direct sunlight. Size is relative to upper canopy and varies by community. If this size is upper canopy, count as intermediate trees.
*Big Trees (mature, dominant and co-dominant, upper canopy)*—Larger than intermediate trees, occupy upper canopy, receive direct sunlight; tallest may extend above average big-tree level.
*Understory*—Substrates, herbs/low shrubs+trees, tall shrubs+trees.
*Overstory*—Intermediate and big trees.
*Total Plot, or Overall*—All strata of the plot combined.

## GENERAL

**Prefire Exposed Soil/Rock** is considered unburned if there is no sign of overlying substrates or vegetation that burned. Avoid sites with >50% exposed prefire soil/rock, see guidelines.
**Rehab Site**—Mulch or other does not count, estimate as if that was not present. Planted, growing vegetation can be tallied where appropriate, but not as new colonizers. A specific factor may not be rated if is not relevant, shows inconsequential presence or insignificant indication of severity (write in N/A for not applicable), or when effects are unclear and cannot be reasonably judged (write in UC for uncertain).
**Percent Plot Area Burned**—Record the percent surface area (burned substrates and low-growing plants) showing any impact from fire for the 98-ft (30-m) diameter plot, and for the nested 66-ft (20-m) plot, if that is used for the understory.
**Prefire Variables**—Report cover (percent area), depth (inches) and density (number of trees) plot-wide as if before fire. Consider burned evidence + unburned areas within plot or nearby; reasonable approximation of prefire conditions. If too difficult to estimate, write in UC for uncertain.
**Enhanced Growth Factors**—100 percent + percent productivity above that, judged to be fire-enhanced; regard amount of green biomass in terms of cover, volume and density. If plots show about the same or less productivity than before fire, then enter as not applicable (N/A). If plot shows enhanced growth, then enter the percent productivity that is augmented by fire, with 100 percent being the same postfire productivity as prefire (for example, 200 percent represents double the estimated prefire productivity); write in UC if uncertain.

## SUBSTRATE RATING FACTORS
### Do not count litter or fuels built up after fire.

**Litter/Light Fuel**—Relative amount consumed of leaves, needles, and < 3-inches (<7.6-cm) diameter wood on the ground at time of fire. Not new litter-fall. Count litter/light fuel even if it occurs under living plants.
**Duff condition**—Relative amount consumed and charring of decomposed organic material lying below the litter. Not fine root mass. Count duff even if it occurs under living plants.
**Medium Fuel**—Consumption of down woody fuel between 3–8 inches (7.6–20.3 cm).
**Large Fuel**—Loss and charcoal from down woody fuel >8-inch diameter (20.3 cm). Base both classes on change to fuel load. Omit or join as one if either fuel class < 5 percent plot cover, see text. Include stumps in appropriate size class, if relevant.
**Soil Cover/Color**—New exposed soil and color change; lightening at moderate to high, ~10 percent red at high severity—overlook ash. Consider soil or rock surface *not* covered by litter, duff or low herbaceous cover less than about 30 cm. If such occurs under taller shrubs and trees, count it.

## HERBS, LOW SHRUBS AND TREES LESS THAN 3 FEET (1 METER) RATING FACTORS

**Percent Foliage Altered**—Only low shrubs and trees (<3 ft), prefire live or dead cover that are newly brown, black or consumed. Ignore resprout.
**Frequency Percent Living**—Percent of prefire vegetation that is still alive after fire, based on number plot-wide; survivorship, not cover, not new seedlings. Include unburned as well as burned, resprouting perennial herbs, low shrubs and trees (<3 ft) pot-wide. Include all green vegetation as well as burned plants that have not had enough time to resprout but remain viable. Burned plants may need to be examined for viable growth points. Do not include new plants from seed or suckers.
**Colonizers**—Potential dominance 2–3 years postfire of new (native or exotic) plants from seed; includes herbs and tree seedlings, plus aspen or other tree-to-shrub suckers, and nonvascular plants (for example, thistle, fireweed, pokeweed, ferns, moss, fungi, seedlings of lodgepole pine, slash pine, western larch, many weedy spp.). Rate only if spp. response to fire is known.

# FIREMON LA CBI Cheat Sheet (cont.)

**Species Composition/Relative Abundance**—Change in spp. and/or relative abundance of spp. anticipated 2–3 years postfire. How much does postfire spp. composition resemble prefire stratum? Consider presence of new or absence of old spp., plus how dominance is spread across spp.

## TALL SHRUBS AND TREES 3 TO 16 FEET (1 TO 5 METERS) RATING FACTORS

**Percent Foliage Altered**—Percent prefire live-or-dead crown volume (leaves, stems) newly brown, black or consumed. Ignore new resprout; it does *not* lessen the amount of prefire foliage altered.

**Frequency Percent Living**—Percent of prefire tall shrubs/trees that are still alive after fire. This is a measure of survivorship based on numbers of individuals. Include unburned as well as burned but viable tall shrubs/trees 3–6 ft (1–5 m) tall plot wide; examine growth points for viability if needed. Do not include new plants from seed or suckers. Account for potential mortality that could occur up to 2 years postfire.

**Percent Change in Cover**—Overall *decrease* in cover of tall shrubs/trees between 3 and 16 ft tall (1 and 5 m), relative to the area occupied by those plants before fire. Count resprouting from plants that burned, plus the unburned plants as cover that lessens the amount of decrease in cover. Do not include suckers or plants newly germinating from seeds.

**Species Composition/Relative Abundance**—Change in spp. composition and/or relative abundance of spp. anticipated 2 to 3 years postfire.

## INTERMEDIATE AND BIG TREE RATING FACTORS (COMBINED)

**Percent Unaltered (green)**—Percent prefire live-or-dead crown volume unaltered by fire. Include new resprout from burned crowns, not from bases.

**Percent Black (torch)**—Percent prefire live-or-dead crown volume that actually caught fire (black or consumed stems, leaves). May or may not be viable postfire; resprout from black crowns does not lessen percent black. At high severity, consumption of fine branching is evident. Include deciduous blackened crowns.

**Percent Brown (scorch)**—Percent prefire live crown volume affected by scorch or girdle without direct flame contact. Brown is due to proximal heating, where foliage did not catch fire. Includes delayed mortality, insect damage, and brown foliage that has fallen to ground.

**Percent Canopy Mortality**—Percent prefire live canopy volume made up by trees killed directly or indirectly by fire within 1–2 years. Proportion of a plot's total once-living canopy lost to dead trees (include insect/disease kill) in relation to total prefire canopy volume.

**Char Height**—Mean char height from ground flames averaged over all trees. The mean is halfway between upper and lower heights on a tree. Include unburned (char height = 0) and burned trees *only* when char height is discernable. Do *not* include black from crown fire; enter N/A for most crown fire burns.

## RECORD FOR EACH OVERSTORY STRATUM, BUT DO NOT COUNT IN CBI SCORES

**Percent Girdled (at root or lower bole)**—Percent of trees effectively killed by heat through the lower bark, sufficient to kill cambium around lower boles or buttress roots. Include trees either dead or likely to die within 1–2 years. Do not include trees killed by torch or scorch to crown. May or may not char through bark and into the wood; may have loose sloughing bark in 1–2 years.

**Percent Felled (downed)**—Percent live-or-dead trees, that were standing before fire but now are on the ground. Usually from wind throw after fire, they exhibit fresh up-turned root masses, and different charring patterns than trees that were down when fire occurred.

**Percent Tree Mortality**—Percent of once living trees on the plot that were killed by the fire, based on number of trees. Suspected insect and disease effects also may be included, if such contributed to killing whole trees relatively soon after fire (for example, within 1–2 years).

## RATING ADVICE

Factors that are not applicable or cannot be resolved in a plot are not rated; they are omitted from that plot's composite ratings. Moreover, if there is much uncertainty about how a specific factor should be rated, or whether it is even relevant to the plot, then that factor should be left unranked. Only the number of rated factors is used to compute averages. If a factor is not rated, enter not applicable (N/A) or uncertain (UC) on the CBI data form. Do not just leave the field blank; such factors are not part of the CBI average, but one wants to know whether these factors were actually assessed and it was decided not to rate them, or just accidentally overlooked and skipped. Zeros, on the other hand, are valid entries and do get averaged into composite scores. Zeros should be used when a rating factor is applicable and exhibits an unburned condition. A zero represents no detected change in an observable factor.

 FIREMON LA Form

## BURN SEVERITY -- COMPOSITE BURN INDEX (BI)

| PD - Abridged | Examiners: | | | Fire Name: | |
|---|---|---|---|---|---|
| Registration Code | | Project Code | | Plot Number | |
| Field Date mmddyyyy | / / | Fire Date mmyyyy | / | | |
| Plot Aspect | | Plot % Slope | | UTM Zone | |
| Plot Diameter Overstory | | UTM E plot center | | GPS Datum | |
| Plot Diameter Understory | | UTM N plot center | | GPS Error (m) | |
| Number of Plot Photos | | Plot Photo IDs | | | |

| BI Long Form | % Burned 100 feet (30 m) diameter from center of plot | | | | | Fuel Photo Series | |
|---|---|---|---|---|---|---|---|
| STRATA RATING FACTORS | BURN SEVERITY SCALE | | | | | | FACTOR SCORES |
| | No Effect | Low | | Moderate | | High | |
| | 0.0 | 0.5 | 1.0 | 1.5 | 2.0 | 2.5 | 3.0 |

### A. SUBSTRATES

| % Pre-Fire Cover: Litter = | Duff = | Soil/Rock = | Pre-Fire Depth (inches): Litter = | Duff = | Fuel Bed = | | |
|---|---|---|---|---|---|---|---|
| Litter/Light Fuel Consumed | Unchanged | -- | 50% litter | -- | 100% litter | >80% light fuel | 98% Light Fuel |
| Duff | Unchanged | -- | Light char | -- | 50% loss deep char | -- | Consumed |
| Medium Fuel, 3-8 in. | Unchanged | -- | 20% consumed | -- | 40% consumed | -- | >60% loss, deep ch |
| Heavy Fuel, > 8 in. | Unchanged | -- | 10% loss | -- | 25% loss, deep char | -- | >40% loss, deep ch |
| Soil & Rock Cover/Color | Unchanged | -- | 10% change | -- | 40% change | -- | >80% change |

Σ  N  $\overline{X}$

### B. HERBS, LOW SHRUBS AND TREES LESS THAN 3 FEET (1 METER):

| Pre-Fire Cover = | | % Enhanced Growth | | | | | |
|---|---|---|---|---|---|---|---|
| % Foliage Altered (blk-brn) | Unchanged | -- | 30% | -- | 80% | 95% | 100% + branch loss |
| Frequency % Living | 100% | -- | 90% | -- | 50% | < 20% | None |
| Colonizers | Unchanged | -- | Low | -- | Moderate | High-Low | Low to None |
| Spp. Comp. - Rel. Abund. | Unchanged | -- | Little change | -- | Moderate change | -- | High change |

Σ  N  $\overline{X}$

### C. TALL SHRUBS AND TREES 3 to 16 FEET (1 TO 5 METERS):

| Pre-Fire Cover | | % Enhanced Growth | | | | | |
|---|---|---|---|---|---|---|---|
| % Foliage Altered (blk-brn) | 0% | -- | 20% | -- | 60-90% | > 95% | Signifent branch loss |
| Frequency % Living | 100% | -- | 90% | -- | 30% | < 15% | < 1% |
| % Change in Cover | Unchanged | -- | 15% | -- | 70% | 90% | 100% |
| Spp. Comp. - Rel. Abund. | Unchanged | -- | Little change | -- | Moderate change | -- | High Change |

Σ  N  $\overline{X}$

### D. INTERMEDIATE TREES (SUBCANOPY, POLE-SIZED TREES)

| Pre-Fire % Cover | Pre-Fire Number Living | | | Pre-Fire Number Dead | | | |
|---|---|---|---|---|---|---|---|
| % Green (Unaltered) | 100% | -- | 80% | -- | 40% | < 10% | None |
| % Black (Torch) | None | -- | 5-20% | -- | 60% | > 85% | 100% + branch loss |
| % Brown (Scorch/Girdle) | None | -- | 5-20% | -- | 40-80% | < 40 or > 80% | None due to torch |
| % Canopy Mortality | None | -- | 15% | -- | 60% | 80% | %100 |
| Char Height | None | -- | 1.5 m | -- | 2.8 m | -- | > 5 m |

Σ  N  $\overline{X}$

Post Fire:  %Girdled     %Felled     %Tree Mortality

### E. BIG TREES (UPPER CANOPY, DOMINANT, CODOMNANT TREES)

| Pre-Fire % Cover | Pre-Fire Number Living | | | Pre-Fire Number Dead | | | |
|---|---|---|---|---|---|---|---|
| % Green (Unaltered) | 100% | -- | 95% | -- | 50% | < 10% | None |
| % Black (Torch) | None | -- | 5-10% | -- | 50% | > 80% | 100% + branch loss |
| % Brown (Scorch/Girdle) | None | -- | 5-10% | -- | 30-70% | < 30 or > 70% | None due to torch |
| % Canopy Mortality | None | -- | 10% | -- | 50% | 70% | %100 |
| Char Height | None | -- | 1.8 m | -- | 4 m | -- | > 7 m |

Σ  N  $\overline{X}$

Post Fire:  %Girdled     %Felled     %Tree Mortality

| Community Notes/Comments: | CBI = Sum of Scores / N Rated: | Sum of Scores | N Rated | CBI |
|---|---|---|---|---|
| | Understory (A+B+C) | | | |
| | Overstory (D+E) | | | |
| | Total Plot (A+B+C+D+E) | | | |

% Estimators:  **20 m Plot:** 314 m²  1% = 1x3 m   5% = 3x5 m   10% = 5x6 m     *After, Key and Benson 1999, USGS NRMSC, Glacier Field Station*
              **30 m Plot:** 707 m²  1% = 1x7 m (<2x4 m)  5% = 5x7 m   10% = 7x10 m    Version 4.0 8 27, 2004

Strata and Factors are defined in FIREMON Landscape Assessment, Chapter 2, and on accompanying BI "cheatsheet."    www.fire.org/firemon/lc.htm

# FIREMON Database

## User Manual

John F. Caratti

## SUMMARY

The FIREMON database software allows users to enter data, store, analyze, and summarize plot data, photos, and related documents. The FIREMON database software consists of a Java application and a Microsoft® Access database. The Java application provides the user interface with FIREMON data through data entry forms, data summary reports, and other data management tools. The Microsoft Access database contains the tables that store the actual FIREMON data and the standard lookup codes used in various FIREMON fields.

## INTRODUCTION

Fire effects monitoring is defined by two tasks: field data collection and data evaluation. Field data collection is discussed in detail in the **Integrated Sampling Strategy** and **Field Assessment** documentation. FIREMON data entry and data summary procedures are outlined here.

Managing and maintaining the FIREMON database encompasses three major tasks: 1) data entry, 2) data management, and 3) data summary. Data entry is accomplished by physically entering the collected field data into standardized Microsoft® Access database tables. Data management includes populating a plant species list for the database, adding additional codes used in the vegetation sampling methods (for example, life forms such as forb, grass, or shrub), and adding user-specific codes to the FIREMON lookup code tables. Data summary reports are generated by a C program (sum.exe) developed specifically for the FIREMON data summary display reports.

Data stewards can perform a variety of tasks using the FIREMON database. Plot data are entered through data entry forms for each FIREMON sampling method. All the FIREMON plot-sampling methods are accessible by clicking the tabs above entry forms. A separate data entry form is provided for the **Metadata** and **Fire Behavior** data because these methods apply to one or more FIREMON plots. **Metadata** and **Fire Behavior** data are linked to FIREMON plots through the MDID and FireID fields, respectively, on the **Plot Description** form.

Basic plot data summaries are displayed through data summary reports. These reports display basic summary data for a FIREMON plot. The FIREMON Analysis Tools (FMAT), which calculate plot statistics, can be executed from the FIREMON database or as a stand-alone program. The NRCS plants database is provided with the FIREMON database, and a data entry form allows users to build a species list for their data. FIREMON also provides a data entry form for adding customized sampling codes. A Simple Query Builder tool allows users to query data from any of the FIREMON tables. However, more sophisticated queries must be developed by FIREMON users and stored in the FIREMON *Data* database. The FIREMON database also includes a program to generate random transect starting points along a baseline and random quadrat starting points along transects.

## FIREMON DATABASE INSTALLATION AND CONFIGURATION

### Installing the FIREMON Software

Log on to the FIREMON Web site (www.fire.org/firemon), navigate to the FIREMON software link and download the FIREMON installation file. Run the installation file, which installs the Java application, FIREMON database, and data analysis software. The default installation directory and FIREMON software configuration files are set to c:\firemon; however, the FIREMON software may be installed in any directory. When the files are successfully installed, the FIREMON directory should appear similar to the directory in figure DB-1.

The c:\firemon directory includes the FIREMON Java application, database, analysis software, and subdirectories for plot photos and metadata documents. The Java application (JFiremon.exe) contains data entry forms, data summary reports, and data management tools. The C program (sum.exe) is used by the JFiremon.exe program to summarize data for the data summary reports. The database (firemondata.mdb) is an empty database containing all the FIREMON data tables and lookup code tables (such as the NRCS species codes and landform codes). The FIREMON training database (firemondata_training.mdb) contains fictional data for testing and training purposes. The FIREMON Analysis Tools include the data analysis program (fmat.exe) and associated help files (firemon.hlp, roboex32.dll). The photo and documents subdirectories are for storing photos and documents; they contain sample photos and a document used with the training data set.

The JFiremon program is the Java application in which users interact with the FIREMON database. The FIREMON Java application may be linked to any FIREMON database (for example, firemondata.mdb or firemon_training.mdb). The *File → Save As...* option on the main FIREMON toolbar allows users to rename a FIREMON database. The firemondata.mdb database is read only and must be renamed after the first installation of FIREMON to preserve an empty set of FIREMON tables for creating new

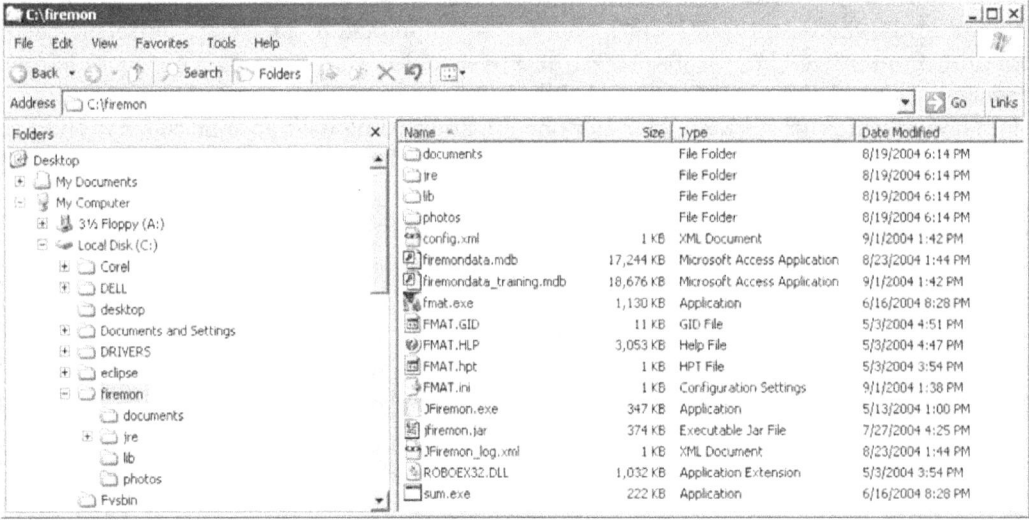

**Figure DB-1**—Files contained in the FIREMON directory.

FIREMON Database Guide

FIREMON databases. Because the JFiremon application is separate from the FIREMON database, users are able to keep their data intact when new versions of JFiremon.exe are released.

## Configuring the FIREMON Database Application

Open the FIREMON application (JFiremon.exe) using the desktop icon or Start menu. The main FIREMON form appears along with the plot data entry forms on the screen (fig. DB-2) after the FIREMON splash screen closes. The database name is displayed in the upper left corner of the JFiremon application.

Next, close the plot data entry forms window and select *File → Settings…* from the FIREMON toolbar. This will display the FIREMON *Configuration and Settings* form (fig. DB-3).

The *Configuration and Settings* form allows users to set the default directories for plot photos and documents, the FIREMON analysis software, and the FIREMON data tables. Once these directories are set, the FIREMON application uses these settings until they are changed again. Click on the *Save Settings* button to save the current settings.

The *Photos Base Directory* and *Documents Base Directory* settings set the hyperlinks for plot photos and metadata documents to the directories storing those items.

## Populating the Plant Species Codes Lookup Table

FIREMON uses the NRCS Plants database codes as defaults, although local codes may be used. If plant species data are collected using the FIREMON sampling methods, then users must populate the plant

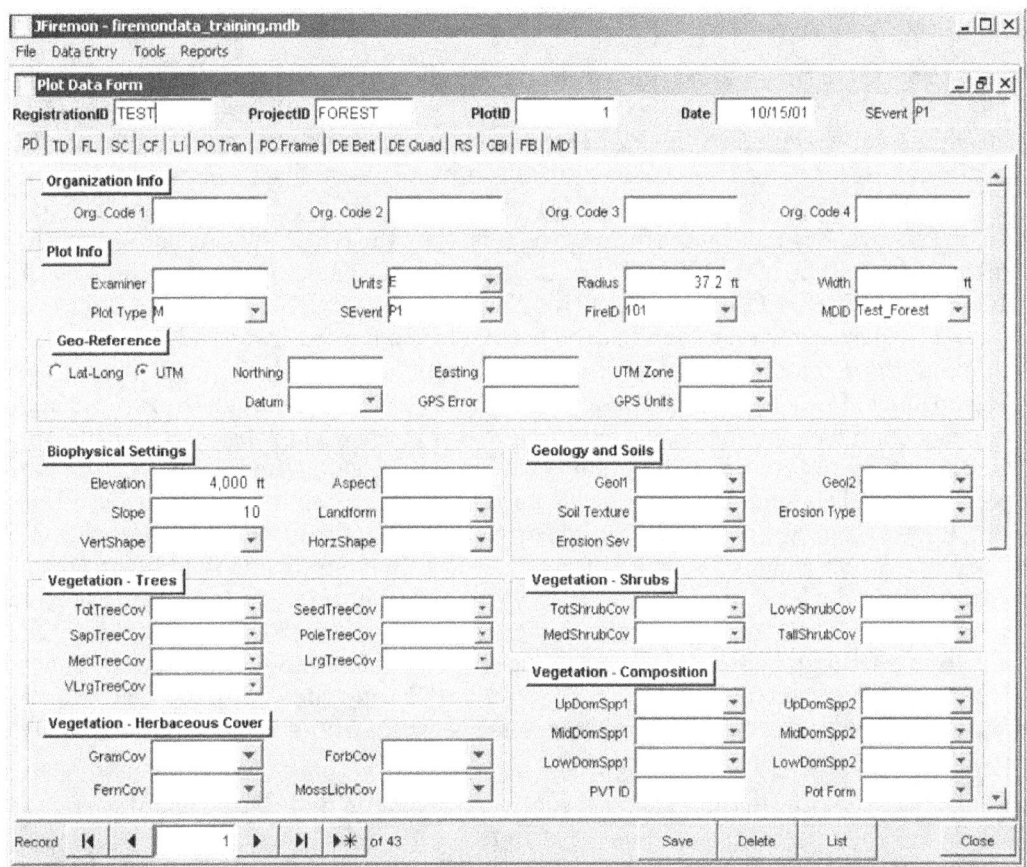

**Figure DB-2**—Main form for the FIREMON ap-plication database.

FIREMON Database Guide

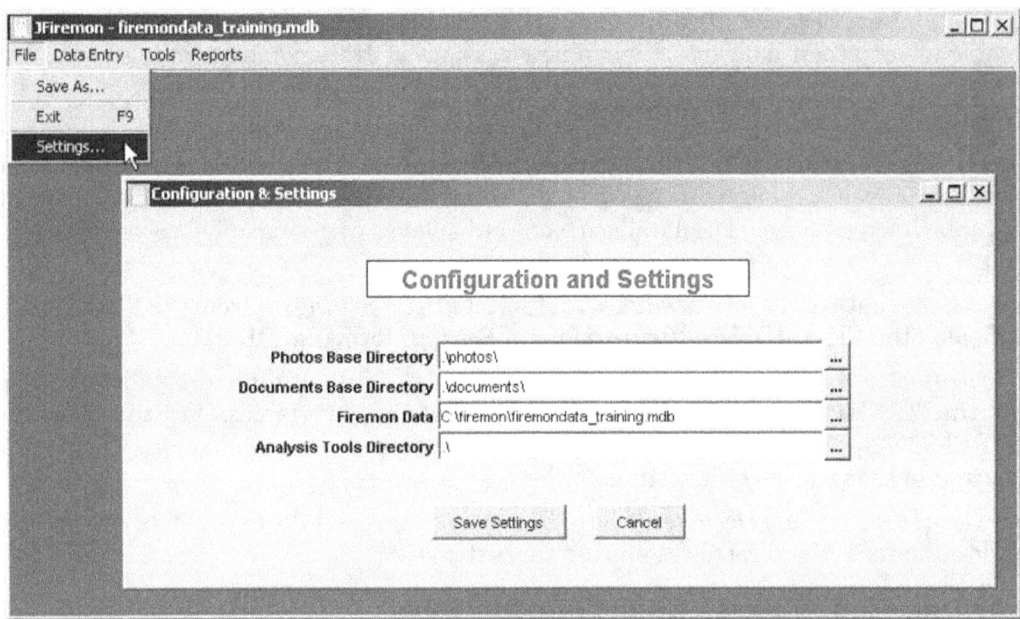

**Figure DB-3**—FIREMON configuration settings

species code table with a list of plant species for their project area. Although FIREMON provides the entire list of species from the NRCS Plants database, this list is much too large to display effectively in a dropdown list for data entry.

Prepare to add species to the lookup table by closing the data entry forms. Otherwise, the data entry forms will need to be closed and reopened before the species you add will show in the species drop down list. To add species, select *Data Entry → Plant Species Codes* from the FIREMON toolbar to display the NRCS Plants database form (fig. DB-4). This form allows users to search for a plant species by the NRCS Plants code, scientific name, or common name. Users must find each plant species for their project area and populate the *Local Code* field on the form. To do this, open the NRCS species list window by clicking the *List* button at the bottom of the NRCS Plant Codes window. A table will appear showing the default search fields: Symbol, Symbol Code, and Scientific Name. More search fields may be added by using the *Add Field* menu and dropdown list found at the bottom of the Record List window. Search fields may be deleted (except the defaults) by right clicking on a column header and clicking *Remove this Column*. Use the *Find* button at the bottom of the Record List window to search through the records in the NRCS Plants database. Search on any field selected in the Record List window by selecting the appropriate Search Column, typing your search in the Find What box, and clicking the *Find Next* button. When the desired species is found and highlighted in the Record List window, click the *Add to Local Code* button on the NRCS Plants Codes window. For example, in figure DB-5, the Common Name column has been added to the Record List window (shown lower right), the Common Name field has been selected to search for grand fir in the Find window (lower left), and ABGR has been entered into the lookup table by clicking *Add to Local Code* (top). Users can change the local code from the one provide by NRCS. For example, deleting ABGR and entering GF in the Local Code field on the NRCS Plants Codes window links the NRCS code for grand fir (ABGR) to the local code (GF) and adds GF to the species lookup table. Any open data entry forms must be closed and reopened before newly added species are displayed in the species dropdown list.

The FIREMON data entry forms build a plant species code lookup table based on all species that have the *Local Code* field populated. If a species is not in this list, the data entry forms will not allow that species to be entered in the database.

FIREMON Database Guide

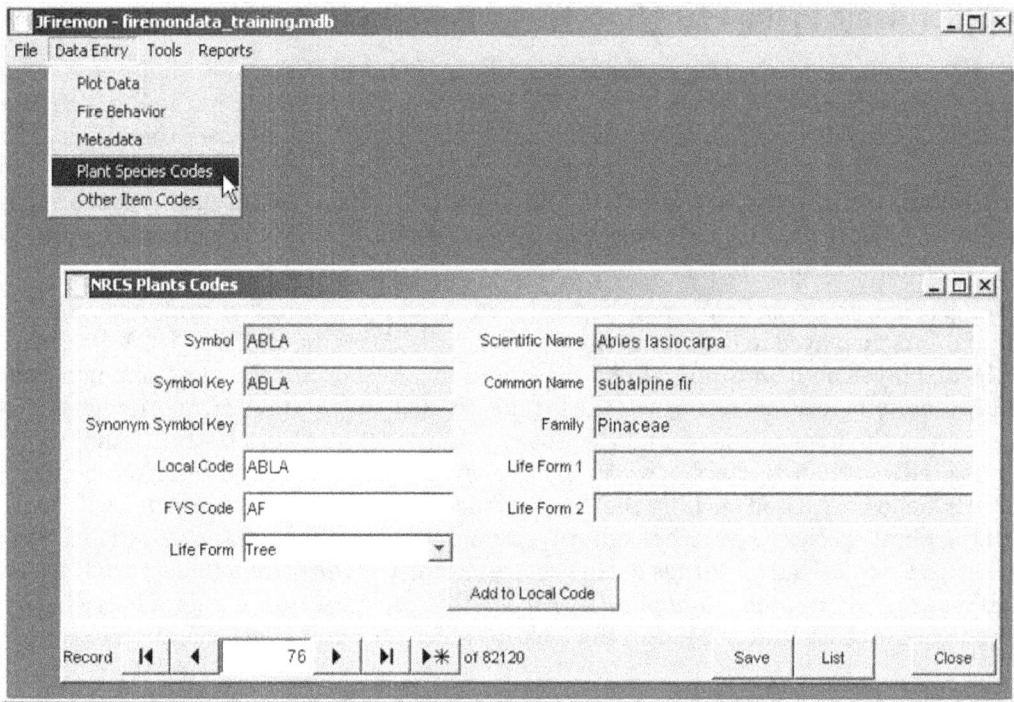

**Figure DB-4**—NRCS Plants database form.

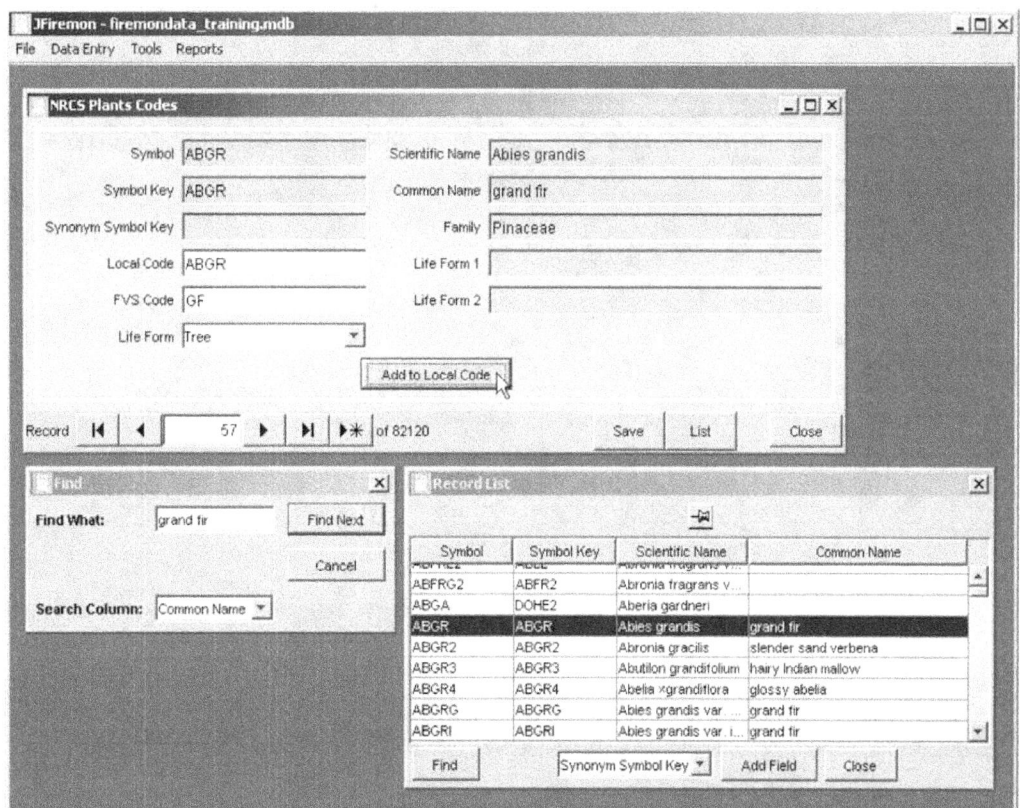

**Figure DB-5**—Refreshing the records for the data entry forms.

USDA Forest Service Gen. Tech. Rep. RMRS GTR 164 CD. 2006                                                                                                         DB-5

## Adding Other Items to the Plant Species Lookup Table

The FIREMON vegetation sampling methods (SC, CF, LI, PO, and DE) allow other items to be entered in the plant species (*Item*) field. These item codes, however, are not stored in the Plant Species Codes table. FIREMON provides another table, which users may customize, to store these fields. Examples of other items include ground cover codes used with the point intercept method, plant life form codes (shrub, grass, forb, and so forth), and density counts of items other than plant species (such as elk pellets, gopher mounds). Select *Data Entry → Other Item Codes* from the FIREMON toolbar to display the *Other Item Codes* form (fig. DB-6).

The FIREMON *Ground Cover Codes* table is displayed in the upper window, and the *FIREMON Other Item Codes* table is displayed in the lower window. Codes displayed in the *FIREMON Other Item Codes* table will be displayed on *Item* dropdown lists. Ground cover codes are the most common codes to add to the list of other items, and users may use these codes to easily populate the *Other Item* list by selecting the desired row in the upper window and clicking the *Add to Other Codes* button. Other codes may be manually entered into the lower table (*Other Item Codes*) by typing a code and description in the Selected Ground Cover Code boxes, and clicking the *Add to Other Codes* button. These codes will be displayed, along with the plant species codes, when entering data in the *Item* field on the SC, CF, LI, PO, and DE sampling forms. Click the *Save Settings* button before closing the *Other Item Codes* form. It is important to enter all codes for items other than plant species in this table because the data entry forms will not allow users to enter codes that are not in the lookup tables for plant species and other items.

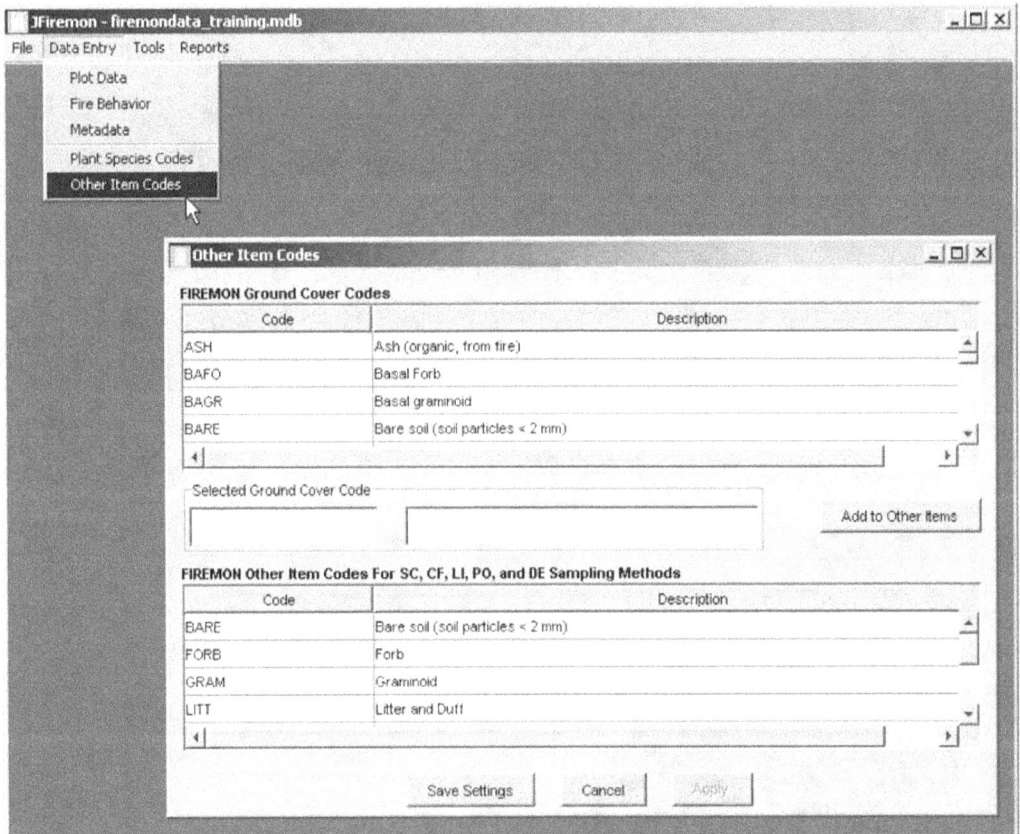

**Figure DB-6**—Other Item Codes form.

# FIREMON DATA ENTRY FORMS

Once the plant species codes and other item codes are populated, users may enter plot data for each of the FIREMON sampling methods. Select *Data Entry → Plot Data* on the FIREMON toolbar to display the Plot Data entry forms (fig. DB-7). When the Plot Description data entry form opens, Record 1 will be selected. If the form is empty (first time use) start entering your data, otherwise click on the *New Record* button (right facing black triangle and asterisk) on the Access record navigation bar at the bottom of the form. This advances beyond the last record in the database and allows users to enter new records. Similarly, the *New Record* button can be used to enter new data on the other data entry forms. When editing existing records, use the record selectors or the *List* button to find the desired record and begin editing the record (fig. DB-8).

Attempting to enter a new record before valid data are entered in all the plot key fields (RegID, ProjID, PlotID, and Date) returns an error message stating there are missing fields (fig. DB-9). You must either delete the current record or fill in the required key fields. The current record may be deleted by right clicking on the record and pressing the *Delete This Record* button (fig. DB-10).

## Important Instructions Common to all Data Entry Forms

Each data entry form has required "key" fields necessary to link and organize database entries. These fields are identified in the paragraphs that follow. Understanding these requirements before collecting data may be useful in speeding up the data entry process.

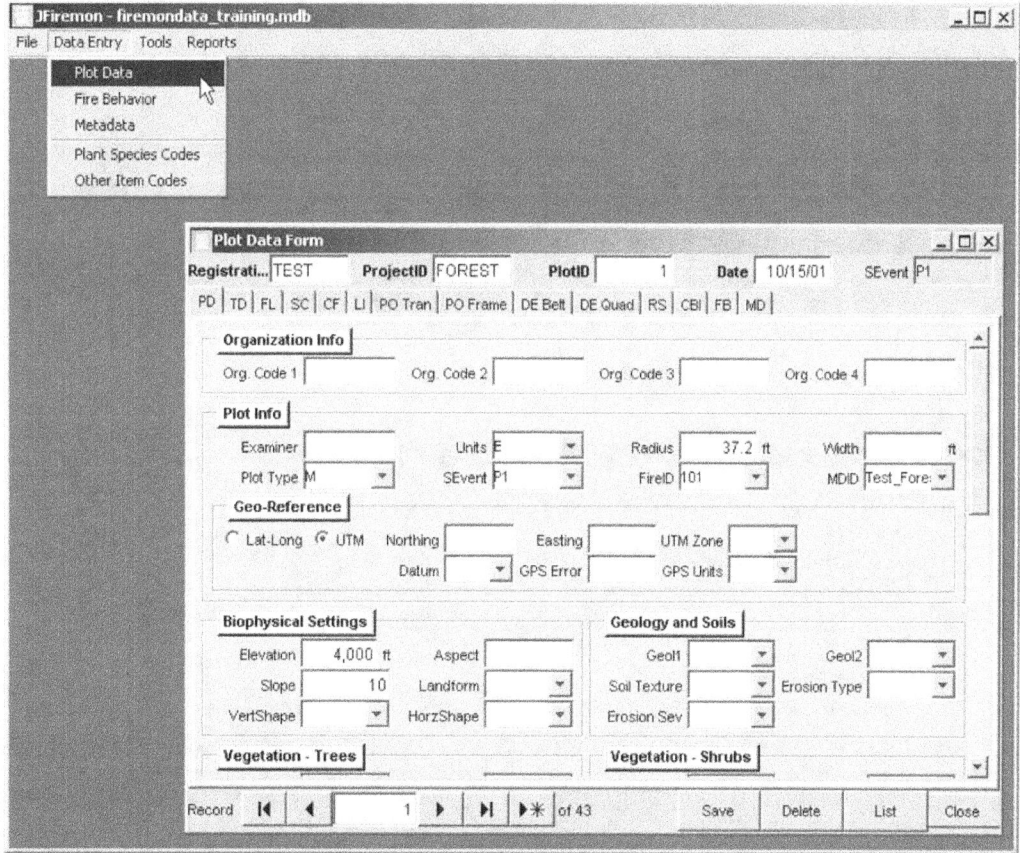

**Figure DB-7**—Adding a new record to the FIREMON database.

FIREMON Database Guide

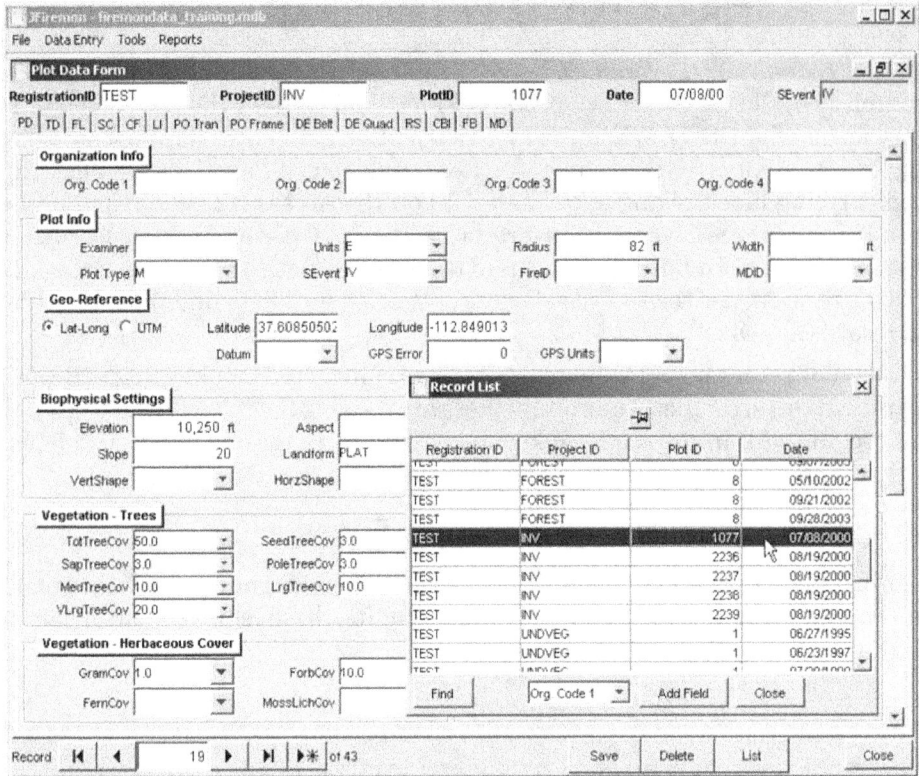

**Figure DB-8**—Editing an existing record in the FIREMON database.

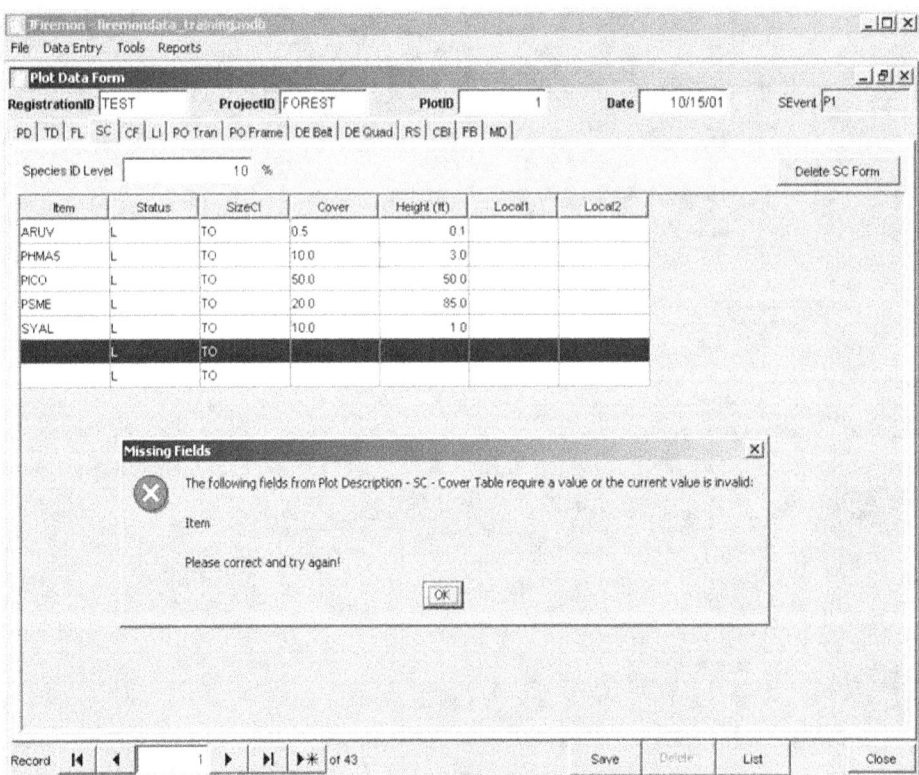

**Figure DB-9**—Example of adding a null record.

FIREMON Database Guide

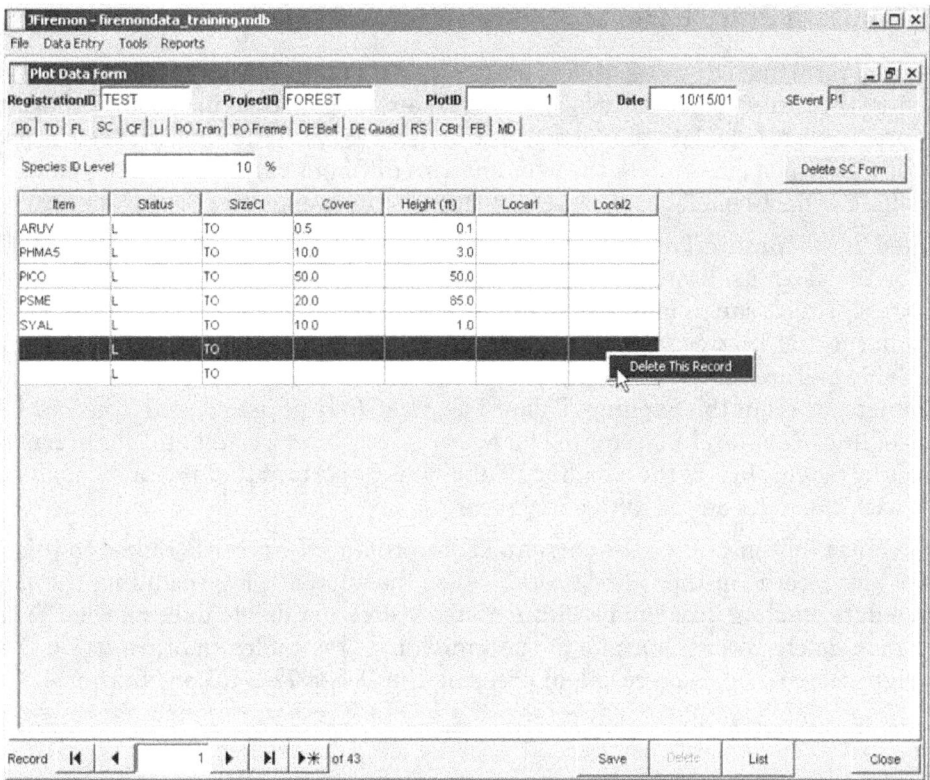

**Figure DB-10**—Null record error message.

Many data entry forms will require transect numbers, lengths, points, plot size, or other overall sample design data. It is important to provide accurate information as these entries are used to calculate plot summaries including items such as cover, density, frequency, and tons per acre.

The Plot Data (PD) form asks users to specify the units of the entered data as English or metric and, once entered, is used to specify units on all forms linked to that PD. The chosen units dictate how density, cover, and frequency values are calculated when summary reports are requested. Therefore, it is important to enter all data in the units that were specified on the PD. The units are described in the description of each sampling method. When summary reports are generated, users may view them in English or metric units.

## Plot Description (PD) Data Entry Form

The first data entry form is for the PD sampling method (fig. DB-8). Users are required to enter data in this form first for each plot. The FIREMON key fields (*RegID*, *ProjID*, *PlotID*, and *Date*) are required and automatically populate these fields in the other sampling method data entry forms. Users must also enter the units in which the FIREMON data are collected (E = English, M = metric), the plot type (C = control, M = measured), and the sampling event (P$n$ = pretreatment, R$n$ = remeasurement, IV = inventory).

All fields on the PD form relate to the general description of the plot. General categories of these fields include location, biophysical setting, geology and soils, vegetation cover, vegetation composition, ground cover, fuels, fire behavior and effects, plot photo IDs, and comments.

The *Delete* button on the PD form deletes the current PD record and any records linked to this record. Because all tables are linked to the PD table, this button deletes all records in the database for the current plot.

FIREMON Database Guide

## Tree Data (TD) Data Entry Form

The TD data entry form (fig. DB-11) is used to enter tree data for mature trees, saplings, and seedlings. Users must enter the correct plot sizes used in the TD sampling method in order for the data summary and analysis programs to calculate the correct tree density values. The snag plot size is assumed to be the same as the macro plot size, unless the user enters a different value in the snag plot size field. The snag plot size field cannot be left blank. Click on the TD tab to open the Tree Data form.

The *TagNo* field in the Mature Trees table is a key field and is required for each record. Each mature tree recorded on the plot must have a unique tag number entered in the database. If you did not use tags to mark your trees, *TagNo* can be numbered sequentially (1, 2, 3, 4…). Enter a tree species code and one or more fields in the mature tree table. The *SizeCL_Dia*, *Species*, and *TreeStat* fields are key fields in the Saplings Table and are required for each record. Enter a count and one or more other fields for each species by diameter class in the Saplings Table. The *SizeCL_Ht*, *Species*, and *TreeStat* fields are key fields in the Seedlings Table and are required for each record. Enter a count and one or more other fields for each species by height class in the Seedlings Table. It is important that the units of all fields entered be consistent with the units on the data entry form.

The *Delete TD Form* button deletes the current TD record and the records linked to this record. This button deletes any records in the tree data tables for the current plot, including the plot size data, individual tree data, sapling data, and seedling data. It does not delete data on the PD or other data forms. Users may delete records specific to the individual tree tables (mature trees, seedlings, and saplings) by right clicking on the record and pressing the *Delete This Record* button.

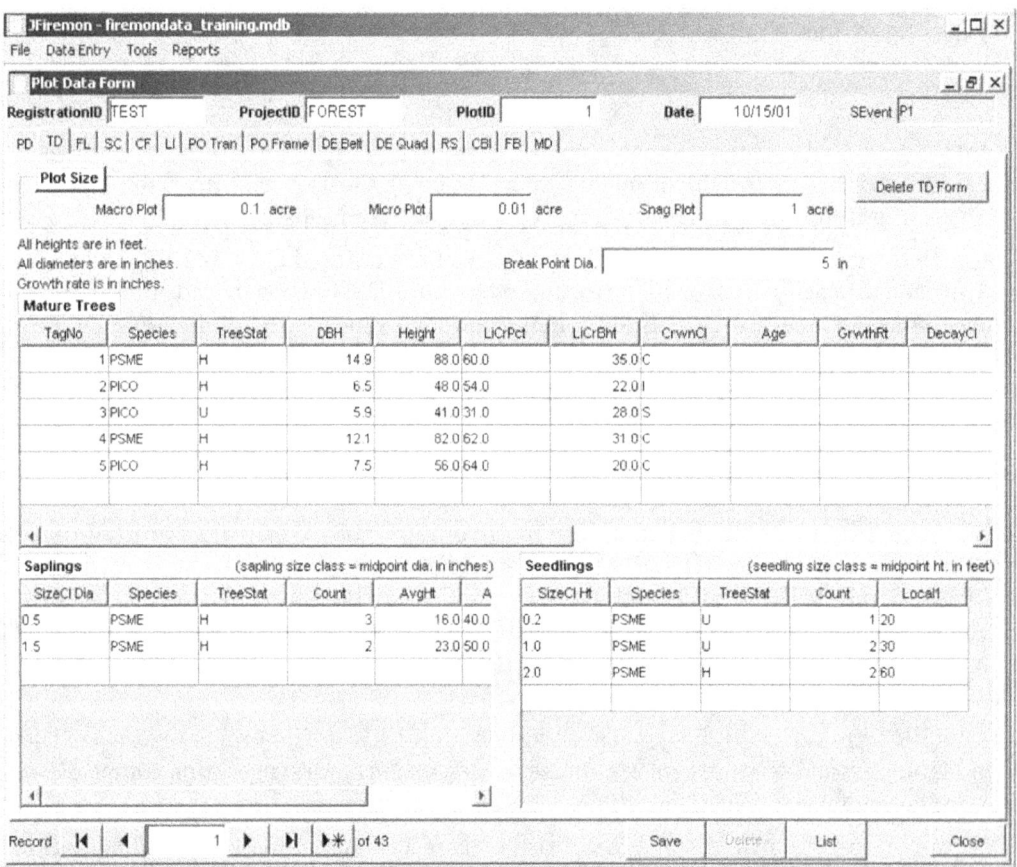

**Figure DB-11**—Tree data entry form.

## Fuel Loading (FL) Data Entry Form

The FL data entry form (fig. DB-12) is used to enter fuel counts for fine and coarse fuels, duff and litter depths, and vegetation cover and height. Users must enter the correct transect lengths, slope values, fuel counts, duff/litter depths, litter percent, log diameters, and decay classes in order for the fuel loading calculations to provide the correct values.

The *Transect* field is a key field in the *Fine Woody Debris* table and is required for each record. Number the sampling planes (transects) ascending and sequentially starting with transect number 1. Enter the slope of the transect, fuel counts, duff/litter depths, and litter percentages for each transect. Zero should be entered when a fine woody fuel class is not counted on a transect, although the fuel loading calculations assume a zero value if these fields are left blank. The *Transect* and *LogNum* fields are key fields in the *Coarse Woody Debris* table and are required for data entry. As with fine woody debris, number transects sequentially starting with number 1. Number logs sequentially for each transect or for the entire plot starting with number "1." Enter the diameter and decay class for each piece of coarse woody debris. If no pieces of coarse woody debris are located on a transect, we suggest you enter a transect and log number and enter "0" for the diameter field and "X" in the decay class field. This is not required but helps with recordkeeping. You will receive a warning message about zero diameter and invalid decay class fields from the data summary and analysis programs, but the biomass values will be correct. The *Transect* field is a key field in the *Vegetation* table and is required for each record. Enter shrub and/or herbaceous cover and height for each transect. If vegetation cover and height are collected at only one point along the transect, the data summary calculations include only this one point when

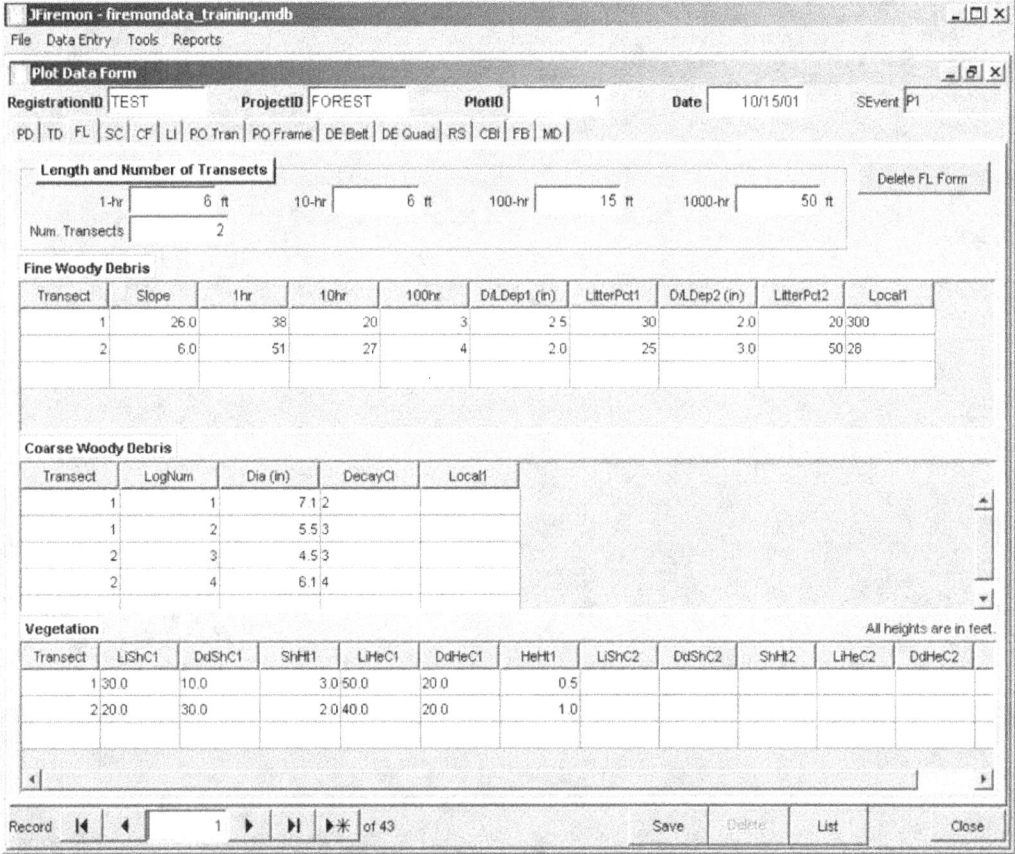

**Figure DB-12**—Fuel loading data entry form.

averaging these values across the transect. If no cover is noted for one of the cover components, enter "0" for that component. It is important that the units for all fields entered be consistent with the units on the data entry form.

The *Delete FL Form* button deletes the current FL record and any records linked to this record. This button deletes any records in the fuel data tables for the current plot including the transect length data, fine fuels data, coarse fuels data, and vegetation cover data. It does not delete data on the PD or other data forms. Users may delete records specific to the individual fuel tables (fine woody debris, coarse woody debris, vegetation) by right clicking on the record and pressing the *Delete This Record* button.

## Species Composition (SC) Data Entry Form

The SC data entry form (fig. DB-13) is used for ocular estimates of plant species cover and height for a plot. The *Species ID Level* field is required for data entry. It is important to enter the correct species identification level on this form so future users know whether a full or reduced plant species list was collected. The *Item*, *Status*, and *Size Class* are key fields and are required for each record. Enter a cover and/or height value for each item by status and size class. It is important that the units for all fields entered be consistent with the units on the data entry form.

The *Delete SC Form* button deletes the current SC record and any records linked to this record. This button deletes any records in the species composition tables for the current plot including the *Species ID Level* and all species on the plot. It does not delete data on the PD or other data forms. Users may delete individual species records by right clicking on the record and pressing the *Delete This Record* button.

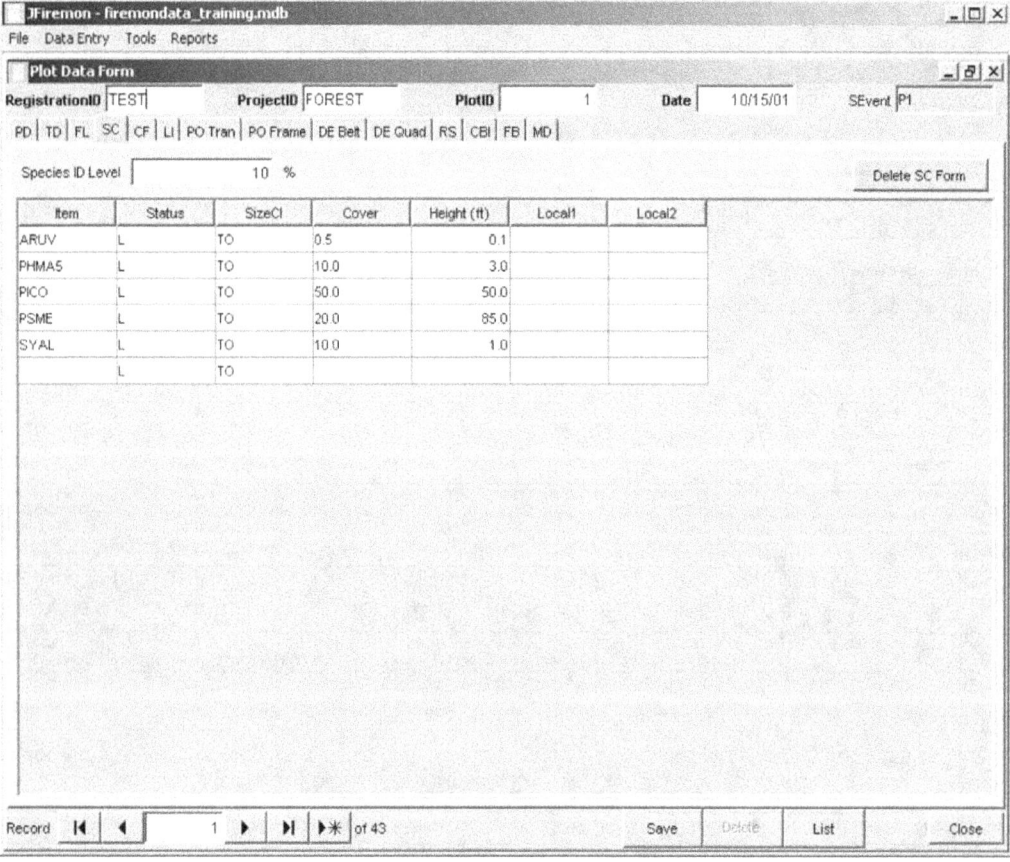

**Figure DB-13**—Plant species composition data entry form.

FIREMON Database Guide

## Cover/Frequency (CF) Data Entry Form

The CF data entry form (fig. DB-14) is for plant species cover and/or frequency data. Users must enter the correct number of transects and quadrats per transect in order for the data summary and analysis programs to calculate the correct average cover and frequency.

The *Transect*, *Item*, and *Status* fields are key fields and are required for each record. Enter one or more of the following fields: canopy cover, frequency, and height for each item by status on each quadrat. The data entry form allows a maximum of 20 quadrats per transect. The fields for each quadrat are numbered sequentially (CC1, NRF1, Ht1 for the first quadrat). It is important that the units for all fields be consistent with the units on the data entry form.

The *Delete CF Form* button deletes the current CF record and any records linked to this record. This button deletes any records in the cover/frequency tables for the current plot, including the transect/quadrat size data and all the species data for the plot. It does not delete data on the PD or other data forms. Users may delete individual species records by right clicking on the record and pressing the *Delete This Record* button.

## Line Intercept (LI) Data Entry Form

The LI data entry form (fig. DB-15) is for line intercept data for plant species. Users must enter the correct number of transects, transect length, and start/stop points for each intercept in order for the data summary and analysis programs to calculate the correct cover values.

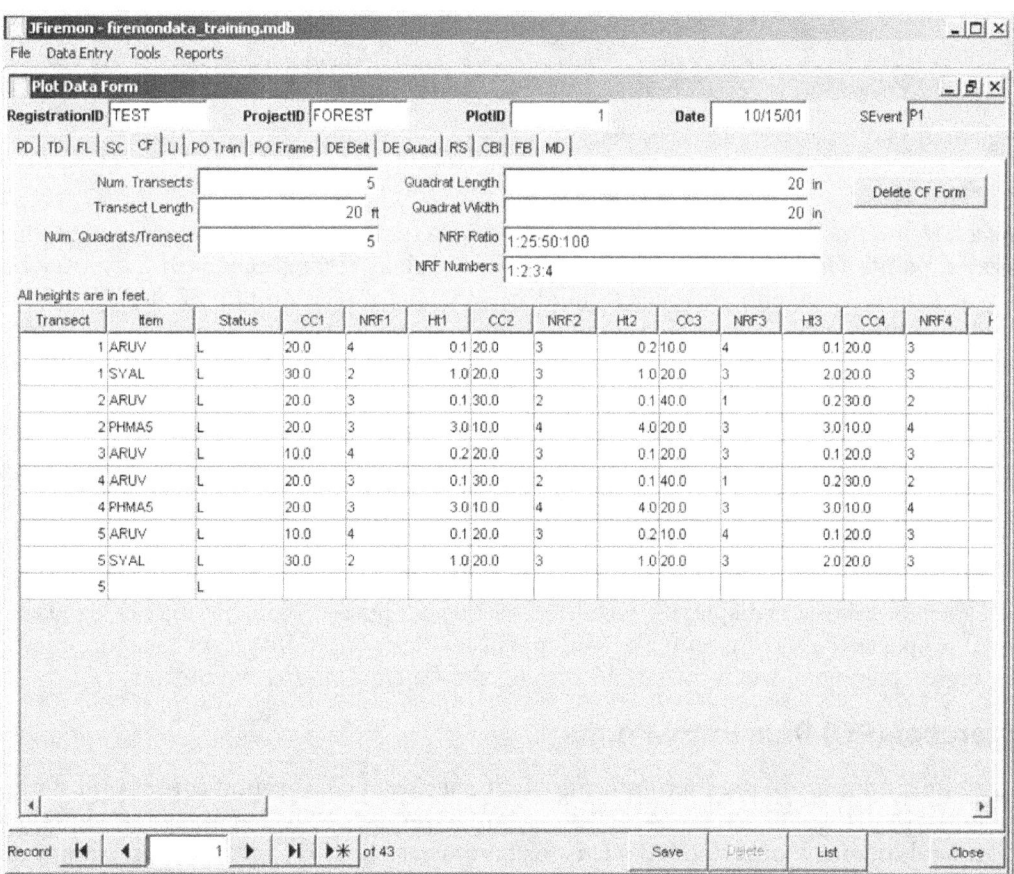

**Figure DB-14**—Cover/frequency data entry form.

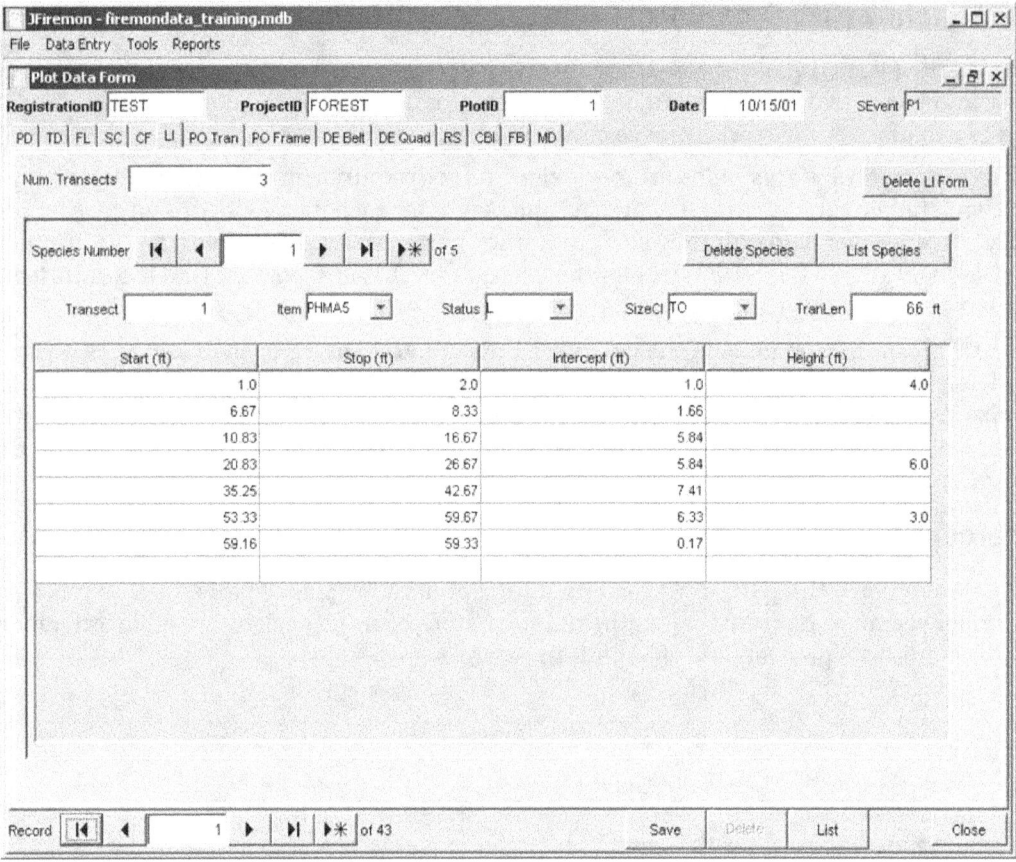

**Figure DB-15**—Line intercept data entry form.

The *Transect*, *Item*, *Status*, *SizeCl*, and *TranLen* fields are required for each record. Enter the transect start and stop points for each species by status and size class. Transect length may vary by species sampled. For example, sagebrush species could be sampled in the middle 30 meters of a 60-meter transect, while juniper could be sampled along the entire transect. Height values may be entered at one or more intercepts. The average height calculations for the line intercept data include only intercepts where heights are entered. Click the (right facing black triangle and asterisk) button on the *Species* navigation bar at the top of the LI data entry form to enter data for a new species.

The *Delete LI Form* button deletes the current LI record and any records linked to this record. This button deletes any records in the line intercept tables for the current plot, including the number of transects, all the species, and all the intercept measurements for the plot. It does not delete data on the PD or other data forms. Users may delete a single species from the plot using the *Delete Species* button.

A list of all species entered is displayed with the *List Species* button. This list may be used to navigate to a particular species on a transect for editing purposes. Individual intercept measurements may be deleted by right clicking on the record and pressing the *Delete This Record* button

## Point Intercept (PO) Data Entry Forms

The PO data entry forms are used for entering plant species and/or ground cover point data. The *PO Tran* data entry form (fig. DB-16) is used to enter point cover data collected along individual transects. The *PO Frame* data entry form (fig. DB-17) is for point cover data collected within frames (groups of points) along individual transects.

FIREMON Database Guide

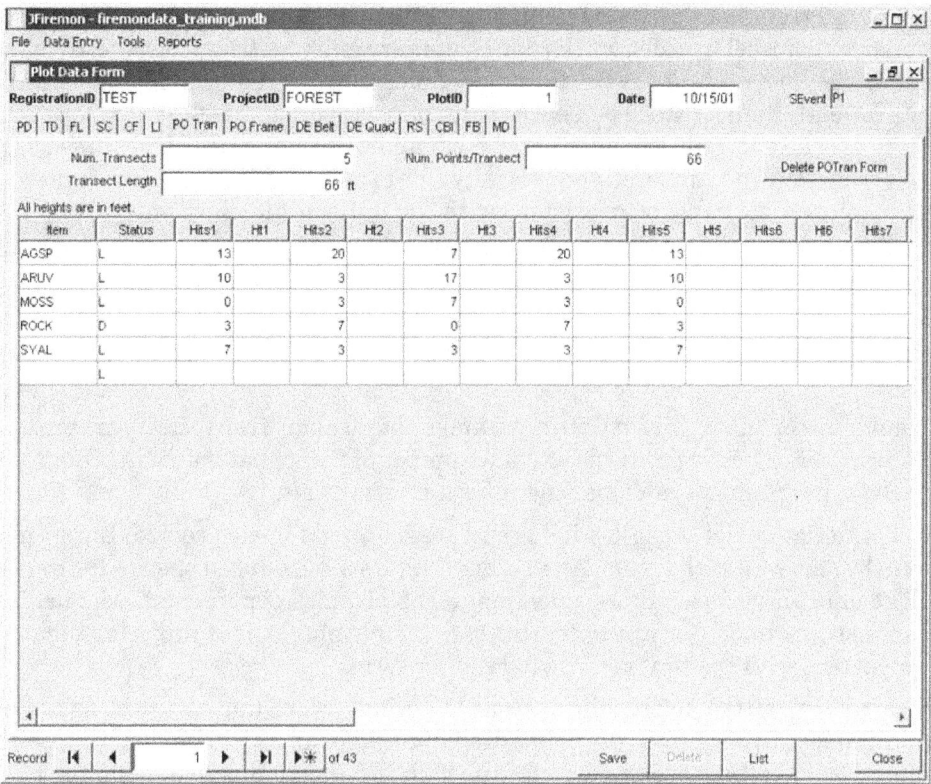

Figure DB-16—Point intercept transect data entry form.

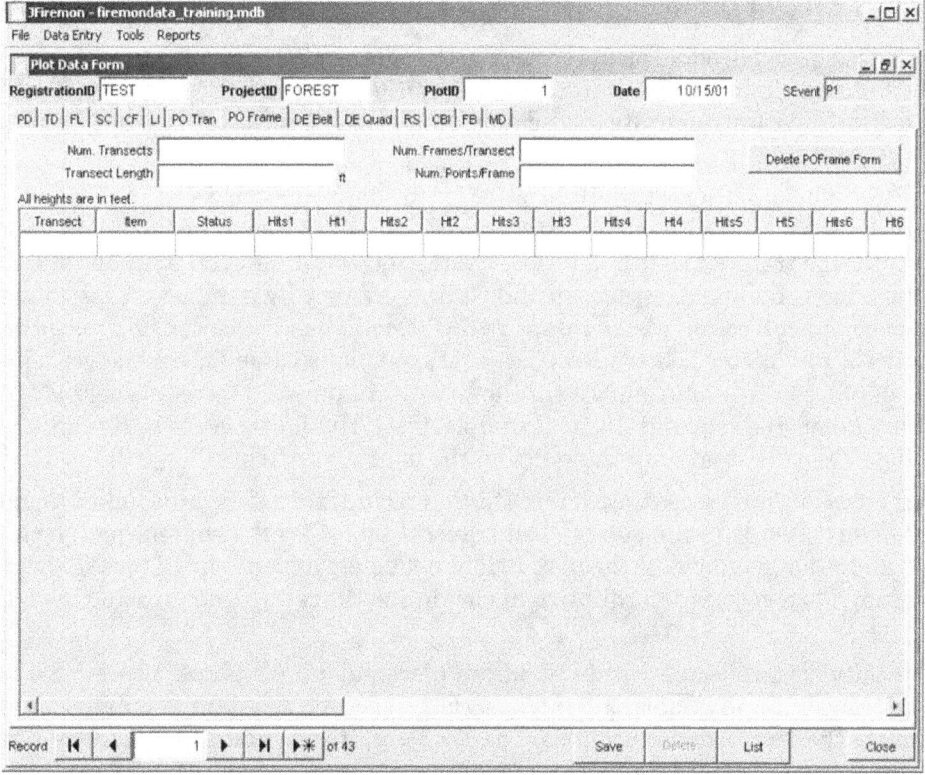

Figure DB-17—Point intercept frame data entry form.

When entering point intercept data collected along transects, users must enter the correct number of transects and points per transect in order for the data summary and analysis programs to calculate the correct cover values. The *Item* and *Status* fields are key fields and are required for each record. Enter the number of hits for each item by status on each transect. An average height value for each species on the transect may also be entered. The data entry form allows a maximum of 20 transects per plot. The fields for each transect are numbered sequentially (for example, Hits1, Ht1 for transect 1). It is important that the units for all fields be consistent with the units on the data entry form.

The *Delete POTran Form* button deletes the current PO transect record and any records linked to this record. This button deletes any records in the point intercept transect tables for the current plot, including the transect data and all the species data for the plot. It does not delete data on the PD or other data forms. Users may delete individual species by right clicking on the record and pressing the *Delete This Record* button.

When entering point intercept data collected within frames placed along transects, users must enter the correct number of transects, frames per transect, and number of points per frame in order for the data summary and analysis programs to calculate the correct cover values.

The *Transect*, *Item*, and *Status* fields are key fields and are required for each record. Enter the number of hits for each item by status in each frame. An average height value for each species in the frame may also be entered. The data entry form allows a maximum of 20 frames per transect. The fields for each frame are numbered sequentially (for example, Hits1, Ht1 for frame 1). It is important that the units for all fields be consistent with the units on the data entry form.

The *Delete POFrame Form* button deletes the current PO frame record and any records linked to this record. This button deletes any records in the point intercept frame tables for the current plot, including the transect/frame data and all the species data for the plot. It does not delete data on the PD or other data forms. Users may delete individual species records by right clicking on the record and pressing the *Delete This Record* button.

## Density (DE) Data Entry Forms

The DE forms are for counts of individual plants or other items. The *DEBelt* data entry form (fig. DB-18) is for density data collected along individual belt transects. The *DEQuad* data entry form (fig. DB-19) is used for density data collected within quadrats placed along individual transects.

When entering density data collected within belt transects, users must enter the correct number of transects, transect length, and transect width in order for the data summary and analysis programs to calculate the correct density values. The *Item*, *Status*, and *SizeCl* fields are key fields and are required for each record. Enter the transect length, transect width, and count for each item by status and size class in each belt transect. Belt transect length and width may vary by item, according to the sample design. For example, juniper trees may be counted in a 6 x 60 meter belt transect while sagebrush plants may be counted in a 2 x 60 meter belt transect. An average height value for each species in the belt transect may also be entered. The data entry form allows a maximum of 20 transects per plot. The fields for each transect are numbered sequentially (for example, Cnt1, Ht1 for transect 1). It is important that the units for all fields be consistent with the units on the data entry form.

The *Delete DEBelt Form* button deletes the current DE Belt record and any records linked to this record. This button deletes any records in the density belt transect tables for the current plot, including the transect data and all the species data for the plot. It does not delete data on the PD or other data forms. Users may delete individual species records by right clicking on the record and pressing the *Delete This Record* button.

When entering density data collected within quadrats placed along transects, users must enter the correct number of transects and quadrats per transect in order for the data summary and analysis programs to calculate the correct density values. The *Transect*, *Item*, *Status*, and *SizeCl* fields are key fields and are required for each record. Enter the quadrat length, quadrat width, and count for each item

# FIREMON Database Guide

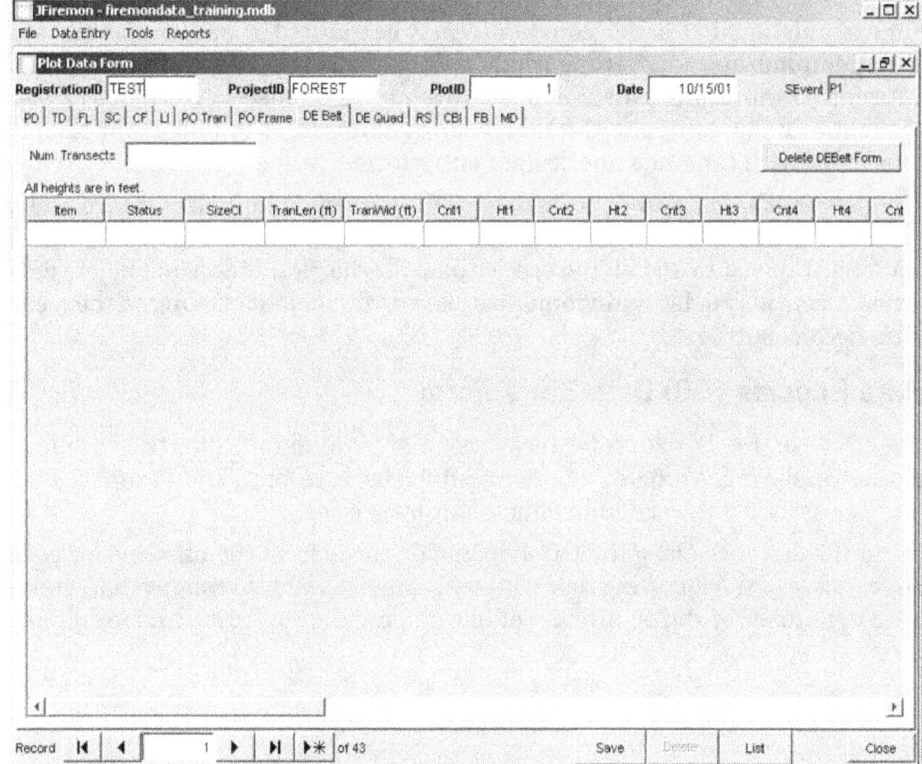

**Figure DB-18**—Density belt transect data entry form.

**Figure DB-19**—Density quadrat data entry form.

by status and size class in a quadrat. Quadrat size may vary by species according to the sample design. For example, smaller herbaceous plants may be counted in smaller quadrats and larger shrubs counted in larger quadrats. An average height value for each species in the quadrat may also be entered. The data entry form allows a maximum of 20 frames per transect. The fields for each quadrat are numbered sequentially (for example, Cnt1, Ht1 for quadrat 1). It is important that the units for all fields be consistent with the units on the data entry form.

The *Delete DEQuad Form* button deletes the current DE quadrat record and any records linked to this record. This button deletes any records in the density quadrat tables for the current plot, including the transect/frame data and all the species data for the plot. It does not delete data on the PD or other data forms. Users may delete individual species records by right clicking on the record and pressing the *Delete This Record* button.

## Rare Species (RS) Data Entry Form

The RS form (fig. DB-20) is for rare perennial plant data. Enter the baseline length for each plot. The *Species* and *PlantNo* fields are required for each record. The *PlantNo* field is a unique, sequential number assigned to each individual plant by species.

Enter the distance along the baseline and distance from the baseline for each individual plant. Enter one or more of the following fields: status, stage, maximum canopy diameter, second canopy diameter, height, number of stems, number of flowers, number of fruits, and local fields 1, 2, and 3.

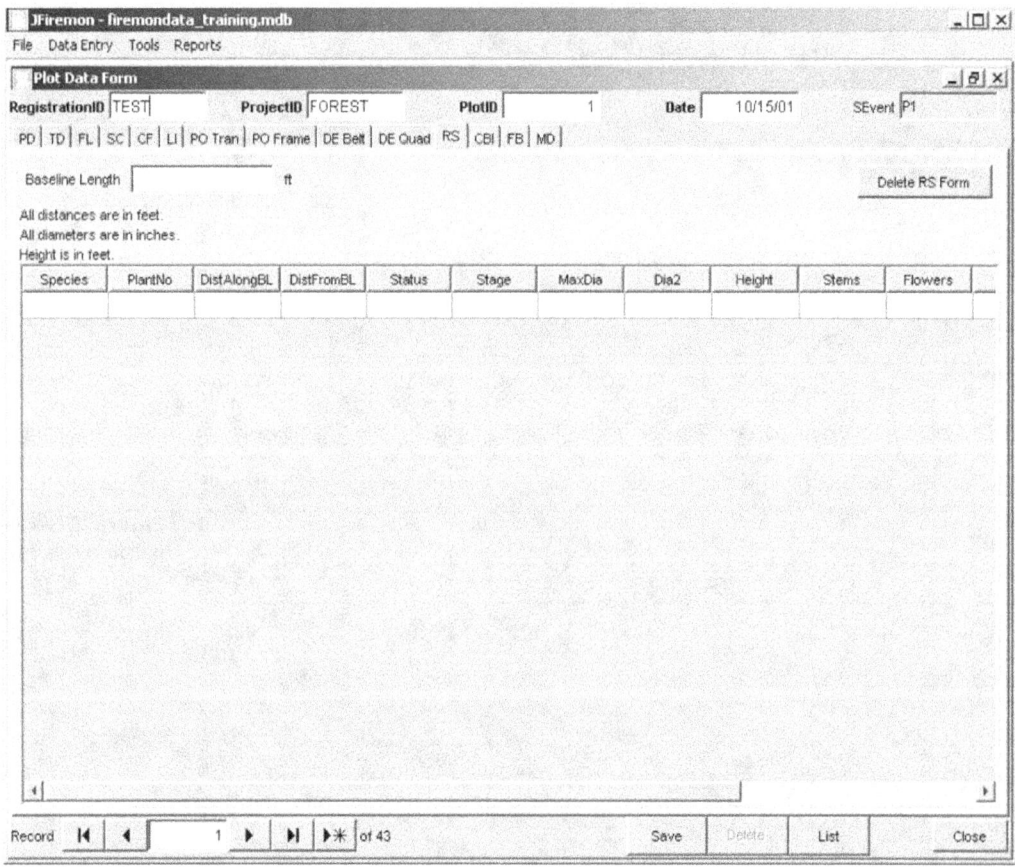

**Figure DB-20**—Rare species data entry form.

FIREMON Database Guide

The *Delete RS Form* button deletes the current RS record and any records linked to this record. This button deletes any records in the RS tables for the current plot, including all individual plant records for the plot. It does not delete data on the PD or other data forms. Users may delete one or more individual plant records by right clicking on the record and pressing the *Delete This Record* button.

## Composite Burn Index (CBI) Data Entry Form

The CBI form (fig. DB-21) is used for the calculation of Composite Burn Index value by plot. Enter the effects of fire, ranked from 1 to 3, for each of the vegetation strata on the form. The CBI values for each stratum and for the total plot are updated as the data are entered or edited. The *Delete CBI Form* button deletes the current CBI record. It does not delete data on the PD or other data forms.

## Fire Behavior (FB) Data Entry Form

The FB data entry form (fig. DB-22) is for fire behavior data. The *Fire Behavior* form is separate from other data entry forms because it is not entered for every plot. The records in the *Fire Behavior* table are linked to the plot data through the *FireID* field on the PD form. One Fire Behavior record may be linked to many plots. Fire Behavior data may be entered for different times during the same fire. To add a new fire behavior record using the current *FireID*, click the *New Record* button (right facing black triangle and asterisk) on the *FBData* record navigator bar in the upper part of the *Fire Behavior Entry* form. The *Delete* button deletes the current FB record and all data recorded for the fire. It does not delete data on the PD or other data forms. The *Delete FBData* button deletes observations recorded at one time

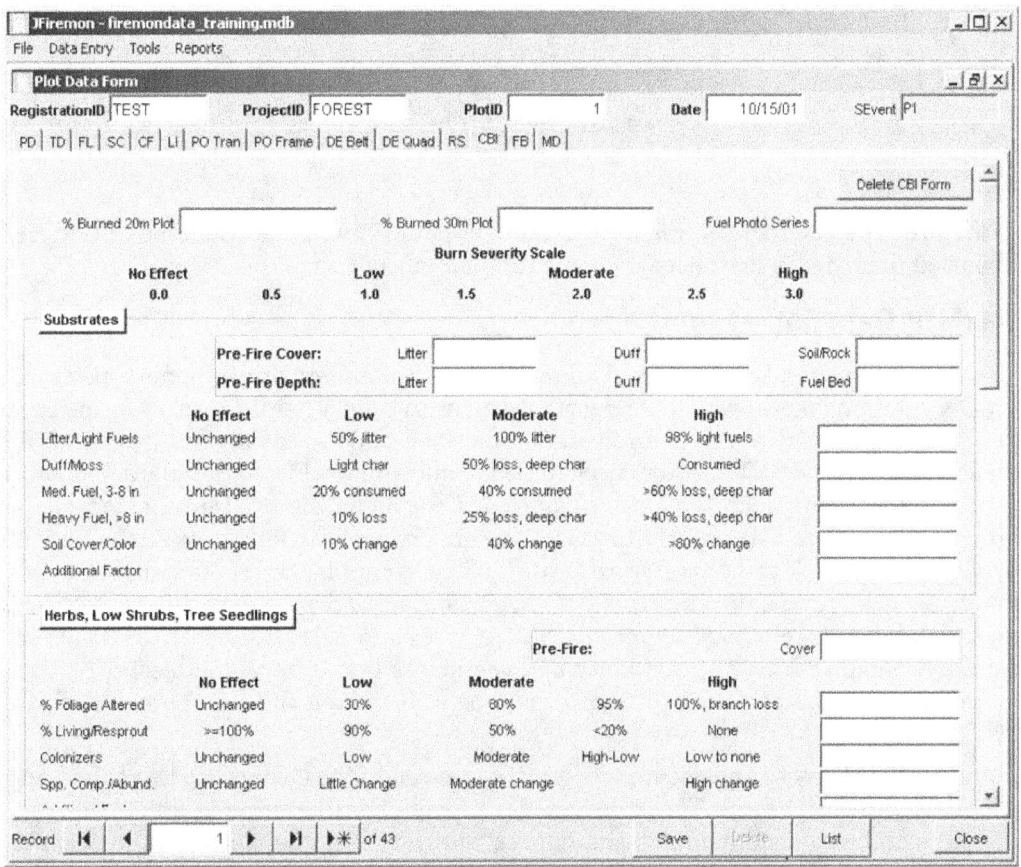

**Figure DB-21**—Composite Burn Index data entry form.

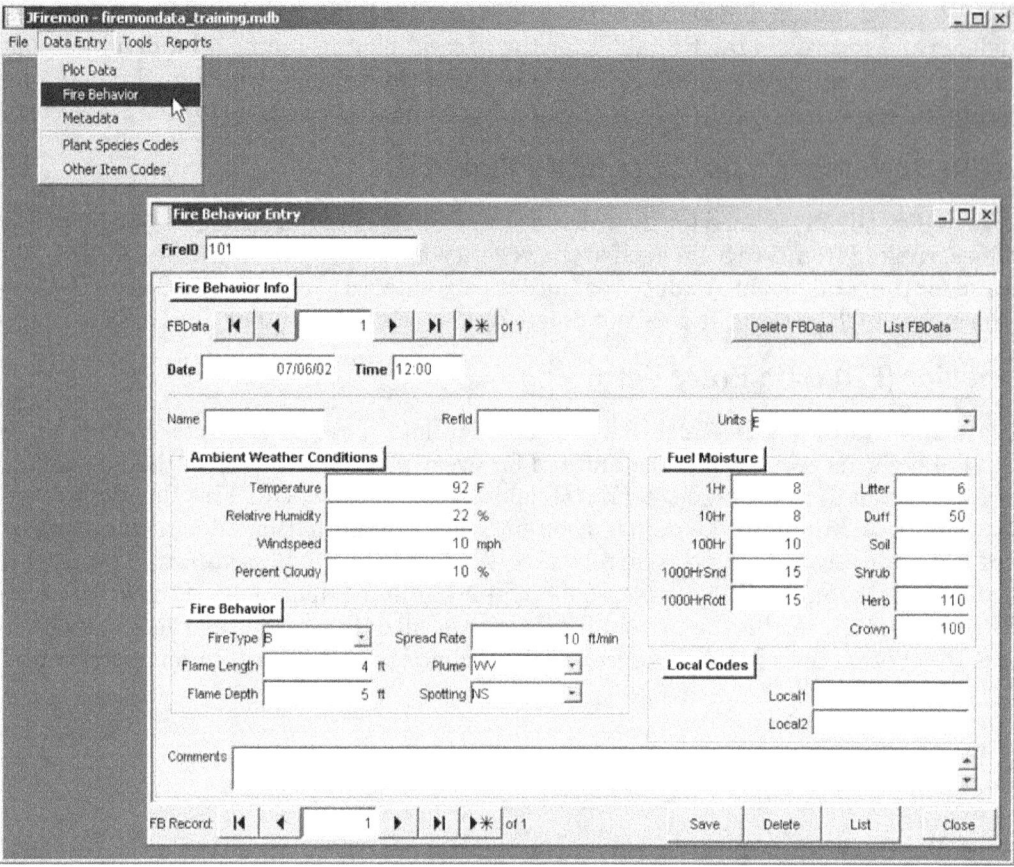

**Figure DB-22**—Fire Behavior data entry form.

period for the fire. The *List FBData* displays a list of all observations recorded for the current *FireID* and may be used to navigate to a specific observation for editing purposes.

## Metadata (MD) Data Entry Form

The MD data entry form (fig. DB-23) is used for entering any metadata and/or general comments about the FIREMON sampling methods. The MD form is separate from other data entry forms because it is not entered for every plot. The records in the MD table are linked to the plot data through the *MDID* field on the PD form. One MDID record may be linked to many plots. Enter metadata and/or comments by subject in the comments field. Click the *New Record* button on the *MDData* navigation bar in the upper part of the *Metadata Entry Form* to start recording comments for a new subject related to the current *Metadata ID*. Click the *New Record* button (right facing black triangle and asterisk) on the *MDData* navigation bar in the upper part of the *Metadata Entry Form* to start recording comments for a new subject related to the current *Metadata ID*. The *Comment* field can store up to 65,536 characters. If more text is needed, split up the metadata and comments into different subjects. Text from word processor documents (such as Microsoft Word) may be copied from the Windows clipboard into the *Comments* field.

The file name for Word processing documents may be entered in the *Documents Link* field and opened via hyperlink from the MD form. These documents provide additional metadata that are linked to FIREMON plots and also display tables and figures, which can not be stored in the comments field. These documents should be stored in the documents subdirectory or the user selected directory listed in the *Document Base Directory* field in the FIREMON Configuration and Settings form (fig. DB-3).

FIREMON Database Guide

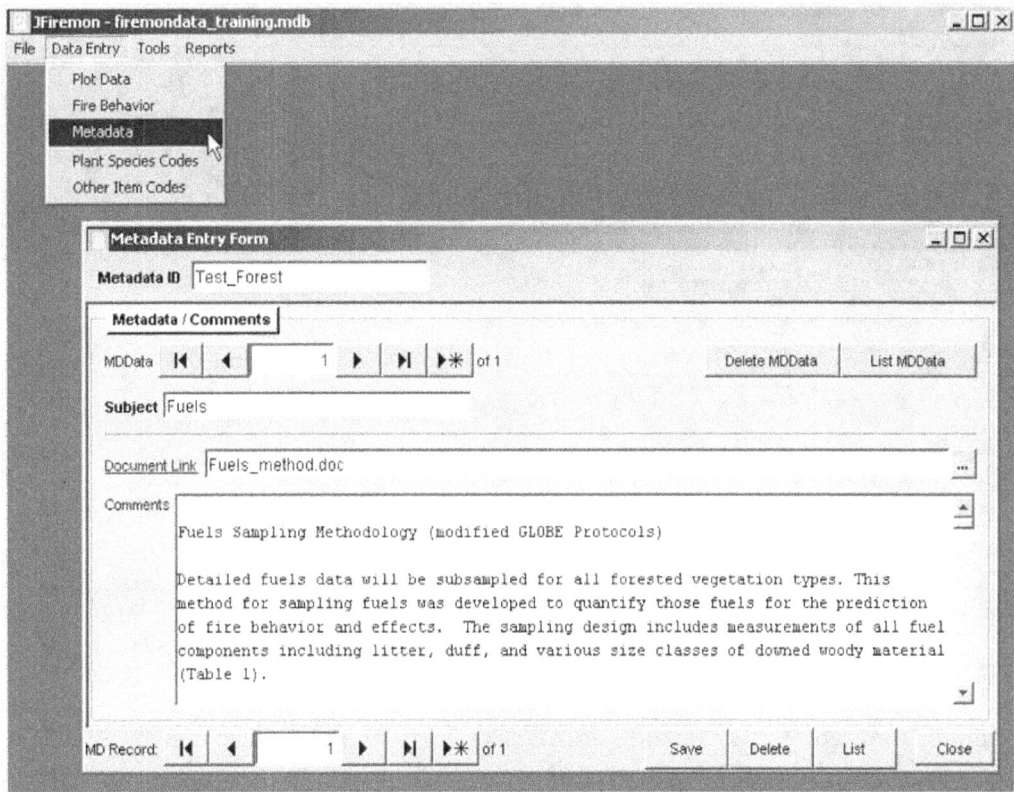

Figure DB-23—Metadata data entry form.

The *Delete button* deletes the current MD record and associated MD records. It does not delete data on the PD or other data forms. The *Delete MDData* button deletes metadata or comments for one subject only. The *List MDData* displays a list of all subject records for an MDID record and may be used to navigate to a specific subject for editing purposes.

## FIREMON DATA SUMMARY REPORTS

FIREMON provides reports to display summary data for each plot in the FIREMON database. Reports may be printed or exported as .pdf, .csv, or .html files. Select *Reports → Data Summary Reports* from the FIREMON toolbar to display the *Data Summary Reports* form (fig. DB-24). Users may select the output units for the reports (English or metric) independent of input units. The data summary report is selected using radio buttons. The plots displayed on the report may be filtered by any of the FIREMON plot key fields and the sample event (*SEvent*). For example, users may report plots for only a specific project or select only pretreatment plots. Click on the *Generate Report* button to display the report. Use the *Print* button to print the report or the *Export* button to save the report in the desired file format.

### Tree Data (TD) Summary Reports

The TD data summary reports (fig. DB-25 and DB-26) display summary data for all trees on a plot and by tree species on a plot. Tree density, basal area, average live crown base height (crown fuel base height), average height, and quadratic mean diameter (QMD) are calculated for mature trees. Seedling, sapling, and snag density are also calculated. Density is calculated per acre (ha), average heights are displayed in feet (m), and QMD is displayed in inches (cm).

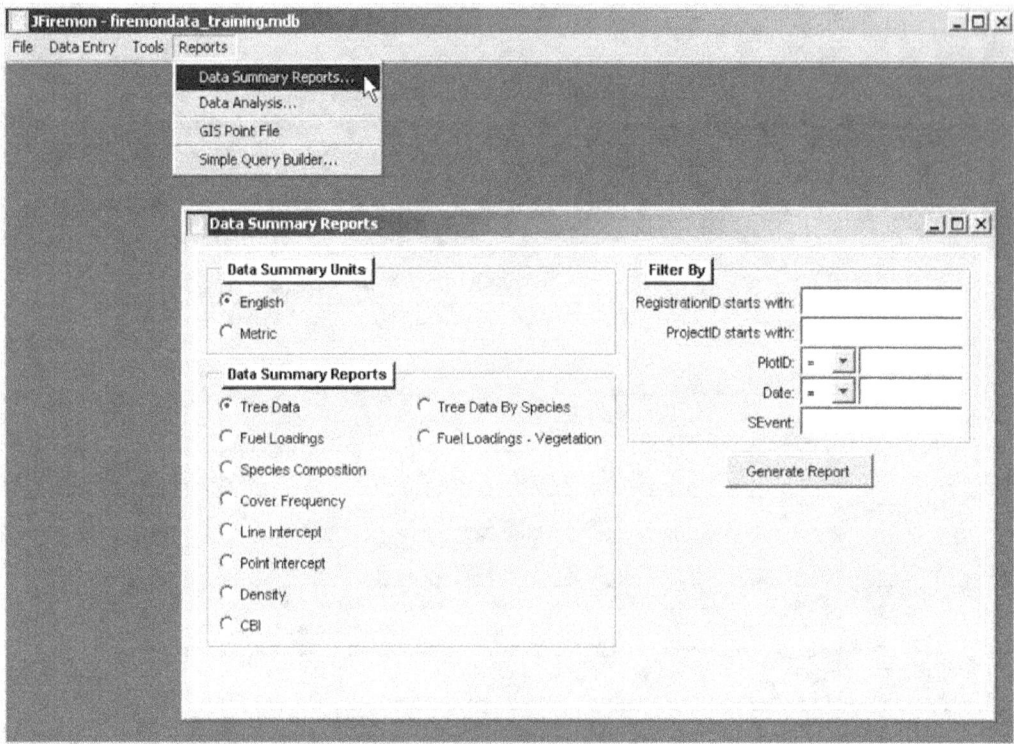

**Figure DB-24**—Data summary units form.

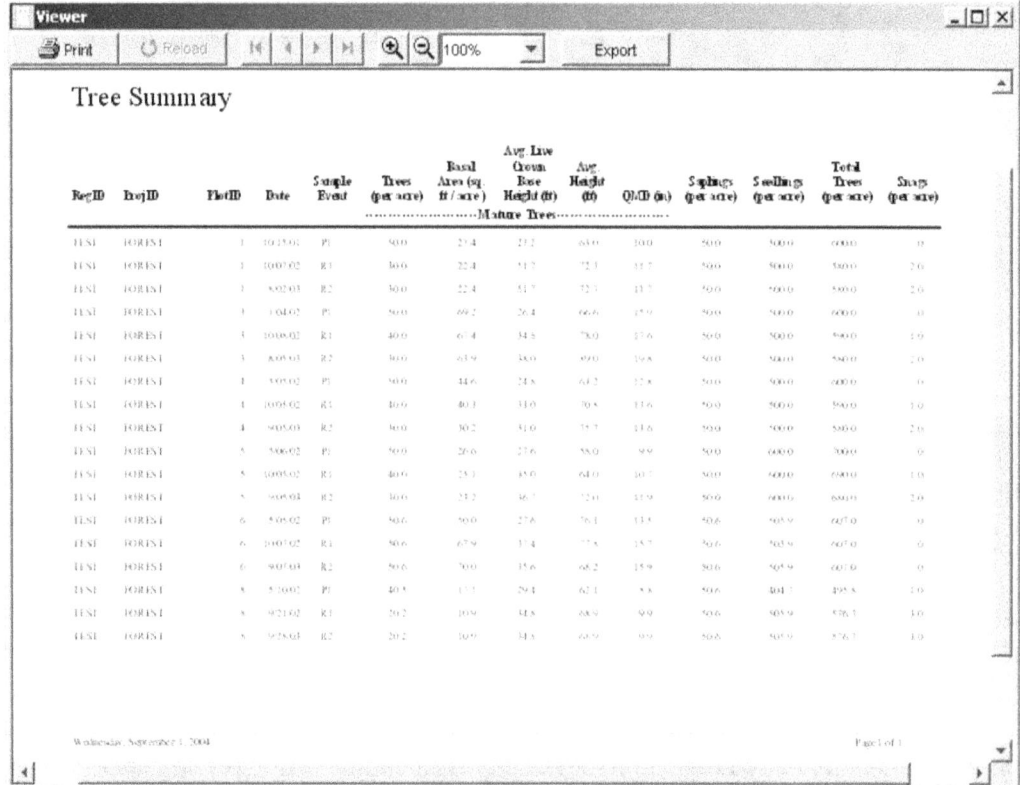

**Figure DB-25**—Tree data summary form.

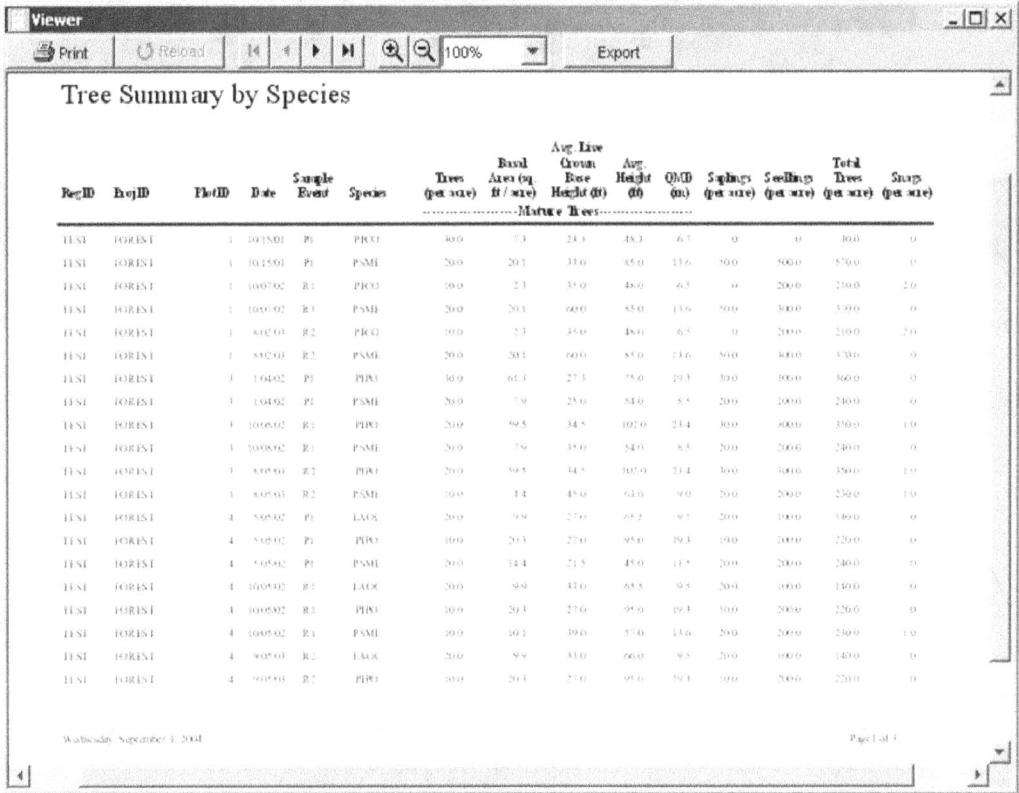

**Figure DB-26**—Tree species data summary form.

## Fuel Loading (FL) Summary Reports

The FL summary reports (figs. DB-27 and DB-28) display fuel loadings in tons per acre (kg per m$^2$) calculated from fuel counts and duff / litter measurements on the fuels transects along with average vegetation cover and heights. Down woody loadings are calculated for 1-hour, 10-hour, 100-hour, 1 to 100-hour, 1,000-hour sound, 1,000-hour rotten, and 1 to 1,000 hour fuels. Biomass is calculated based on the equations presented in the *Handbook for Inventorying Downed Woody Material* (Brown 1974). Nonslash, composite values are used for quadratic mean diameter, nonhorizontal correction, and specific gravity of fine woody debris. Decay class 1, 2, and 3 pieces of coarse woody debris are considered sound and assigned a specific gravity of 0.40. Decay class 4 and 5 pieces are considered rotten and assigned a specific gravity of 0.30. Loading of litter and duff is calculated using bulk densities of 2.75 lb per ft$^3$ and 5.5 lb per ft$^3$, respectively. Duff and litter depth summaries are provided. Averages for live shrub cover, dead shrub cover, live herbaceous cover, and dead herbaceous cover are also calculated. Average shrub height and herbaceous height are calculated in feet (m). Biomass of live and dead shrubs and biomass of live and dead herbaceous plants are calculated using the equation,

Equation DB-1 $\qquad\qquad B = H*C*BD$

where, B is biomass (kg per m$^2$), H is height (m), C is percent cover/100, and BD is bulk density (kg per m$^3$)

Bulk density used for the herbaceous and shrub components are 0.8 kg per m$^3$ and 1.8 per m$^3$, respectively. Vegetation heights entered in English units on the FL form are automatically converted to metric units for the biomass calculations.

FIREMON Database Guide

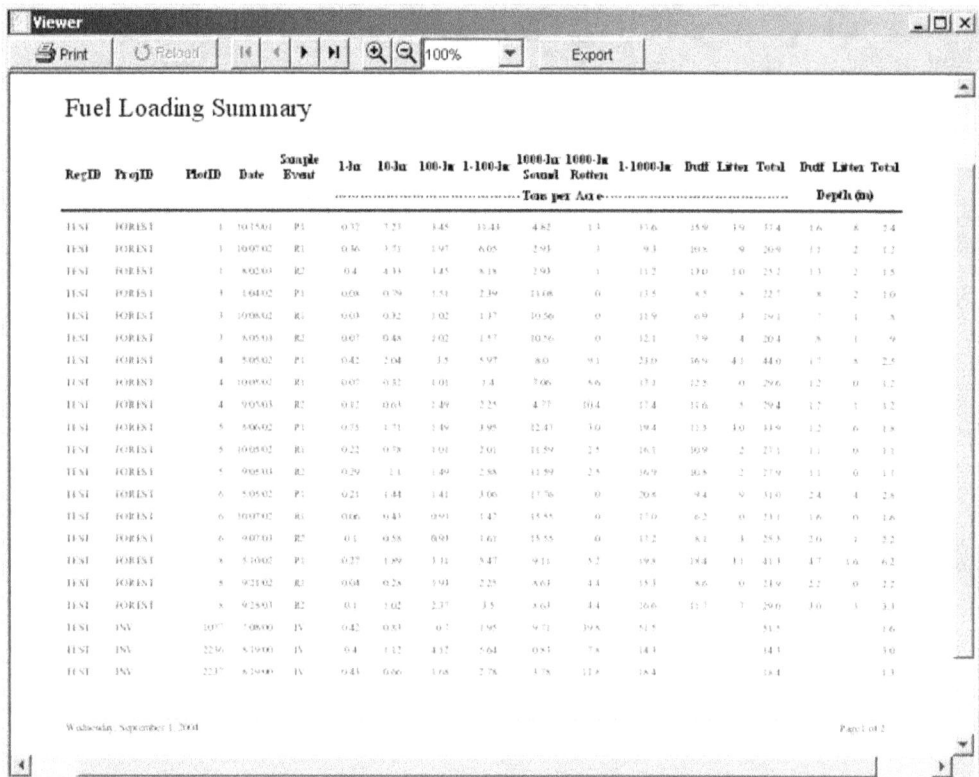

Figure DB-27—Fuel loading data summary report.

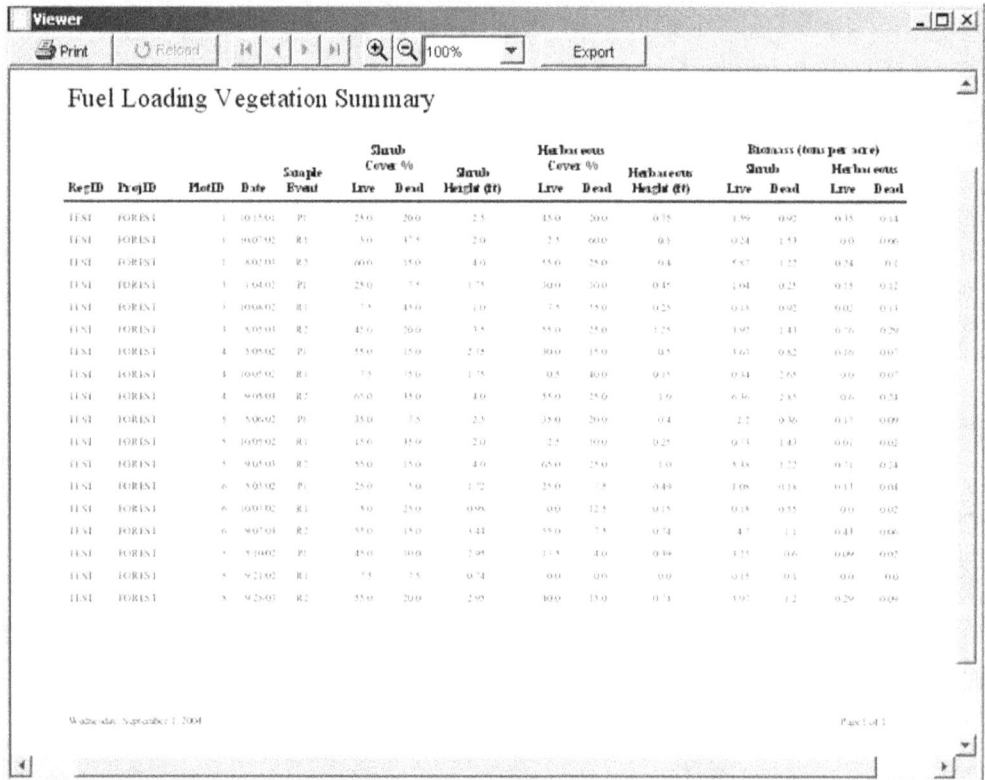

Figure DB-28—Fuel loading vegetation summary report.

## Species Composition (SC) Summary Report

The SC summary report (fig. DB-29) displays the same information as the data entry form because no calculations are required to calculate average cover or height for a plot. Average height is displayed in feet or meters depending on the output units selected.

## Cover/Frequency (CF) Summary Report

The CF summary report (fig. DB-30) displays average plant species cover and frequencies for the different frame sizes used in the CF method. Average cover is calculated for each plant species by status. The frequency of occurrence is calculated for the different frame sizes used to collect the frequency data. The Summary Report allows up to four frame sizes. Frequency is calculated by dividing the number of quadrats in which a species is present by the total number of sampled quadrats.

## Line Intercept (LI) Summary Report

The LI summary report (fig. DB-31) displays average cover and height values for plant species sampled using the line intercept method. Plant species cover on a transect is calculated by dividing the total intercept for each species on a transect by the total length of the transect. Average cover is then calculated for the plot by averaging the transect cover values. Average height is calculated by averaging all the height measurements for a species by status and size class. Average height is displayed in feet or meters depending on the output units selected.

## Point Cover (PO) Summary Report

The PO summary report (fig. DB-32) displays average cover and height values for plant species sampled using the point cover method. Plant species and ground cover values on a transect are calculated by

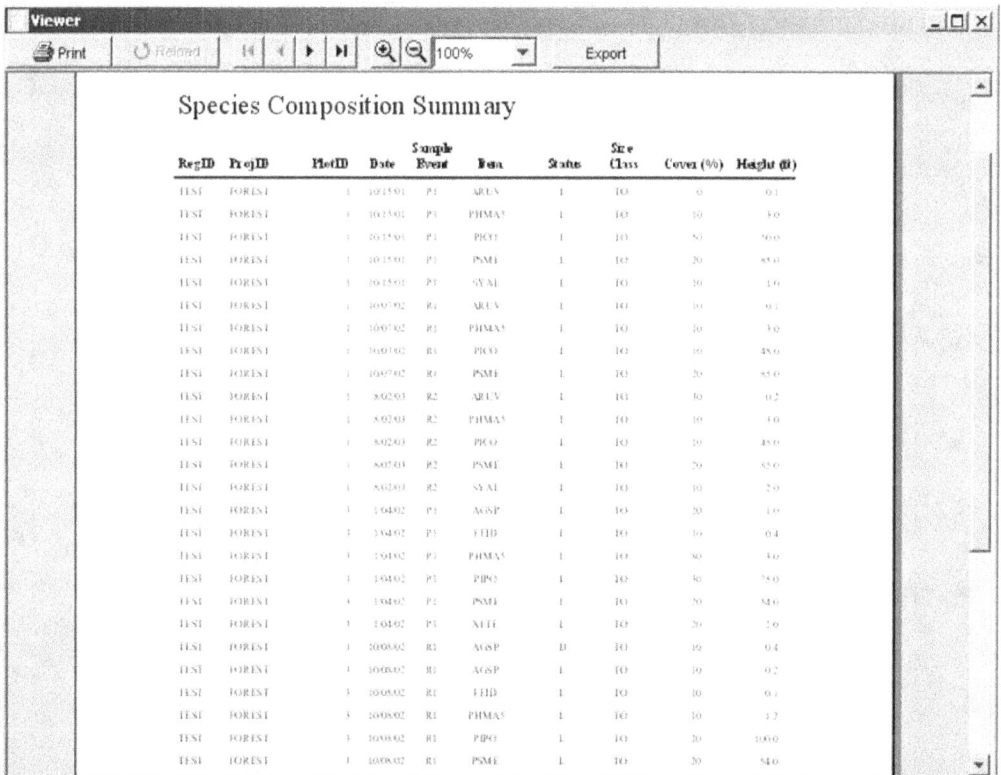

**Figure DB-29**—Species composition summary report.

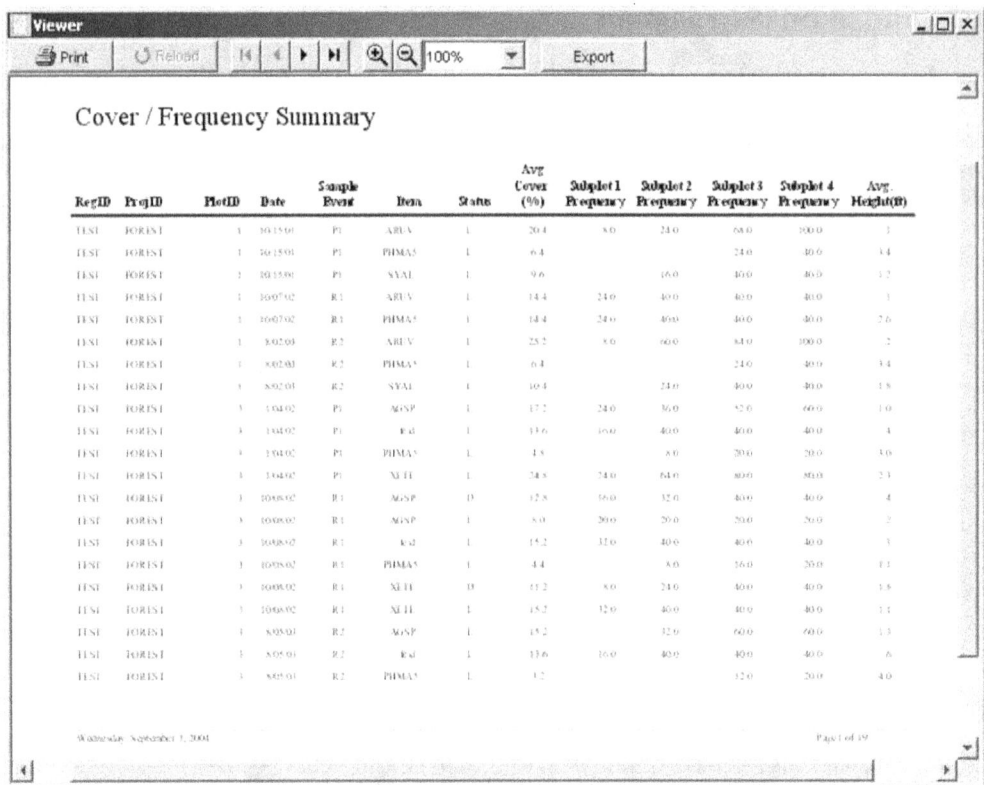

**Figure DB-30**—Cover/frequency summary report.

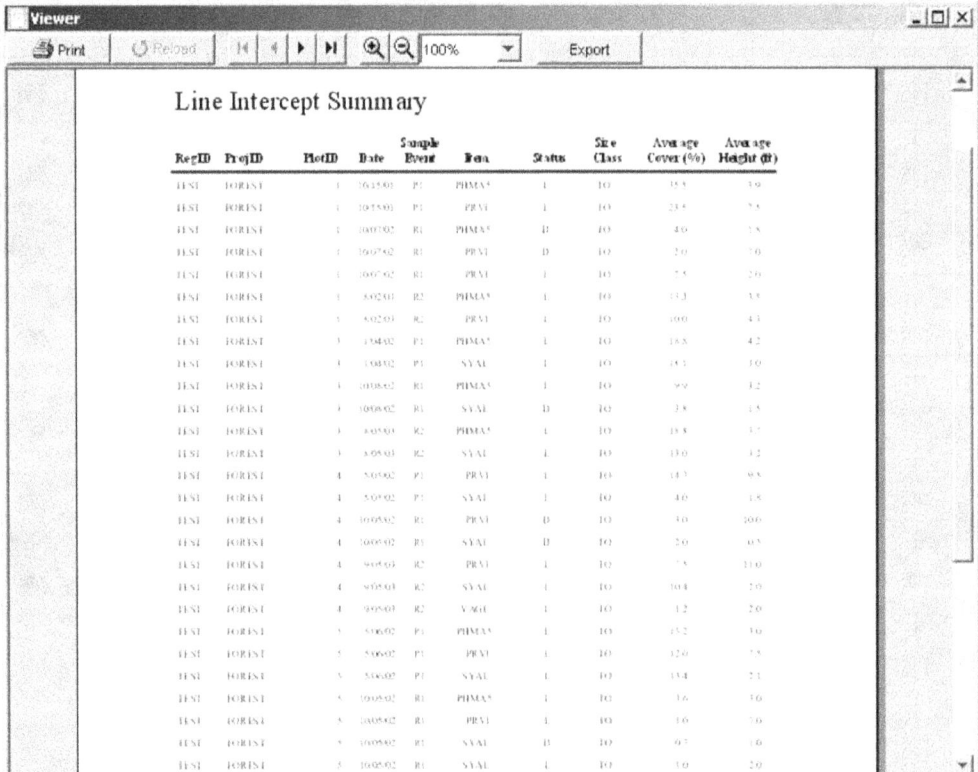

**Figure DB-31**—Line intercept summary report.

FIREMON Database Guide

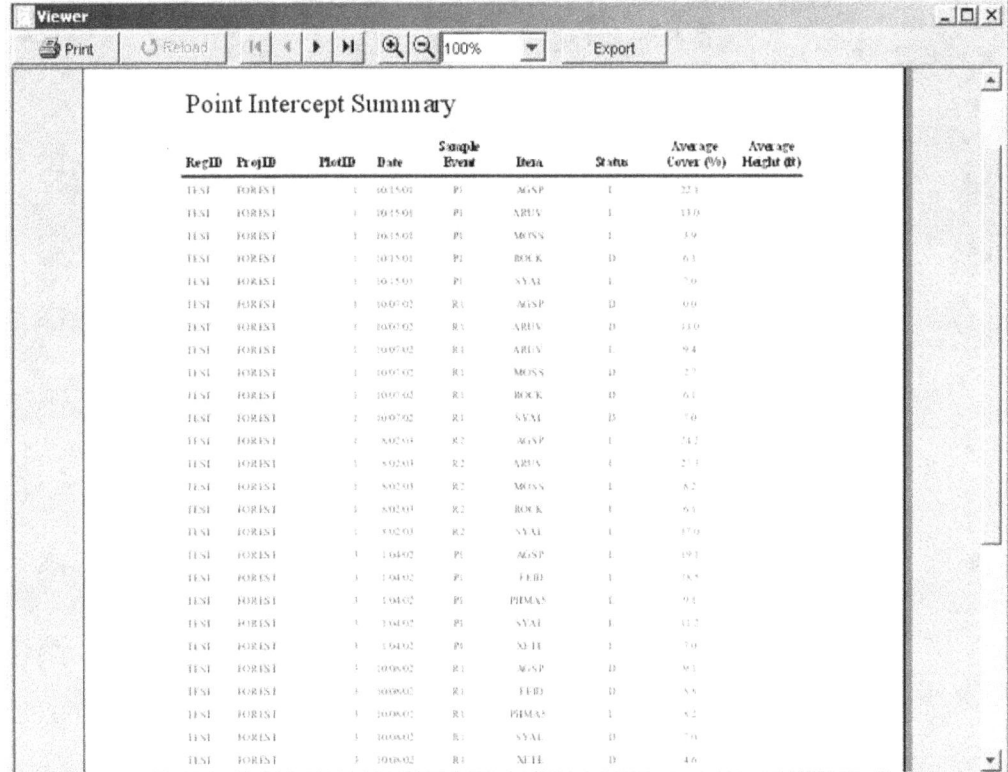

**Figure DB-32**—Point intercept summary report.

dividing the number of hits for an item divided by the total number of points sampled on the transect. The transect cover values are then averaged to provide an average cover value for the plot. Average height values in feet (m) are also calculated. Plant species and ground cover values for point frames are calculated by dividing the number of hits for an item divided by the total number of points per frame. The frame cover values are used to calculate an average cover value for the plot.

## Density (DE) Summary Report

The DE summary report (fig. DB-33) displays density summaries for plant species and other items sampled using the density method. The average number of items per quadrat or belt transect is calculated. The average number of items per $ft^2$ ($m^2$) and per acre (hectare) are also calculated. The average height in feet (m) is calculated for each item.

## Composite Burn Index (CBI) Summary Report

For each plot the CBI summary report (fig. DB-34) displays the Composite Burn Index for each stratum and the summarized CBI values for understory, overstory, and the total plot.

# FIREMON DATA ANALYSIS PROGRAM

The FIREMON Analysis Tools (FMAT) program is opened by selecting *Reports → Data Analysis* from the FIREMON toolbar. When FMAT opens, it may not be connected to a database or may not be connected to the same database you are using in FIREMON. Open a database by selecting *Options → Settings* from the main toolbar (fig. DB-35), click *Open* and double click on the database you want to use. Databases have an .mdb extension and are found in the installation folder (default: c:\firemon). Select the *RegID* and *ProjectID* for the plots you want to examine (fig. DB-36).

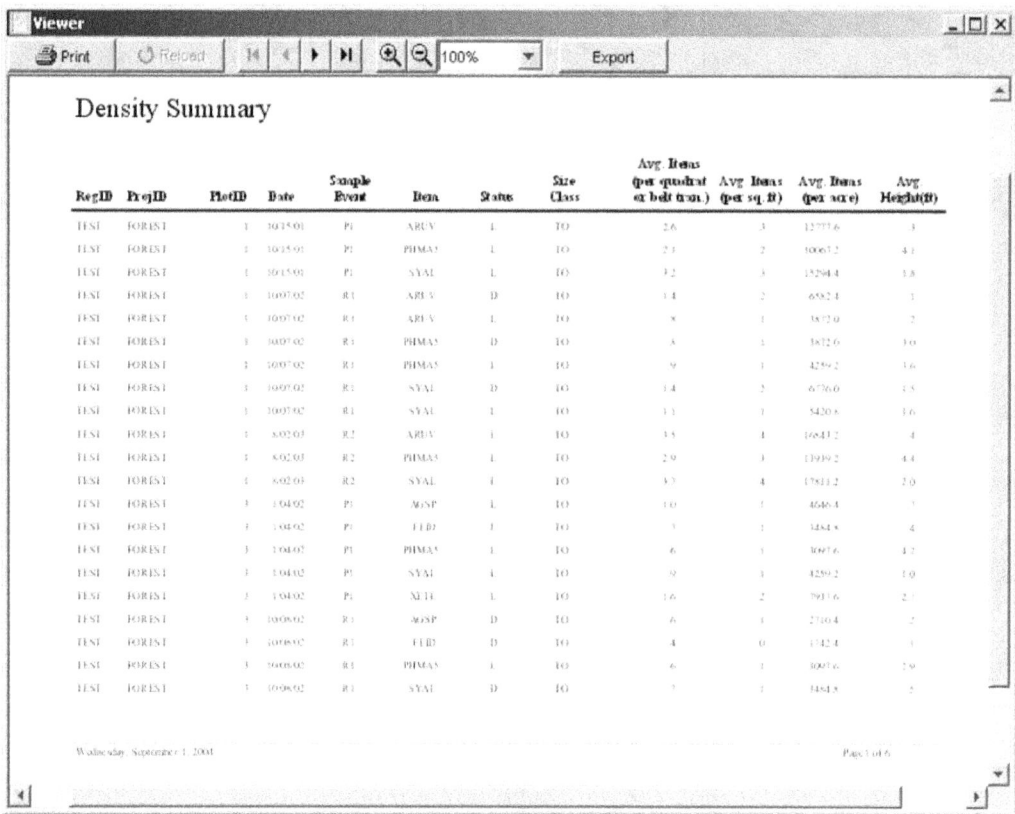

Figure DB-33—Density summary report.

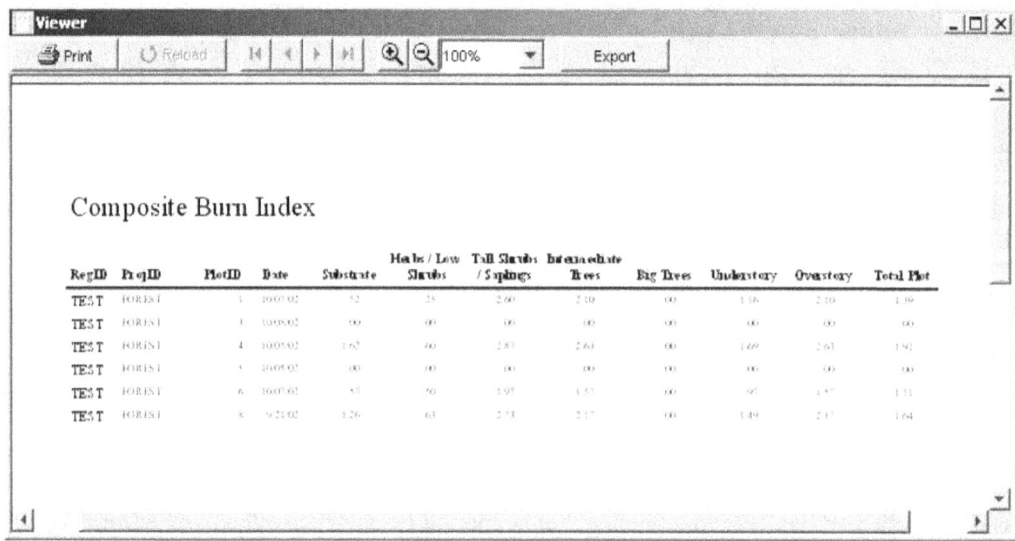

Figure DB-34—Composite burn index summary report.

FIREMON Database Guide

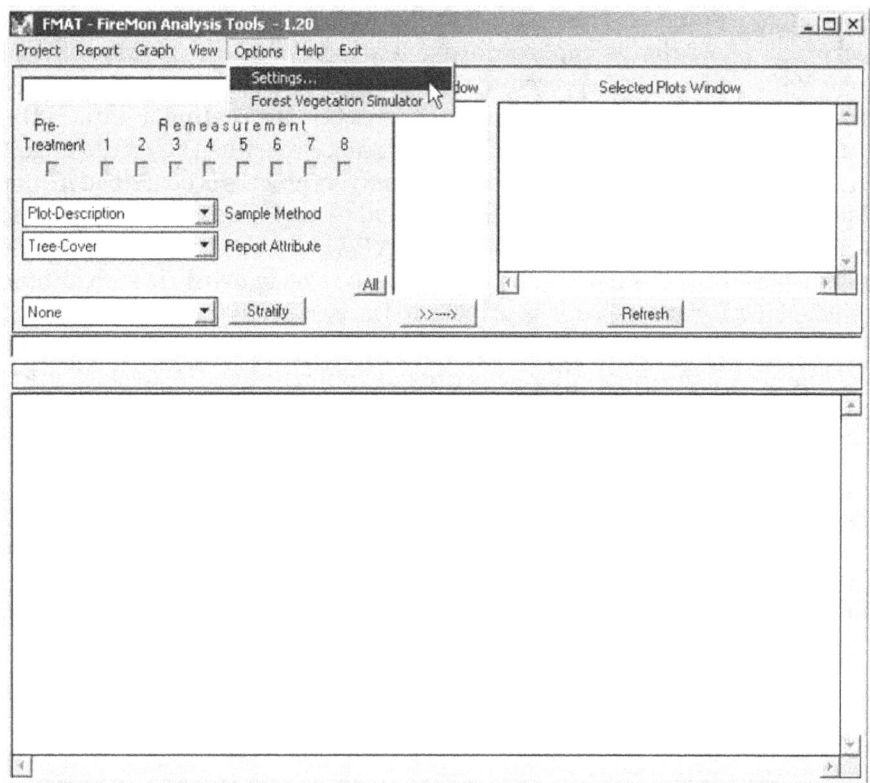

**Figure DB-35**—FIREMON data analysis program screen.

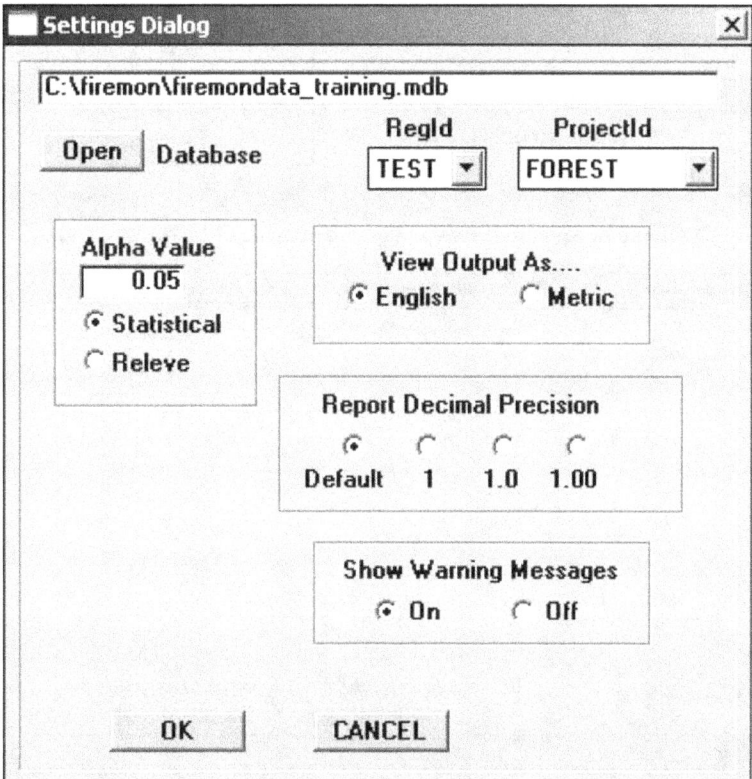

**Figure DB-36**—Settings dialog box.

Four additional settings are found in the Settings Dialog box. The *Alpha Value* field lets users select the significance level of the F-test that is applied during analysis. When the p-value of the F-test is below the selected *Alpha Value*, a Dunnett's procedure for multiple comparisons with a control is performed to identify attribute(s) that are significantly different from the pretreatment data. If the p-value of the F-test is greater than the *Alpha Value*, no significant differences between attribute means are noted and the Dunnett's procedure is not applied. The statistical analysis process is described in more detail below. Selecting the Releve radio button eliminates all statistical tests. Users can select the decimal precision and choose to see output in metric or English units. FMAT warns users of possible problems during the analysis process (such as if the diameter of a piece of coarse woody debris is zero or blank). If warning messages are selected (*On*), each message is printed to the screen and users must click *OK* to continue processing. If warning messages are not selected (*Off*), they are printed to a file (error.txt) that can be viewed when the analysis is finished. After selecting all options, click *OK* to close the Settings Dialog box. Plots with the *RegID* and *ProjectID* you selected in the Settings Dialog are displayed in the *Plot Window*.

Select plots for analysis by clicking the plot number(s) and clicking the arrow button below the *Plot Window*. The selected plots will move to the *Selected Plots Window* (fig. DB-37). You can move all the plots by clicking the *All* button then clicking the arrow button. Select the *Sampling Events* you want to analyze (for example, pretreatment, first remeasurement, second remeasurement, and so forth) by marking the appropriate check boxes. Select the method and attribute for analysis using the *Sample*

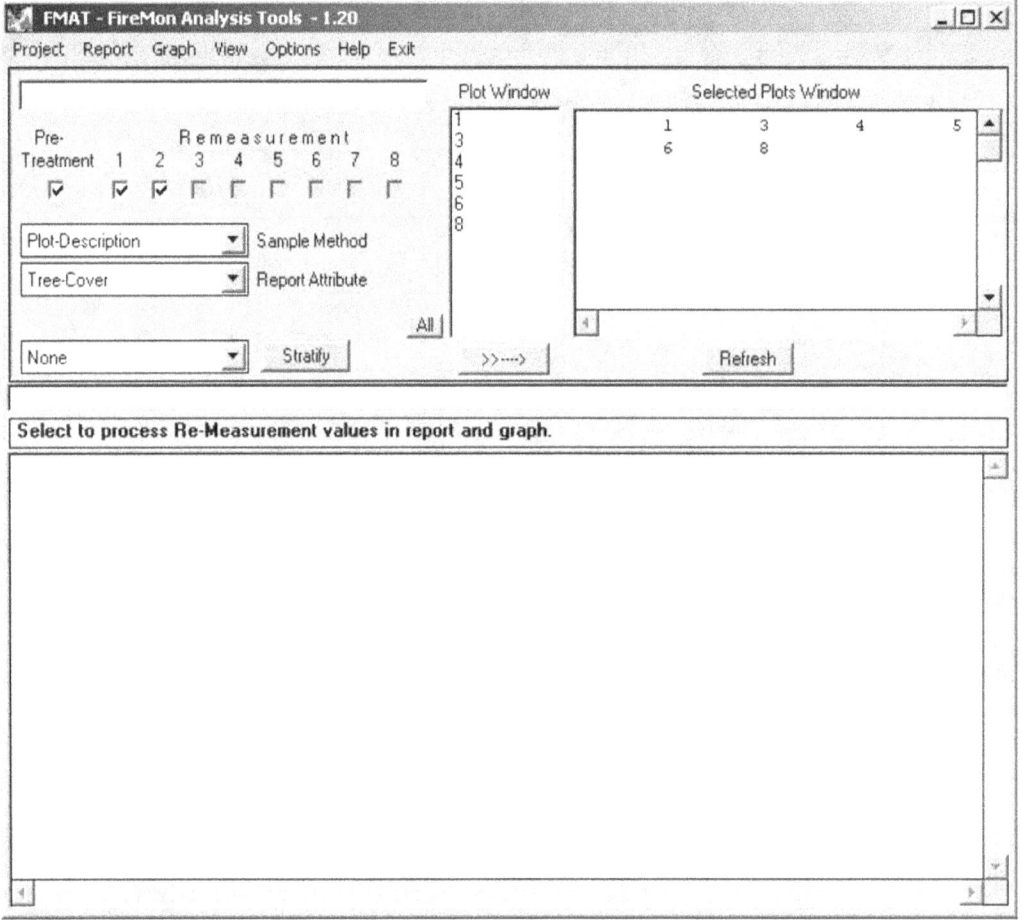

**Figure DB-37**—FIREMON analysis plot screen showing plots that have been selected for analysis (Selected Plots Window).

FIREMON Database Guide

*Method* and *Report Attribute* combo boxes. Generate a report by selecting *Report* → *Create Report* from the main toolbar (fig. DB-38). Reports can be printed directly from the window, saved to a file, or moved to a word processor file using the Windows *Copy* and *Paste* commands.

The FMAT program uses a randomized block analysis of variance (ANOVA) procedure for a single attribute by two or more sampling events. ANOVA is used to test the hypothesis that several means are equal. As described above, in addition to determining that differences exist among the means, FMAT determines which means differ using a Dunnett's procedure for multiple comparisons to compare a set of treatments against a single control mean. FMAT uses the first sampling event (most recent pretreatment or first remeasurement) as the control and the subsequent sampling events as the set of

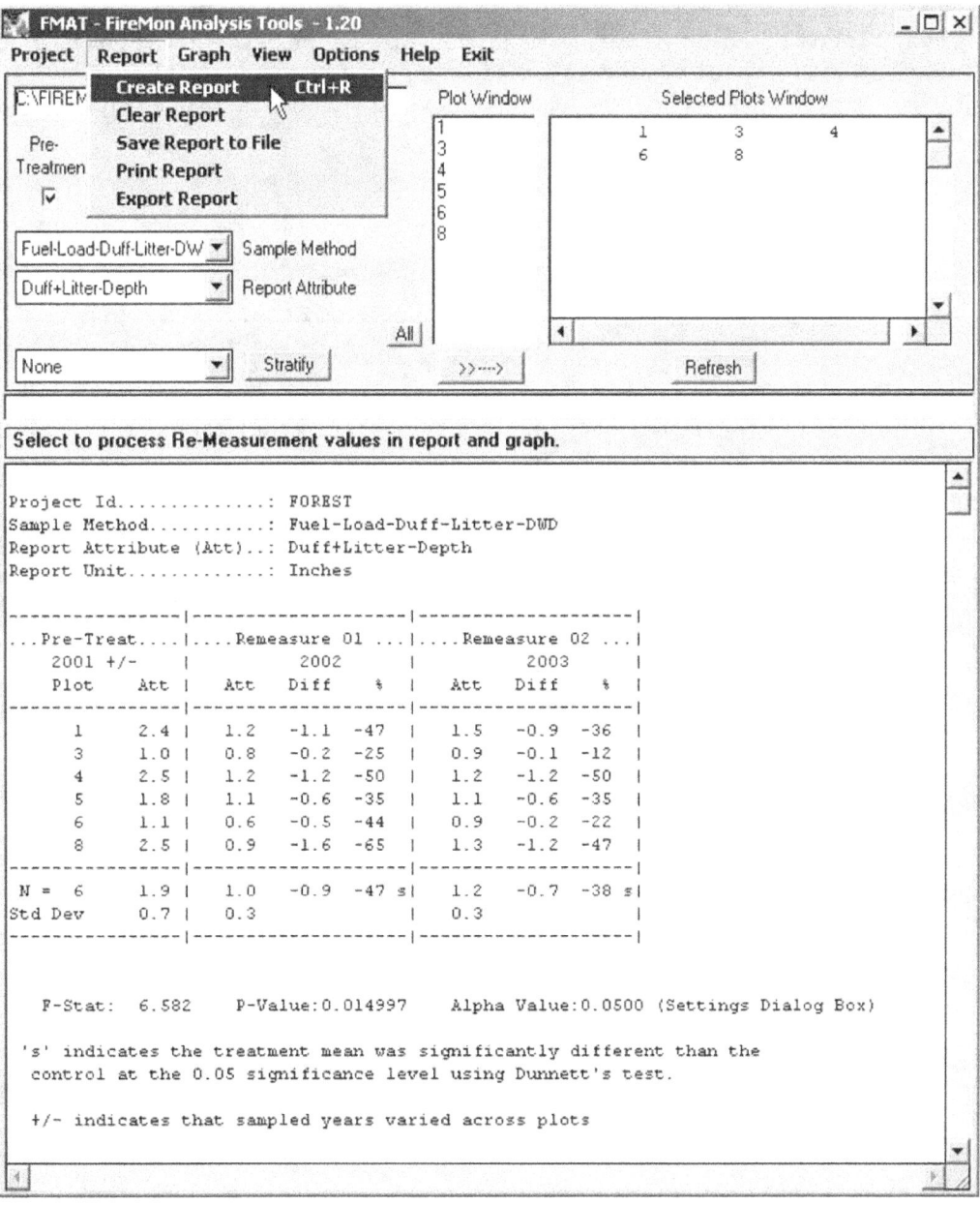

**Figure DB-38**—FIREMON analysis program report display.

treatments. The FIREMON analysis report displays the attribute value for each plot and the absolute and percent difference for each value relative to the first sampling event. The mean values, standard deviation, and differences in mean values are displayed at the bottom of the plot attribute value table. The F-statistic and associated p-value for the ANOVA table are displayed below the plot attribute table (fig. DB-38). An "S" next to an attribute mean indicates a significant difference when compared to the first sampling event using Dunnett's procedure at the 0.01 significance level. An "s" denotes significance at the 0.05 level.

FMAT assumes the data meets the assumptions of parametric tests (for example, normally distributed observations, homogenous variance, independent observations, and so on). While the tests used in FMAT are robust to these assumptions, validity of results may be adversely affected if data do not meet the assumptions of the tests used. If needed, data should be analyzed in more a comprehensive program such as SAS.

Generate a graph by selecting *Graph → Create Graph* from the main toolbar (fig. DB-39). The FIREMON analysis graph displays the mean values for the attribute by sampling event. Graphs may be printed directly from the window, saved to a file, or copied to the clipboard and pasted into word processing documents. Graphs may be copied from the report display text box and then pasted into documents. For additional details, see the **Analysis Tools Guide.**

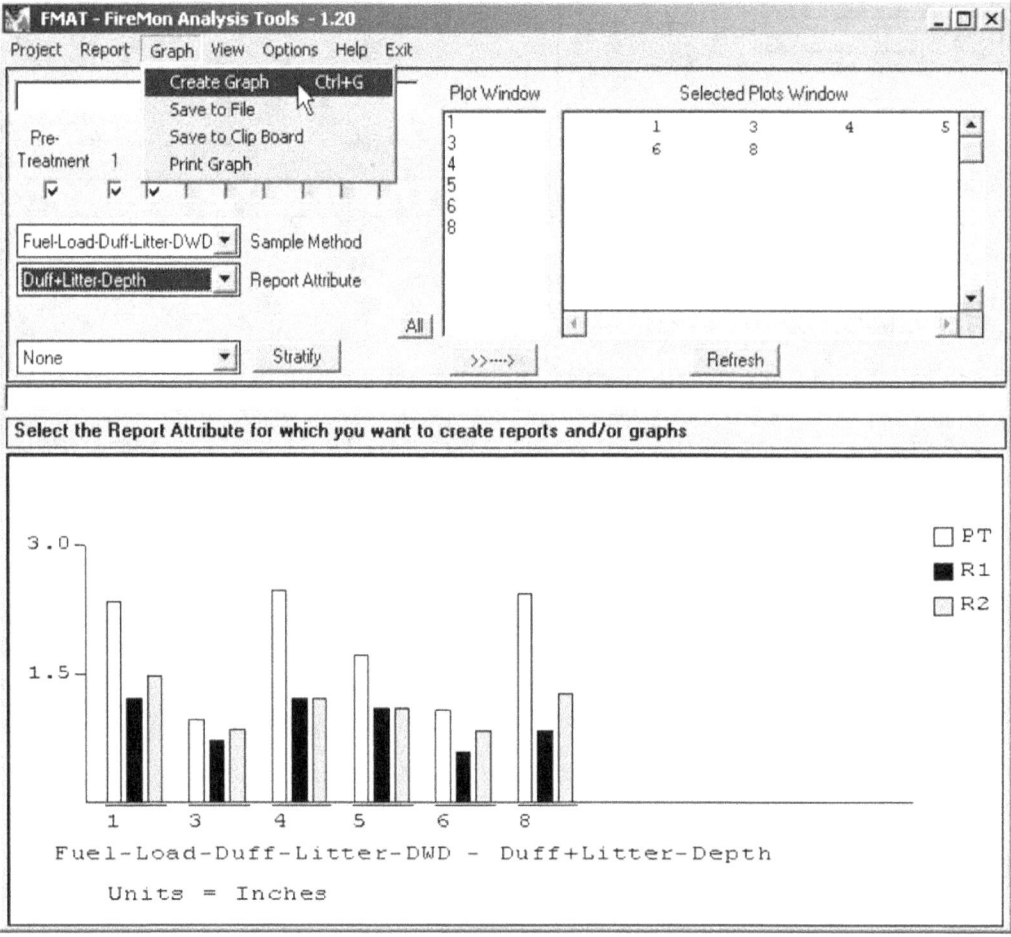

**Figure DB-39**—FIREMON analysis program graph display.

FIREMON Database Guide

# SIMPLE QUERY BUILDER

Select *Reports → Simple Query Builder* from the FIREMON toolbar to display the *Simple Query Builder* Form. The FIREMON Simple Query Builder is a form that allows users to query the FIREMON tables. This tool provides a quick way to view data in the FIREMON data tables and has only a few query options. Users may select any fields from one or more FIREMON tables. For example, users may select the transect length fields from the FLMacro table and the fine fuel counts plus duff/litter depths from the FLFineDL table (fig. DB-40). The query results may be filtered by Registration ID, Project ID, Plot ID, and Date. The query results are displayed in form view for easier viewing on the screen and can be exported in .csv file format to import into spreadsheets or statistical analysis software (fig. DB-41). All FIREMON plot data tables, fields, and field descriptions are listed in appendix E as a reference for building these simple queries. See **Customizing the FIREMON Database** in this chapter for a brief tutorial on designing more complex queries in Access.

# GIS POINT FILE

Select *Reports → GIS Point File* to display all the plot locations in the database (fig. DB-42). This tool provides a quick way to export all the plot locations to create a point coverage in a GIS. All location fields in the PD table are displayed for all plots in the database. Location fields include latitude, longitude, UTM zone, northing, easting, datum, GPS error, and GPS error units. This file may be printed or exported in .csv, .pdf, or .html file formats.

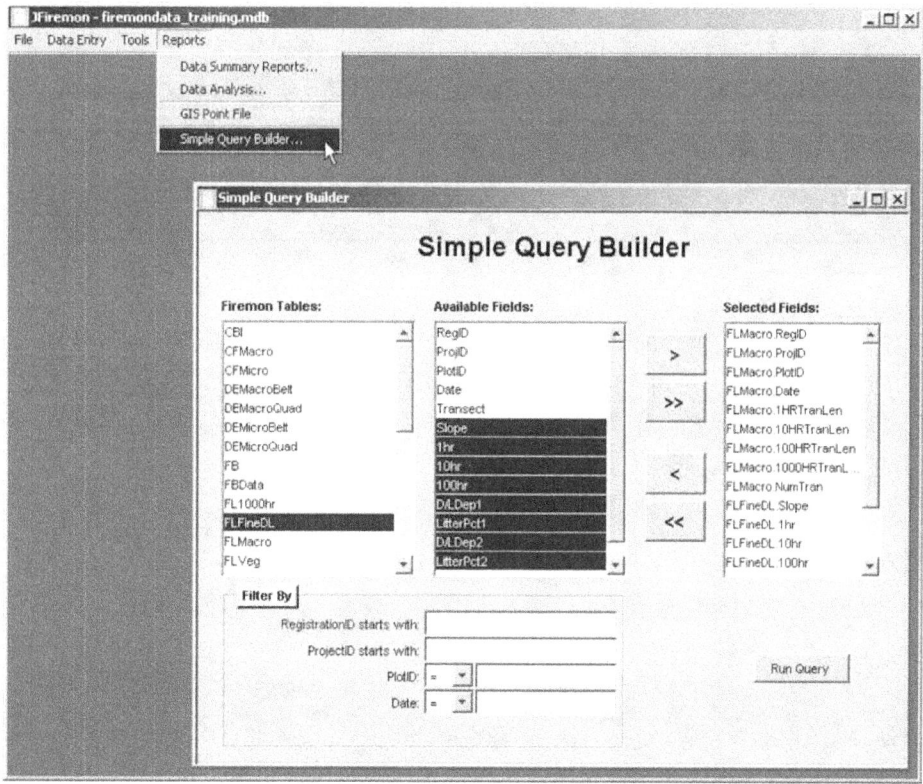

**Figure DB-40**—Simple query builder screen.

FIREMON Database Guide

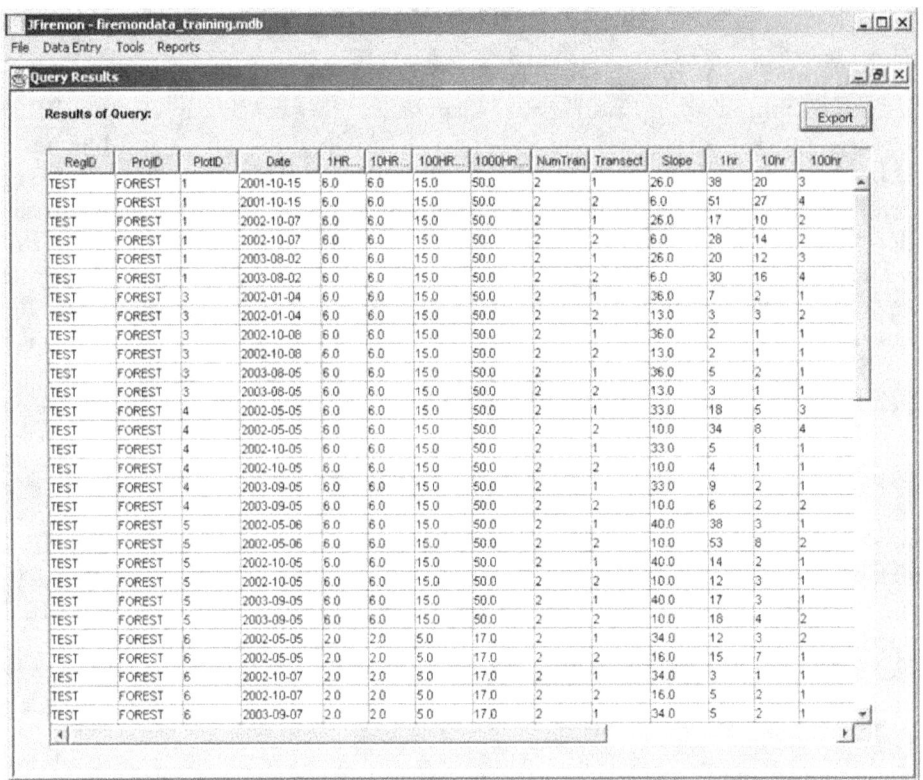

Figure DB-41—Query results display form view.

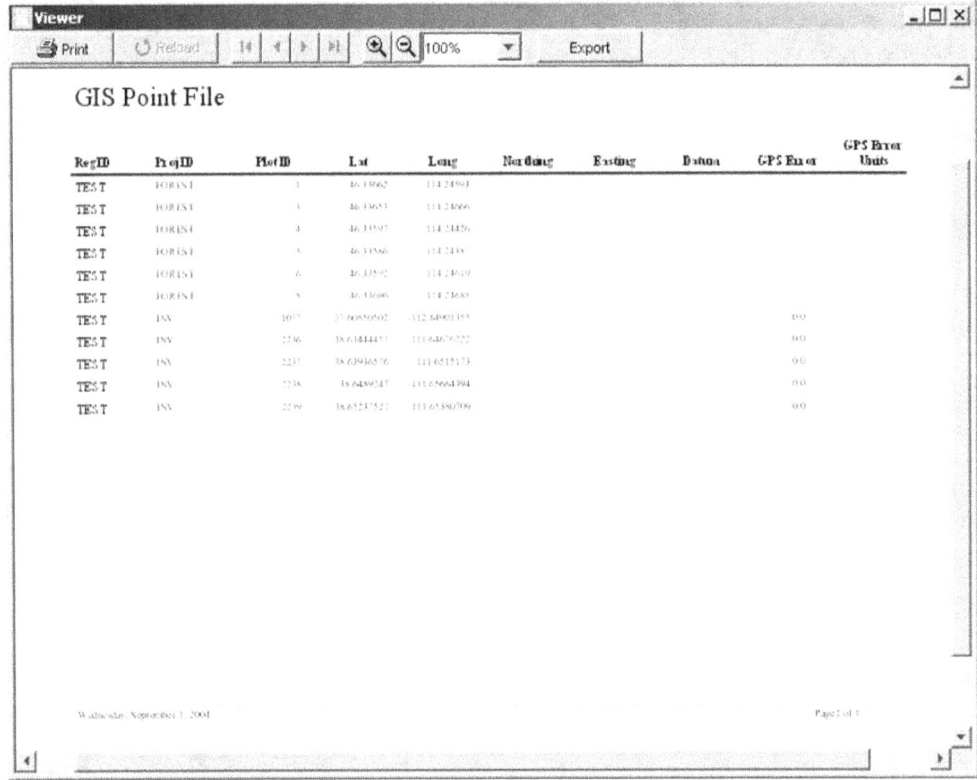

Figure DB-42—GIS point file screen.

FIREMON Database Guide

# RANDOM TRANSECT LOCATOR PROGRAM

Select *Tools → Transect Layout* from the FIREMON toolbar to display the *Random Transect Locator* form (fig. DB-43). This tool generates random starting points for transects placed along a baseline. It also generates random starting points for placing quadrats along a transect. Enter the number of transects, transect length, and the maximum distance from the baseline to place the first quadrat. The maximum distance from the baseline for placement of the first quadrat depends on transect length, number of quadrats, spacing of quadrats, and quadrat size. Click on the *Transect Locations* button to generate the transect locations, and click on the *Quadrat Locations* button to generate the starting points for the first quadrat.

# CUSTOMIZING THE FIREMON DATABASE

The FIREMON database may be tailored to local users by editing the lookup code tables to provide user specific codes for various FIREMON fields. Customizing the FIREMON code tables is generally not recommended for users who will share data with other FIREMON users. The more FIREMON is customized, the more difficult it is to share data with other FIREMON users. FIREMON user groups should coordinate their use of custom codes. Unique queries may also be developed for customized reports or for exporting data to spreadsheets and statistical software packages.

All customized code tables and user-designed queries should reside in the FIREMON database where users are storing their FIREMON plot data. Users with customized FIREMON databases will not lose their customized version of FIREMON code tables and queries when they install updated versions of the JFiremon application.

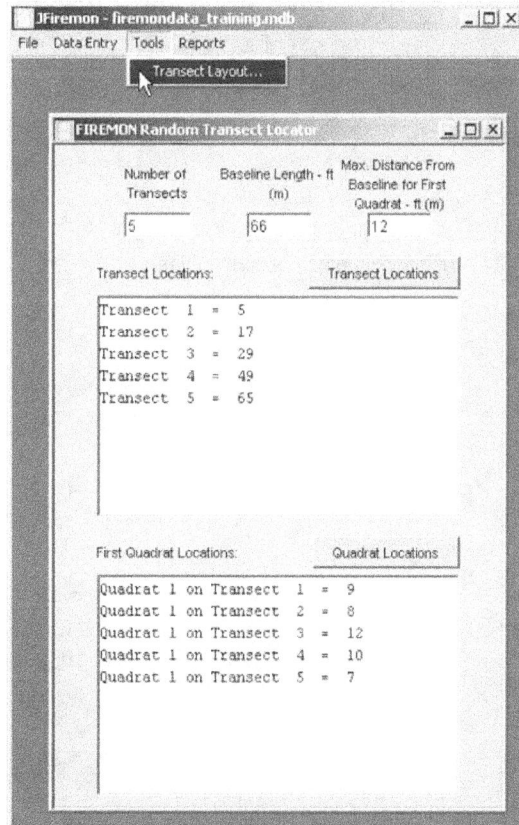

**Figure DB-43**—Random transect locator program.

# FIREMON Database Guide

## Customizing FIREMON Codes

The FIREMON code tables may be modified to include locally specific codes. Users may edit these tables in the FIREMON *Data* database (.mdb). The FIREMON code tables are named _codes*.* (fig. DB-44). The code tables are systematically named by the table name and field name for which they store codes. For example, _codesTD_CrwnClass is the table with the crown class codes (fig. DB-45) used in the TD sampling method and data entry form.

The code tables generally have three fields. The first field is the code, the second field is the description of the code, and the third field is a number that sorts the order of the codes. The sorting number is used to display fields in the dropdown list in an order other than alphabetic. Codes may be added, modified, or deleted. However, if longer codes are added, the field size must be lengthened in the FIREMON table. Table DB-1 lists all the code tables in FIREMON, the FIREMON tables and forms that use the codes, and a brief description of the type of codes stored in each table.

## Developing Custom Queries in FIREMON

The JFiremon application has a *Simple Query Builder* tool that allows users to display and export data from FIREMON tables. However, this form only builds relatively simple queries and can filter only

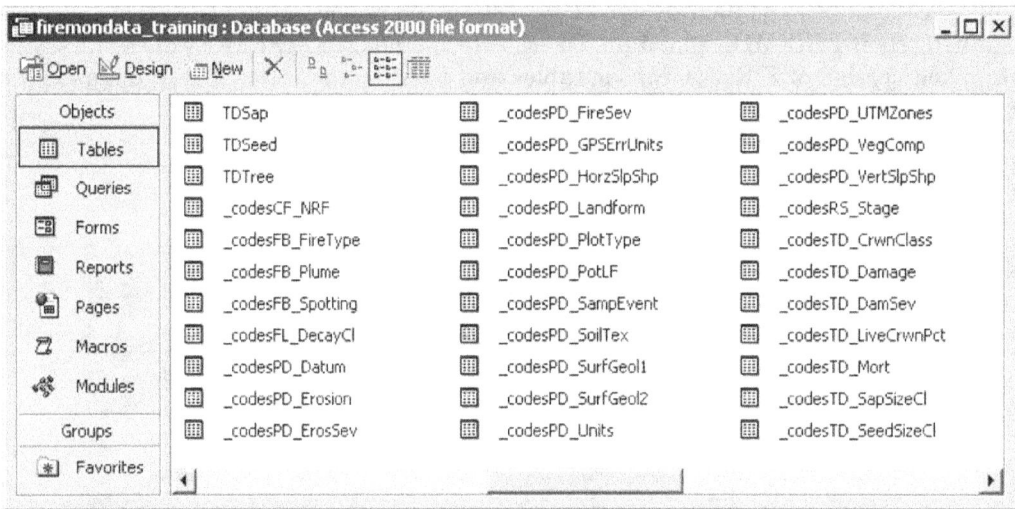

**Figure DB-44**—FIREMON database code tables.

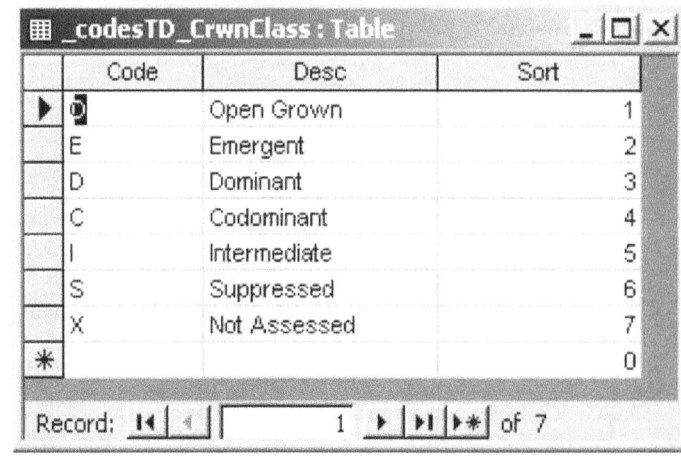

**Figure DB-45**—Crown class codes.

FIREMON Database Guide

Table DB-1 FIREMON code tables.

| Table name | FIREMON tables/forms | FIREMON code description |
|---|---|---|
| codes CanopyCover | PD, FL, SC, CF | Canopy cover classes |
| codes GroundCover | SC, CF, PO | Ground cover |
| codes OtherItems | SC, CF, LI, PO, DE | Other items |
| codes SizeCl | SC, LI, DE | Tree/shrub size classes |
| codes Status | SC, CF, PO, LI, DE, RS | Plant species status |
| codesCF NRF | CF | Nested rooted frequency |
| codesFB FireType | FB | Fire type |
| codesFB Plume | FB | Plume |
| codesFB Spotting | FB | Spotting |
| codesFL DecayCl | FL | Log decay class |
| codesPD Datum | PD | Datum |
| codesPD Erosion | PD | Erosion type |
| codesPD ErosSev | PD | Erosion severity |
| codesPD FireSev | PD | Fire severity |
| codesPD GPSErrUnits | PD | GPS error units |
| codesPD HorzSlpShp | PD | Horizontal slope shape |
| codesPD Landform | PD | Landform |
| codesPD PlotType | PD | Plot type |
| codesPD PotLF | PD | Potential life form |
| codesPD SampEvent | PD | Sampling Event |
| codesPD SoilTex | PD | Soil texture |
| codesPD SurfGeol1 | PD | Primary surficial geology |
| codesPD SurfGeol2 | PD | Secondary surficial geology |
| codesPD Units | PD | Measuring units |
| codesPD UTMZones | PD | UTM zone |
| codesPD VegComp | PD | Non species vegetation composition codes |
| codesPD VertSlpShp | PD | Vertical slope shape |
| codesRS Stage | RS | Plant species stage |
| codesTD CrwnClass | TD | Crown class |
| codesTD Damage | TD | Damage |
| codesTD DamSev | TD | Damage severity |
| codesTD LiveCrwnPct | TD | Live crown percent |
| codesTD Mort | TD | Mortality |
| codesTD SapSizeCl | TD | Sapling size class |
| codesTD SeedSizeCl | TD | Seedling size class |
| codesTD Snag | TD | Snag decay class |
| codesTD TreeStatus | TD | Tree status |

records based on the four primary key fields in FIREMON (RegID, ProjID, PlotID, and Date). More complex queries require the use of the *Access Query Design* window described below. Users may design their own queries and store them in the FIREMON database. Click on *Queries → New → Design* view to display the *Access Query Design* window (fig. DB-46). Customized queries are useful for exporting data into statistical software packages or for generating custom reports. All FIREMON tables, fields, and field descriptions are listed in appendix E as a reference for building customized queries.

Access prompts users to click on tables to add to the *Access Query Design* window. Selected tables and their relationships are added to the design window. Users may select the desired fields along with any criteria for specific fields. For example, the query in figure DB-46 joins the PD table with the FLMacro and FLFineDL tables. This query selects the FireID, PlotID, transect number, slope, and fine fuel counts for all plots in the FIREMON database that were located inside the fire assigned to FireID = 101. The resulting query is displayed in figure DB-47.

The FIREMON database relationships are presented in appendix F.

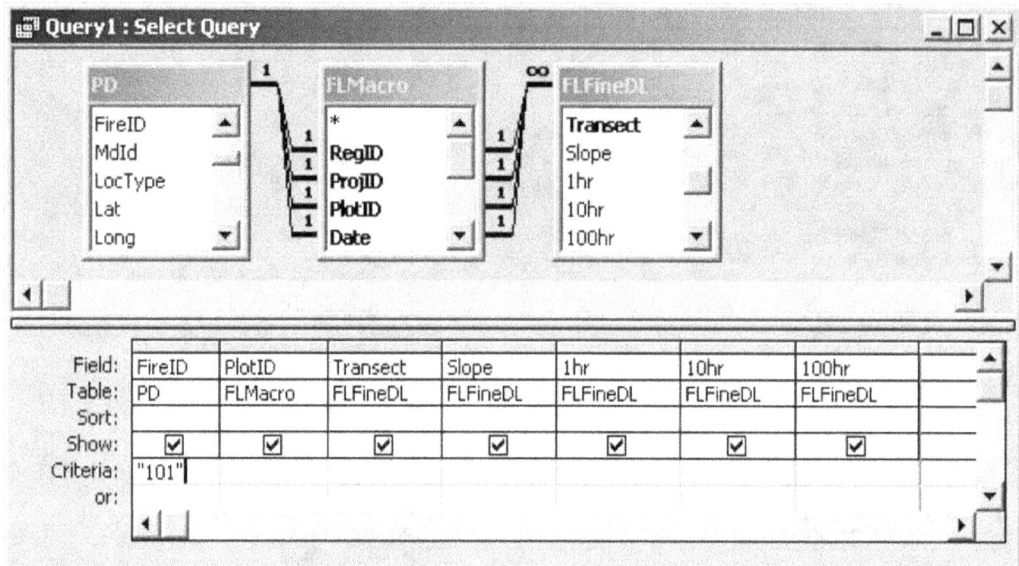

**Figure DB-46**—Access query design window.

**Figure DB-47**—Query results displaying fine fuel counts on transects by FireID.

# Analysis Tools (AT)

## User Manual

Larry J. Gangi

## SUMMARY

The FIREMON Analysis Tools program is designed to let the user perform grouped or ungrouped summary calculations of single measurement plot data, or statistical comparisons of grouped or ungrouped plot data taken at different sampling periods. The program allows the user to create reports and graphs, save and print them, or cut and paste them into a word processor.

## INTRODUCTION

Fire effects monitoring is defined by two tasks: field data collection and evaluation. Field data collection has been discussed in detail in the **Integrated Sampling Strategy** and FIREMON sampling documentation. Discussed here are the software and procedures for the evaluation of the field data to assess fire effects.

The FIREMON analysis component encompasses two major tasks: 1) data entry and 2) data analysis. Data entry is accomplished by physically entering the collected field data into a set of standardized FIREMON Microsoft® Access databases. Data analysis is accomplished using data summary routines and the FIREMON Analysis Tools computer program developed specifically for FIREMON. The data summary programs are discussed in the **FIREMON Database User Manual.**

The FIREMON Analysis Tools program imports the data from the database and performs common calculations, then presents results in report or graphical form. For example, fuels data are entered by transect. The Analysis Tools program calculates the fuel loading from the raw transect data and summarizes fuel load by size class for each plot. The analysis program summarizes all database entries into one report for storage as a computer file or paper report.

The statistics program performs temporal monitoring analyses using Dunnett's procedure for multiple comparisons with a control and provides output reports showing the results of statistical tests. The summary can be stratified by any macroplot-level field or by the user.

FMAT assumes the data meets the assumptions of parametric tests (for example, normally distributed observations, homogenous variance, independent observations, and so on). While the tests used in FMAT are robust to these assumptions, validity of results may be adversely affected if data do not meet the assumptions of the tests used. If needed, data should be analyzed in more a comprehensive program such as SAS.

The analysis package can also produce the necessary files to run the Forest Vegetation Simulator and the associated Fire and Fuels Extension. Lastly, data can be exported in comma-delimited format for input into spreadsheets or statistical programs.

# FIREMON ANALYSIS

## Getting Help

FIREMON Analysis has three levels of help that are available while it is running:

1. Single line help is always shown on the main program window, just above the report window. Whenever you click on a window item such as a list box, you will see a simple one-line message explaining what the list box does. You'll notice it changing as you click on different items.
2. F1 help is also available. Click on a screen item then hit the F1 function key. A small window will pop up giving you a somewhat detailed explanation of the item you have clicked on.
3. Full help is available from the main menu. This is a complete and detailed help system.

## Getting Started

The first time you start up the FIREMON Analysis Tools (FMAT) program you will get a message instructing you to set a database and a project file. The message window that pops up will give you a list of short instructions to guide you.

A project file (.prj file extension) saves all of your screen settings. You can have as many project files as you like. For example, if you have built a particular list of plots that you want to work with, save them to a project file. Later you can use them again by opening the project file, and you won't have to create the list again.

Project files are also helpful in that they save your place when you leave FIREMON Analysis Tools program. The next time you enter the program your settings will be restored.

Once you have set your project file, the name and full directory (folder) path will be shown in the project window (just under the main menu). The path that is shown will be used as a default path whenever you go to open or save a file. You can set the path by selecting *Project → New* from the main menu.

## The Settings Dialog Box

The *Settings Dialog* box (fig. AT-1) is available by selecting *Options → Settings* from the main menu. Choose the database that includes the plot data you want to analyze by clicking the *Open* button and navigating to the appropriate Access database. The FMAT program runs independently from FIREMON. Changing the database in FMAT does not change the database you are connected to in FIREMON, or vice versa. Choose a database and select the *Registration ID* and *Project ID* from the list boxes. Only plots with the selected *Registration ID* and *Project ID* will be shown in the *Plot Window* on the main screen.

Clicking the *Statistical* radio button on the *Settings Dialog* will provide users with statistical results. The alpha (a) value is the critical value used in the F-test. When the probability of the calculated F statistic is lower than the alpha value, it indicates there are significant differences between at least two of the means. If that is true, Dunnett's procedure is used to identify which treatment means are significantly different than the control mean. The alpha value should be set to match the $Z_a$ level used when the sample size was calculated in the **Integrated Sampling Strategy**. This value should be recorded in the project notebook or Metadata table. If the *Relevé* radio button is selected, no statistical tests are performed.

You can elect to see your output (report/graph) in English or metric units, regardless of how the data were entered in the database, by clicking the appropriate radio button on the *Settings Dialog*.

Four radio buttons are provided to set the decimal precision you want to see on your reports. Although you can leave it set to *Default*, there may be times when you find that you need more or less precision.

Analysis Tools (AT) Guide

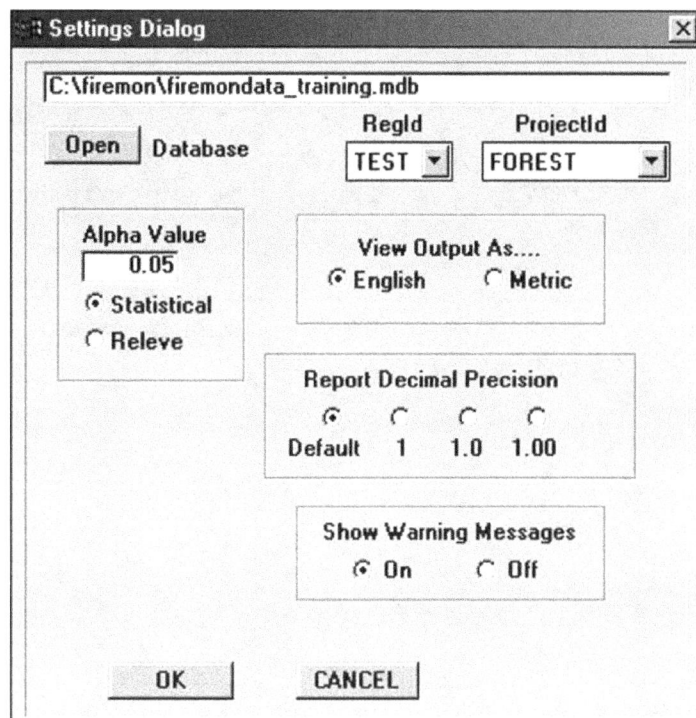

Figure AT-1—The Settings Dialog screen.

You can also turn the program's warning messages on or off. Warning messages occur for invalid or missing data when you create a report or a graph. When warnings are "on" a popup box is seen every time a problem is encountered. When set to "off," the program directs the messages to a text file that you can open for viewing or printing.

## How to Run—A Brief Overview

The program is set up such that you can create reports and graphs, then save and print, or cut and paste into a word processor such as Microsoft Word.

Once you have selected the database you want to use and saved the settings to a project file, you are ready to do analysis. Select the plots you want to analyze by clicking them in the *Plot Window*, then clicking the arrow button. The plot numbers will appear in the *Selected Plots Window*. You can remove plots from the *Selected Plots Window* by highlighting with your mouse and hitting the delete key.

Select the *Pretreatment* and *Remeasurement* check boxes for the sampling events you want analyze. At least one box needs to be checked. Grayed out boxes indicate there are no data for that sampling event. Select the *Sample Method* and *Report Attribute* you want. Lastly, select *Create Report* or *Create Graph* from the *Report* or *Graph* menu dropdown lists. The results will appear in the lower window. Figure AT-2 shows the opening screen of the FIREMON Analysis Tools program with the report for 10-hour biomass for six plots and three sampling events in the report window.

### Example report

The FMAT uses a randomized block Analysis of Variance (ANOVA) procedure to compare a single attribute between two or more sampling events. ANOVA is used to test the alternative hypothesis that at least one of the means is significantly different from at least one of the other means at the significance level set in the *Alpha Value* window on the *Settings Dialog*.

When significant differences of means are found, the FMAT will determine which means differ from the control using Dunnett's procedure multiple comparison t-test. This test is used to compare a set of

Analysis Tools (AT) Guide

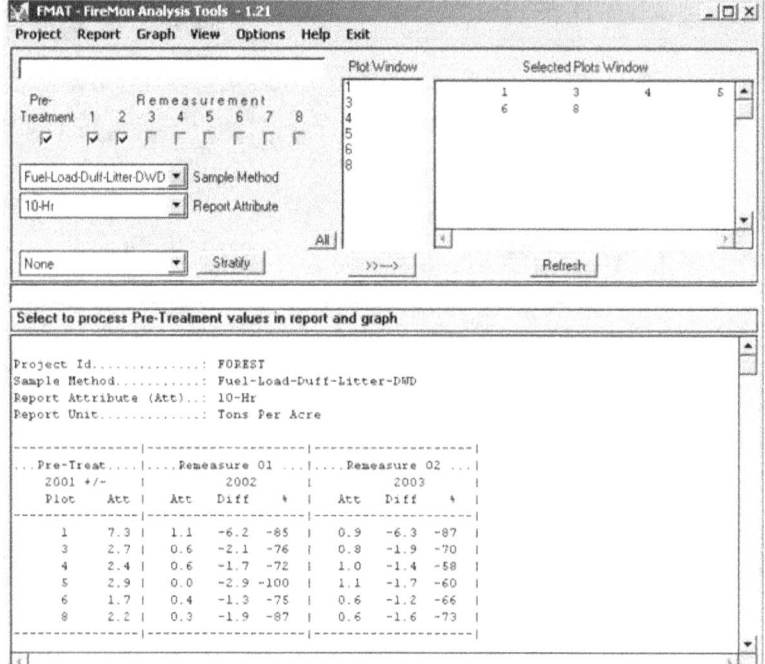

**Figure AT-2**—The FMAT main screen.

treatment means against a single control mean (fig. AT-3). FMAT uses the first sampling event (most recent pretreatment or first remeasurement) as the control and the subsequent sampling events as the set of treatments.

The FIREMON analysis report displays the attribute value for each plot and the absolute and percent difference for each value, relative to the first sampling event. The mean values and differences in mean values are displayed at the bottom of the plot attribute value table. When users select the *Statistical*

```
Project Id..............: FOREST
Sample Method...........: Fuel-Load-Duff-Litter-DWD
Report Attribute (Att)..: 10-Hr
Report Unit.............: Tons Per Acre

----------------|--------------------|--------------------|
...Pre-Treat....|....Remeasure 01 ...|....Remeasure 02 ...|
   2001 +/-     |       2002         |       2003         |
   Plot   Att   |   Att   Diff    %  |   Att   Diff    %  |
----------------|--------------------|--------------------|
      1   7.3   |   1.1   -6.2  -85  |   0.9   -6.3  -87  |
      3   2.7   |   0.6   -2.1  -76  |   0.8   -1.9  -70  |
      4   2.4   |   0.6   -1.7  -72  |   1.0   -1.4  -58  |
      5   2.9   |   0.0   -2.9 -100  |   1.1   -1.7  -60  |
      6   1.7   |   0.4   -1.3  -75  |   0.6   -1.2  -66  |
      8   2.2   |   0.3   -1.9  -87  |   0.6   -1.6  -73  |
----------------|--------------------|--------------------|
  N = 6   3.2   |   0.5   -2.7  -83 S|   0.8   -2.4  -73 s|
  Std Dev 2.0   |   0.4              |   0.2              |
----------------|--------------------|--------------------|

   F-Stat:  8.932    P-Value:0.005954    Alpha Value:0.0500 (Settings Dialog Box)

   's' indicates the treatment mean was significantly different than the
       control at the 0.05 significance level using Dunnett's test.

   'S' indicates the treatment mean was significantly different than the
       control at the 0.01 significance level using Dunnett's test.
```

**Figure AT-3**—Example of report output.

radio button on the *Settings Dialog*, the F-statistic and associated P-value for the ANOVA table are displayed below the plot attribute table.

If the F-statistic is significant, the mean values that are significantly different (from the control) are noted with an *S* or *s*, indicating significance at the 0.01 and 0.05 levels, respectively.

Summary reports similar to those produced by FIREMON can be produced by selecting *Summary-Report* in the *Sample Method* drop down list box (fig. AT-4). You can copy and paste the text from these reports, for example, into an Excel spread sheet. In addition there is an *Export Report* menu feature that exports comma delimited files. The file will export as *Export.txt*.

## Example graph

The FIREMON analysis graph simply displays the mean values for the attribute by sampling event. Graphs may be copied to the clipboard and then pasted into word processing documents (fig. AT-5).

The *Graphing Option* dialog box lets you set options to arrange the overall look of the graph (fig. AT-6). Multiple bar graphs are displayed in the graph window. The settings in the dialog box help you control how the graph window will look.

If you have elected to group your plots, then a list box will be shown in the upper left of the dialog box so you can select a group to display, or *All Strata* (show a graph for each stratification group). The list box will not appear if you have not grouped your plots.

The *Rows* and *Columns* list boxes control how the individual graphs will be arranged in the Graph Window. For example, if the dialog box informs you that 12 graphs will be created, you may want to select 2 Rows and 6 Columns. A little experimentation with the rows and columns options will help you get familiar with how they work.

Use the *Y Axis* edit text box to set the limit of the y-axis in your graphs.

```
                                    Tree Data

Project Id: FOREST
........Trees Per Acre
BA......Basal Area - Square Feet
LCBH....Average Live Crown Base Height - Feet
Height..Average Height - Feet
QMD.....Quadradic Mean Diameter inches
            Mature    BA    LCBH   Height   QMD   Saplng  Seedlng  Total   Snags
   1    P1   50.0    27.4   27.2    63.0   10.0    50.0    500.0   600.0   0.0
   1    R1   30.0    22.4   51.7    72.7   11.7    50.0    500.0   580.0   2.0
   1    R2   30.0    22.4   51.7    72.7   11.7    50.0    500.0   580.0   2.0
   3    P1   50.0    69.2   26.4    66.6   15.9    50.0    500.0   600.0   0.0
   3    R1   40.0    67.4   34.8    78.0   17.6    50.0    500.0   590.0   1.0
   3    R2   30.0    63.9   38.0    89.0   19.8    50.0    500.0   580.0   2.0
   4    P1   50.0    44.6   24.8    63.2   12.8    50.0    500.0   600.0   0.0
   4    R1   40.0    40.3   33.0    70.8   13.6    50.0    500.0   590.0   1.0
   4    R2   30.0    30.2   31.0    75.7   13.6    50.0    500.0   580.0   2.0
   5    P1   50.0    26.6   27.6    58.0    9.9    50.0    600.0   700.0   0.0
   5    R1   40.0    25.1   35.0    64.0   10.7    50.0    600.0   690.0   1.0
   5    R2   30.0    23.2   36.7    72.0   11.9    50.0    600.0   680.0   2.0
   6    P1   50.6    50.0   27.6    76.1   13.5    50.6    505.9   607.0   0.0
   6    R1   50.6    67.9   37.4    77.8   15.7    50.6    505.9   607.0   0.0
   6    R2   50.6    70.0   35.6    68.2   15.9    50.6    505.9   607.0   0.0
   8    P1   40.5    17.1   29.4    62.1    8.8    50.6    404.7   495.8   1.0
   8    R1   20.2    10.9   34.8    68.9    9.9    50.6    505.9   576.7   3.0
   8    R2   20.2    10.9   34.8    68.9    9.9    50.6    505.9   576.7   3.0
```

**Figure AT-4**—Example of summary report output.

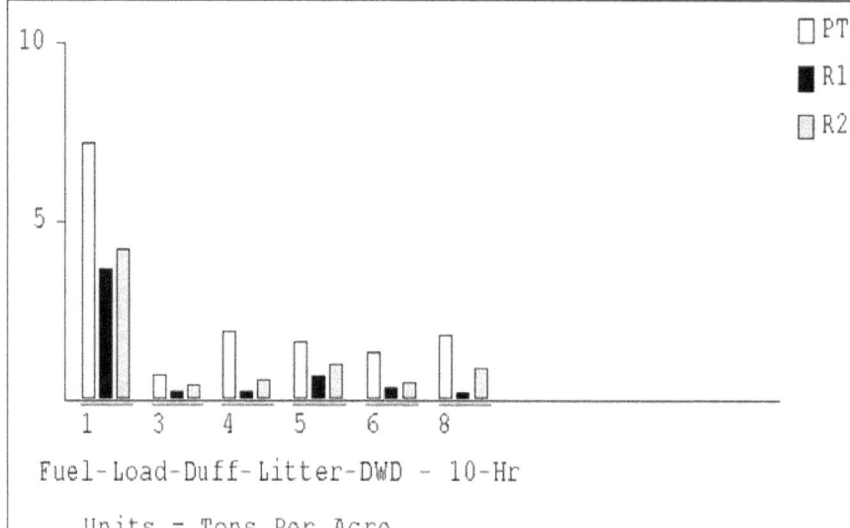

**Figure AT-5**—Example of graph output.

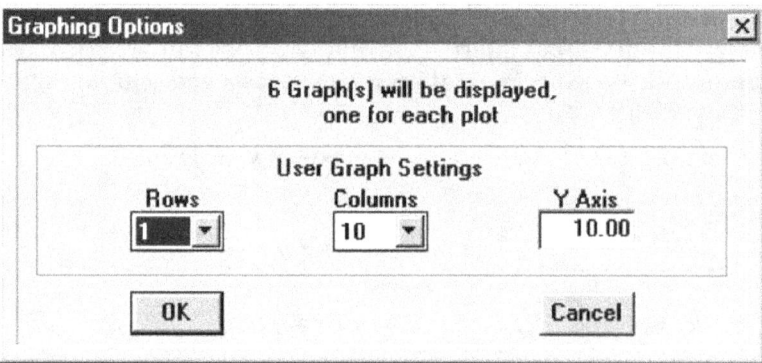

Figure AT-6—The Graphing Options dialog box.

## Grouping Your Plots

You can group your plots by selecting an attribute from the stratification list box and then clicking the *Stratify* button. The items in the list box correspond to the fields in the Plot Description database table.

When you click the *Stratify* button you will see your selected plots in the *Selected Plots Window* move into groups. For example, if you stratified by Landform, you might see your plots grouped under #VAL and #HIL. The "#" prefix distinguishes group codes from plot numbers.

The *Selected Plots Window* is a typical text edit window. You can delete, paste, and build your own stratification groups manually, if needed. To manually stratify plots, move the plots you want to stratify into the *Selected Plots Window* by clicking on them in the *Plot Window* then clicking the arrow button. Optionally, you can just type plot numbers in the *Selected Plots Window*. Next, put your cursor in the *Selected Plots Window* and type # *name* to the left of each group, where *name* is a descriptive label for the group. For example, you could label your groups by fuel load: #High, #Med, #Low. To organize your groups click the *Refresh* button. Do not click the *Stratify* button after manually stratifying plots or your group labels will be erased.

Analysis Tools (AT) Guide

Some points to remember:
1. There is a "None" item in the strata list box. Selecting it will ungroup your plots.
2. All group labels must have a "#" prefix.
3. If missing stratification codes are found, the associated plots will be labeled #Missing.

## Forest Vegetation Simulator Option

FMAT provides a means for users to convert and move tree and fuels data into the Forest Vegetation Simulator (FVS) file formats. This option is available under *Options* → *Forest Vegetation Simulator* on the main menu. For more information on using the FVS, see Reinhardt and Crookston (2003) or the FVS Web page, www.fs.fed.us/fmsc/fvs/.

FMAT will only process one set of FVS files at a time. Before you choose this option you will need to move the plot you want to make the FVS files for into the *Selected Plots Window* and mark the *Pretreatment* or one *Remeasurement* check box on the main window. The current version of FMAT processes only one plot and one sampling event at a time.

When you select *Forest Vegetation Simulator* option, a dialog box will open (fig. AT-7). The default plot number is the same as that used in FIREMON, although you can modify it. Select the *Variant-Region*, forest, and habitat type or plant association from the dropdown lists, and then select the filename you want to use. If you want the FVS files stored with a name other than the default, click the *Set File* button and enter the name you want to use. When you click *OK*, all of the FVS files will be built using the same base name. For example, if you change the name to plot001.tre the other files will be plot001.loc and plot001.slf. If there are fuels data, the file will be plot001.kcp. A popup will notify you the files have been saved. Files with the same name will be overwritten without warning.

## Some Hints

Always hit the *Refresh* button each time you select plots or click a *Pretreatment* or *Remeasurement* check box. Doing so will force the species list and remeasurement check boxes to update, reflecting the new plots you have selected. In addition, the *Refresh* button will realign your plots in the *Selected Plots Window*.

If you have your database open while you are running FMAT and you make a change to your data in the FIREMON data entry application, you will need to close and reopen the data entry screen so that your changes will be detected by the FMAT program.

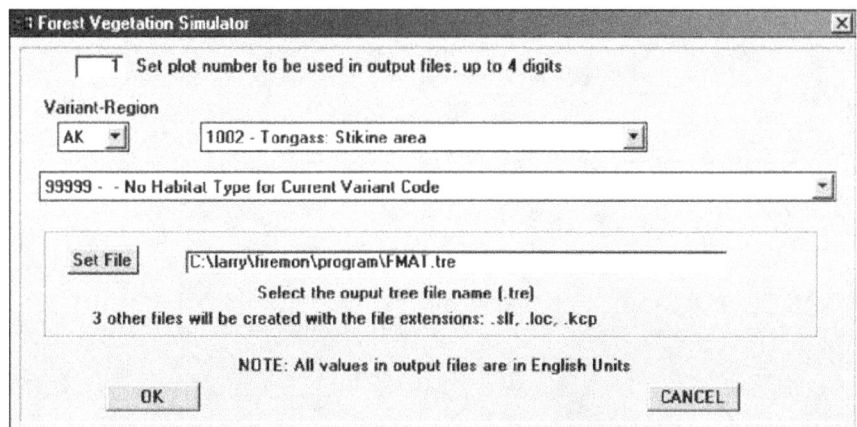

**Figure AT-7**—Example of a Forest Vegetation Simulator dialog box.

The report window on the lower half of the FMAT screen is a text edit window. You can cut, copy, and paste the results that are presented there. Any new reports will be placed at the current cursor position.

If the *Pretreatment* and/or *Remeasurement* check boxes are grayed out, that means there are no Plot data for that sample event.

The *Species* list box will appear only when it is appropriate for the *Sample Method*.

Whenever you create a report or graph, the program will gather and check the needed data from the database. The program will report any problems or inconsistencies; for example, a missing transect length, no data for a selected remeasurement, or data values out of limits. In most cases you will need to correct the problems before analysis can be completed. The major exception to this is with the Fuel Load data where many people add logs with no diameter or decay class just to let users know that no logs intersected on a particular sampling plane. FMAT will present a warning that it found a log with zero diameter and decay class, but will still produce a report with the correct biomass. Similarly, if some heights are missing in the FIREMON Fuel Load Vegetation table, a warning will inform you they are missing, but the mean height will be calculated from all the nonmissing values.

# Metadata (MD)

## Description

**Robert E. Keane**

## SUMMARY

The Metadata (MD) table in the FIREMON database is used to record any information about the sampling strategy or data collected using the FIREMON sampling procedures. The MD method records metadata pertaining to a group of FIREMON plots, such as all plots in a specific FIREMON project. FIREMON plots are linked to metadata using a unique metadata identifier that is entered in the MD table and in the PD data for each FIREMON plot. Metadata pertaining to a single plot are recorded in the comments field on the FIREMON PD data form.

## INTRODUCTION

The Metadata (MD) method is used to record metadata, information about data, for FIREMON sampling strategies and sampling methods. Metadata are recorded for multiple FIREMON plots. The data are linked to individual plots using a unique metadata identifier (MDID) that is entered in the MD table and in the Plot Description (PD) table for each plot. Specific comments on a single FIREMON plot are entered in the comments field on the PD data form. The MD table is for commenting on a number of plots, usually in one project.

Metadata are entered in the MD table by subject. A subject may be a FIREMON sampling method, the FIREMON Integrated Sampling Strategy, or any other aspect of a FIREMON project. The MD table allows FIREMON users the flexibility to record metadata on any subject related to a FIREMON project. Any text-based information that can be copied to the Windows clipboard, such as Word documents, Excel spreadsheets, or Access databases, can be pasted into the MD Comments field. The metadata field can store up to 64,000 characters in each Comments field. Images and documents with figures can be linked using the Document Link field.

Some examples of FIREMON metadata include:

1) General information about how FIREMON Integrated Sampling Strategy was applied to the plots in the project.
2) Specific metadata for each FIREMON sampling method such as the transect locations and quadrat placement for the FIREMON replicated vegetation sampling methods (CF, LI, PO, RS, and DE).
3) User-defined codes for fields, which differ from the standard FIREMON codes.
4) Units of measure for fields, which differ from the standard FIREMON units of measure.

## SAMPLING PROCEDURE

**Field 1: Metadata ID**—Enter a unique 15-character code to identify the metadata records. This code is also entered in the PD table to link FIREMON plots to these metadata.

**Field 2: Subject**—Enter a unique 20-character code for the metadata. Use the Subject field to divide your comments into logical groups. For example, if you entered "TD" in the Subject field you would organize all of the notes about tree sampling in the Comments field below. Then you could enter "FL-veg" in the Subject field and compile your notes in the Comments field on how vegetation was sampled when you were doing the FL method. Other subject codes may be entered by the user to record metadata on any other aspect of a FIREMON project.

**Field 3: Document Link**—The MD Comments field will not store images or figures. The Document Link field allows you to hyperlink to documents, figures, or images that cannot be stored in the Comments field. If the file you are linking to is in the same folder as you have entered in the Documents Base Directory field on the Configuration and Setting menu, then you need only enter the file name and extension. Otherwise you must specify the entire path to the file.

**Field 4: Comments**—Memo field. 60,000+ characters used to record any information pertinent to project plots. Text information can be pasted from Word documents, Access databases, Excel spreadsheets, or any other software that can copy text to the windows clipboard.

# Metadata (MD) Form

MD Page _ _ of _ _

**Metadata Information**

| Field 1 | Field 2 | Field 3 |
|---|---|---|
| Metadata ID | Subject | Document Link |
|  |  |  |

Field 4: Comments

# Fire Behavior (FB)

## Sampling Method

**Robert E. Keane**

## SUMMARY

The Fire Behavior (FB) method is used to describe the behavior of the fire and the ambient weather and fuel conditions that influence the fire behavior. Fire behavior methods are not plot based and are collected by fire event and time-date. In general, the fire behavior data are used to interpret the fire effects documented in the plot-level sampling. Unlike the other plot-level sampling methods, the Fire Behavior methods are documented observations taken over the course of a fire event. The FireID field in the PD method links the FB database table to the plot-level data. The Flame Length, Spread Rate, and Fire Behavior Picture fields in the Plot Description (PD) method allow plot specific fire behavior data.

## INTRODUCTION

Fire managers achieve desired fire effects by burning under narrow environmental conditions that create desired fire behavior. Two examples of fire effects are, constraining mortality to 50 percent or achieving 60 percent fuel reduction. Fire behavior is the only direct observation that fire managers can use to judge the outcome for desired fire effect. Fire managers must juggle weather conditions such as temperature, humidity, and wind, with fuel moistures and topography to get a flame length or fireline intensity that will ultimately satisfy the burn objectives by generating the desired fire effects. The only way the fire manager can successfully perform this complex balancing act is to get extensive experience burning across the wide range of environmental conditions *and* to review the results of others who burn under the same conditions. One way that fire managers can document their experience is to record environmental conditions and resultant fire behavior in a database so that others can reference the conditions and then compare them with the actual results of the burn or the fire effects.

The FIREMON Fire Behavior (FB) methods were designed to document the weather and fuel conditions at the time of the fire and the fire behavior that resulted from those conditions. The data are linked to the plot-level sampling by a common field so that they can then be used to interpret the effects of that burn. For example, it is informative to know that the 50 percent fuel consumption measured using the FL methods was achieved with 70 °F temperature, 30 percent relative humidity, and 12 percent 1,000-hour fuel moistures.

The Fire Behavior methods are unlike the other FIREMON methods in that the FB methods are recorded by fire rather than plot. For example, if 50 FIREMON macroplots located in a 500-acre prescribed fire unit are all burned with the same fire, then they all will be linked to the same information in the FB table. The Fire Behavior method and database table are used to document the fire behavior at one or more times during a burn. Each macroplot is linked to this fire behavior information via the Fire ID in the PD form. The fire manager can query the Fire Behavior data to determine the burning conditions, fire behavior, and resultant fire effects on each of the burned plots.

The project objectives will determine which fields on the FB form need to be sampled.

## SAMPLING PROCEDURE

The sampling procedures for all FB procedures are documented in many other publications, most notably the Fire-Weather Observer's Handbook (Fischer and Hardy 1976), and will not be described here. This method is presented in the order of the fields in the FB data table.

### Fire Information

**Field 1: Fire ID**—Enter a Fire ID of up to 15 characters. The ID number or name relates the fire to plots in the PD table. This field links this fire scale data with the plot scale data in the PD method.

**Field 2: Fire Date**—Enter the date of fire as an 8-digit number in the MM/DD/YY format where MM is the month number, DD is the day of the month, and YY is the last two digits of the year. For example, April 01, 2001, would be entered 04/01/01.

**Field 3: Fire Time**—Enter the time of day that these observations were recorded. Use 24-hour time. For example, if it is 8 a.m. enter 08:00; if it is 8 p.m. enter 20:00.

**Field 4: Fire Name**—The name of the fire is entered in this field as a 25-character (or less) code in this field. This is a nonstandardized field so anything can be entered here, but we suggest the name follow the convention used by fire management where it is derived from the drainage or major landmark where the fire starts.

**Field 5: Reference Fire ID**—Enter a unique 20-character fire code taken from the database of other fire management agencies. Record the source of this reference ID in the FIREMON metadata table.

**Field 6: Units** (E/M)—Units of measure. E—English or M—Metric.

### Recording Ambient Weather Conditions

The next set of fields allows the fire manager to store the weather conditions at various times during the burn. These weather data can be measured onsite or downloaded from a RAWS station or other weather station near the burn. The source of these data is recorded in the FIREMON metadata table.

**Field 7: Temperature** (degrees F/degrees C)—Enter the temperature at the time and date listed in Fields 2 and 3.

**Field 8: Relative Humidity** (percent)—Enter the relative humidity measured at the specified date and time.

**Field 9: Windspeed** (miles/hour or km/hour)—Enter the typical windspeed recorded at the specified date and time.

**Field 10: Percent Cloudy** (percent)—Enter the percent cloudiness at the specified date and time.

### Recording Fuel Moistures

The next set of fields describes the measured fuel moistures at the date and time of burn. Standard fuel moisture measurement techniques should be employed. Basically there are two methods of measuring fuel moistures. The oven-dry method requires that multiple samples of all the fuels class sizes be collected in the field, stored in airtight containers (zip-close bags work well), and brought back promptly to be weighed and dried. The mass of the individual samples is measured first, and then the samples are put in an oven at 100 °C. The 1- and 10-hour fuels, and litter and duff should be dry in 24 hours. Weight a few selected samples of the larger fuels every 24 hours until they reach equilibrium. When the piece weights of a class, for example the 100-hour fuel class, reach equilibrium then you can make a final weight of all of the pieces in the class. Calculate the percent moisture (by weight) for each piece using equation FB-1, then average the moistures for each class.

Equation FB-1
$$M = \left( \frac{WW - DW}{DW} \right) *100$$

where, $WW$ = wet weight of the piece and $DW$ = dry weight of the piece.

The fuel moisture for a class is the average moisture measured across all of the samples. When cutting pieces off logs for fuel samples, you do not need to cut them thicker than 3 inches (7.5 cm). Thicker pieces will unnecessarily extend the drying time. If you use the oven-dry method you will not be able to enter the fuel moisture data the day of the fire.

The second method involves indirect measurements of fuel moisture using probes or other instrumentation. The moisture estimates using these methods are generally not as accurate as the oven-dry method, but they allow quick moisture estimates and, depending on the project objectives, may be sufficiently accurate.

Record the method of fuel moisture measurement in the FB Comments field.

**Field 11: 1-Hour Fuel Class Moisture** (percent)—Enter the fuel moisture of the 1-hour downed dead woody fuel class (less than 0.25 inch or 1 cm in diameter).

**Field 12: 10-Hour Fuel Class Moisture** (percent)—Enter the fuel moisture of the 10-hour downed dead woody fuel class (0.25 to 1.0 inch or 1 to 2.5 cm in diameter).

**Field 13: 100-Hour Fuel Class Moisture** (percent)—Enter the fuel moisture of the 100-hour downed dead woody fuel class (1.0 to 3.0 inches or 2.5 to 8 cm in diameter).

**Field 14: 1000-Hour Fuel Sound Class Moisture** (percent)—Enter the fuel moisture of the sound 1,000-hour downed dead woody fuel class (greater than 3.0 inches or 8 cm in diameter). Sound fuels are decay class 1, 2, and 3 pieces. See the **Fuel Load (FL)** sampling method for more information.

**Field 15: 1,000-Hour Fuel Rotten Class Moisture** (percent)—Enter the fuel moisture in percent of the rotten 1,000-hour downed dead woody fuel class (greater than 3.0 inches or 8 cm in diameter). Rotten fuels are decay class 4 and 5 pieces. See the **Fuel Load (FL)** sampling method for more information.

**Field 16: Litter Moisture** (percent)—Enter the moisture of the litter layer. This is the layer that contains recognizable needles, cone scales, and leaves.

**Field 17: Duff Moisture** (percent)—Enter the moisture of the duff layer. This is the layer that contains unrecognizable decomposing organic material. Try to get a moisture estimate for the entire duff profile.

**Field 18: Soil Moisture** (percent)—Enter the moisture of the uppermost soil layer. This is the top 4 inches (10 cm) of mineral soil just below the duff layer.

**Field 19: Live Shrub Moisture** (percent)—Enter the moisture of the live shrubs.

**Field 20: Live Herb Moisture** (percent)—Enter the moisture of the live herbaceous plants.

**Field 21: Live Crown Moisture** (percent)—Enter the moisture of the live tree crown foliage. If possible, take samples from all parts of the tree crowns.

## Recording Fire Behavior

The last set of fields describes the measured or observed fire behavior of the fire at the selected time and date. Fire behavior is often observed rather than measured. Follow the directions in the Fire Observers Handbook (Fischer and Hardy 1976) when estimating these standard fire behavior characteristics.

**Field 22: Fire Type**—Enter the code that best describes the fire that is described by the following observations in Fields 23 through 27.

>   **F**—Flanking
>   **B**—Backing
>   **H**—Head
>   **C**—Crown

**Field 23: Flame Length** (ft/m)—Estimate flame length at this time and date of this. Precision: $\pm 1$ ft/0.3 m

**Field 24: Flame Depth** (ft/m)—Estimate flame depth at this time and date of this fire. Precision: $\pm 1$ ft/0.3 m

**Field 25: Spread Rate** (ft/min or m/min)—Estimate the average speed of the fire at this time and date. To estimate this parameter, using a watch, note the time it takes the flame front to go 30 feet (10 m) and then divide 30 by the number of minutes (or fraction of) to get an answer in ft/min, or divide 10 by the number of minutes (or fraction of) to get an answer in m/min.

**Field 26: Plume Behavior**—Estimate the dynamics of the plume using the following codes.

>   **WV**—Plume well ventilated, rising, and dispersing high above burn
>   **US**—Plume unstable and its behavior is erratic.
>   **PD**—Plume is dropping and going downhill into the valleys

**Field 27: Spotting Observations**—Estimate the spotting behavior of the fire at this time and date using the following codes.

>   **SD**—Spotting downslope or downwind
>   **SU**—Spotting upslope or upwind
>   **SE**—Spotting is erratic and random
>   **NS**—No spotting observed
>   **NA**—Difficult to determine spotting due to smoke or obstruction

## Local Codes and Comments

**Field 28: Local Code 1**—Enter a user-designed code that is up to 20 characters in length and uniquely describes some condition on the FIREMON plot. Do not embed blanks in your codes to avoid confusion and database problems. Document your coding method in the Comments field.

**Field 29: Local Code 2**—Enter a user-designed code that is up to 20 characters in length and uniquely describes some condition on the FIREMON plot. Do not embed blanks in your codes to avoid confusion and database problems. Document your coding method in the Comments field.

**Field 30: Comments**—Memo field. 60,000+ characters may be used to record any information pertinent to the FB information. Text information can be pasted from word documents, Access databases, Excel spreadsheets, or any other software that can copy text to the windows clipboard.

## FIRE BEHAVIOR (FB) EQUIPMENT LIST

Camera with film
Clipboard
FB field forms
Field notebook
Flagging
Hammer (2)
Hatchet (1)
Chain saw or hand saw
Labels
Lead pencils with lead refills

Maps, charts, and directions
Map protector or plastic bag
Masking tape
Pocket calculator
Plot sheet protector or plastic bag
Plot sheets for plots that will be burned (to fill in Fire ID)
Watch with second hand
Weather kit
Zip-close bags or other plastic containers

# Fire Behavior (FB) Form

FB Page _ _ of _ _

## Fire Information

| Field 1 | Field 2 | Field 3 | Field 4 | Field 5 | Field 6 |
|---|---|---|---|---|---|
| FireID | Fire Date | Fire Time | Fire Name | Reference Fire ID | Units (E/M) |
|  |  |  |  |  |  |

## Ambient Weather Conditions

| Field 7 | Field 8 | Field 9 | Field 10 |
|---|---|---|---|
| Temp. (F/C) | Rel. Humidity | Wind (mi/hr or km/hr) | Percent Cloudy |
|  |  |  |  |

## Fuel Moistures

| Field 11 | Field 12 | Field 13 | Field 14 | Field 15 | Field 16 | Field 17 | Field 18 | Field 19 | Field 20 | Field 21 |
|---|---|---|---|---|---|---|---|---|---|---|
| 1-hr Moisture | 10-hr Moisture | 100-hr Moisture | 1000-hr Sound Moisture | 1000-hr Rotten Moisture | Litter Moisture | Duff Moisture | Soil Moisture | Shrub Moisture | Herb Moisture | Crown Moisture |
|  |  |  |  |  |  |  |  |  |  |  |

## Fire Behavior

| Field 22 | Field 23 | Field 24 | Field 25 | Field 26 | Field 27 |
|---|---|---|---|---|---|
| Fire Type | Flame Length (ft/m) | Flame Depth (ft/m) | Spread Rate (ft/min. or m/min.) | Plume | Spotting |
|  |  |  |  |  |  |

## Local Codes

| Field 28 | Field 29 |
|---|---|
| Local Code 1 | Local Code 2 |
|  |  |

Field 30: Comments

# How To...
## Guide

Duncan C. Lutes
Robert E. Keane
John F. Caratti
Carl H. Key
Nathan C. Benson

## HOW TO CONSTRUCT A UNIQUE PLOT IDENTIFIER

This is probably the most critical phase of FIREMON sampling because this plot ID must be unique across all plots that will be entered in the FIREMON database. The plot identifier is made up of three parts: Registration Code, Project Code, and Plot Number.

The FIREMON Analysis Tools program will allow summarization and comparison of plots only if they have the same Registration and Project Codes. This restriction is set because typically each monitoring project has unique objectives with the sample size and monitoring methods developed for specific reasons intimately related to each project. Comparisons made between projects with dissimilar methods may not be appropriate.

**Registration Code ID**—The Registration Code is a four-character code determined by you or assigned to you. The Registration Code should be used to identify a large group of people, such as all the people at one District of a National Forest or a number of people working under one monitoring leader. You are required to use all four characters. Chose your Registration Code so that the letters and numbers are related to your business or organization. For example:

  MFSL = Missoula Fire Sciences Lab
  MTSW = Montana DNRC, Southwest Land Office
  CHRC = Chippewa National Forest, Revegetation Crew
  RMJD = Rocky Mountain Research Station, John Doe

**Project Code**—The Project Code is an eight-charcter code used to identify project work that is done within the group. You are not required to use all eight characters. Some examples of Project Codes are:

  TCRESTOR = Tenderfoot Creek Restoration
  BurntFk = Burnt Fork Project
  SCF1 = Swan Creek Prescribed Fire, Monitoring Crew 1
  BoxCkDem = Box Creek Demonstration Project

It will be easier to read the sorted results if you do not include digits in the left-most position of the project code. For instance, if two of your projects are 22Lolo and 9Lolo, then when sorted 22Lolo will come before 9Lolo. The preferred option would be to name the projects Lolo09 and Lolo22, although Lolo9 and Lolo22 will sort in the proper order, also.

**Plot Number**—Identifier that corresponds to the site where sampling methods are applied. Integer value.

## HOW TO LOCATE A FIREMON PLOT

The FIREMON plot describes the area where the FIREMON methods are applied. Each plot location is determined by the sampling approach—relevé or statistical—then its location is identified with written directions to the plot center and, finally, monumented with a permanent marker.

### Identifying the Appropriate Place for a FIREMON Plot

FIREMON plot location procedures differ by sampling approach. If a relevé approach is used, then FIREMON plots are located by traversing the stand or strata to be sampled to find the range of vegetation and biophysical conditions that exist within the strata. When a location is found that has the conditions that comprehensively represent conditions across the entire strata, the crew boss will identify the plot center. Representative conditions should be assessed from a wide range of ecological attributes with the most important decision criteria keyed to the project objectives. First, the macroplot should represent the vegetation conditions of the strata. This includes species composition, vertical stand structure (canopy layers), plant size (such as, diameter and height), and plant health. Next, use the biophysical environment to judge representativeness of the macroplot. The macroplot should represent modal topography conditions or average slope, aspect, slope position, and elevation attributes. The disturbance history should then be taken into account by making sure that the disturbance evidence (insect, disease, fire, browsing, and so forth) on the macroplot represents the entire strata. Lastly, make sure the fuel characteristics are representative of the macroplot. Be sure to judge fuels individually (fine woody debris, coarse woody debris, and vegetation) and as a group (fire behavior fuel model) to make sure the macroplot is not located in an atypical fuel condition.

Being able to locate a plot in a representative portion of the stand without preconceived bias is somewhat unrealistic. Most plot locations will contain some element of sampler bias. However, in complex ecosystems with high spatial and temporal variability when many attributes must be measured, relevé appears to be the most simple, efficient, and tenable sampling approach available. One way to minimize bias and subjectivity is to mark a plot location, then randomly choose a direction (you can use the second hand on a watch) and place the plot center 100 ft (30 m) away along the randomly selected direction.

There are a few rules that should be followed for relevé plot location. First, establish the plot at least 150 ft (50 m) from any major change in vegetation or ecosystem conditions, such as a roadway, ecotone, or watercourse, and 150 ft (50 m) from the edge of the strata. Next, be sure the macroplot does not contain any atypical features such as brush piles, trails, or camp spots.

If a statistical approach is used, you must randomly locate the plots within the strata or across the landscape using one of three distribution techniques: systematic, random, or cluster. The systematic method is usually preferred because most of the FIREMON plots are located a set distance and azimuth from one another, making them easier to relocate. The random method uses some process that allows plots to be located at any point within the strata with equal probability. This means that two or more plots may be directly adjacent to one another. Plot locations can be picked by overlaying a map of the strata with a clear plastic sheet marked with random points or using randomly determined x- and y-coordinates. Clustered sampling is used to randomly locate plots around a point of origin. In FIREMON we suggest placing the point of origin near the intersection of multiple strata. This will reduce travel and sampling times and may allow you to increase the number of plots per strata. Unfortunately, cluster sampling this way usually involves some sort of bias because areas near the point of origin are more likely to be sampled than parts farther away. Also, the plots are, more often than not, near the boundary edges and may not be located to give the best description of typical conditions. These biases may lead to debate over the validity of statistical results when using clustered plots. There is a more extensive description of the statistical, random, and clumped plot location techniques in the **Integrated Sampling Strategy** chapter.

Once the plots have been marked on a map, the distance and azimuth to each one must be calculated. This can be done using the map scale and plotter or, more easily, using a GIS.

## FIREMON Plot Center

The center of the FIREMON macroplot is a marked, discreet point around which the sampling methods are applied. FIREMON plots that will be resampled must have this point permanently identified. The plot center should be identified by two means: 1) written directions including, at least, latitude/longitude or UTM coordinates, and 2) a physical monument. Additionally, plot photos can be quite useful in relocating the FIREMON plot center.

## Written Directions to the FIREMON Plot

Directions should be kept in the FIREMON notebook and/or the Metadata Table in the FIREMON database. They should carefully describe the directions to the plot from some well-known location that is not likely to change. Instead of "Travel 2 miles from the big boulder in Pat Firemon's pasture..." use "Turn right at the junction of Highway 87 and Forest Service Road 829, then travel 2 miles...". Give more specific directions as you get closer to the plot. The final leg should include distance and azimuth to the plot center. For instance, "...travel 2 miles up FS Rd 829. At Orion Park turn left on FS Rd 73 and go 3.1 miles. On left (north) side of road are two blazed DF (10.3" and 14.4")—if you go over culvert for Johnson's Springs you've gone 0.1 mile too far. Trees should also be marked with flagging. Between the two trees is a red stake. The first plot is 245 ft at 330 degrees true north (17.5 degrees E declination) from the stake. Plot center is marked with flagging and 1-ft tall, 2" x 2" red wood stake. The plot stake is 18.2 ft@155 degrees from a 12.4" DF (tag number #33), 25.8 ft@200 degrees from a 9.0" DF (#34), and 23.0 ft@050 degrees from a 18.1" PP (#35). Tags are at 1-ft height facing plot center. The other five plots are in a line spaced two chains apart on an azimuth of 330 true north, starting from the first plot, and marked with red stakes and flagging." The written directions should always include latitude/longitude or UTM coordinates for plot locations, averaged over 200 readings along with the associated location error.

The directions should be accompanied by a USGS 7.5 minute quadrangle map or aerial photo showing all of the plot locations and, if needed, a hand drawn map showing features not on the quad map such as side roads and log landings. Be sure to record any unique characteristics that might help locate the plot such as a group of blow down trees or a spring.

## Monumenting the FIREMON Plot

The actual method of marking the plot center depends on the land use, distance from roads, vegetation, and the type of treatment that will be applied at the plot. We suggest using the most permanent marker possible for the situation. In areas not frequented by the public, domestic livestock, or hoofed wildlife, or in areas that will not see the use of rubber tired skidders, brightly painted steel fence posts or reinforcing bar (rebar) that extend above the understory vegetation make good choices. Rebar is heavy so you don't want to use it if you are traveling long distances on foot. In such cases, metal electrical conduit might be a better choice. If the unit is going to be burned, the best identification is by stamping plot markers or using casket tags. At the other end of the spectrum, a short wooden 2 x 2 inch (5 x 5 cm) post pounded down to ground level may be best. Relocation of wooden stakes can be made more easily if two or three bridge spikes are pounded around the stake, below ground level, so that the plot center can be relocated with a metal detector. Generally, spikes below ground level are not disturbed by or do damage to rubber tired skidders. Identify the plot center marker in some way (paint color, tag, stamped identification) so that it is not confused with other markers on the plot such as the ones used to identify the fuels transects or the vegetation baseline. Each project is unique, and it is up to the crew boss to determine the best method for permanently marking the plot center. When used in combination—good written directions, GPS locations, flagging, and tree tags—you should be able to locate all plot centers with minimal effort.

## HOW TO PERMANENTLY ESTABLISH A FIREMON PLOT

FIREMON macroplots can be located permanently by driving a 3-ft (1-m) long 1-inch (2.5 cm) diameter piece of concrete reinforcement steel bar (rebar) down 1 to 2 ft (0.3 to 0.5 m) into the ground. Use a heavy hammer if possible. Sometimes it's not possible to drive the rebar into the ground because of rocks or hard soil. If so, drive the rebar in as deep as possible and then hacksaw off the top leaving about 6 inches (15 cm) of rebar showing. Tie a tag around the rebar about 4 inches (10 cm) from the top. As aluminum tags may melt, tags should be hard gauge steel (casket steel) if the plot will be burned in the near future. Use aluminum tags if the plot has already been burned. This tag should have an ID number stamped on it. Write the ID tag down in your notebook. Try to make the ID number part of the plot number. It is highly recommended that an orange or colorful cap be put on the exposed rebar for two reasons. First, it will be easy to relocate and find on the ground. And, more important, it will be highly visible so that no one gets hurt by tripping or falling on it.

The location of the plot must be documented using three methods. First, stand over the rebar and estimate the longitude and latitude of the plot location using a GPS unit. Average at least 200 instantaneous readings to get the most accurate geo-position. Second, take photos of the plot by following the recommendations in **How To Take Plot Photos.** Third, the rebar should be benchmarked by referencing it to at least three semipermanent monuments. A monument can be a large (greater than 8 inches diameter breast height), healthy tree, large rock (greater than 6 ft diameter), or stump. Don't use logs, snags, or objects that can easily be moved. Measure the distance and direction (degrees azimuth) from the plot center to the monument and write in a field notebook. Be sure to describe the monument in detail including unique attributes of the monument. For example, record the approximate species, diameter breast height, and height of a tree monument, or describe the type and size of a rock monument. If possible, permanently mark a monument so that it is more easily identified. For example, blaze a tree monument or scar a rock monument.

## HOW TO DEFINE THE BOUNDARIES OF A MACROPLOT

The boundaries of a macroplot are defined by tying flagging on branches or vegetation at a fixed distance away from the plot monuments accounting for slope of the plot. For circular plots a spring loaded logger's tape makes flagging the plot boundaries easier because it automatically rewinds itself as you walk back toward plot center to get around trees. Tie or hook a cloth or logger's tape to the plot center monument, walk out a distance equal to the plot radius, and tie a flag to a semipermanent structure such as a branch, grass bunch, or downed log (fig. HT-1). Flagging should be placed near eye level along the boundary of the macroplot at intervals that allow the field crew accurate measures of plant cover or tree measurement. We suggest that at least eight points along the plot perimeter be flagged, but dense undergrowth, high tree densities, or severe topography might require additional flagging. Err on the conservative side and put more flag perimeter points when in doubt. It is important that the distances be adjusted for the slope of the ground by multiplying the fixed distances by a slope correction factor. See **How to Adjust for Slope** for the correction factors. The flagging should clearly identify which plants are inside or outside the macroplot boundaries.

If the macroplot is rectangular, then cloth tapes are stretched to form the boundaries with all of the corners marked with stakes or rebar. Tie flagging on branches that cross over the tape. The tape may not be visible from all parts of the plot, and the flags will make it easier to identify the plot boundaries.

How To ... Guide

**Figure HT-1**—Mark the macroplot boundary with flagging.

While selection of flag color may seem trivial, it can be an important phase in the sampling effort. Select a color that is easy to see and different from what other resource groups at your agency or district office may be using. For monitoring plots, it is often helpful if the flags stay up for at least 3 to 5 years so that the same boundaries can possibly be used for the second measurement. It is also good practice to write the plot number on the flagging.

# HOW TO ESTABLISH PLOTS WITH MULTIPLE METHODS

## Plot Layout

Typically, the FIREMON sampling will be done using more than one method. If so, this may mean more than one plot type will be used to gather all of the data. In FIREMON there are four general recommended plot types used for sampling: 1) a 0.1 acre (0.04 ha) circular plot for the Plot Description (PD), Species Composition (SC), and Tree Data (TD) methods; 2) a 66 x 66 ft (20 x 20 m) vegetation sampling plot for the Cover/Frequency (CF), Point Intercept (PO), Line Intercept (LI), Density (DE), and Rare Species (RS) methods; 3) a hexagonal path defined by the 75-ft (25-m) Fuel Loading (FL) sampling planes; and 4) a 50-ft (0.18-acre) or 15-m radius (0.07-ha) sampling plot for the Composite Burn Index (CB) method. FIREMON allows you to use almost any plot size and shape, but the point is that when you lay out the different plot types they should, as much as possible, cover the same area (fig. HT-2).

The FIREMON plot center coincides with the start of the first sampling plane of the FL method as well as the center of the PD, SC, TD, and CBI methods so it is not difficult to lay them out. The CF, PO, LI, DE, and RS methods use transects and quadrats oriented perpendicular to a baseline, and it may not be clear where to start the baseline in order for the vegetation plot to be positioned appropriately. If you are using the recommended plot size for the CF, PO, LI, DE, or RS methods, then from the FIREMON

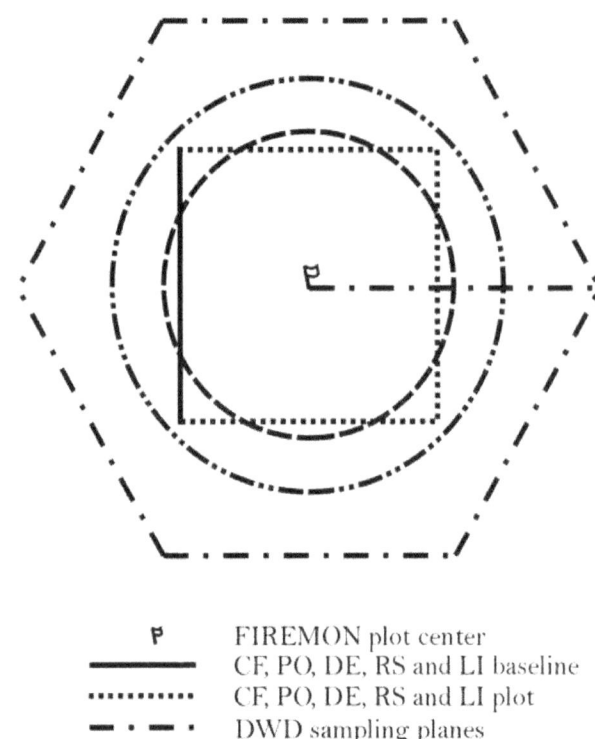

**Figure HT-2**—Four general plot shapes are used when sampling with the recommended protocols. Often more than one of these plots will be used. If so, the different plots should be laid out so they overlay each other as much as possible.

plot center measure 33 ft (10 m) down slope, then 33 ft (10 m) across slope to the right, and locate the start of the baseline at that point. If the plot is located on flat ground, the starting point for the baseline is located 33 ft (10 m) true south and 33 ft (10 m) due west from the plot center. To locate the baseline when using a different plot size, measure one-half the plot width down slope from plot center, then one-half the plot height across the slope, and locate the start of the transect there. If you are using a rectangular plot the longer dimension will generally go across the slope.

## Sampling Order

Each sampling method will impact the sampling area to some extent, and the impact from one method may negatively influence the ability to sample with another method. For instance, tree sampling often leads to lots of trampling because the samplers are moving back and forth across the plot to mark the boundary, then sampling small and mature trees. If you try to measure herbaceous vegetation heights after tree sampling you won't get a true representation because of the trampling.

Use table HT-1 to order the sampling on your plot. Look for the method(s) that you will be using on each FIREMON plot and sample them in the order you find them in the table. For example, if you are using the TD and CF methods, complete the CF protocol before moving on to the TD protocol. The first three methods are equally sensitive to plot disturbance. However, rarely would more than one be used on a particular plot. Modify the order if needed so that plot measurements reflect the actual plot condition, not the condition caused by sampling (such as trampling).

**Table HT-1**—Suggested order for sampling when using multiple methods. Modify as necessary for your project.

| Order | Method |
|---|---|
| 1 | RS |
| 2 | SC |
| 3 | PO |
| 4 | CF |
| 5 | DE |
| 6 | LI |
| 7 | TD |
| 8 | FL |
| 9 | CB |
| 10 | PD |

# HOW TO DETERMINE SAMPLE SIZE

## Plotting Graphs of Mean Values for Varying Sample Sizes

It may be necessary to sample more than the recommended number of transects or quadrats in order to sufficiently capture the plant species variation within the macroplot. The FIREMON Line Intercept, Point Intercept-Transect, and Density-Belt Transect sampling methods are transect-based methods that may require adding more transects or making the existing transect length longer in order to capture the variability of the attribute of interest. The FIREMON Cover/Frequency (CF), Point Intercept-Frame, and Density-Quadrat methods are quadrat-based sampling methods that may require installing more quadrats on longer transects or installing more quadrats on additional transects in order to capture the variability of the attribute of interest.

The following example uses the FIREMON CF method to determine a sufficient sample size for estimating plant species canopy cover. Begin by laying out the minimum number of transects and quadrats. See **How to Locate Transects and Quadrats** for more details. Then record plant species canopy cover for each quadrat. Using a calculator and graph paper or a spreadsheet program such as Microsoft® Excel, plot the average canopy cover of selected plant species for varying number of quadrats (table HT-2). Start with averaging canopy cover for the first quadrat and end with averaging canopy cover over all quadrats. It may be necessary to plot average plant species cover values on more than one plot if some plant species have low cover values and some plant species have high cover values. Plotting cover values at different scales will allow you to see fluctuations in the graphs for all species, regardless of the absolute cover values. In this example, four species with high relative cover values were plotted on one graph (fig. HT-3), and two species with low relative cover values were plotted on a second graph (fig. HT-4).

Fluctuations in the graph will level out at a sufficient number of quadrats for sampling the attribute plotted—canopy cover, in this example. More quadrats should be sampled if the line graphs do not level out. If more quadrats need to be sampled, add more transects with the same number of quadrats as the other transects, or make the existing quadrats longer and add more quadrats. A third option would be to place additional quadrats on the existing transects. However, this can lead to questions about the spatial correlation of some attributes, so we recommend using this option only if the first two are not feasible. Record plant species canopy cover for the new quadrats and then plot the graphs again with the additional cover values. These graphs can be plotted for other attributes such as frequency, density, and height. The basic idea is to plot a graph for the attribute you are interested in measuring and adjust the sample size appropriately. This method can be used to plot mean attribute values by transect for the transect sampling methods. The number of transects sampled may be adjusted accordingly.

Table HT-2  Average canopy cover values for selected plant species. Average values are calculated for successively larger numbers of quadrats.

| Agropyron spicatum | Agropyron smithii | Bromus japonicus | Achillea millefolium | Koelaria cristata |
|---|---|---|---|---|
| 0 | 0 | 0 | 0 | 0 |
| 20.5 | 20.4 | 0.5 | 0 | 0 |
| 23.3 | 23.3 | 1.3 | 0 | 1 |
| 23.4 | 23.3 | 6 | 0.25 | 1.5 |
| 18.4 | 18.5 | 5.4 | 0.2 | 1.2 |
| 15.2 | 14.9 | 5 | 0.17 | 1 |
| 13.6 | 14.5 | 5.7 | 0.14 | 1 |
| 11.1 | 14.4 | 5.4 | 0.25 | 1 |
| 10.7 | 13.4 | 5.1 | 0.22 | 1.22 |
| 9.5 | 12.5 | 5.6 | 0.3 | 1.1 |
| 9.1 | 12.5 | 4.5 | 0.27 | 1 |
| 8.4 | 11.2 | 5.2 | 0.25 | 0.92 |
| 7.4 | 10 | 5 | 0.46 | 1.1 |
| 7.3 | 9.2 | 4.8 | 0.5 | 1.2 |
| 7.2 | 9.2 | 4.5 | 0.47 | 1.1 |
| 6.3 | 9.2 | 4.4 | 0.63 | 1.06 |
| 6.3 | 7.8 | 4.3 | 0.59 | 1.06 |
| 6.3 | 7.8 | 4.1 | 0.56 | 1.17 |
| 5.6 | 7.1 | 4 | 0.68 | 1.26 |
| 5.4 | 7.1 | 4 | 0.65 | 1.35 |
| 5.3 | 7.1 | 3.8 | 0.67 | 1.29 |
| 5.4 | 7.1 | 3.8 | 0.64 | 1.3 |
| 4.3 | 6.2 | 3.5 | 0.62 | 1.3 |
| 4.3 | 6.2 | 3.5 | 0.62 | 1.3 |
| 4.3 | 6.2 | 3.5 | 0.62 | 1.3 |

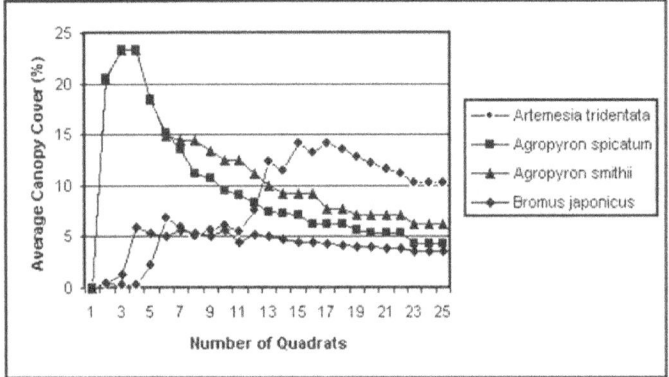

**Figure HT-3**—Plot of average canopy cover versus number of quadrats for four plant species on a plot having approximately 5 percent or more cover.

**Figure HT-4**—Plot of average canopy cover versus number of quadrats for two plant species on a plot having approximately 1 percent or less cover.

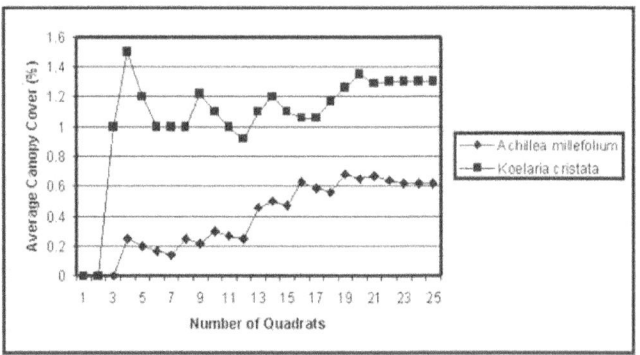

# HOW TO ESTIMATE COVER

## General Cover Estimation Techniques

Depending on the objectives of your monitoring project you may be estimating cover using the Plot Description (PD), Fuel Load (FL), Cover/Frequency (CF), and Species Composition (SC) methods. In FIREMON cover estimation is usually made at one of two general scales. The cover estimation is made over the entire macroplot with the PD and SC methods and, usually, on a 20 inch (0.50 cm) square or 6-ft (2-m) diameter plot with the CF and FL methods, respectively. Cover estimation is much more straightforward with the LI method and is described in the LI documentation.

Cover is usually defined one of two ways. First, cover may be the outside edge or drip line of the plant crown being assessed, with all of the spaces within the crown included in the estimate. Although there is no standard definition, this cover assessment is sometimes called canopy cover, or if tree cover is being estimated, crown cover. Second, cover may be estimated as the vertical projection of the plant foliage and supporting parts with all of the spaces within the crown excluded from the estimate. Again, there is no standard definition; however, this cover assessment is sometimes called foliar cover. Since the foliage of overstory trees is dense, the two assessments are nearly synonymous for that component. Apart from FIREMON, it is common for cover of overstory trees to be estimated using the canopy cover definition, and the cover of other components to be estimated using foliar cover definition.

In FIREMON, all the methods suggest using foliar cover for the cover estimates. Some figures in this section were developed to help users estimate foliar cover. Other figures are presented to help users visualize concepts, such as subdividing and grouping, to make better cover estimations. If users want to estimate canopy cover using a method that is different than foliar cover, it should be noted in the Metadata table.

Of the two types, foliar cover is a better measure of vegetation change over time. If you estimate cover based on plant perimeter you are purposely disregarding the open spaces in the canopy. These spaces are likely locations for new growth to occur so it would be possible for foliar cover to increase over time without any associated increase in canopy cover. Changes in foliar cover are often important to fire monitoring projects, so we suggest sampling that characteristic rather than just canopy cover.

The biomass equations used in the FIREMON Analysis Tools package are based on foliar cover, but this is not always the case. If you are collecting species-specific cover data to derive biomass outside of the FIREMON Analysis Tools, be sure that you sample cover the same way that it was sampled when the equations were developed or your biomass estimations will be incorrect.

You may need to make cover estimates for a number of ecosystem components. The most common estimates are for living vegetation, such as individual species, structural layer, or life form. Other components include cover of dead vegetation such as fine and coarse woody debris and dead herbaceous material. Finally, bare ground, rocks, and ash are examples of nonvegetation components that may be sampled.

In FIREMON, "cover" is the vertically projected cover of the component being sampled. Vertically projected cover is best described as the cover of the sampling entity if it were compressed straight to the ground. To make good estimations of cover, field samplers will need to visualize this compression for each component. This might not be hard to visualize if samplers are estimating the cover of logs, lichen, or some other low entity, but as the vegetation gets taller and occupies more layers, the task becomes more difficult. Experience will help samplers be less intimidated by this task. Sometimes a plant that is rooted outside the sampling area has vegetation—branches and leaves primarily—growing within the sampling area. This vegetation should be included in the cover estimates you make.

If a particular species was sampled sufficiently early or late in the season that you believe its cover at peak phonological development would be one cover class greater or lower, enter the estimated peak cover. For example, if leaves have fallen off the plant and are on the ground, mentally reconstruct the plant with leaves attached and estimate its cover.

The cover classes used in FIREMON are relatively broad, typically 10 percent (table HT-3), so the precision of cover estimates are secondary to accuracy.

**Table HT-3**  Cover classes used in FIREMON.

| Code | Cover |
|---|---|
| | *percent* |
| 0 | Zero |
| 0.5 | 0  1 |
| 3 | >1  5 |
| 10 | >5  15 |
| 20 | >15  25 |
| 30 | >25  35 |
| 40 | >35  45 |
| 50 | >45  55 |
| 60 | >55  65 |
| 70 | >65  75 |
| 80 | >75  85 |
| 90 | >85  95 |
| 98 | >95  100 |

The easiest way to get a cover estimate in the field is through an iterative process where you first note that cover is between two cover classes, then use the midpoint of those two classes and try to determine which half the cover is in. Continue this until you have narrowed cover down to one class. For instance, say you are looking at a sampling area where you know for certain that cover is between 15 and 55 percent (cover classes 20 and 60). Next, try and determine if cover is between 15 and 35 percent or between 35 and 55 percent. If you think cover is lower than 35 percent, then try to determine if cover is between 15 and 25 percent or of it is between 25 and 35 percent. Cover class will be 20 if you choose the lower half or 30 if you choose the upper half.

Experienced field samplers usually get accurate estimations of cover using two methods: grouping or subdividing. On smaller plots many samplers mentally group the plants to one corner of the sampling area and then estimate the cover. Cover estimates are easier when you group using a marked quadrat such as the one described in the CF methods. The subdivision method uses our natural ability to estimate cover better on small areas than large areas. It is typically used on large plots, such as the FIREMON macroplot. Subdivision also helps get accurate cover estimates when the entire plot cannot be seen from one location.

When using the subdivision method, divide the sampling area into quadrants or some other easily determined area, and estimate the cover in each part. It may be useful to use the grouping technique on each quadrant also. In figure HT-5, A, illustrates how plants, represented by the circles, might be distributed across a plot. B shows the plot divided into quadrants and plants being mentally combined within the quadrant with the combined cover shown using circles in C or as squares in D. The areas of all the circles in A make up 10 percent cover. In C, percent cover for quadrants 1, 2, 3, and 4 is 12, 4, 16, and 8 percent, respectively. Percent cover, then, is (12 + 4 + 16 + 8)/4 = 10 percent. Usually you will also be developing a species list for the plot, so as you walk around the perimeter looking at the cover within each quadrant you can record the species at the same time.

The dimensions of the imaginary item groups can be used by the field sampler to estimate cover on the entire plot. For instance, if you visually group all of the herbaceous cover on the macroplot and find that it would fit in a circle about 24 ft across, that group would constitute a cover of about 10 percent on the typical 0.10 acre macroplot (table HT-4). This method can be used at the individual species level for estimating cover in layers or for a composite estimation of cover.

Regardless of the technique used for estimating cover, samplers will need to calibrate their eye in order to make an accurate assessment of cover. Field crews should develop a plan so that samplers can calibrate their eyes periodically throughout the field season. The best way to do this is by visually estimating cover on a FIREMON plot-sized area where the "true" value has been verified using either PO for ground cover, CF for herbaceous cover, and/or LI for shrub cover. Below are some illustrations that are designed to help samplers as they begin to calibrate their eyes.

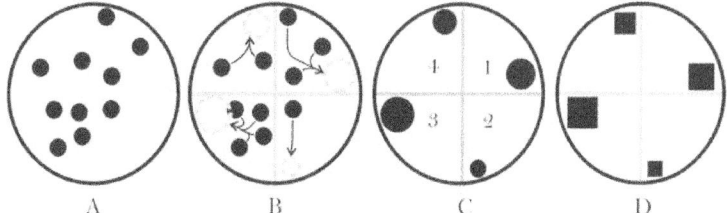

**Figure HT-5**—The subdivision and grouping technique for estimating cover. Average the cover in quadrants to get the cover class estimation.

**Table HT-4** Percent cover for different radius areas on a 0.10 acre (400 m²) circular macroplot.

| Radius | Radius | Percent of 0.10 acre (400 m²) macroplot |
|---|---|---|
| ft | m | |
| 37.2 | 11.28 | 100 |
| 26.3 | 7.98 | 50 |
| 18.6 | 5.64 | 25 |
| 11.8 | 3.57 | 10 |
| 8.3 | 2.52 | 5 |
| 3.7 | 1.13 | 1 |

The illustrations in figure HT-6 represent a circular plot of any size. Before reading the figure caption try to estimate the area inside each large circle that is covered by the smaller circles. As a hint, people tend to overestimate.

The illustrations in figure HT-7 represent a 6-ft diameter sampling area with 1 inch and 0.25 inch diameter pieces scattered inside. Try to estimate the percent of vertically projected cover before reading the caption.

Notice that in figure HT-7 the illustrations of branches are two-dimensional representations of three-dimensional entities—that is each illustration shows how much ground the dead branches would cover if they were compressed straight to the ground. In the field, samplers will need to get comfortable with imagining all of the suspended pieces moved to the ground in order to estimate vertically projected cover.

The illustrations in figure HT-8 represent live cover on a 6-ft diameter sample plot. This cover estimate includes both the branches and leaves together. The percent cover is listed below each illustration. Remember, some of the vegetation in the sampling area might be rooted outside the sampling area; however, it is included in your cover estimate.

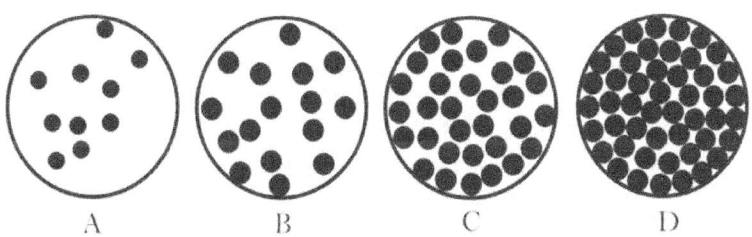

**Figure HT-6**—Cover illustrations showing different levels of cover. Cover in A, B, C, and D is 10, 25, 50, and 75 percent, respectively.

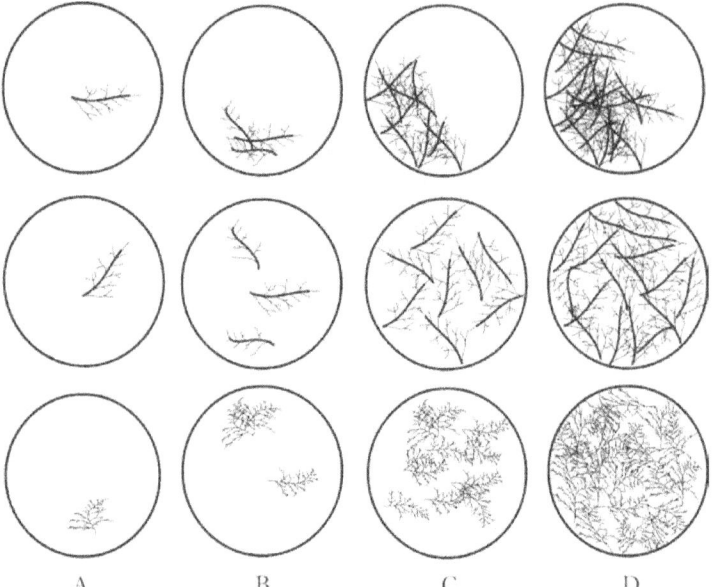

**Figure HT-7**—Cover illustrations showing levels of dead vegetation. Each circle represents a 6-ft diameter sampling area. Thick lines represent pieces 1 inch diameter and thin lines 0.25 inch diameter. The cover in columns A, B, C, and D is 1, 3, 10, and 20 percent, respectively.

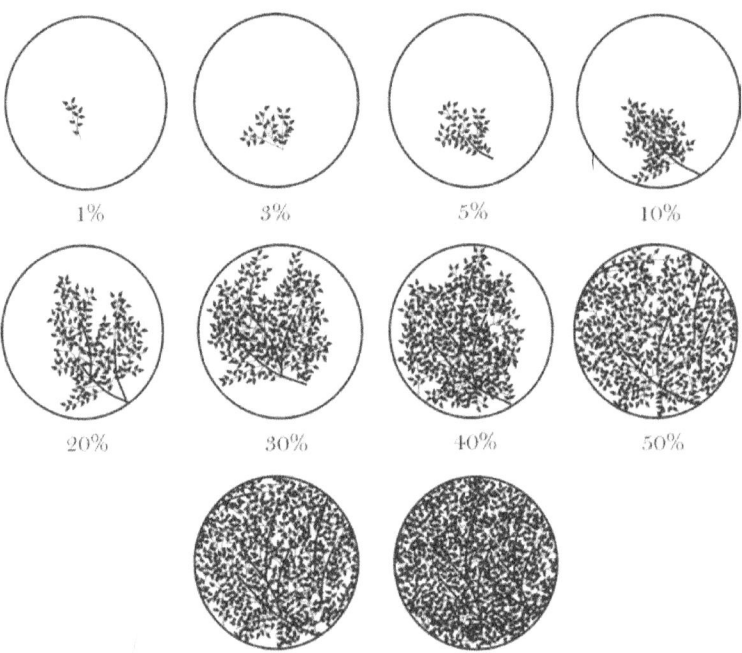

**Figure HT-8**—Cover of live vegetation on a 6-ft diameter sampling area. Percent cover is indicated below each illustration.

Estimating the cover of multiple entities makes the estimation task more difficult because you have to mentally separate each entity. It is easiest to first make an estimate of the total vertically projected cover on the sampling area, and then estimate cover of the entities from greatest cover to least cover. Figure HT-9 shows two entities, woody and nonwoody vegetation, being sampled on the same sampling area. First, the total cover (A) would be estimated, then nonwoody cover (B), and finally, woody cover (C). Because of overlap between entities, the sum of the entities may be greater than the total cover and may sum to be greater than 100 percent.

How To ... Guide

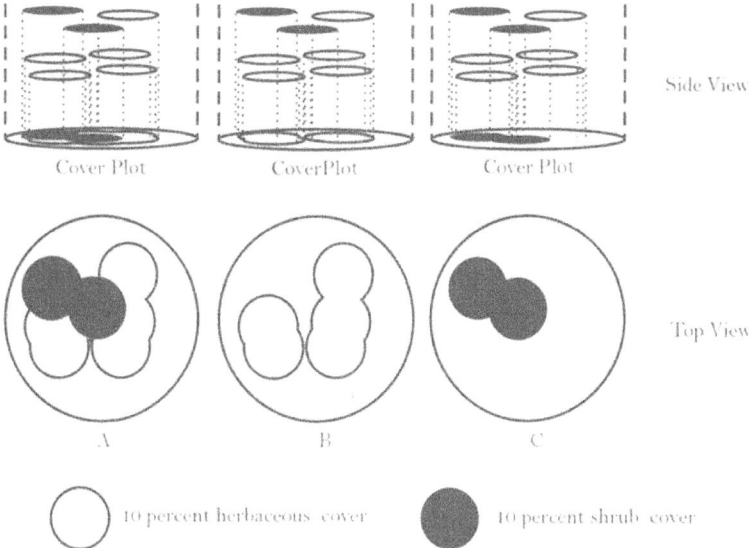

**Figure HT-9**—In these illustrations each circle, individually, represents 10 percent cover. A) Total cover within the sampling area is about 40 percent. B) Cover of nonwoody vegetation is about 30 percent. C) Cover of woody vegetation is about 15 percent. Note that, because of overlap, the sum of cover for the entities will always be greater than or equal to the total cover, and may sum to greater than 100 percent cover. In terms of field data, cover in A, B, and C would be recorded as 40, 30, and 20, respectively.

## Additional Hints for Estimating Cover When Using the Species Composition (SC) Method

Use the techniques provided above to estimate cover on the SC sampling area. Both the subdivision and grouping techniques may be helpful on a large macroplot. If you are sampling species cover, do not include overlap between canopies of the same plant species (fig. HT-10). If cover is measured by size class for a plant species, estimate the cover for each size class and include canopy overlap between different size classes (fig. HT-11).

**Figure HT-10**—In this figure, the small trees underneath the canopy of the larger trees are the same plant species. If cover estimates are being made for total cover by species then cover is estimated as the projection of the large tree canopy onto the ground, which overlaps the canopy of the smaller trees.

**Figure HT-11**—In this figure the small trees underneath the canopy of the larger trees are the same plant species but a different size class (seedlings versus saplings). If cover estimates are being made by species and size class, then one cover estimate should be made for the seedlings and one estimate for the saplings.

## Additional Hints for Estimating Cover When Using Quadrats

Estimating cover within quadrats is made easier by marking the quadrat frame to indicate subplot sizes and knowing the percent of quadrat area each subplot represents (fig. HT-12). Subplots are used to estimate cover for a plant species by mentally grouping cover for all individuals of a plant species into one of the subplots. The percent size of that subplot, in relation to the size of the quadrat being sampled, is used to make a cover class estimate for the species. Cover estimates should include plant cover over the plot even if a plant is rooted outside the plot (fig. HT-13).

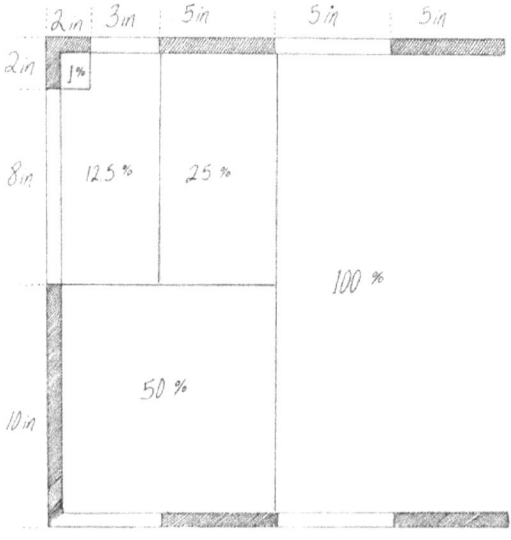

**Figure HT-12**—Subplot dimensions for the standard 20 x 20 inch (50 x 50 cm) quadrat used for cover/frequency sampling.

How To ... Guide

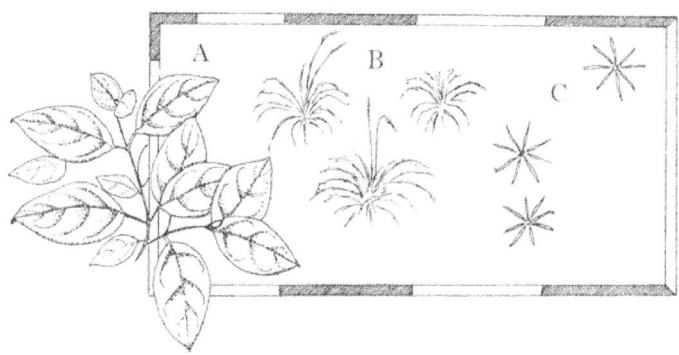

**Figure HT-13**—Cover from species A is estimated even though this species is not actually rooted within the quadrat.

## HOW TO USE A COMPASS—SIGHTING AND SETTING DECLINATION

Compass headings are used to navigate to plots, determine plot aspect, orient sampling transects, and other field tasks. In FIREMON we call the compass heading the *azimuth*. It is extremely important that the field crews are familiar with using a compass so they can walk a course with a known azimuth and find the azimuth of a course where the azimuth is unknown. For example, when describing the directions to a new FIREMON plot, the crew will have to determine what distances and azimuth to record so that sampling crews will be able to find the plot at the next sampling time. For subsequent sampling visits, the crew will follow the directions provided by the crew that sampled the plot initially. When you are determining a course, compass use is different than when you are following a course. We recommend that you use a compass that has a sighting mirror and declination adjustment. The parts of a compass are shown in figure HT-14.

If you want to determine a course between two points, say a flagged tree and a sample plot, stand at the tree, set up your compass so that you can see the compass face in the mirror, with the sighting line in the mirror and the sight on the compass lined up. Now, with the compass level, hold it out in front of you at eye level so that you can see the plot center in the sight on top of the compass. If the compass is not held level, the compass needle will not rotate freely and the azimuth will be wrong. Sometimes there are bubbles in the compass and you can use the bubble's reflection in the mirror to let you know if the compass is not level. With the plot center in the sight and the scribed line on the mirror lined up with the sighting marks on the compass face, hold the compass still and while looking at the compass needle in the mirror, use your free hand to turn the compass housing until the compass needle is parallel to the orienting arrow. Make sure that the north end of the compass needle (usually marked red) is pointing the same direction as the orienting arrow or your azimuth will be off 180 degrees. Now, lower the compass and record the azimuth from the sighting mark closest to the mirror. If the declination is set to zero, you will be recording the azimuth in magnetic degrees. If the declination is set to the declination for your sampling area, then you will be recording the azimuth in degrees true north.

If the bearing between two points is known and you want to follow that course, then set the known azimuth at the sighting mark closest to the mirror. Next, hold the compass out in front of you, level and at eye level. Keep the compass in the same position relative to your eye and rotate your body until you see in the mirror that the compass needle is parallel with the orienting needle. Make sure the scribed line on the mirror is lined up with the sights marks or you will be going the wrong direction. With the needles parallel, use the sight on the mirror to pick an obvious object, such as a rock or tree, that falls in line with the direction you want to go. Pick something distinct that you will be able to see for the entire distance as you are walking. For example, don't pick just any tree as many look alike. Instead, pick a tree with a forked crown or other distinctive feature. Once you get to the object you will repeat the procedure. For long distances or in dense vegetation you may have to repeat this sight-and-walk procedure many times before you get to your destination. When laying out the sampling planes for the FL method you use this technique to guide the sampler pulling the tape. For instance, for the first

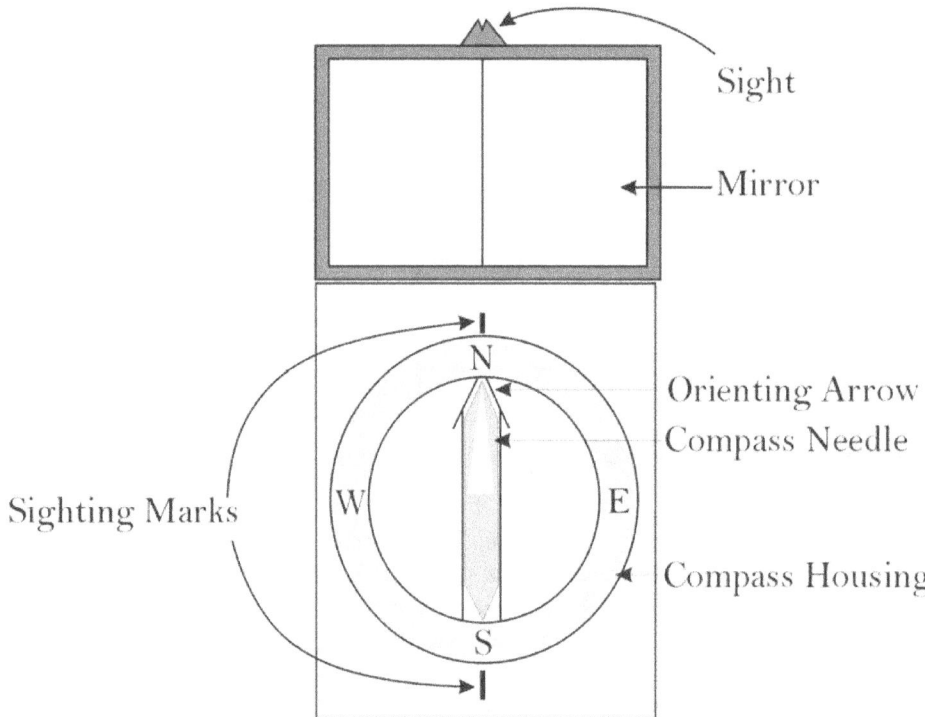

**Figure HT-14**—The parts of a compass. Azimuth numbers have been left off for readability.

sampling plane you set the azimuth 090 degrees at the sighting mark closest to the mirror, then hold the compass in front of you and guide the other sampler along making sure to keep that person on the 090 degree azimuth.

Magnetic declination is the difference between true north (pointing to the North Pole) and magnetic north (the direction the needle on your compass points). We suggest always using true north degrees, which requires that you set the declination on your compass. The declination for the area you are sampling should be set the same on all the compasses in the project and recorded in the Comments field of the Plot Data table or in the Metadata table. Declination can change substantially as you move from place to place, especially in the Northern United States and Canada, so be sure that you look up the declination at each of your sampling sites. Probably the best declination values are available from aircraft sectional charts. However, those maps may not be immediately available. Ask a pilot friend to give you an old sectional chart when it has gone out of date (usually once a year) and keep it on hand for reference. Declination changes over time, so do not use sectional charts or declination maps greater than 10 years old to get the declination value. Other sources for declination information are the World Wide Web or people familiar with the sampling area. For example, the USGS has a coarse scale map of declination on its Web site (http://geomag.usgs.gov/usimages.html).

Different compasses have different mechanisms for changing the declination but most use a small screw on the compass housing. Use a screwdriver (usually supplied with a new compass) and turn the declination screw left or right until the mark in the orienting arrow lines up with the proper declination. Be sure that you use the correct declination direction, east or west. In figure HT-15 declination is set to 6 degrees east declination.

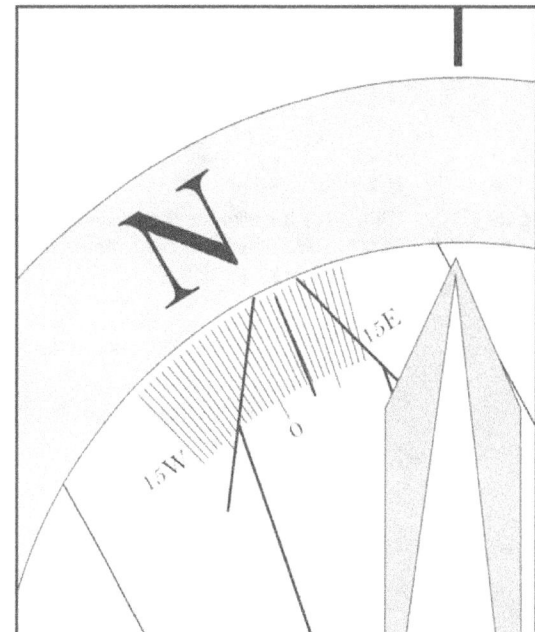

**Figure HT-15**—The mark in the orienting arrow indicates the compass is set to 6 degrees east declination.

## HOW TO ESTABLISH A BASELINE FOR TRANSECTS

The suggested baseline is 66 ft (20 m) long and is oriented upslope with the 0-ft (0-m) mark at the lower point (fig. HT-16). On flat areas, the baseline runs from south to north with the 0-ft (0-m) mark on the south end of the baseline. Transects are placed perpendicular to the baseline (across the slope) and are sampled starting at the baseline. Running transects across the slope will reduce the possibility of erosion along paths taken by samplers as they move along the transect. The greatest concern for erosion is on areas with sparse ground vegetation. On flat areas, transects are located west to east.

The baseline is established by stretching a tape measure the desired distance between two stakes. Permanently mark the baseline with four markers. In the written description of the plot location record the foot (meter) location of each marker along the baseline. For a 66-ft (20-m) baseline markers are placed at 0 ft (0 m), 3 ft (1 m), 63 ft (19 m), and 66 ft (20 m). Locating a transect nearer that 3 ft (1 m) to the end of the baseline may mean that portions of quadrats will lie outside the vegetation plot. It is not always necessary to permanently mark the 3-ft and 63-ft points, but doing so provides a backup in case any markers are disturbed. The markers at the start and end of the baseline should be brightly painted or marked with flagging so that they are easy to relocate for subsequent sampling.

In general you should locate the baseline using the most permanent marker available for the situation. In areas not frequented by the public, domestic livestock, hoofed wildlife, or areas that won't see the use of rubber tired skidders, brightly painted steel fence posts or reinforcing bar (rebar) that extend above the understory vegetation make good choices. Rebar is heavy so you don't want to use it if you are traveling long distances on foot. In these cases, metal electrical conduit might be a better choice. If the unit is going to be burned, the best identification is by stamping the plot marker or using casket tags. At the other end of the spectrum, a short wooden 2 x 2 inch (5 x 5 cm) post pounded down to ground level may be best. Relocation of wooden stakes is easier if two or three bridge spikes are pounded around the stake, below ground level, so that the plot center can be relocated with a metal detector. Generally, spikes below ground level are not disturbed by or do damage to rubber tired skidders. Identify the baseline markers in some way (for example, paint color, tag, or stamped identification) so that they are not confused with other markers on the plot such as the ones used to identify the fuels transects or the plot center. Use a file or other type of permanent marker to display 1, 2, 3, or 4 notches in the appropriate

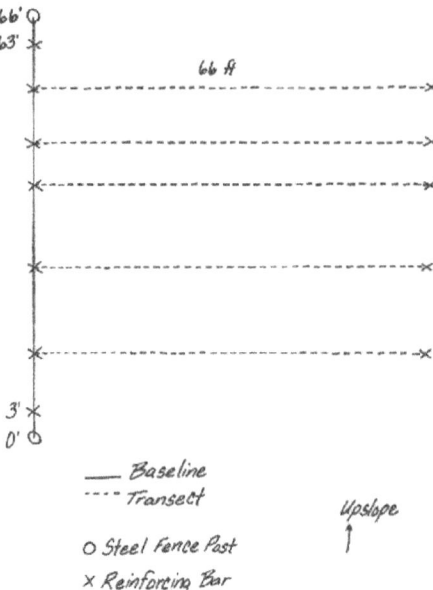

**Figure HT-16**—Orient the baseline upslope with the zero end positioned at the lower point. Transects are oriented across the slope to limit the opportunity for erosion.

marker with one notch denoting the starting marker and four notches denoting the ending marker along the baseline. Each project is unique, and it is up to the crew boss to determine the best method for permanently marking the baseline.

From the starting and ending markers, take a compass bearing (azimuth) and distance to the plot center and record the information on the plot location. Take a compass bearing from the starting marker to the ending marker and record this on the map and on the plot location description. Determine the compass bearing perpendicular to the baseline for each of the transects and record them on the general description map. If permanent transects are established, mark the beginning and ending of each transect with reinforcing bar.

Transect locations should be recorded on the general description map and in a written description of the plot layout design used in the project. Transects are placed perpendicular to the baseline with a compass and tape to ensure they are parallel to each other and can be relocated.

## HOW TO LOCATE TRANSECTS AND QUADRATS

The FIREMON Random Transect and Quadrat Locator program (fig. HT-17) selects random starting points for transects along the macroplot baseline. The program also selects random starting points for systematic placement of quadrats along a transect. The numbers are unit-less, so you can enter them in feet or in meters and the result will be in feet or meters.

Enter a seed number for the random number generator. Then enter the number of transects and the length of the baseline in feet (meters). If two or more transects start at the same point or are too close to each other (less than the quadrat width), press the run button again to generate another set of transect locations. If you are placing quadrats along the transects, you can also generate random starting points for placement of the first quadrat. Enter the maximum distance from the baseline in feet (meters) in which the first quadrat on each transect could be placed. For example, if you are placing five 20 x 20 inch (50 x 50 cm) quadrats along a 66-ft (20-m) transect and placing them 12 ft (4 m) apart, the maximum distance would be 12 ft (4 m). If the random starting point equals 10 ft, your quadrats would start at 10, 22, 34, 46, and 58 ft.

How To ... Guide

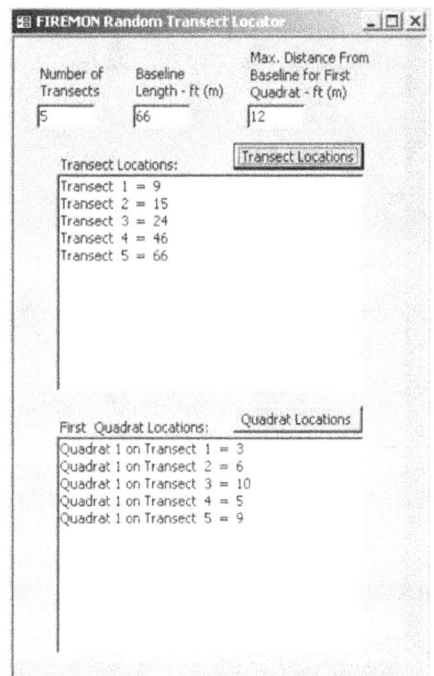

**Figure HT-17**—FIREMON random transect and quadrat locator.

Press the Transect Locations button to generate random transect starting points along the baseline. Press the Quadrat Placement button to generate random starting points for quadrat placement along each transect. Tables HT-5 and HT-6 present five sampling schemes for randomly locating five transects along a 66-ft (20-m) baseline. If you were using the FIREMON Line Intercept, Point Intercept, or Density method with only three transects along the baseline, you would use transect locations 1, 3, and 5 in tables HT-5 and HT-6.

Figure HT-18 illustrates example transect locations and quadrat placement for the recommended FIREMON Cover/Frequency sampling method. Five transects are located along a 66-ft baseline using sampling scheme 3 from table HT-5. Five quadrats are placed systematically along each transect starting 12 ft from the baseline and placed at 12-ft intervals.

Figure HT-19 illustrates example transect locations, quadrat placement, and belt transect placement for the recommended FIREMON Density sampling method. Five transects are located along a 66-ft baseline using sampling scheme 3 from table HT-5. Five 3 x 3 ft quadrats for sampling herbaceous plants are placed systematically along each transect starting 12 ft from the baseline and placed at 12-ft intervals. Three 6-ft (2-m) belt transects for sampling shrubs and trees are placed along each transect.

**Table HT-5** Sample schemes for five randomly placed transects in feet.

| Sample scheme | Transect number | | | | |
|---|---|---|---|---|---|
| | 1 | 2 | 3 | 4 | 5 |
| | Distance along baseline | | | | |
| | *ft* | | | | |
| 1 | 3 | 8 | 14 | 49 | 56 |
| 2 | 3 | 16 | 20 | 22 | 46 |
| 3 | 8 | 24 | 34 | 45 | 57 |
| 4 | 12 | 18 | 22 | 34 | 64 |
| 5 | 17 | 26 | 30 | 42 | 58 |

**Table HT-6** Sample schemes for five randomly placed transects in meters.

| Sample scheme | Transect number | | | | |
|---|---|---|---|---|---|
| | 1 | 2 | 3 | 4 | 5 |
| | Distance along baseline | | | | |
| | *m* | | | | |
| 1 | 2 | 7 | 9 | 16 | 19 |
| 2 | 1 | 8 | 12 | 16 | 19 |
| 3 | 4 | 6 | 9 | 16 | 18 |
| 4 | 2 | 4 | 9 | 15 | 20 |
| 5 | 1 | 7 | 10 | 14 | 20 |

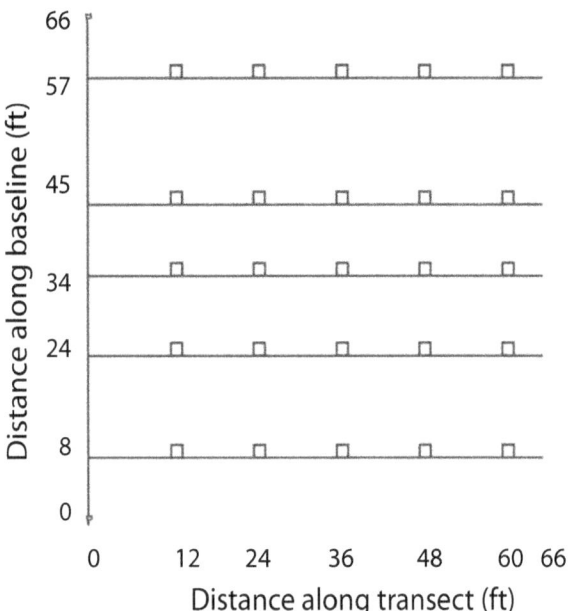

**Figure HT-18**—Transect and quadrat layout for the recommended FIREMON Cover/Frequency sampling method.

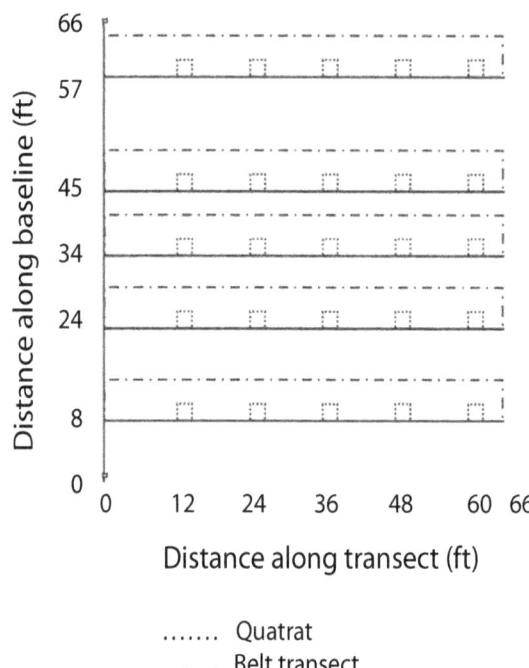

....... Quatrat
_._. Belt transect

**Figure HT-19**—Example belt transect (for large shrubs) and quadrat layout for the recommended FIREMON Density sampling method.

## HOW TO OFFSET A TRANSECT

If an obstacle, such as a large rock or tree, is encountered along a transect, use the following procedure to lay the transect around it (fig. HT-20). First, run the measuring tape from the baseline to the obstacle and place a permanent marker, such as concrete reinforcing bar (rebar), at this point. Note the distance on the measuring tape. Choose the direction to deviate from the transect, left or right, that gives the shortest offset from the original transect. Next, deviate 90 degrees from the transect until the obstacle is cleared and place a temporary marker at this point. Then run a tape from this point in the same azimuth as the original transect until the obstacle is cleared and place a temporary marker at this point. Next, deviate 90 degrees from this point back to the location of the original transect, and place a permanent marker at this spot. Add the distance between the two temporary markers to the desired

**Figure HT-20**—If samplers encounter an obstruction, offset the transect line around the obstruction and correct for the lost distance by adding on the appropriate amount of transect to the end of the same transect.

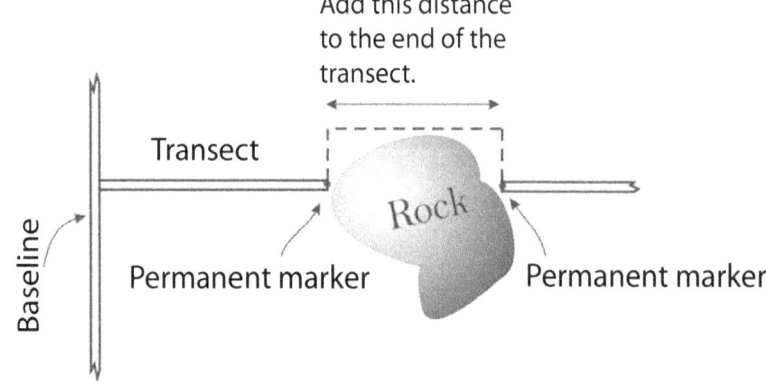

transect length. For example, if you offset 15 ft around a large rock, do not sample on the 15-ft offset; instead, extend the transect out to 81 ft (66 ft +15 ft). Use the Comments field on the Plot Description (PD) form to note the transect modification on the plot.

# HOW TO CONSTRUCT A QUADRAT FRAME

## Cover/Frequency Frames

Quadrats are used to provide estimates of canopy cover and frequency by plant species. Subplots within a quadrat frame are used to improve estimates of canopy cover and provide multiple quadrat sizes to measure frequency. Frequency estimates based on subplots nested within a quadrat are commonly referred to as Nested Rooted Frequency. Various types of materials (wood, metal, or plastic) are suitable for plot frame construction. Five-eighths inch outside diameter PVC pipe works well. Cement two corners in place but leave the remaining two free so that one side can be removed. This will make it easier to slide the quadrat under plant crowns. Large diameter materials should be avoided because they create greater error in estimates of cover and nested frequency than smaller diameter materials. Construction of the standard 20 x 20 inch (50 x 50 cm) quadrat frame for recording canopy cover and nested rooted frequency is described here. Frames having other dimensions can be constructed in a similar fashion.

The 20 x 20 inch (50 x 50 cm) frame used in Cover/Frequency sampling (fig. HT-21) has painted sections with alternating colors (red and white work well). The alternating colors delineate different sized subplots within the quadrat and are used to help estimate cover. Delineation of these sections is made from the inside of the plot frame. A list of subplot sizes used to estimate cover with the standard 20 x 20 inch (50 x 50 cm) quadrat frame described here is displayed in table HT-7. The four subplots used to record nested rooted frequency are illustrated and described in figure HT-22 and table HT-8.

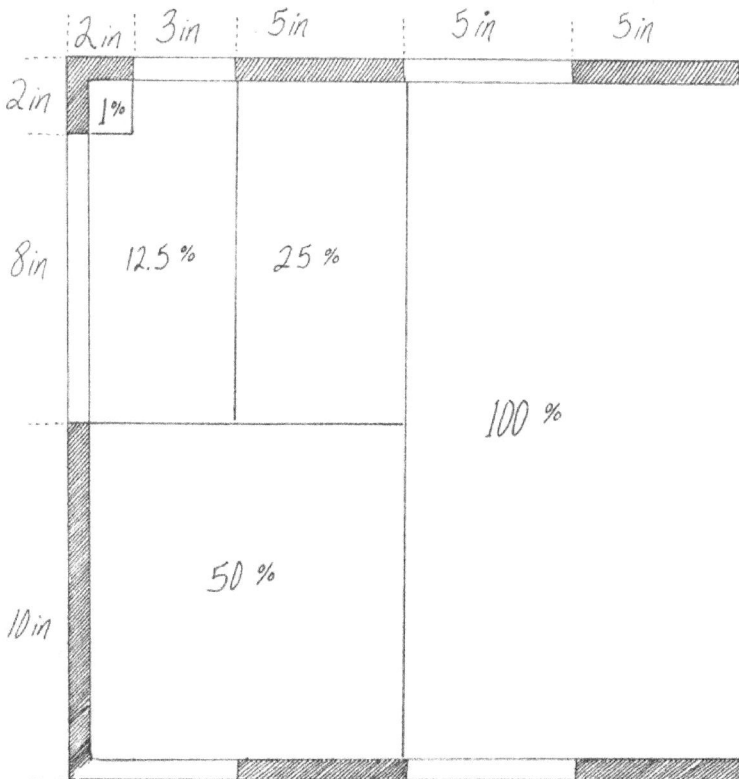

**Figure HT-21**—Dimensions and color coding conventions for a 20 x 20 inch (50 x 50 cm) quadrat frame. Various subplots are shown with their corresponding percent of the total quadrat.

**Table HT-7** Percent of quadrat represented by various subplots within the standard 20 x 20 inch or 50 x 50 cm quadrat.

| Size of subplot | Percent of quadrat |
|---|---|
| *inches (cm)* | |
| 2 x 2 (5 x 5) | 1 |
| 5 x 10 (12.5 x 25) | 12.5 |
| 10 x 10 (25 x 25) | 25 |
| 10 x 20 (25 x 50) | 50 |
| 20 x 20 (50 x 50) | 100 |

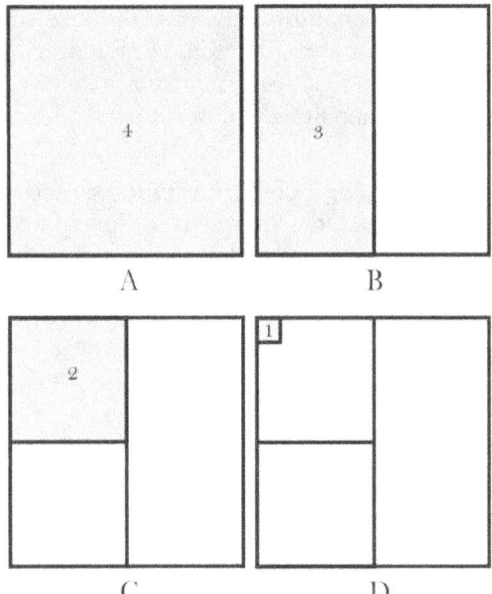

**Figure HT-22**—The numbers inside the plot frame denote the value recorded if a plant is present in that area of the frame. The number 4 corresponds to the entire quadrat (A). The sampling area for number 3 is the entire top half of the quadrat (B). The sampling areas for the numbers 2 and 1 are the upper left quarter and the upper left corner (1 percent) of the quadrat, respectively (C and D). Each larger subplot contains all smaller subplots. Subplots aid the sampler in estimating canopy cover by mentally grouping canopy cover for all individuals of a plant species into one of the subplots.

**Table HT-8** Percent of quadrat represented by the four subplots used to record nested rooted frequency within the standard 20 x 20 inch (50 x 50 cm) quadrat. Each subplot number includes all subplots with smaller numbers. For example, subplot 3 includes subplots 2 and 1 and is 50 percent of the entire quadrat.

| Subplot number for rooted frequency | Size of subplot | Percent quadrat |
|---|---|---|
| | *inches (cm)* | |
| 1 | 2 x 2 (5 x 5) | 1 |
| 2 | 10 x 10 (25 x 25) | 25 |
| 3 | 10 x 20 (25 x 50) | 50 |
| 4 | 20 x 20 (50 x 50) | 100 |

## Density Frames and Belt Transects

Quadrats and belt transects are used to provide estimates of density for plant species. Quadrats are commonly used to estimate density for grasses and forbs, while belt transects are commonly used to estimate shrub and small tree density. A belt transect is essentially a quadrat with a long side, bounded along the transect, and a narrow side.

For density quadrats we suggest using three folding rulers 6 ft (2 m) in length. Folding rulers work well for delineating density quadrats because they allow for varying quadrat sizes and are easily carried in the field. Depending on the quadrat size and shape, density quadrats are constructed using one to three rulers. The sampling area can be designated by placing two folding rulers at right angles to the transect tape at a distance equal to the length of the quadrat. The remaining side of the quadrat can be closed with the third folding ruler. If 3 x 3 ft (1 x 1 m) quadrats are used, then one ruler is folded at right angles and a second ruler used to close the open end of the quadrat. If you anticipate constructing density quadrats longer than 6 ft (2 m) in length, a long piece of rope may be used to close the long end of the quadrat (fig. HT-23).

Belt transects can be constructed by placing two measuring tapes along the length of the transect and at a distance apart equal to the width of the belt (fig. HT-24). The width of the belt is measured at both ends with a folding ruler, 6 ft (1 m) in length. Instead of stretching parallel tapes to delineate the belt transect, observers may walk along the transect with a ruler the width of the belt transect. The ruler is oriented perpendicular to the measuring tape with one end adjacent to the tape. Observers can then count individual plants or other items that are within the belt as they walk the length of the transect.

**Figure HT-23**—Use a folding frame or folding ruler to lay out three sides of the sampling quadrat; slide the ruler under any vegetation and close the quadrat with another yardstick.

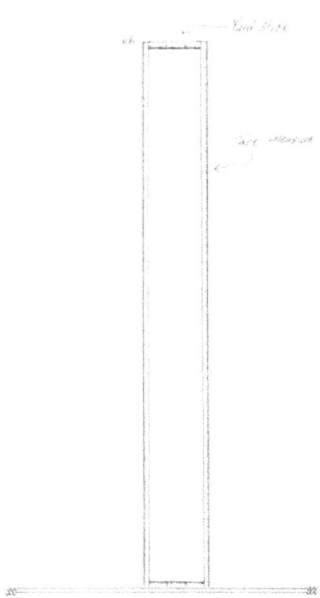

**Figure HT-24**—Density belt transects can be constructed easily using two measuring tapes and two yardsticks.

# HOW TO CONSTRUCT POINT FRAMES AND GRID FRAMES

Point frames are used to provide estimates of plant species cover and ground cover using the Point Intercept (PO) sampling method. Point frames can be constructed out of plastic or metal tubing or wood. The basic concept is to design a free-standing frame with 10 holes so that the pins can be lowered vertically (fig. HT-25). The point frame should stand far enough above the ground to sample vegetation 3 to 4.5 ft (1 to 1.5 m) tall. Pins are commonly placed 3 to 4 inches (7.5 to 10 cm) apart, but spacing is dependent on the size and spacing of the vegetation being sampled.

The PO method may be used in conjunction with the Cover/Frequency (CF) method to sample ground cover by using the CF sampling quadrat as a point frame. A pencil or pen is used to record ground cover "hits" at the four corners and the four midpoints on each side of the quadrat for a total of eight points per quadrat.

Grid frames can be constructed out of metal or plastic tubing or wood. The basic design is to build two square or rectangular frames with wires stretched across the length and width of each frame to form a grid. The grid must have identical spacing for both frames. The frames are placed on top of each other a small distance apart. The entire sighting frame has three to four legs, preferably adjustable type tripod legs (fig. HT-26). The two sets of wire grids allow the observer to always have a vertical line of sight when recording point data. The point frame should stand far enough above the ground to sample the vegetation 3 to 4.5 ft (1 to 1.5 m) tall. Crosshairs are commonly placed 3 to 4 inches (7.5 to 10 cm) apart, but spacing is dependent on the size and spacing of the vegetation being sampled.

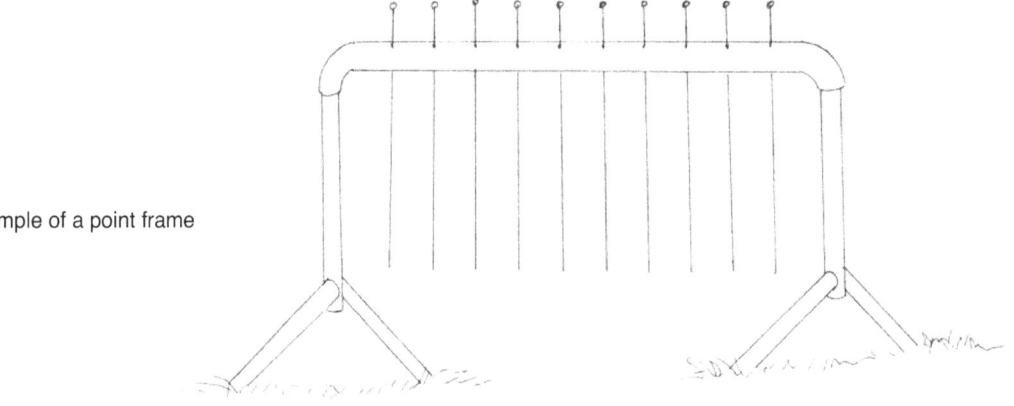

**Figure HT-25**—Example of a point frame with 10 pins.

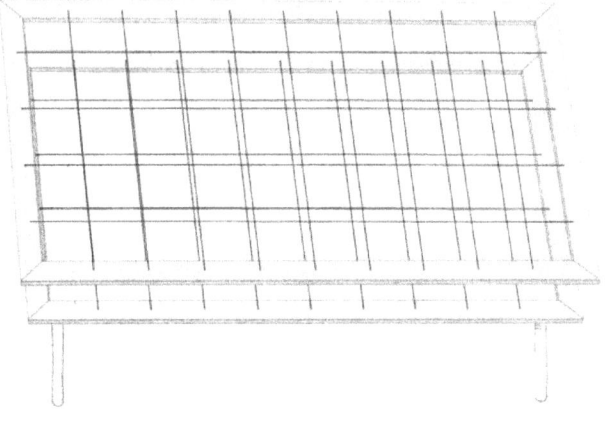

**Figure HT-26**—Example of a grid frame with 36 points (4 x 9).

# HOW TO COUNT BOUNDARY PLANTS

Boundary plants are defined as plants that have a portion of basal vegetation intersecting the sampling boundary. The boundary could be a flagged macroplot such as used in the Tree Density (TD) sampling method but is more likely to be a quadrat like the one used in the Density (DE) method. For different life forms the basal area may be defined differently. For example the basal area of a tree is measured 4.5 ft (1.37 m) above the ground. For shrub species the basal portion is the area beneath the plant where stems grow out of the ground at some predefined density. The basal area of bunchgrasses is where the aerial portions grow out of the ground and is somewhat intuitive (fig. HT-27).

Decisions on which boundary plants to count and which ones to exclude must be consistently applied to each quadrat. For example, it is fairly easy to determine if thin, single-stemmed plants are in or out of the quadrat, but plants with larger basal diameters (such as bunchgrasses) could be partly in and partly out of the quadrat. There are a several ways to count boundary plants, but one easy way to consistently apply boundary decisions is to count all boundary plants "in" on two adjacent sides of a quadrat and "out" on the other two sides. Figure HT-28 is a top view representation of plants around a quadrat. Some of the plants only have their aerial portion overlapping the quadrat so they are not

**Figure HT-27**—Bunchgrasses (A), shrubs (B), saplings (C) and trees (D) all have their basal area defined differently. Before any sampling, the crew leader should determine how the basal area will be defined for the different life forms.

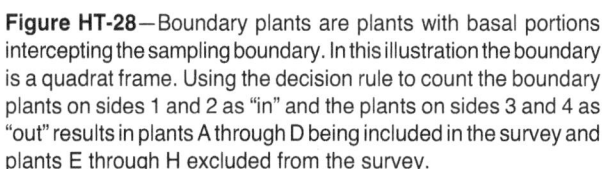

**Figure HT-28**—Boundary plants are plants with basal portions intercepting the sampling boundary. In this illustration the boundary is a quadrat frame. Using the decision rule to count the boundary plants on sides 1 and 2 as "in" and the plants on sides 3 and 4 as "out" results in plants A through D being included in the survey and plants E through H excluded from the survey.

boundary plants. The plants with their basal portion touching the quadrat are boundary plants. These are the plants that will have the decision rules applied to them. For example, if the rules are that boundary plants on sides 1 and 2 are "in" and the plants on sides 3 and 4 are "out," then four of the boundary plants in figure HT-28 would be "in" (plants A through D) and four would be "out" (plants E through H).

## HOW TO DOT TALLY

Instead of counting items in your head or inefficiently tallying items in groups of five using lines, try using a box tally method where you put a dot for each item. Dots and lines are used to record counts as shown in figure HT-29. Each completed box represents 10 items. You will find that this is an efficient and quick method of tallying large numbers of items such as fuel intersects or tree seedlings.

## HOW TO MEASURE PLANT HEIGHT

In FIREMON, tree or plant height measurement is accomplished using a clinometer for taller plants, or a yardstick for shorter plants.

### Plants Greater Than 20 Feet (6 m)

To measure the height of taller vegetation using a clinometer, first, attach a cloth or logger's tape to the tree or plant at breast height and walk away from the tree, on the slope contour, a distance a little more than the tree is tall. For example, if you think a tree is about 30 ft tall, walk out about 40 ft. Once you are in position, read the angle to the top of the tree from the *percent scale* in the clinometer. Next, take the percent reading from your position to bottom of the tree, where it enters the ground. This angle will usually be negative, which is okay (fig. HT-30).

Calculate height using the equation:

Equation HT-1
$$HT = \theta_1 \left( D/100 \right) - \theta_2 \left( D/100 \right)$$

where, HT is the tree height in the same units as D.

$\theta_1$ is the angle from the sampling position to the top of the tree, measured in percent slope.

$\theta_2$ is the angle from the sampling position to the bottom of the tree, measured in percent slope.

D is the horizontal distance from the tree in feet or meters.

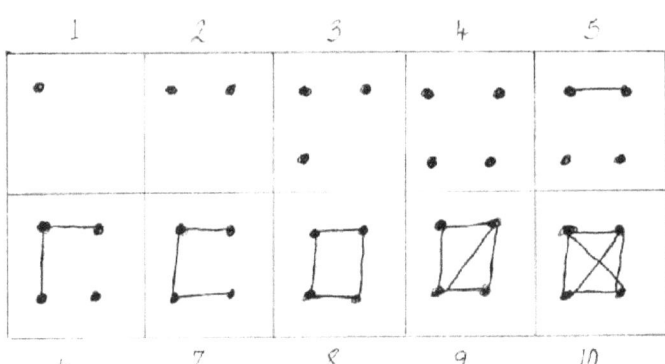

**Figure HT-29**—Use the dot tally system to record counts quickly and accurately.

**Figure HT-30**—Measure the height of tall trees by moving away from the tree a known distance (slope corrected) and measure the percent slope to the top and the bottom of the tree. Use equation HT-1 to calculate height. Units of tree height will be the same as the units of the distance (D) measurement.

Note that when the $\theta_2$ is negative you actually add the two angles. If either $\theta_1$ or $\theta_2$ are greater than 100 percent, go out another 20 ft and recheck the angles. Measuring high angles can cause a lot of error. It may be helpful to make a table of tree height as a function of tree angles (total angle from top to bottom of the tree) and distances, to reduce the time spent making height calculations in the field.

Sometimes it is difficult to see the top of the tree because of obstructions or other tree tops. In these cases, have someone shake the tree at DBH which will cause the top to move and perhaps make it more visible. The sampler also might have to move uphill or downhill one or two paces, making sure the distance from the tree is kept constant, to get a better view of the tree top. If you need to move up or downslope, correct for slope using the correction factors in **How to Adjust for Slope.**

Often, the notetaker on the FIREMON plot will be too busy with other duties to hear or record the tree height measurement. Therefore, we recommend that the sampler measuring tree heights take a field notebook into the field and write down the measured heights along with the tree and plot numbers to make sampling go faster.

Depending on the project objective(s), tree height is recorded to the top of the stem or to the highest live foliage; sometimes dead tops are not considered part of the height of the tree. This subject is discussed in detail in the **Tree Data (TD) Sampling Method** chapter. The highest foliage might not be directly above the tree bole, as in hardwoods, so the sampler must be sure to check all lateral branches to make sure the highest foliage is being measured.

If the tree is leaning at an angle measure tree height using the procedure described above even though the tree would measure taller if it were standing up straight.

## Plants Between 10 and 20 Feet (3 to 6 m) Tall

Depending on the precision requirements of the monitoring project, these plants can either be measured using a clinometer, for more precise estimates, or they may be estimated. The clinometer method is described in the section immediately preceding this one. Faster, less precise measurements can be made by looking at a tree and moving your eyes up in increments. Many people are about 6 ft tall so it is easy to think in that scale. Starting at the ground level simply move your eyes up 6 ft at a time, keeping a mental tally. At first, you should check your estimate using a clinometer, but with a little practice you will be able to estimate heights of these plants to between 1 and 2 ft, consistently.

## Plants Less Than 10 Feet (3 m) Tall

Measure these plant heights with a yardstick (meterstick) for the highest precision. You can also measure the height of the top of your boots, knees, hips, head, and raised hand, then use those measurements to estimate the plant heights.

## Other Height Measuring Tips

Often you will be able to estimate the height of tall trees based on the measured height of another tree that is close by. This can substantially reduce the sampling time because you only need to make four or five height measurements per plot. If the project objectives require high precision, then the height of each tree taller than 20 ft should be measured with a clinometer.

Sometimes the sampling methods call for an average plant height across species or life form. When the plants vary greatly in height, this can be tricky because it is hard to estimate an average. One way is to imagine a piece of plastic draped over the plants you are interested in then estimate what the average height of the plastic sheet would be.

When measuring the height of herbaceous plants, measure only to the point that includes approximately 80 percent of the plant biomass. For example, inflorescence height in graminoids is not typically measured.

# HOW TO MEASURE DBH

The diameter of a tree or shrub is conventionally measured at exactly 4.5 ft (1.37 m) above the ground surface, measured on the uphill side of the tree if it is on a slope. Wrap a diameter tape around the bole or stem of the plant, without twists or bends, and without dead or live branches caught between the tape and the stem.

When making the diameter measurement, the diameter tape should always be positioned so that it is perpendicular to the tree stem at the point of measurement. If the tree splits above breast height, record as one tree with the diameter measurement made at a representative area below the swell caused by the separation. If the tree splits below breast height then record two trees with diameter measured as close as possible to breast height while still getting a representative measure. If there is a stem deformity at breast height, measure the diameter at the closest location above or below that will allow the most representative diameter measurement (fig. HT-31).

There may be times when it is necessary to remove problem branches to thread the DBH tape around the tree bole. If so, carefully remove just enough of the unwanted branches so you do not threaten the survival of the tree. Sometimes it is impossible to measure diameter at breast height with a tape because of protruding branches and other trees. In these rare cases, use a plastic ruler to obtain a diameter (see **How To Measure Diameter with a Ruler**).

Some tree species (juniper, for instance) have multiple stems at breast height. If the project objectives for your project indicate, you should measure the stem diameter of these individuals at ground level. This is called the diameter at root crown or DRC. Use the DRC measurement to classify the tree as a mature tree or sapling and mark the Local Code field so you know that DRC was measured. Some studies may indicate it is also important to count the number of stems at breast height. The down side to DRC measurements is that they may not give an accurate representation of mortality (counting the number of stems at breast height can overcome this), and basal diameters may not adequately portray canopy fuels for fire modeling.

How To ... Guide

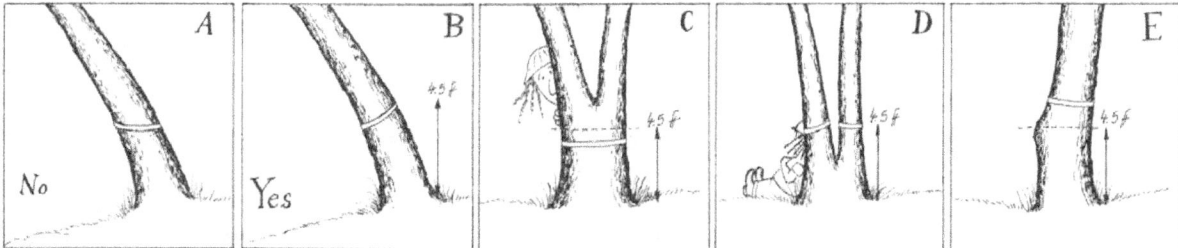

**Figure HT-31**—DBH measurements. A) Diameter tape is not perpendicular to the tree stem. B) Correct way to measure tree diameter is with tape perpendicular to tree stem. C) If the tree splits above breast height measure tree diameter below any swell cause by the separation. D) If the tree splits below breast height measure as two trees. E) Measure the most representative diameter above or below any deformity.

## HOW TO MEASURE DIAMETER WITH A RULER

Using a ruler to measure tree or log diameter can be easier and quicker than with a diameter tape. However, ruler measurements can give biased estimates of diameter if done incorrectly. To measure diameter correctly first, hold one end of the ruler so that it is aligned with the one edge of the tree (at breast height) or log, then, while keeping the ruler in the same position, move your head to the other end of the ruler so that you are looking at the other side of the tree or log at a perpendicular angle to the ruler (fig. HT-32). Estimate the diameter to the appropriate precision. With practice you will only need to use the ruler to measure trees that are on the boundary between two classes.

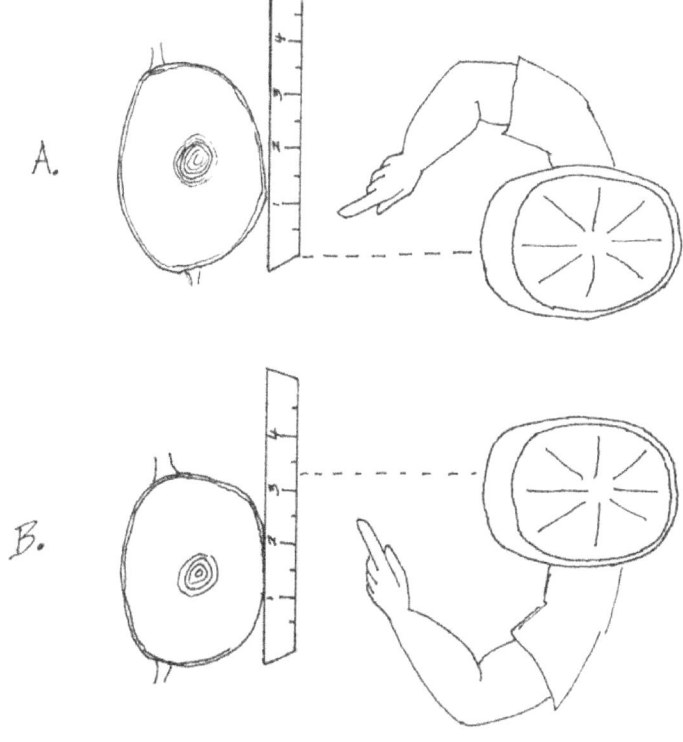

**Figure HT-32**—To accurately measure tree diameter using a ruler, A) align the left side of the ruler with the left side of the tree, then B) without moving the ruler, move your head so that it is aligned with the right side of the tree. The same method can be used for logs.

# HOW TO AGE A TREE

Tree age is estimated by extracting a core from the tree at stump height (about 1 ft above ground line) on the downhill side of the tree using an increment borer. Stump height is used instead of the conventional height at DBH because it is difficult to estimate the time it takes the tree to grow to DBH, especially in severe environments such as the upper subalpine and woodland ecosystems. In fact, the lower down on the tree, the better the estimate of total age. The increment borer is positioned on the downhill side of the tree at an angle that will ensure that the pith of the tree will be struck as the increment core is drilled into the tree. The pith is required to absolutely estimate the age of the tree at stump height. Tree coring requires a certain degree of experience to consistently obtain the pith. To extract a core, screw the increment corer into the tree just deep enough to hit the pith, and then insert the spoon of the corer on the top part of the core (spoon is facing downward). Next, unscrew the borer one-half turn so that the spoon is facing upward. Tug gently on the spoon, and if the planets are in alignment, the core will be extracted in one piece. Be especially careful to collect all of the pieces of broken cores so that the rings can be accurately counted.

Age is not simply the count of the tree rings from cambium to pith. You will have to add an estimate of the number of years it took the tree to grow to stump height to get total tree age. Sometimes this can be done by counting branch whorls on the bole from the ground to stump height, but often branch whorls are difficult to identify and count. Usually you will have make a best-guess estimate on the years it took the tree to grow to the height where the core was taken. Sometimes, the pith is rotten or difficult to extract, so the sampler must also estimate the number of rings it would take to get to the pith. In that case, three counts must be added to obtain true tree age: (total ring count) + (years to stump height) + (years to pith) = Tree Age.

Once the core is extracted, the sampler can either count the age of the core while it is still in the spoon, or store the core for ring counting later in a laboratory or office. If trees are young (<100 years old) and growth rates are not important, then the sampler can probably count the tree rings in the field and record the age estimate in the appropriate field on the forms. A small magnifying glass is often helpful for counting rings. Rings on some cores can be difficult to read because they are tightly packed or difficult to distinguish. It is advisable to take these cores back to the office where the rings can be counted with the proper equipment. Cores that are to be taken back to the office can be stored in drinking straws to reduce breakage, but the straws should be slit so the core can dry and not get moldy. Or cores can be mounted on planks. Wood planks or sections of plywood can be grooved with a router using a diameter that corresponds to the increment corer. Then, in the field, the core can be glued into the groove with wood glue (fig. HT-33). The advantage of using the wood mount is that once the glue has dried, the cores

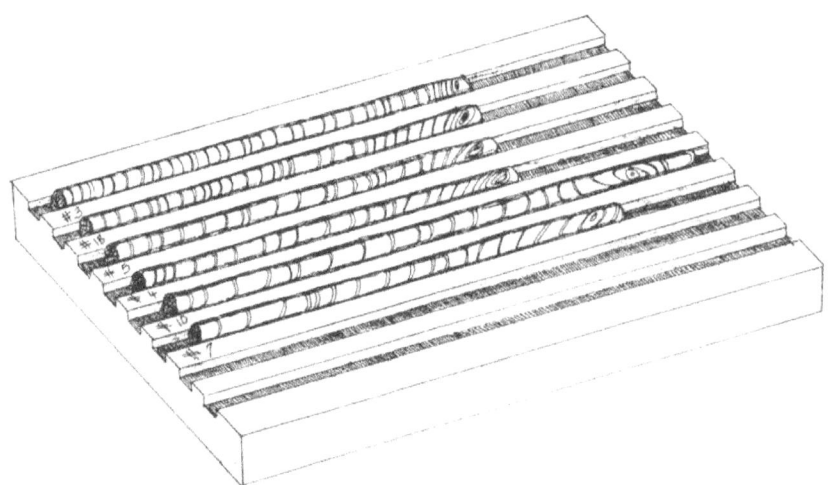

**Figure HT-33**—A core board is used to permanently store tree cores.

How To ... Guide

can be sanded. This will heighten the contrast between rings and make the age and growth rate determination easier and more accurate. The disadvantage is that the glue and wood mounts are difficult to transport in the field. It's best to record plot ID, tree tag number, date, years to stump height, and years to pith directly on the mount or straw.

## HOW TO MEASURE SLOPE

Slope measurements are made so that a correction factor can be applied to slope distance to get the horizontal distance. To find the percent slope, aim the clinometer at the eye level of sampler at the other end of the line (fig. HT-34). Be sure to read the slope off the percent scale in the clinometer. For slope correction use the absolute value of percent slope. If there is a height difference of the samplers, adjust the height where you are aiming so that the slope reading is accurate.

## HOW TO ADJUST FOR SLOPE

Some distance measurements in FIREMON must be corrected to horizontal distance. Examples are macroplot radius and the distance you measure out from a tree when measuring tree height. Table HT-9 shows the correction factor for slope by 10-percent class. Note that the correction factor on slopes less

**Figure HT-34**—Measure the slope of each line by aiming the clinometer at eye level on the sampler at the opposite end of the measuring tape, then reading and recording the percent slope seen on the scale in the instrument.

Table HT-9 Correct for slope by multiplying the horizontal distance you need to travel by the appropriate correction factor listed.

| Slope | | Correction | Slope | | Correction |
|---|---|---|---|---|---|
| Percent | Degrees | factor | Percent | Degrees | factor |
| 10 | 5.71 | 1.00 | 100 | 45.00 | 1.41 |
| 20 | 11.31 | 1.02 | 110 | 47.73 | 1.49 |
| 30 | 16.70 | 1.04 | 120 | 50.19 | 1.56 |
| 40 | 21.80 | 1.08 | 130 | 52.43 | 1.64 |
| 50 | 26.57 | 1.12 | 140 | 54.46 | 1.72 |
| 60 | 30.96 | 1.17 | 150 | 56.31 | 1.80 |
| 70 | 34.99 | 1.22 | 160 | 57.99 | 1.89 |
| 80 | 38.66 | 1.28 | 170 | 59.53 | 1.97 |
| 90 | 41.99 | 1.35 | 180 | 60.95 | 2.06 |

than 30 percent is negligible. To find corrected distance, multiply the horizontal distance you need by the appropriate correction factor. For example, when marking the boundary of a macroplot with a radius of 37.2 ft on a site with 50-percent slope, you would have to measure the upslope and downslope radius out to a distance of 37.2 x 1.12 = 41.67 ft.

## HOW TO DOCUMENT PLOT LOCATION AND FIRE EFFECTS WITH PHOTOS

Photographs, conventional or digital, are a useful means to document the FIREMON plot. They provide a unique opportunity to visually assess fire effects and document plot location. Previously established FIREMON plots can be found by orienting the landmarks in photos to visual cues in the field. Photos can be compared to determine important changes after a fire. Photos provide excellent communication tools for describing fire effects to the public and forest professionals.

Document the FIREMON macroplot location using two photographs taken facing north and east. For the north-facing photo move about 10 ft south of the FIREMON macroplot center then take the photo facing north, being sure that the plot center stake or rebar will be visible in the picture (fig. HT-35). Then move west of the plot center about 10 ft and take a photo facing east, again being sure that the plot center stake or rebar will be visible in the picture. For these pictures be sure that the camera is focused on the environment surrounding the plot, not the distance or foreground, and that the camera is set for the correct exposure and aperture for existing light conditions. A flash might be needed in low-light conditions.

Photos taken with conventional film can be identified by assigning a code that integrates the roll number and/or name and the picture number. For example, picture 8 taken on roll John Smith Roll 1 might be called JSR01P08. Use a consistent system so the plot photos do not get mixed up.

You must label the roll so that you will be able to find the correct photos after the film has been developed. One way is to take a picture of a card with the roll information on it, as your first photo. Or, you can write the roll information on the film canister before you load it into the camera. The first method is the most foolproof. For digital cameras, write the plot identification on a whiteboard and place the board in the plot scene or enter the file name of the digital picture on the PD form. Film photos will need to be scanned once they are developed and stored on your computer in digital format.

**Figure HT-35**—Take your plot photos so that they show the plot center and the general plot conditions.

# HOW TO RECORD A GPS LOCATION

Basically, there are a number of options today for GPS receivers and ways to acquire locational data. Here are recommend protocols on only a few issues.

## Acceptable Accuracy

The two-dimensional accuracy of X, Y coordinates reported by the GPS should be less that 10 meters, preferably less than 7 meters. There is no procedural requirement on elevation, or the Z coordinate.

## Geodetic Datum

The more accurate and more recent NAD83 should be selected, unless there is strong need to use the data predominantly within a local GIS, *and* the local standard is for some other datum. For example, many National Park Service sites still use NAD27 in order to reference data taken from their older base maps. These were some of the first areas mapped with USGS 1:24,000 quadrangles. In any event, it is important to note the datum used for plot location so conversions can be made if necessary.

## When Digitizing a Single Point

Y-Code receivers such as Rockwell's PLGR should be set to an averaging mode and allowed to log a number of points until the coordinates stabilize, at which time the plot center is either jotted down or saved in memory. These receivers are only available to approved Federal government employees.

P-Code receivers such as Trimble's Pathfinder or GeoExplorer should be treated similarly *only* if "selective availability" is off. If "selective availability" is on, a differential correction should be used. Set the receiver to log and save a hundred or so points for each plot center. There then are options to do differential correction "on the fly," or later by receiving suitable reference control data from a surveyed base station. You will have to check with GPS-knowledgeable people in your area to find how to access local base station data.

# FIREMON

## Appendices

## Appendix A: NRIS Damage Categories, Agents, Severity Ratings, and Tree Parts

| Damage categories code | Description |
|---|---|
| 00 | None (default) |
| 10 | General insects |
| 11 | Bark beetles |
| 12 | Defoliators |
| 13 | Chewing insects |
| 14 | Sucking insects |
| 15 | Boring insects |
| 16 | Seed/cone/flower/fruit insects |
| 17 | Gallmaker insects |
| 18 | Insect predators |
| 19 | General diseases |
| 20 | Biotic damage |
| 21 | Root/butt diseases |
| 22 | Stem decays/cankers |
| 23 | Parasitic/epiphytic plants |
| 24 | Decline complexes/dieback/wilts |
| 25 | Foliage diseases |
| 26 | Stem rusts |
| 27 | Broom rusts |
| 30 | Fire |
| 41 | Wild animals |
| 42 | Domestic animals |
| 50 | Abiotic damage |
| 60 | Competition |
| 70 | Human activities |
| 71 | Harvest |
| 80 | Multidamage (insect/disease) |
| 90 | Unknown |
| 99 | Physical effects |

**Damage agents**: General insects

| Category | Agent | Common name |
|---|---|---|
| 10 | 000 | General insects |
|  | 001 | Thrips |
|  | 007 | Clerid beetle |
|  | 009 | Green rose chafer |
|  | 017 | Bagworm moth |
|  | 019 | Scarab |
|  | 021 | *Steremnius carinatus* |
|  | 023 | Wood wasps |

Severity rating: 101 = minor; 102 = severe.

# Appendix A: (Continued)

**Damage agents**: Bark beetles.

| Category | Agent | Common name |
| --- | --- | --- |
| 11 | 000 | Bark Beetles |
|  | 001 | Roundheaded pine beetle |
|  | 002 | Western pine beetle |
|  | 005 | Lodgepole pine beetle |
|  | 006 | Mountain pine beetle |
|  | 007 | Douglas fir beetle |
|  | 009 | Spruce beetle |
|  | 012 | Red turpentine beetle |
|  | 013 | *Dryocoetes affaber* |
|  | 015 | Western balsam bark beetle |
|  | 016 | *Dryocoetes sechelti* |
|  | 017 | Ash bark beetles |
|  | 018 | Native elm bark beetle |
|  | 021 | Sixspined ips |
|  | 022 | Emarginate ips |
|  | 024 | *Ips latidens* |
|  | 026 | Monterey pine ips |
|  | 028 | Northern spruce engraver beetle |
|  | 029 | Pine engraver |
|  | 030 | Ips engraver beetles |
|  | 031 | *Ips tridens* |
|  | 032 | Western ash bark beetle |
|  | 034 | *Orthotomicus caelatus* |
|  | 035 | Cedar bark beetles |
|  | 036 | Western cedar bark beetle |
|  | 037 | Tip beetles |
|  | 038 | Douglas fir twig beetle |
|  | 039 | Twig beetles |
|  | 040 | Foureyed spruce beetle |
|  | 041 | Fir root bark beetle |
|  | 042 | *Pseudohylesinus dispar* |
|  | 043 | Douglas fir pole beetle |
|  | 044 | Silver fir beetle |
|  | 045 | Small European elm bark beetle |
|  | 046 | Spruce engraver |
|  | 048 | True fir bark beetles |
|  | 049 | Douglas fir engraver |
|  | 050 | Fir engraver |
|  | 053 | Four eyed bark beetle |
|  | 054 | Hemlock beetle |

Severity rating: 111 = unsuccessful bole attack: pitchout and beetle brood absent; 112 = strip attacks: galleries and brood present; 113 = successful bole attack: galleries and brood present; 114 = topkill.

## Appendix A: (Continued)

**Damage agents**: Defoliators.

| Category | Agent | Common name | Agent | Common name |
|---|---|---|---|---|
| 12 | 000 | Defoliators | 088 | Aspen blotchminer |
| | 001 | Casebearer | 089 | Gypsy moth |
| | 003 | Looper | 090 | Cottonwood leafminers |
| | 005 | Sawfly | 094 | Western tent caterpillar |
| | 007 | Larger elm leaf beetle | 096 | Forest tent caterpillar |
| | 008 | Spanworm | 098 | Leafcutting bees |
| | 011 | Western blackheaded budworm | 099 | Blister beetle |
| | 013 | Whitefly | 102 | Willow sawfly |
| | 014 | Fall cankerworm | 104 | Lodgepole sawfly |
| | 015 | Alder flea beetle | 106 | Pine infesting sawflies |
| | 016 | Mountain mahogany looper | 109 | Ponderosa pine sawfly |
| | 018 | Oak worms | 115 | Hemlock sawfly |
| | 020 | Western larch sawfly | 116 | Pine butterfly |
| | 021 | Fruit tree leafroller | 117 | False hemlock looper |
| | 022 | Uglynest caterpillar | 118 | California tortoiseshell |
| | 023 | Boxelder defoliator | 120 | Bruce spanworm |
| | 030 | Pear sawfly | 121 | Rusty tussock moth |
| | 033 | Boxelder leafroller | 122 | Whitemarked tussock moth |
| | 035 | Spruce webspinning sawfly | 123 | Douglas fir tussock moth |
| | 036 | Two year budworm | 124 | Western tussock moth |
| | 037 | Large aspen tortrix | 125 | Spring cankerworm |
| | 039 | Sugar pine tortrix | 135 | Aspen leafminer |
| | 040 | Western spruce budworm | 136 | Yellowheaded spruce sawfly |
| | 043 | Aspen leaf beetle | 137 | Tenlined June beetle |
| | 044 | Cottonwood leaf beetle | 138 | Japanese beetle |
| | 045 | Leafhopper | 139 | Larch sawfly |
| | 046 | Poplar tentmaker | 140 | Mountain ash sawfly |
| | 047 | Larch casebearer | 141 | Elm leaf beetle |
| | 049 | Lodgepole needleminer | 142 | Spearmarked black moth |
| | 050 | Ponderosa needleminer | 143 | Giant silkworm moth |
| | 051 | Black Hills Pandora moth | 144 | Redhumped caterpillar |
| | 052 | Pandora moth | 146 | Larch looper |
| | 053 | Sycamore lace bug | 150 | Spruce needleminer (west) |
| | 054 | Lace bugs | 154 | *Thryridopteryx ephemeraeformis* |
| | 055 | Oak leaftier | 155 | Leafroller/seed moth |
| | 058 | Yellownecked caterpillar | 156 | Willow defoliation |
| | 059 | Walkingstick | 157 | Euonymus caterpillar |
| | 060 | Spruce coneworm | 159 | Larch bud moth |
| | 061 | Introduced pine sawfly | 160 | Pine needle sheathminer |
| | 066 | White fir needleminer | 162 | Cottonwood leaf beetle |
| | 071 | Elm leafminer | 164 | Saddle backed looper |
| | 072 | Geometrid moth | 165 | Leaf roller |
| | 073 | Leafblotch miner | 168 | Green striped looper |
| | 074 | Spotted tussock moth | 174 | Pine looper |
| | 077 | Brown day moth | 176 | *Zadiprion townsendi* |
| | 082 | Fall webworm | 177 | Douglas fir budmoth |
| | 083 | Hemlock looper | 179 | Phantom hemlock looper |
| | 085 | Tent caterpillar moth | 180 | Tent caterpillar |
| | 086 | Satin moth | 188 | Elm sawfly |
| | 087 | Willow leafblotch miner | 189 | June beetles/leaf chafers |

Severity rating: 121 = Light defoliation (1 25%), no topkill.
122 = Light defoliation (1 25%), topkill ≤10%.
123 = Light defoliation (1 25%), topkill >10%.
124 = Moderate defoliation (26 75%), no topkill.
125 = Moderate defoliation (26 75%), topkill ≤10%.
126 = Moderate defoliation (26 75%), topkill >10%.
127 = Heavy defoliation (76 100%), no topkill.
128 = Heavy defoliation (76 100%), topkill ≤10%.
129 = Heavy defoliation (76 100%), topkill >10%.

# Appendix A: (Continued)

**Damage agents**: Chewing insects.

| Category | Agent | Common name |
|---|---|---|
| 13 | 000 | Chewing Insects |
|  | 001 | Grasshopper |
|  | 002 | Shorthorn grasshoppers |
|  | 005 | Clearwinged grasshopper |
|  | 006 | Cicadas |
|  | 007 | Eurytomids |
|  | 008 | Cutworms |
|  | 010 | Pales weevil |
|  | 012 | Periodical cicada |
|  | 013 | Migratory grasshopper |
|  | 014 | Valley grasshopper |
|  | 015 | Strawberry root weevil |
|  | 020 | Northern pitch twig moth |
|  | 021 | Ponderosa pine tip moth |
|  | 022 | Pine needle weevil |
|  | 025 | *Thrips madronii* |
|  | 026 | Ash plant bug |
|  | 028 | Pitch eating weevil |

Severity rating: 131 = minor; 132 = severe.

**Damage agents**: Sucking insects.

| Category | Agent | Common name | Agent | Common name |
|---|---|---|---|---|
| 14 | 000 | Sucking Insects | 035 | Treehoopers |
|  | 001 | Scale insect | 039 | Black pineleaf scale |
|  | 002 | Western larch woolly aphid | 040 | Spruce spider mite |
|  | 003 | Balsam woolly adelgid | 043 | Maple aphids |
|  | 004 | Hemlock woolly adelgid | 044 | Spruce bud scale |
|  | 006 | Aphid | 046 | Pine leaf adelgid |
|  | 008 | Western pine spittlebug | 047 | White pine adelgid |
|  | 010 | Spittlebug | 048 | Pine bark adelgid |
|  | 012 | Pine needle scale | 049 | Root aphid |
|  | 014 | Giant conifer aphids | 050 | Mealybug |
|  | 017 | Spruce aphid | 051 | Cottony maple scale |
|  | 018 | Wooly apple aphid | 052 | Fir mealybug |
|  | 022 | Pine thrips | 053 | Douglas fir mealybug |
|  | 026 | Lecanium scale | 061 | Pine tortoise scale |
|  | 028 | Oystershell scale | 063 | Birch aphid |
|  | 029 | Pinyon needle scale | 068 | European elm scale |
|  | 030 | Ponderosa pine twig scale |  |  |

Severity rating: 141 = minor; 142 = severe.

**Damage agents**: Boring insects.

| Category | Agent | Common name | Agent | Common name |
|---|---|---|---|---|
| 15 | 000 | Boring insects | 041 | Flatheaded fir borer |
|  | 001 | Shoot borer | 042 | Whitespotted sawyer |
|  | 002 | Termite | 043 | Redheaded ash borer |
|  | 003 | Ponderosa pine bark borer | 045 | Oberea shoot borers |
|  | 004 | Bronze birch borer | 048 | *Pissodes dubius* |
|  | 006 | Bronze poplar borer | 050 | White pine weevil |
|  | 007 | Carpenter bees | 051 | Lodgepole terminal weevil |
|  | 008 | Flatheaded borer | 052 | Ambrosia beetles |
|  | 009 | Golden buprestid | 053 | Cottonwood borer |
|  | 010 | Carpenter ants | 056 | Ash borer |
|  | 011 | Gouty pitch midge | 057 | Lilac borer |
|  | 012 | Shootboring sawflies | 058 | *Prionoxystus robiniae* |
|  | 013 | Roundheaded borer | 059 | Maple shoot borers |
|  | 014 | Flatheaded apple tree borer | 060 | Western subterranean termite |
|  | 017 | Pitted ambrosia beetle | 063 | European pine shoot moth |
|  | 018 | Carpenterworm moths | 064 | Western pine tip moth |
|  | 019 | Poplar and willow borer | 065 | Nantucket pine tip moth |
|  | 020 | Pine reproduction weevil | 066 | Lodgepole pine tip moth |
|  | 021 | Douglas fir twig weevil | 067 | Southwestern pine tip moth |
|  | 027 | Ponderous borer | 070 | Saperda shoot borer |
|  | 029 | Western pine shoot borer | 071 | Clearwing moths |
|  | 030 | Eucosma species | 073 | Roundheaded fir borer |
|  | 034 | Warren's collar weevil | 074 | Western larch borer |
|  | 035 | Powderpost beetle | 075 | Western cedar borer |
|  | 036 | Tarnished plant bug | 076 | Douglas fir pitch moth |
|  | 037 | *Magdalis* spp. | 077 | Sequoia pitch moth |
|  | 038 | White pine bark miner | 083 | Ottonwood twig borer |
|  | 039 | Locust borer | 085 | Banded ash borer |
|  | 040 | California flathead borer |  |  |

Severity rating: 151 = minor; 152 = severe.

## Appendix A: (Continued)

**Damage agents**: Seed/cone/flower/fruit insects.

| Category | Agent | Common name |
|---|---|---|
| 16 | 000 | Seed/cone/flower/fruit insects |
|  | 001 | Douglas fir cone moth |
|  | 002 | Lodgepole cone beetle |
|  | 003 | Limber pine cone beetle |
|  | 004 | Mountain pine cone beetle |
|  | 005 | Ponderosa pine cone beetle |
|  | 010 | Douglas fir cone midge |
|  | 011 | Cone scale midge |
|  | 012 | Pecan |
|  | 015 | Fir coneworm |
|  | 017 | Pine coneworm |
|  | 019 | Ponderosa twig moth |
|  | 020 | *Dioryctria pseudotsugella* |
|  | 021 | Dioryctria moths |
|  | 022 | Lodgepole cone moth |
|  | 023 | Seed chalcid |
|  | 025 | Cone maggot |
|  | 027 | Ponderosa pine seed worm/moth |
|  | 028 | Spruce seed moth |
|  | 029 | Boxelder bug |
|  | 031 | Western conifer seed bug |
|  | 033 | *Magastigmus lasiocarpae* |
|  | 034 | Spruce seed chalcid |
|  | 035 | Ponderosa pine seed chalcid |
|  | 036 | Fir seed chalcid |
|  | 037 | Douglas fir seed chalcid |
|  | 040 | Roundheaded cone borer |
|  | 042 | Coneworm |
|  | 043 | Harvester ants |
|  | 048 | Coneworm |
|  | 049 | Prairie tent caterpillar |

Severity rating: 161 = minor; 162 = severe.

**Damage agents**: Gallmaker insects.

| Category | Agent | Common name |
|---|---|---|
| 17 | 000 | Gallmaker insects |
|  | 003 | Cooley spruce gall adelgid |
|  | 006 | Gall midge |
|  | 007 | Douglas fir needle gall midge |
|  | 008 | Gall mite |
|  | 009 | Spruce gall midge |
|  | 013 | Gall aphid |
|  | 014 | Alder gall mite |
|  | 015 | Psyllid |
|  | 018 | Gouty pitch midge |
|  | 019 | Spider mites |

Severity rating: 171 = minor; 172 = severe.

**Damage agents**: Insect predators.

| Category | Agent | Common name |
|---|---|---|
| 18 | 000 | Insect predators |
|  | 001 | Lacewing |
|  | 002 | Blackbellied clerid |
|  | 003 | Redbellied clerid |
|  | 005 | Western yellowjacket |

Severity rating: 181 = minor; 182 = severe.

**Damage agents**: General diseases.

| Category | Agent | Common name |
|---|---|---|
| 19 | 000 | General diseases |

Severity rating: 191 = minor; 192 = severe.

**Damage agents**: Biotic damage.

| Category | Agent | Common name |
|---|---|---|
| 20 | 000 | Biotic damage |
|  | 001 | Damping off |
|  | 002 | Gray mold |

Severity rating: 201 = minor; 202 = severe.

**Damage agents**: Root/butt diseases.

| Category | Agent | Common name |
|---|---|---|
| 21 | 000 | Root/butt diseases |
|  | 001 | Armillaria root disease |
|  | 003 | Cylindrocladium root disease |
|  | 004 | Brown crumbly rot |
|  | 006 | Fusarium root rot |
|  | 007 | White mottled rot |
|  | 009 | Ganoderma rot of conifers |
|  | 010 | Annosus root disease |
|  | 012 | Tomentosus root disease |
|  | 014 | Black stain root disease |
|  | 015 | Schweinitzii butt rot |
|  | 017 | Laminated root rot |
|  | 022 | Pythium root rot |
|  | 026 | Yellow pitted rot |

Severity rating for trees:
211 = tree within 30 ft of tree with deteriorating crown, tree with diagnostic symptoms or signs, or tree killed by root disease.
212 = pathogen (sign) or diagnostic symptom detected no crown deterioration.
213 = crown deterioration detected no diagnostic symptoms or signs.
214 = both crown deterioration and diagnostic signs symptoms detected.

# Appendix A: (Continued)

**Damage agents**: Stem decays.

| Category | Agent | Common name |
|---|---|---|
| 22 | 000 | Stem decays |
|  | 001 | Heart rot |
|  | 002 | Stem rot |
|  | 003 | Sap rot |
|  | 006 | Black knot of cherry |
|  | 007 | Atropellis canker |
|  | 012 | Black canker of aspen |
|  | 024 | Gray brown saprot |
|  | 025 | Cryoptosphaeria canker of aspen |
|  | 026 | Cytospora canker of fir |
|  | 027 | Western red rot |
|  | 028 | Rust red stringy rot |
|  | 029 | Sooty bark canker |
|  | 035 | Amelanchier rust |
|  | 036 | Cedar apple rust |
|  | 038 | Hypoxylon canker of aspen |
|  | 040 | Sterile conk trunk rot of birch |
|  | 047 | Red ring rot |
|  | 048 | Aspen trunk rot |
|  | 051 | Phomopsis canker |
|  | 057 | Cytospora canker of aspen |
|  | 059 | Red belt fungus |
|  | 062 | Brown heartrot |
|  | 063 | *Coniophora puteana* |
|  | 064 | Tinder fungus |
|  | 065 | Purple conk |
|  | 066 | *Leptographium wagnerii* |
|  | 067 | *Phellimus hartigii* |
|  | 068 | False tinder fungus |
|  | 070 | Yellow cap fungus |
|  | 071 | Oyster mushroom |
|  | 074 | Cedar brown pocket rot |
|  | 075 | Lanchnellula canker |
|  | 077 | Phomopsis blight |

Severity rating: 220 = 0 4% rotten.   225 = 46 55% rotten.
221 = 5 15% rotten.   226 = 56 65% rotten.
222 = 16 25% rotten.   227 = 66 75% rotten.
223 = 26 35% rotten.   228 = 76 85% rotten.
224 = 36 45% rotten.   229 = 86 100% rotten.

**Damage agents**: Parasitic.

| Category | Agent | Common name |
|---|---|---|
| 23 | 000 | Parasitic |
|  | 001 | Mistletoe |
|  | 003 | Vine damage |
|  | 006 | Lodgepole pine dwarf mistletoe |
|  | 008 | Western dwarf mistletoe |
|  | 009 | Limber pine dwarf mistletoe |
|  | 011 | Douglas fir dwarf mistletoe |
|  | 013 | Larch dwarf mistletoe |

Severity rating: 231 = Hawksworth tree DMR rating 1.
232 = Hawksworth tree DMR rating 2.
233 = Hawksworth tree DMR rating 3.
234 = Hawksworth tree DMR rating 4.
235 = Hawksworth tree DMR rating 5.
236 = Hawksworth tree DMR rating 6.

**Damage agents**: Decline complexes/dieback/wilts.

| Category | Agent | Common name |
|---|---|---|
| 24 | 000 | Decline complexes/dieback/wilts |
|  | 004 | Ash decline/yellow |
|  | 022 | Dutch elm disease |

Severity rating: 241 = minor; 242 = severe.

**Damage agents**: Foilage diseases.

| Category | Agent | Common name |
|---|---|---|
| 25 | 000 | Foliage diseases |
|  | 001 | Blight |
|  | 002 | Broom rust |
|  | 003 | Juniper blights |
|  | 004 | Leaf spots |
|  | 005 | Needlecast |
|  | 006 | Powdery mildew |
|  | 009 | True fir needlecast |
|  | 013 | Large pored spruce laborador tea rust |
|  | 014 | Ink spot of aspen |
|  | 015 | Pine needle rust |
|  | 019 | Cedar leaf blight |
|  | 020 | Dogwood anthracnose |
|  | 022 | Elytroderma disease |
|  | 023 | Fire blight |
|  | 027 | Brown felt blight |
|  | 028 | Larch needle blight |
|  | 031 | Spruce needle cast |
|  | 032 | Fir needle cast |
|  | 033 | White pine needle cast |
|  | 034 | Lophodermella needle cast |
|  | 035 | Lophodermium needle cast |
|  | 036 | Marssonina blight |
|  | 037 | Melampsora rusts |
|  | 039 | Larch needle cast |
|  | 040 | Dothistroma needle blight |
|  | 041 | Brown felt blight of pines |
|  | 042 | Snow blight |
|  | 043 | Swiss needle cast |
|  | 049 | Fir needle rust |
|  | 050 | Douglas fir needle cast |
|  | 052 | Rhizophaeria needle cast |
|  | 054 | Brown spot needle blight |
|  | 056 | Septoria leaf spot and canker |
|  | 058 | Diplodia blight |
|  | 061 | Shepherd's crook |
|  | 062 | Dothistroma needle blight |
|  | 064 | Broom rust |
|  | 065 | Spruce needle rust |
|  | 067 | Spruce needle cast |
|  | 068 | Hardwood leaf rusts |
|  | 072 | Sirococcus shoot blight |
|  | 073 | Shepherds crook |
|  | 074 | Delphinella shoot blight |

Severity rating: 251 = minor; 252 = severe.

# Appendix A: (Continued)

**Damage agents**: Stem rusts.

| Category | Agent | Common name |
|---|---|---|
| 26 | 000 | Stem rusts |
|  | 001 | White pine blister rust |
|  | 002 | Western gall rust |
|  | 003 | Stalactiform blister rust |
|  | 004 | Comandra blister rust |
|  | 011 | Bethuli rust |

Severity rating: 261 = branch infections located greater than 2 ft from tree bole.
262 = branch infections located between 6 inches and 2 ft from tree bole.
263 = bole infections or branch infections located within 6 inches of bole.
264 = topkill.

**Damage agents**: Broom rusts.

| Category | Agent | Common name |
|---|---|---|
| 27 | 000 | Broom rusts |
|  | 001 | Spruce broom rust |
|  | 003 | Juniper broom rust |
|  | 004 | Fir broom rust |

Severity rating: 271 = minor; 272 = severe.

**Damage agents**: Fire.

| Category | Agent | Common name |
|---|---|---|
| 30 | 000 | Fire |

Severity rating: 301 = minor; 302 = severe.

**Damage agents**: Wild animals.

| Category | Agent | Common name |
|---|---|---|
| 41 | 000 | Wild animals |
|  | 001 | Bear |
|  | 002 | Beaver |
|  | 003 | Big game |
|  | 004 | Mice or voles |
|  | 005 | Pocket gophers |
|  | 006 | Porcupines |
|  | 007 | Rabbits or hares |
|  | 008 | Sapsucker |
|  | 009 | Squirrels |
|  | 010 | Woodpeckers |
|  | 011 | Moose |
|  | 012 | Elk |
|  | 013 | Deer |
|  | 014 | Feral pigs |

Severity rating: 411 = minor; 412 = severe.

**Damage agents**: Domestic animals.

| Category | Agent | Common name |
|---|---|---|
| 42 | 000 | Domestic animals |
|  | 001 | Cattle |
|  | 002 | Goats |
|  | 003 | Horses |
|  | 004 | Sheep |

Severity rating: 421 = minor; 422 = severe.

**Damage agents**: Abiotic damage.

| Category | Agent | Common name |
|---|---|---|
| 50 | 000 | Abiotic damage |
|  | 001 | Air pollutants |
|  | 002 | Chemical |
|  | 003 | Drought |
|  | 004 | Flooding/high water |
|  | 005 | Frost |
|  | 006 | Hail |
|  | 007 | Heat |
|  | 008 | Lightning |
|  | 009 | Nutrient imbalances |
|  | 010 | Radiation |
|  | 011 | Snow/ice |
|  | 012 | Wild fire |
|  | 013 | Wind/tornado |
|  | 014 | Winter injury |
|  | 015 | Avalanche |
|  | 016 | Mud/land slide |

Severity rating: 501 = minor; 502 = severe.

**Damage agents**: Competition.

| Category | Agent | Common name |
|---|---|---|
| 60 | 000 | Competition |

Severity rating: 601 = minor; 602 = severe.

**Damage agents**: Human activities.

| Category | Agent | Common name |
|---|---|---|
| 70 | 000 | Human activities |
|  | 001 | Herbicides |
|  | 002 | Human caused fire |
|  | 003 | Imbedded objects |
|  | 004 | Improper planting technique |
|  | 005 | Land clearing |
|  | 006 | Land use conversion |
|  | 007 | Logging damage |
|  | 008 | Mechanical |
|  | 009 | Pesticides |
|  | 010 | Roads |
|  | 011 | Soil compaction |
|  | 012 | Suppression |
|  | 013 | Vehicle damage |
|  | 014 | Road salt |

Severity rating: 701 = minor; 702 = severe.

# Appendix A: (Continued)

**Damage agents:** Harvest.

| Category | Agent | Common name |
|---|---|---|
| 71 | 000 | Harvest |

Severity rating: 711 = minor; 712 = severe.

**Damage agents:** Unknown.

| Category | Agent | Common name |
|---|---|---|
| 90 | 000 | Unknown |

Severity rating: 900 = 0 9% affected.
901 = 10 19% affected.
902 = 20 29% affected.
903 = 30 39% affected.
904 = 40 49% affected.
905 = 50 59% affected.
906 = 60 69% affected.
907 = 70 79% affected.
908 = 80 89% affected.
909 = 90 100% affected

**Damage agents:** Multidamage (insect/disease).

| Category | Agent | Common name |
|---|---|---|
| 80 | 000 | Multidamage (insect/disease) |
|  | 001 | Aspen defoliation (12037,12096, 25036, and 25037) |
|  | 002 | Subalpine fir mortality (11015, 21001, 21010, 50014) disturbances |

Severity rating: 801 = minor; 802 = severe.

**Tree parts.**

| Code | Description |
|---|---|
| UN | Unspecified |
| TO | Top |
| FO | Foliar (crown) |
| LI | Limb |
| BO | Bole, other than top or base |
| BA | Base |
| RO | Roots |
| WT | Whole tree |
| TT | Top third of crown |
| MT | Middle third of crown |
| BT | Bottom third of crown |
| TA | Above merch top |
| TB | Below merch top |

**Damage agents:** Physical effects.

| Category | Agent | Physical effects | How to code severity (actual percent) |
|---|---|---|---|
| 99 | 000 | Physical effects |  |
|  | 000 | Unknown |  |
|  | 001 | Broken top | Percent of missing height |
|  | 002 | Dead top | Percent of dead height |
|  | 003 | Limby (large limbs top to bottom) | Percent of bole with many limbs/knots |
|  | 004 | Forked top | Percent of height above fork |
|  | 005 | Forked below merch top | Percent of bole affected |
|  | 006 | Crook or sweep | Percent of bole affected |
|  | 007 | Checks, bole cracks | Percent of bole affected |
|  | 008 | Foliage discoloration | Percent of foliage discolored |
|  | 010 | Lack of seed source | NA |
|  | 011 | Poor planting stock | NA |
|  | 012 | Poor growth | NA |
|  | 013 | Total board ft volume loss |  |
|  | 014 | Total cubic ft volume loss |  |
|  | 015 | Bark removal |  |
|  | 016 | Foliage loss |  |
|  | 017 | Sunscald |  |
|  | 018 | Uproot |  |
|  | 019 | Scorched foliage |  |
|  | 020 | Scorched bark |  |
|  | 021 | Dieback |  |
|  | 022 | Poor crown form |  |

These severities **do not** need to be proceeded by the category code of 99. Only the actual percentage needs to be recorded.

FIREMON Appendices

# Appendix B. NRIS Lithology Codes

| Primary code | Primary lithology | Secondary code | Secondary description |
|---|---|---|---|
| IGEX | Igneous extrusive | ANDE | Andesite |
| | Igneous extrusive | BASA | Basalt |
| | Igneous extrusive | LATI | Latite |
| | Igneous extrusive | RHYO | Rhyolite |
| | Igneous extrusive | SCOR | Scoria |
| | Igneous extrusive | TRAC | Trachyte |
| IGIN | Igneous intrusive | DIOR | Diorite |
| | Igneous intrusive | GABB | Gabbro |
| | Igneous intrusive | GRAN | Granite |
| | Igneous intrusive | QUMO | Quartz monzonite |
| | Igneous intrusive | SYEN | Syenite |
| | Metamorphic | GNEI | Gneiss |
| | Metamorphic | PHYL | Phyllite |
| | Metamorphic | QUAR | Quartzite |
| | Metamorphic | SCHI | Schist |
| | Metamorphic | SLAT | Slate |
| SEDI | Sedimentary | ARGI | Argillite |
| | Sedimentary | CONG | Conglomerate |
| | Sedimentary | DOLO | Dolomite |
| | Sedimentary | LIME | Limestone |
| | Sedimentary | SANS | Sandstone |
| | Sedimentary | SHAL | Shale |
| | Sedimentary | SILS | Siltstone |
| | Sedimentary | TUFA | Tufa |
| UNDI | Undifferentiated | MIEXME | Mixed extrusive and metamorphic |
| | Undifferentiated | MIEXSE | Mixed extrusive and sedimentary |
| | Undifferentiated | MIIG | Mixed igneous (extrusive and intrusive) |
| | Undifferentiated | MIIGME | Mixed igneous and metamorphic |
| | Undifferentiated | MIIGSE | Mixed igneous and sedimentary |
| | Undifferentiated | MIINME | Mixed intrusive and metamorphic |
| | Undifferentiated | MIINSE | Mixed intrusive and sedimentary |
| | Undifferentiated | MIMESE | Mixed metamorphic and sedimentary |

# Appendix C: NRIS Landform Codes

| Landform | Code | Landform | Code |
|---|---|---|---|
| Alluvial fan | ALFA | Nivation hollow | NIHO |
| Alluvial flat | ALFL | Plateau | PLAT |
| Avalanche talus | AVTA | Ridge | RIDG |
| Break | BREA | Stream | STRE |
| Cirque | CIRQ | Stream terrace | STTE |
| Dip slope | DISL | Structural | STRU |
| Drumlin | DRUM | Structural bench | STBE |
| Kame | KAME | Terracette | TERR |
| Kettle | KETT | Trough floor | TRFL |
| Landslide | LAND | Trough wall | TRWA |
| Moraine | MORA | Upland | UPLA |
| Mountain slope | MOSL | | |

# Appendix D: Rick Miller Method for Sampling Shrub-Dominated Systems

Derived from Miller, R. F.; Svejcar, T. J.; Rose, J. A. 2000. Impacts of western juniper on plant community composition and structure. Journal of Range Management. 53: 574–585.

*Macroplot:* 40 x 60 m (fig. AP-1)

## Plot layout

1. Permanent macroplot is staked in each corner and the lower left corner GPS'ed.
2. Three 60-m transects are located at the 0, 20, and 40 m points along the 40-m line.

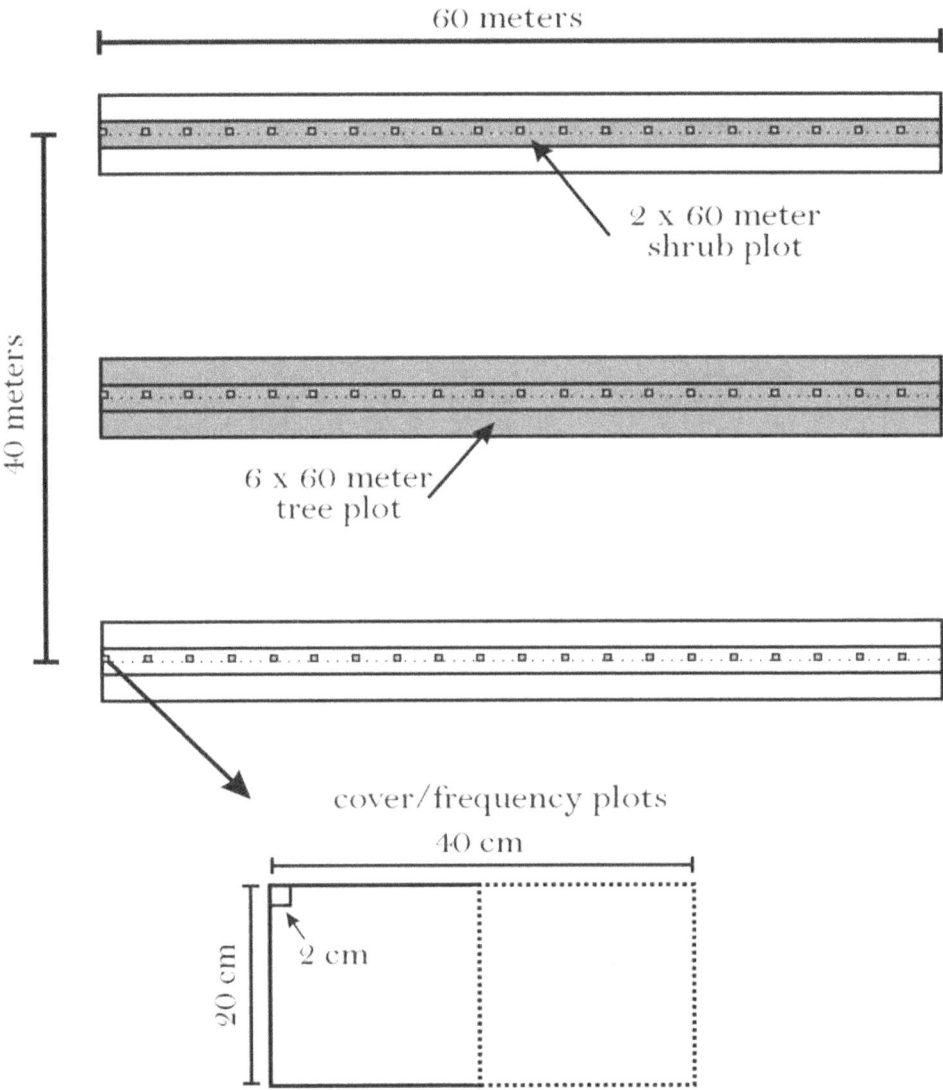

**Figure AP-1**—Rick Miller plot design for sampling in shrub-dominated systems.

## Tree measurements

1. Juniper tree counts are recorded along each 60-m line in a 6 x 60 m belt (a tape or 3-m pole is used to measure the width along the belt).
2. Trees are recorded in the following classes: Old Growth, Dominant (75 percent to maximum height), Subcanopy (3 m to 75 percent of maximum canopy height), Sapling (1 to 3 m), Juveniles (30 cm to 1 m). Trees <30 cm are measured in the 2 x 60 m shrub plot. All old growth trees (separated from younger trees by differences in bark and canopy morphology) occurring with the 40 x 60 m macroplot are recorded. (Note: If there are less than 20 trees >1 m in height in the macroplot, we count all trees within the macroplot.)
3. Tree canopy cover is measured along each 60-m line using the line intercept technique (we use the 3-m PVC pole to mark the canopy edge where it intercepts the line).
4. If trees are to be aged, cores are collected at 30 cm height for trees >6 to 10 cm in diameter. Trees <6 cm diameter are cut at ground level and a disk collected. The disk is always labeled on the top.
5. Parameters measured for cored trees are:
    a. Basal diameter
    b. Tree height (estimated using the 3-m pole)

## Shrub measurements

1. Shrub density is measured along each 60-m line in a 2 x 60 m belt (a tape or 1-m pole is used to measure the width along the belt).
2. Shrub cover is measured along each 60-m line using the line intercept. Live and dead cover are recorded separately using the 15-cm rule (canopy gaps >15 cm within a single shrub canopy or dead foliage >15 cm within a live canopy are reported as such.

## Herbaceous and ground measurements

Nested rooted frequency and cover by functional groups. Optionally, also measure herbaceous cover by species. However, we always do nested frequency and cover by functional groups.

Functional Groups:

1. Deep rooted perennial grasses (PG)
2. Shallow rooted perennial grasses (such as, *Poa sandbergii*)
3. Annual grasses
4. Perennial forbs
5. Annual forbs
6. Bare ground
7. Rock
8. Biological crusts and mosses

Nested frequency is measured in 0.02-, 0.2-, and 0.4-$m^2$ plots every 3 m along each 60-m line (n = 60 plots/macroplot).

Herbaceous cover is estimated in 20, 0.2-$m^2$ plots spaced every 3 m along each 60-m line.

Functional group cover is estimated in 20, 0.2-$m^2$ plots spaced every 3 m along each 60-m line.

# Appendix E: FIREMON Table Names, Field Names, and Field Descriptions

| Table name | Field name | Field description |
|---|---|---|
| CBI | RegID | Registration ID |
| CBI | ProjID | Project ID |
| CBI | PlotID | Plot ID |
| CBI | Date | Date |
| CBI | Percent 20m Plot Burned | Percent of 20 m plot burned |
| CBI | Percent 30m Plot Burned | Percent of 30 m plot burned |
| CBI | FuelPhoto | Fuel photo |
| CBI | PreFire Cov Litter | Litter cover before the fire |
| CBI | PreFire Cov Duff | Duff cover before the fire |
| CBI | PreFire Cov Soil | Soil cover before the fire |
| CBI | PreFire Depth Litter | Litter depth before the fire |
| CBI | PreFire Depth Duff | Duff depth before the fire |
| CBI | PreFire Depth FuelBed | Fuelbed depth before the fire |
| CBI | Litter | Litter score |
| CBI | Duff | Duff score |
| CBI | Med Fuels | Medium fuels score |
| CBI | Heavy Fuels | Heavy fuels score |
| CBI | Soil Cover | Soil cover score |
| CBI | Add Factor Substrates | Additional substrate factor score |
| CBI | PreFire Cov Herbs | Herbaceous cover before the fire |
| CBI | Percent Foliage Alt Herbs | Percent altered herbaceous foliage score |
| CBI | Percent Living Herbs | Percent live herbaceous score |
| CBI | Colonizers Herbs | Herbaceous colonizers score |
| CBI | Species Diversity Herbs | Herbaceous species diversity score |
| CBI | Add Factor Herbs | Additional herbaceous factor score |
| CBI | Enhanced Growth Fact Herbs | Enhanced growth factor for herbaceous plants |
| CBI | PreFire Cov Tall Shrubs | Tall shrub cover before the fire |
| CBI | Percent Foliage Alt Tall Shrubs | Percent altered shrub foliage score |
| CBI | Percent Green Tall Shrubs | Percent green shrub score |
| CBI | Percent Living Tall Shrubs | Percent live shrub score |
| CBI | Species Diversity Tall Shrubs | Shrub species diversity score |
| CBI | Add Factor Tall Shrubs | Additional shrub factor score |
| CBI | Enhanced Growth Fact Tall Shrubs | Enhanced growth factor for shrubs |
| CBI | PreFire Cov Int Trees | Intermediate tree cover before the fire |
| CBI | PreFire Den Int Trees | Intermediate tree density before the fire |
| CBI | Percent Green Int Trees | Percent green intermediate tree score |
| CBI | Percent Black Int Trees | Percent black intermediate tree score |
| CBI | Percent Brown Int Trees | Percent brown intermediate tree score |
| CBI | Percent Canopy Mort Int Trees | Percent intermediate tree canopy mortality score |
| CBI | Char Height Int Trees | Intermediate tree char height score |
| CBI | Add Factor Int Trees | Additional intermediate tree factor score |
| CBI | Percent Girdled Int Trees | Percent intermediate trees girdled |
| CBI | Percent Felled Int Trees | Percent intermediate trees felled |
| CBI | Percent Tree Mort Int Trees | Percent intermediate tree mortality |
| CBI | PreFire Cov Big Trees | Big tree cover before the fire |
| CBI | PreFire Den Big Trees | Big tree density before the fire |
| CBI | Percent Green Big Trees | Percent green big tree score |
| CBI | Percent Black Big Trees | Percent black big tree score |
| CBI | Percent Brown Big Trees | Percent brown big tree score |
| CBI | Percent Canopy Mort Big Trees | Percent big tree canopy mortality score |
| CBI | Char Height Big Trees | Big tree char height score |
| CBI | Add Factor Big Trees | Additional big tree factor score |
| CBI | Percent Girdled Big Trees | Percent big trees girdled |
| CBI | Percent Felled Big Trees | Percent big trees felled |
| CBI | Percent Tree Mort Big Trees | Percent big tree mortality |
| CBI | Plant Community Notes | Notes on plant species community |
| CBI | Substrate CBI | Substrate composite burn index |
| CBI | Low Shrub CBI | Low shrub composite burn index |
| CBI | Tall Shrub Sapling CBI | Tall shrub composite burn index |

(con.)

# Appendix E: (Continued)

| Table name | Field name | Field description |
|---|---|---|
| CBI | Intermediate Trees CBI | Intermediate tree composite burn index |
| CBI | Big Trees CBI | Big tree composite burn index |
| CBI | Understory CBI | Understory composite burn index |
| CBI | Overstory CBI | Overstory composite burn index |
| CBI | Total Plot CBI | Total plot composite burn index |
| CFMacro | RegID | Registration ID |
| CFMacro | ProjID | Project ID |
| CFMacro | PlotID | Plot ID |
| CFMacro | Date | Date |
| CFMacro | NumTran | Number of transects |
| CFMacro | TranLen | Transect length ft (m) |
| CFMacro | NumQuadTran | Number of quadrats/transect |
| CFMacro | QuadLen | Quadrat length inches (cm) |
| CFMacro | QuadWid | Quadrat width inches (cm) |
| CFMacro | NFRatio | Nested frequency subplot size ratio percent |
| CFMacro | NFNum | Nested frequency subplot numbers |
| CFMicro | RegID | Registration ID |
| CFMicro | ProjID | Project ID |
| CFMicro | PlotID | Plot ID |
| CFMicro | Date | Date |
| CFMicro | Transect | Transect number |
| CFMicro | Item | Item code plant species or other item |
| CFMicro | Status | Health of species (live or dead) |
| CFMicro | CC1 | Canopy cover of item by quadrat percent |
| CFMicro | NRF1 | Nested rooted frequency of item by quadrat |
| CFMicro | Ht1 | Height of item by quadrat |
| CFMicro | CC2 | |
| CFMicro | NRF2 | |
| CFMicro | Ht2 | |
| CFMicro | CC3 | |
| CFMicro | NRF3 | |
| CFMicro | Ht3 | |
| CFMicro | CC4 | |
| CFMicro | NRF4 | |
| CFMicro | Ht4 | |
| CFMicro | CC5 | |
| CFMicro | NRF5 | |
| CFMicro | Ht5 | |
| CFMicro | CC6 | |
| CFMicro | NRF6 | Nested rooted frequency of item by quadrat |
| CFMicro | Ht6 | Height of item by quadrat |
| CFMicro | CC7 | |
| CFMicro | NRF7 | |
| CFMicro | Ht7 | |
| CFMicro | CC8 | |
| CFMicro | NRF8 | |
| CFMicro | Ht8 | |
| CFMicro | CC9 | |
| CFMicro | NRF9 | |
| CFMicro | Ht9 | |
| CFMicro | CC10 | |
| CFMicro | NRF10 | |
| CFMicro | Ht10 | |
| CFMicro | CC11 | |
| CFMicro | NRF11 | |
| CFMicro | Ht11 | |
| CFMicro | CC12 | |
| CFMicro | NRF12 | |
| CFMicro | Ht12 | |

(con.)

# Appendix E: (Continued)

| Table name | Field name | Field description |
|---|---|---|
| CFMicro | CC13 | |
| CFMicro | NRF13 | |
| CFMicro | Ht13 | |
| CFMicro | CC14 | |
| CFMicro | NRF14 | |
| CFMicro | Ht14 | |
| CFMicro | CC15 | |
| CFMicro | NRF15 | |
| CFMicro | Ht15 | |
| CFMicro | CC16 | |
| CFMicro | NRF16 | |
| CFMicro | Ht16 | |
| CFMicro | CC17 | |
| CFMicro | NRF17 | |
| CFMicro | Ht17 | |
| CFMicro | CC18 | |
| CFMicro | NRF18 | |
| CFMicro | Ht18 | |
| CFMicro | CC19 | |
| CFMicro | NRF19 | |
| CFMicro | Ht19 | |
| CFMicro | CC20 | |
| CFMicro | NRF20 | |
| CFMicro | Ht20 | |
| DEMacroBelt | RegID | Registration ID |
| DEMacroBelt | ProjID | Project ID |
| DEMacroBelt | PlotID | Plot ID |
| DEMacroBelt | Date | Date |
| DEMacroBelt | NumTran | Number of transects |
| DEMacroQuad | RegID | Registration ID |
| DEMacroQuad | ProjID | Project ID |
| DEMacroQuad | PlotID | Plot ID |
| DEMacroQuad | Date | Date |
| DEMacroQuad | NumTran | Number of transects |
| DEMacroQuad | NumQuadTran | Number of quadrats/transect |
| DEMicroBelt | RegID | Registration ID |
| DEMicroBelt | ProjID | Project ID |
| DEMicroBelt | PlotID | Plot ID |
| DEMicroBelt | Date | Date |
| DEMicroBelt | Item | Item code; plant species or other item |
| DEMicroBelt | Status | Health of species (live or dead) |
| DEMicroBelt | SizeCl | Size class |
| DEMicroBelt | TranLen | Transect length   ft (m) |
| DEMicroBelt | TranWid | Transect width   ft (m) |
| DEMicroBelt | Cnt1 | Count (number of items) by transect |
| DEMicroBelt | Ht1 | Average height of item by transect   ft (m) |
| DEMicroBelt | Cnt2 | |
| DEMicroBelt | Ht2 | |
| DEMicroBelt | Cnt3 | |
| DEMicroBelt | Ht3 | |
| DEMicroBelt | Cnt4 | |
| DEMicroBelt | Ht4 | |
| DEMicroBelt | Cnt5 | |
| DEMicroBelt | Ht5 | |
| DEMicroBelt | Cnt6 | |
| DEMicroBelt | Ht6 | |
| DEMicroBelt | Cnt7 | |
| DEMicroBelt | Ht7 | |
| DEMicroBelt | Cnt8 | |
| DEMicroBelt | Ht8 | |

(con.)

## Appendix E: (Continued)

| Table name | Field name | Field description |
|---|---|---|
| DEMicroBelt | Cnt9 | |
| DEMicroBelt | Ht9 | |
| DEMicroBelt | Cnt10 | |
| DEMicroBelt | Ht10 | |
| DEMicroBelt | Cnt11 | |
| DEMicroBelt | Ht11 | |
| DEMicroBelt | Cnt12 | |
| DEMicroBelt | Ht12 | |
| DEMicroBelt | Cnt13 | |
| DEMicroBelt | Ht13 | |
| DEMicroBelt | Cnt14 | |
| DEMicroBelt | Ht14 | |
| DEMicroBelt | Cnt15 | |
| DEMicroBelt | Ht15 | |
| DEMicroBelt | Cnt16 | |
| DEMicroBelt | Ht16 | |
| DEMicroBelt | Cnt17 | |
| DEMicroBelt | Ht17 | |
| DEMicroBelt | Cnt18 | |
| DEMicroBelt | Ht18 | |
| DEMicroBelt | Cnt19 | |
| DEMicroBelt | Ht19 | |
| DEMicroBelt | Cnt20 | |
| DEMicroBelt | Ht20 | |
| DEMicroQuad | RegID | Registration ID |
| DEMicroQuad | ProjID | Project ID |
| DEMicroQuad | PlotID | Plot ID |
| DEMicroQuad | Date | Date |
| DEMicroQuad | Transect | Transect number |
| DEMicroQuad | Item | Item code; plant species or other item |
| DEMicroQuad | Status | Health of species (live or dead) |
| DEMicroQuad | SizeCl | Size class |
| DEMicroQuad | QuadLen | Quadrat length   ft (m) |
| DEMicroQuad | QuadWid | Quadrat width   ft (m) |
| DEMicroQuad | Cnt1 | Count (number of items) by quadrat |
| DEMicroQuad | Ht1 | Average height of item by quadrat   ft (m) |
| DEMicroQuad | Cnt2 | |
| DEMicroQuad | Ht2 | |
| DEMicroQuad | Cnt3 | |
| DEMicroQuad | Ht3 | |
| DEMicroQuad | Cnt4 | |
| DEMicroQuad | Ht4 | |
| DEMicroQuad | Cnt5 | |
| DEMicroQuad | Ht5 | |
| DEMicroQuad | Cnt6 | |
| DEMicroQuad | Ht6 | |
| DEMicroQuad | Cnt7 | |
| DEMicroQuad | Ht7 | |
| DEMicroQuad | Cnt8 | |
| DEMicroQuad | Ht8 | |
| DEMicroQuad | Cnt9 | |
| DEMicroQuad | Ht9 | |
| DEMicroQuad | Cnt10 | |
| DEMicroQuad | Ht10 | |
| DEMicroQuad | Cnt11 | |
| DEMicroQuad | Ht11 | |
| DEMicroQuad | Cnt12 | |
| DEMicroQuad | Ht12 | |
| DEMicroQuad | Cnt13 | |
| DEMicroQuad | Ht13 | |

(con.)

# Appendix E: (Continued)

| Table name | Field name | Field description |
|---|---|---|
| DEMicroQuad | Cnt14 | |
| DEMicroQuad | Ht14 | |
| DEMicroQuad | Cnt15 | |
| DEMicroQuad | Ht15 | |
| DEMicroQuad | Cnt16 | |
| DEMicroQuad | Ht16 | |
| DEMicroQuad | Cnt17 | |
| DEMicroQuad | Ht17 | |
| DEMicroQuad | Cnt18 | |
| DEMicroQuad | Ht18 | |
| DEMicroQuad | Cnt19 | |
| DEMicroQuad | Ht19 | |
| DEMicroQuad | Cnt20 | |
| DEMicroQuad | Ht20 | |
| FB | FireID | 12 digit unique fire identifier |
| FBData | FireID | 12 digit unique fire identifier |
| FBData | FDate | Date of the fire; format = DD/MM/YYYY |
| FBData | FTime | Time of day observations recorded; format = 24 hr time |
| FBData | FName | Name of the fire |
| FBData | RefId | Unique fire code |
| FBData | Units | Units of measure; E = English and M = Metric |
| FBData | TObs | Temperature   degrees Fahrenheit (Celsius) |
| FBData | RH | Relative humidity   percent |
| FBData | Wind | Wind speed   miles/hr (meters/sec) |
| FBData | Cloud | Cloudiness   percent |
| FBData | W1Hr | Moisture of 1 hr fuels   percent |
| FBData | W10Hr | Moisture of 10 hr fuels   percent |
| FBData | W100Hr | Moisture of 100 hr fuels   percent |
| FBData | W1000HrSnd | Moisture of 1,000 hr sound fuels   percent |
| FBData | W1000HrRott | Moisture of 1,000 hr rotten fuels   percent |
| FBData | Litter | Moisture of litter layer   percent |
| FBData | Duff | Moisture of duff layer   percent |
| FBData | Soil | Moisture of uppermost soil layer   percent |
| FBData | Shrub | Moisture of live shrubs   percent |
| FBData | Herb | Moisture of live herbaceous plants   percent |
| FBData | Crown | Moisture of tree crown foliage   percent |
| FBData | FireType | Code for type of fire |
| FBData | FLength | Flame length   ft (m) |
| FBData | FDepth | Flame depth   ft (m) |
| FBData | Srate | Average speed of fire   ft/minute (meters/second) |
| FBData | Plume | Dynamic behavior of plume |
| FBData | Spot | Spotting behavior of fire |
| FBData | Local1 | Local field 1 |
| FBData | Local2 | Local field 2 |
| FBData | Comments | Comments |
| FL1000hr | RegID | Registration ID |
| FL1000hr | ProjID | Project ID |
| FL1000hr | PlotID | Plot ID |
| FL1000hr | Date | Date |
| FL1000hr | Transect | Line transect number |
| FL1000hr | LogNum | Log number |
| FL1000hr | Dia | Diameter of log at line intersection   inches (cm) |
| FL1000hr | DecayCl | Log decay class |
| FL1000hr | Local1 | Local field 1 |
| FLFineDL | RegID | Registration ID |
| FLFineDL | ProjID | Project ID |
| FLFineDL | PlotID | Plot ID |
| FLFineDL | Date | Date |
| FLFineDL | Transect | Line tansect number |

(con.)

## Appendix E: (Continued)

| Table name | Field name | Field description |
|---|---|---|
| FLFineDL | Slope | Slope of transect (rise/run)*100  percent |
| FLFineDL | 1hr | Number of pieces 0  0.25 inch (0  0.635 cm) in diameter |
| FLFineDL | 10hr | Number of pieces 0.25  1.0 inch (0.635  2.54 cm) in diameter |
| FLFineDL | 100hr | Number of pieces 1  3 inches (2.54 and 7.62 cm) in diameter |
| FLFineDL | D/LDep1 | Depth of duff/litter profile  inches (cm) |
| FLFineDL | LitterPct1 | Proportion of total profile depth that is litter  percent |
| FLFineDL | D/LDep2 | Depth of duff/litter profile  inches (cm) |
| FLFineDL | LitterPct2 | Proportion of total profile depth that is litter  percent |
| FLFineDL | Local1 | Local field 2 |
| FLMacro | RegID | Registration ID |
| FLMacro | ProjID | Project ID |
| FLMacro | PlotID | Plot ID |
| FLMacro | Date | Date |
| FLMacro | 1HRTranLen | 1 hr transect length  ft (m) |
| FLMacro | 10HRTranLen | 10 hr transect length  ft (m) |
| FLMacro | 100HRTranLen | 100 hr transect length  ft (m) |
| FLMacro | 1000HRTranLen | 1,000 hr transect length  ft (m) |
| FLMacro | NumTran | Number of transects |
| FLVeg | RegID | Registration ID |
| FLVeg | ProjID | Project ID |
| FLVeg | PlotID | Plot ID |
| FLVeg | Date | Date |
| FLVeg | Transect | Line transect number |
| FLVeg | LiShC1 | Live woody cover at point 1 |
| FLVeg | DdShC1 | Dead woody cover at point 1 |
| FLVeg | ShHt1 | Woody height at point 1 |
| FLVeg | LiHeC1 | Live nonwoody cover at point 1 |
| FLVeg | DdHeC1 | Dead nonwoody cover at point 1 |
| FLVeg | HeHt1 | Nonwoody height at point 1 |
| FLVeg | LiShC2 | Live woody cover at point 2 |
| FLVeg | DdShC2 | Dead woody cover at point 2 |
| FLVeg | ShHt2 | Woody height at point 2 |
| FLVeg | LiHeC2 | Live nonwoody cover at point 2 |
| FLVeg | DdHeC2 | Dead nonwoody cover at point 2 |
| FLVeg | HeHt2 | Nonwoody height at point 2 |
| LIMacro | RegID | Registration ID |
| LIMacro | ProjID | Project ID |
| LIMacro | PlotID | Plot ID |
| LIMacro | Date | Date |
| LIMacro | NumTran | Number of transects |
| LIMicroInt | RegID | Registration ID |
| LIMicroInt | ProjID | Project ID |
| LIMicroInt | PlotID | Plot ID |
| LIMicroInt | Date | Date |
| LIMicroInt | Transect | Transect number |
| LIMicroInt | Item | Item code; plant species or other item |
| LIMicroInt | Status | Health of species (live or dead) |
| LIMicroInt | SizeCl | Size class |
| LIMicroInt | Start | Starting point for intercept on measuring tape  inches (cm) |
| LIMicroInt | Stop | Stopping point for intercept on measuring tape  inches (cm) |
| LIMicroInt | Intercept | Intercept length (stop start)  inches (cm) |
| LIMicroInt | Height | Average height for item on transect  ft (m) |
| LIMicroSpp | RegID | Registration ID |
| LIMicroSpp | ProjID | Project ID |
| LIMicroSpp | PlotID | Plot ID |
| LIMicroSpp | Date | Date |
| LIMicroSpp | Transect | Transect number |
| LIMicroSpp | Item | Item code; plant species or other item |
| LIMicroSpp | Status | Health of species (live or dead) |

(con.)

# Appendix E: (Continued)

| Table name | Field name | Field description |
|---|---|---|
| LIMicroSpp | SizeCl | Size class |
| LIMicroSpp | TranLen | Transect length  ft (m) |
| MD | MdId | Metadata key  ID |
| MDComm | MDID | Metadata ID |
| MDComm | Subject | Subject description |
| MDComm | DocLink | Hyperlink to document |
| MDComm | Comments | Metadata or comment text |
| NRCSPlantsDB | Symbol | Plant species symbol |
| NRCSPlantsDB | Symbol Key | Plant species symbol key |
| NRCSPlantsDB | Synonym Symbol Key | Plant species preferred synonym |
| NRCSPlantsDB | Local Code | Local plant species code |
| NRCSPlantsDB | FVS Code | FVS plant species code |
| NRCSPlantsDB | Life Form | Plant species life form |
| NRCSPlantsDB | Scientific Name | Scientific name |
| NRCSPlantsDB | Common Name | Common name |
| NRCSPlantsDB | Family | Plant family |
| NRCSPlantsDB | Life Form 1 | Alternate life form |
| NRCSPlantsDB | Life Form 2 | Second alternate life form |
| PD | RegID | Registration ID |
| PD | ProjID | Project ID |
| PD | PlotID | Plot ID |
| PD | Date | Date |
| PD | OrgCode1 | Organization code 1 |
| PD | OrgCode2 | Organization code 2 |
| PD | OrgCode3 | Organization code 3 |
| PD | OrgCode4 | Organization code 4 |
| PD | Examiner | Name of FIREMON crew boss or lead examiner |
| PD | Units | Units of measurement (English or metric) |
| PD | Radius | Radius/length of the macroplot in ft (m) |
| PD | Width | Width of macroplot in ft (m) |
| PD | PlotType | Type of plot: C = control; M = measured |
| PD | SEvent | Sampling event  reason why plot is being measured at this time |
| PD | FireID | Fire behavior database key  ID |
| PD | MdId | Metadata key  ID |
| PD | LocType | Type of location  L = Lat/Long; U = UTM |
| PD | Lat | Latitude of plot center |
| PD | Long | Longitude of plot center |
| PD | Northing | UTM northing of plot center |
| PD | Easting | UTM easting of plot center |
| PD | Datum | GPS datum |
| PD | GPS Error | GPS error (meters or ft) |
| PD | GPS Err Units | Units for GPS Error: ft = feet; m = meters |
| PD | UTM Zone | UTM zone |
| PD | Elev | Elevation above mean sea level  ft (m) |
| PD | Aspect | Aspect of plot in azimuth  degrees |
| PD | Slope | Average slope (rise/run)*100  percent |
| PD | Landform | Landform code |
| PD | VShape | Shape of plot perpendicular to contour |
| PD | HShape | Shape of plot parallel to contour |
| PD | Geol1 | Primary surficial geology code |
| PD | Geol2 | Secondary surficial geology code |
| PD | SoilTex | Soil texture |
| PD | EType | Erosion type |
| PD | ESev | Erosion severity |
| PD | TreeC | Total tree cover  percent |
| PD | SeedC | Seedling cover  percent |
| PD | SapC | Sapling cover  percent |
| PD | PoleC | Pole cover  percent |
| PD | MedC | Medium tree cover  percent |

(con.)

FIREMON Appendices

## Appendix E: (Continued)

| Table name | Field name | Field description |
|---|---|---|
| PD | LTreeC | Tree cover   percent |
| PD | VLTreeC | Very large tree cover   percent |
| PD | ShrubC | Total shrub cover   percent |
| PD | LShrubC | Low shrub cover   percent |
| PD | MShrubC | Medium shrub cover   percent |
| PD | TShrubC | Tall shrub cover   percent |
| PD | GramC | Graminoid cover   percent |
| PD | ForbC | Forb cover   percent |
| PD | FernC | Fern cover   percent |
| PD | MossC | Moss and lichen cover   percent |
| PD | USpp1 | Most dominant species in upper layer |
| PD | USpp2 | Second most dominant species in upper layer |
| PD | MSpp1 | Most dominant species in middle layer |
| PD | MSpp2 | Second most dominant species in middle layer |
| PD | LSpp1 | Most dominant species in lower layer |
| PD | LSpp2 | Second most dominant species in lower layer |
| PD | PVTId | Potential vegetation type code |
| PD | PotForm | Potential lifeform code |
| PD | BSoilGC | Bare soil ground cover   percent |
| PD | GravelGC | Gravel ground cover   percent |
| PD | RockGC | Rock ground cover   percent |
| PD | DuffGC | Duff and litter ground cover   percent |
| PD | WoodGC | Wood ground cover   percent |
| PD | MossGC | Moss and lichen ground cover   percent |
| PD | CharGC | Charred ground cover   percent |
| PD | AshGC | Ash ground cover   percent |
| PD | BVegGC | Basal vegetation ground cover   percent |
| PD | WaterGC | Water ground cover   percent |
| PD | FModel | Fire behavior model (Anderson 1983) |
| PD | PhotoID | Fuel photo series |
| PD | SHT | Stand height: height of highest stratum that contains at least 10% of canopy cover   ft (m) |
| PD | CBH | Canopy fuel base height   ft (m) |
| PD | CanopyC | Percent canopy cover of forest canopy >6.5 ft   ft (m) |
| PD | FLength | Average flame length   ft (m) |
| PD | SRate | Spread rate; average speed of fire   ft/min (meters/min) |
| PD | FBevPic | Picture code for fire behavior picture |
| PD | FSC | Fire severity code |
| PD | NorthPic | Code for plot photo taken in direction of due north |
| PD | EastPic | Code for plot photo taken in direction of due east |
| PD | Photo1 | Code for plot photo 1 |
| PD | Photo2 | Code for plot photo 2 |
| PD | Local1 | Local code 1 |
| PD | Local2 | Local code 2 |
| PD | Comments | Comments about plot |
| POMacroFrame | RegID | Registration ID |
| POMacroFrame | ProjID | Project ID |
| POMacroFrame | PlotID | Plot ID |
| POMacroFrame | Date | Date |
| POMacroFrame | NumTran | Number of transects |
| POMacroFrame | TranLen | Transect length |
| POMacroFrame | NumFrmTran | Number of frames/transect |
| POMacroFrame | NumPtsFrm | Number of points/frame |
| POMacroTran | RegID | Registration ID |
| POMacroTran | ProjID | Project ID |
| POMacroTran | PlotID | Plot ID |
| POMacroTran | Date | Date |
| POMacroTran | NumTran | Number of transects |
| POMacroTran | TranLen | Transect length |
| POMacroTran | NumPtsTran | Number of points/transect |

(con.)

# Appendix E: (Continued)

| Table name | Field name | Field description |
|---|---|---|
| POMicroFrame | RegID | Registration ID |
| POMicroFrame | ProjID | Project ID |
| POMicroFrame | PlotID | Plot ID |
| POMicroFrame | Date | Date |
| POMicroFrame | Transect | Transect number |
| POMicroFrame | Item | Item code   plant species or other item |
| POMicroFrame | Status | Health of species |
| POMicroFrame | Hits1 | Number of hits for item by frame |
| POMicroFrame | Ht1 | Average height of item by frame   ft (m) |
| POMicroFrame | Hits2 | |
| POMicroFrame | Ht2 | |
| POMicroFrame | Hits3 | |
| POMicroFrame | Ht3 | |
| POMicroFrame | Hits4 | |
| POMicroFrame | Ht4 | |
| POMicroFrame | Hits5 | |
| POMicroFrame | Ht5 | |
| POMicroFrame | Hits6 | |
| POMicroFrame | Ht6 | |
| POMicroFrame | Hits7 | |
| POMicroFrame | Ht7 | |
| POMicroFrame | Hits8 | |
| POMicroFrame | Ht8 | |
| POMicroFrame | Hits9 | |
| POMicroFrame | Ht9 | |
| POMicroFrame | Hits10 | |
| POMicroFrame | Ht10 | |
| POMicroFrame | Hits11 | |
| POMicroFrame | Ht11 | |
| POMicroFrame | Hits12 | |
| POMicroFrame | Ht12 | |
| POMicroFrame | Hits13 | |
| POMicroFrame | Ht13 | |
| POMicroFrame | Hits14 | |
| POMicroFrame | Ht14 | |
| POMicroFrame | Hits15 | |
| POMicroFrame | Ht15 | |
| POMicroFrame | Hits16 | |
| POMicroFrame | Ht16 | |
| POMicroFrame | Hits17 | |
| POMicroFrame | Ht17 | |
| POMicroFrame | Hits18 | |
| POMicroFrame | Ht18 | |
| POMicroFrame | Hits19 | |
| POMicroFrame | Ht19 | |
| POMicroFrame | Hits20 | |
| POMicroFrame | Ht20 | |
| POMicroTran | RegID | Registration ID |
| POMicroTran | ProjID | Project ID |
| POMicroTran | PlotID | Plot ID |
| POMicroTran | Date | Date |
| POMicroTran | Item | Item code; plant species or other item |
| POMicroTran | Status | Health of species (live or dead) |
| POMicroTran | Hits1 | Number of hits for item by transect |
| POMicroTran | Ht1 | Average height of item by transect   ft (m) |
| POMicroTran | Hits2 | |
| POMicroTran | Ht2 | |
| POMicroTran | Hits3 | |
| POMicroTran | Ht3 | |
| POMicroTran | Hits4 | |

(con.)

# Appendix E: (Continued)

| Table name | Field name | Field description |
|---|---|---|
| POMicroTran | Ht4 | |
| POMicroTran | Hits5 | |
| POMicroTran | Ht5 | |
| POMicroTran | Hits6 | |
| POMicroTran | Ht6 | |
| POMicroTran | Hits7 | |
| POMicroTran | Ht7 | |
| POMicroTran | Hits8 | |
| POMicroTran | Ht8 | |
| POMicroTran | Hits9 | |
| POMicroTran | Ht9 | |
| POMicroTran | Hits10 | |
| POMicroTran | Ht10 | |
| POMicroTran | Hits11 | |
| POMicroTran | Ht11 | |
| POMicroTran | Hits12 | |
| POMicroTran | Ht12 | |
| POMicroTran | Hits13 | |
| POMicroTran | Ht13 | |
| POMicroTran | Hits14 | |
| POMicroTran | Ht14 | |
| POMicroTran | Hits15 | |
| POMicroTran | Ht15 | |
| POMicroTran | Hits16 | |
| POMicroTran | Ht16 | |
| POMicroTran | Hits17 | |
| POMicroTran | Ht17 | |
| POMicroTran | Hits18 | |
| POMicroTran | Ht18 | |
| POMicroTran | Hits19 | |
| POMicroTran | Ht19 | |
| POMicroTran | Hits20 | |
| POMicroTran | Ht20 | |
| RSMacro | RegID | Registration ID |
| RSMacro | ProjID | Project ID |
| RSMacro | PlotID | Plot ID |
| RSMacro | Date | Date |
| RSMacro | BLineLen | Length of baseline   ft or meters |
| RSSpp | RegID | Registration ID |
| RSSpp | ProjID | Project ID |
| RSSpp | PlotID | Plot ID |
| RSSpp | Date | Date |
| RSSpp | Species | Plant species code |
| RSSpp | PlantNo | Unique number for each individual plant |
| RSSpp | DistAlongBL | Distance along baseline |
| RSSpp | DistFromBL | Distance from baseline |
| RSSpp | Status | Plant status |
| RSSpp | Stage | Plant stage |
| RSSpp | MaxDia | Canopy diameter measured at maximum diameter |
| RSSpp | Dia2 | Canopy diameter measured at right angles to maximum diameter measurement |
| RSSpp | Height | Plant height |
| RSSpp | Stems | Number of stems |
| RSSpp | Flowers | Number of flowers |
| RSSpp | Fruits | Number of fruits |
| RSSpp | Local1 | Local field 1 |
| RSSpp | Local2 | Local field 2 |
| RSSpp | Local3 | Local field 3 |
| SCCover | RegID | Registration ID |
| SCCover | ProjID | Project ID |

(con.)

# Appendix E: (Continued)

| Table name | Field name | Field description |
|---|---|---|
| SCCover | PlotID | Plot ID |
| SCCover | Date | Date |
| SCCover | Item | Item code |
| SCCover | Status | Heath of species (live or dead) |
| SCCover | SizeCl | Size class |
| SCCover | Cover | Canopy cover percent |
| SCCover | Height | Average height ft (m) |
| SCCover | Local1 | Optional field 1 |
| SCCover | Local2 | Optional field 2 |
| SCMacro | RegID | Registration ID |
| SCMacro | ProjID | Project ID |
| SCMacro | PlotID | Plot ID |
| SCMacro | Date | Date |
| SCMacro | SppIDLevel | Plant species ID level; minimum cover recorded percent |
| TDMacro | RegID | Registration ID |
| TDMacro | ProjID | Project ID |
| TDMacro | PlotID | Plot ID |
| TDMacro | Date | Date |
| TDMacro | MacroPlotSize | Macroplot size acres (square meters) |
| TDMacro | MicroPlotSize | Microplot size acres (square meters) |
| TDMacro | SnagPlotSize | Snagplot size acres (square meters) |
| TDMacro | BreakPntDia | Break point diameter inches (cm) |
| TDSap | RegID | Registration ID |
| TDSap | ProjID | Project ID |
| TDSap | PlotID | Plot ID |
| TDSap | Date | Date |
| TDSap | SizeCl Dia | Size class |
| TDSap | Species | Species code |
| TDSap | TreeStat | Tree status |
| TDSap | Count | Number of trees by species, size class, and status |
| TDSap | AvgHt | Average height ft (m) |
| TDSap | AvgLiCr | Average live crown percent |
| TDSap | Local1 | Local field 1 |
| TDSeed | RegID | Registration ID |
| TDSeed | ProjID | Project ID |
| TDSeed | PlotID | Plot ID |
| TDSeed | Date | Date |
| TDSeed | SizeCl Ht | Size class |
| TDSeed | Species | Species code |
| TDSeed | TreeStat | General health condition of sample tree |
| TDSeed | Count | Number of trees by species, size class, and status |
| TDSeed | Local1 | Local field 1 |
| TDTree | RegID | Registration ID |
| TDTree | ProjID | Project ID |
| TDTree | PlotID | Plot ID |
| TDTree | Date | Date |
| TDTree | TagNo | Tree tag number |
| TDTree | Species | Species code |
| TDTree | TreeStat | Health of tree (live or dead) |
| TDTree | DBH | Diameter breast height inches (cm) |
| TDTree | Height | Tree height ft (m) |
| TDTree | LiCrPct | Live crown percent |
| TDTree | LiCrBHt | Live crown base height ft (m) |
| TDTree | CrwnCl | Crown position class |
| TDTree | Age | Tree age years |
| TDTree | GrwthRt | Tree growth rate (last 10 yrs radial growth) inches (mm) |
| TDTree | DecayCl | Decay cass |

(con.)

## Appendix E: (Continued)

| Table name | Field name | Field description |
|---|---|---|
| TDTree | Mort | Cause of mortality |
| TDTree | DamCd1 | Damage code 1 |
| TDTree | DamSev1 | Severity code 1 |
| TDTree | DamCd2 | Damage code 2 |
| TDTree | DamSev2 | Severity code 2 |
| TDTree | CharHt | Bole char height   ft (m) |
| TDTree | CrScPct | Crown scorch percent |
| TDTree | Local1 | Optional code 1 |

# Appendix F: FIREMON Database Relationships

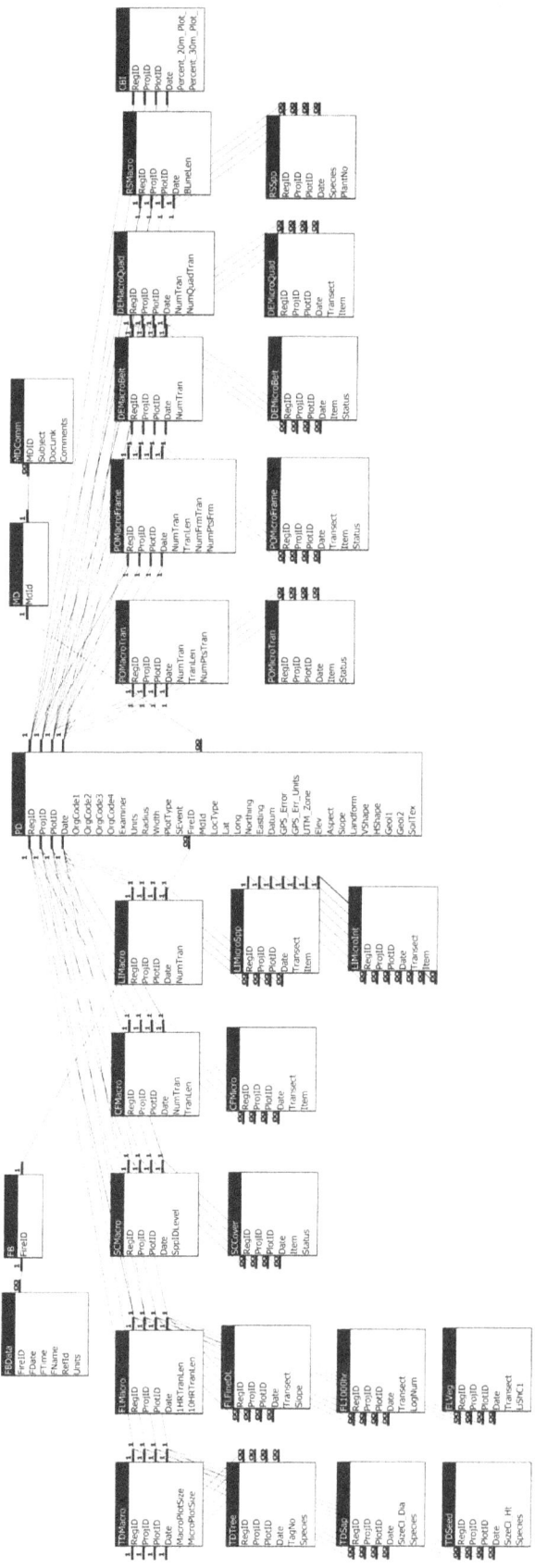

# FIREMON Glossary

The FIREMON Glossary includes terms that are used in the FIREMON protocol but that may be unfamiliar to some readers. The source of each definition is provided in parenthesis at the end of each definition (if applicable). In some cases, words will use the FIREMON documentation as a source. These terms were defined specifically for use in some part of FIREMON, and the definition given may or may not be applicable outside the FIREMON protocol. The term Burn Severity is an example. It was specifically defined for use in the Landscape Assessment methods.

**1,000-hour fuel**: Dead fuels consisting of roundwood 3.0 to 8.0 inches in diameter, estimated to reach 63 percent equilibrium moisture content in 1,000 hours (Fire Effects Guide).

**100-hour fuel**: Dead fuels consisting of roundwood in the size range from 1.0 to 3.0 inches in diameter, estimated to reach 63 percent of equilibrium in 100 hours (Fire Effects Guide).

**10-hour fuel**: Dead fuels consisting of roundwood 0.25 to 1.0 inch in diameter, estimated to reach 63 percent of equilibrium moisture in 10 hours (Fire Effects Guide).

**1-hour fuel**: Dead fuels consisting of dead herbaceous plant material and roundwood less than 0.25 inch in diameter, estimated to reach 63 percent of equilibrium moisture content in 1 hour or less (Fire Effects Guide).

**Abiotic**: Nonliving components of an ecosystem such as air, rocks, soil, water, and so forth. Compare to **biotic** (Wildland Planning Glossary, Glossary of Terms Used In Range Management).

**Aerial plant component**: The upper portion of a plant including branches, leaves, and flowering parts. Compare to **basal vegetation**.

**Alluvium**: A general term for all detrital material deposited or in transit by streams, including gravel, sand, silt, clay, and all variations and mixtures of these (Glossary of Landscape and Vegetation Ecology for Alaska).

**Alpine**: That vegetation occurring between the upper limit of trees (timberline) and the lower limit of snow (snowline) on mountains high enough to possess both of these features (Glossary of Landscape and Vegetation Ecology for Alaska).

**Ash**: The incombustible matter remaining after a substance has burned (Dictionary of Scientific and Technical Terms).

**Aspect**: A position facing or commanding a given direction; exposure. Aspect is the compass direction of the prevailing slope with respect to true north (FSVeg).

**Azimuth**: A horizontal angular measure from true north to an object of interest (FSVeg).

**Basal area**: The area of ground surface covered by the stem or stems of a plant; for trees, measured at 4.5 ft above the ground; for forbs and grasses, measured at the root crown (FSVeg).

**Basal vegetation**: In FIREMON, the area of the cross-section of the plant stem where it enters the ground surface; often expressed as a percent of the plot cover. Compare to **aerial plant component** (FIREMON).

**Baseline**: A permanent line from which all vegetation transects are oriented. Usually, used with the FIREMON CF, PO, LI, DE, and RS sampling methods (FIREMON).

**Belt transect**: A two-dimensional transect with width and length. Compare to **transect** and **quadrat**.

**Biophysical setting**: Describes the physical environment of the FIREMON plot relative to the organisms that grow there. The site characteristics included in a description are topography, geology, soils, and landform (FIREMON).

**Biotic**: Applied to the living components of the biosphere or of an ecosystem, as distinct from the abiotic physical and chemical components. Compare to Abiotic (The Concise Oxford Dictionary of Ecology).

**Biotic plant community**: Any assemblage of populations (plants) living in a prescribed area or physical habitat; an aggregate of organisms that form a distinct ecological unit (Wildland Planning Glossary).

**Bole char height**: A fire severity measurement that is used to quantify potential tree mortality. In FIREMON, it is height of the top of continuous char measured above the ground on the downhill side of the tree, or if on flat ground the top of the lowest point of continuous char (FIREMON).

**Breaklands**: The steep to very broken land at the border of an upland summit that is dissected by ravines (Landforms for Soil Surveys in the Northern Rockies).

**Breakpoint diameter**: In FIREMON the tree diameter above which all trees are tagged and measured individually and below which trees are tallied to species-diameter or species-height classes. Selection of the breakpoint diameter must account for fire monitoring objectives along with sampling limitations and efficiency (FIREMON).

**Broadleaf species**: Deciduous and evergreen trees and shrubs that have seeds within a closed pod or ovary. Compare to **conifer species** (Glossary of Landscape and Vegetation Ecology for Alaska, Webster's New World Dictionary).

**Burn severity**: The degree of environmental change caused by fire, or the result, is the cumulative effect of fire on ecological communities comprising the landscape; the physical and chemical changes to the soil, conversion of vegetation and fuels to inorganic carbon, and structural transformations that bring new microclimates and species assemblages. An analogy to burn severity would be storm severity, which refers to the damage or outcome left in the wake of the storm. Compare to **fire severity** (FIREMON Landscape Assessment Sampling Methods).

**Caliper**: An instrument for determining tree and log diameters by measurement of their rectangular projection on a straight graduated rule via two arms at right angles (and one of them sliding along the rule itself) (Timber Cruising in the Pacific Northwest).

**Canopy cover**: The percentage of ground covered by the vertical projection of the foliage of plant parts. Small openings within the canopy are included. Some sources differentiate canopy cover (with the spaces and small openings included) and foliar cover (with the spaces and small openings excluded). In FIREMON we suggest sampling foliar cover rather than canopy cover. Compare to **foliar cover** (Glossary of Terms Used in Range Management, National Range and Pasture Handbook).

**Canopy fuel base height**: In FIREMON, a subjective assessment of the lowest live or dead fuels attached to the stem of a tree that are sufficient to move fire to the burnable material above. Used to assess crown fire potential (FIREMON).

**Casket tag**: Tags made of high-grade steel that will not melt in a prescribed burn or wildfire. .

**CBI (Composite Burn Index)**: This method uses a field sampling approach on a relatively large plot to determine severity ratings for individual strata, and a synoptic rating for a whole plot area (FIREMON Landscape Assessment Sampling Methods).

**CC (crew costs)**: The cost of outfitting one sampling crew including transportation, supplies, and salary (FIREMON).

**CF (cover frequency)**: Methods used to assess changes in plant species cover and frequency for a macroplot. These methods use quadrats to sample within-plot variation and quantify statistically valid changes in plant species cover, height, and frequency over time. This method is primarily suited for grasses, forbs, and shrubs less than 3 ft in height (FIREMON).

**Char**: The blackened charcoal left from incomplete combustion of organic matter.

**Clay**: Fine-grained material that develops plasticity when mixed with a limited amount of water; composed primarily of silica, alumina, and water, often with iron, alkalies, and alkaline earths. As a soil textural class, soil that is 40 percent or more clay, less than 45 percent sand, and less than 40 percent silt (Dictionary of Scientific and Technical Terms, Glossary of Landscape and Vegetation Ecology for Alaska).

**Clinometer**: An instrument for measuring angles of slope or inclination (Webster's New World Dictionary).

**Cluster sampling**: A method of locating plots in which a group of polygons are sampled in a cluster around an easily accessible point rather than polygons sampled throughout the landscape (FIREMON).

**Conifer species**: A plant belonging to the class Pinatae of cone-bearing gymnospermous trees and shrubs, mostly evergreen, including the pine, spruce, fir, cedar, yew, and cypress. Generally needle-leaved, cone-bearing plants having naked seeds not enclosed in an ovary. Compare to **broadleaf species** (Glossary of Landscape and Vegetation Ecology for Alaska, Webster's New World Dictionary).

**Cover**: See **canopy cover, crown cover**, and **foliar cover**.

**Crown biomass**: The total quantity (weight) of a tree crown including live and dead branches and foliage.

**Crown class**: A code used to describe the position in the canopy of the tree relative to the trees around it and to describe how much light is available to that tree. FIREMON has six categories describing crown class: 1) Open Grown, 2) Emergent, 3) Dominant, 4) Codominant, 5) Intermediate, and 6) Suppressed (based on The Practice of Silviculture).

**Crown cover**: An estimate of tree cover on a plot. Generally, it includes the small openings and spaces in the crown. Sometimes called canopy closure. In FIREMON we suggest sampling foliar cover rather that crown cover; however, project objectives may suggest using crown cover. If crown cover data is collected be sure to note in the Metadata table the fields where it was applied (Forest Measurements).

**Cryptogam**: A plant that reproduces by spores or gametes rather than seeds (an alga, bryophyte, or pteridophyte) (The Concise Oxford Dictionary of Ecology).

**Cryptogamic crust**: Cryptogams such as mosses, algae, lichens, or liverworts growing in a thin crust (The Concise Oxford Dictionary of Ecology).

**Culm groups**: Stalks or stems in grasses (Webster's New World Dictionary).

**CWD (coarse woody debris)**: Generally, pieces greater than 3 inches (10 cm) in diameter, at the point where the piece crosses through the imaginary sampling plane, and longer than 6 ft (2 m). However, definitions vary widely. Dead and down woody debris larger than 3 inches in diameter is often called 1,000-hour (and greater) fuel (Ecology of Coarse Woody Debris in Temperate Ecosystems, FIREMON).

**Damage description**: A code that refers to biotic (insects, disease, or browsing) or abiotic (wind, snow, or fire), damage in trees. This code describes the damaging agent. Compare to **severity description** (FIREMON).

**DBH (diameter at breast height)**: A measure of a tree taken at breast height (4.5 ft), outside the bark of the tree bole, perpendicular to the bole from the uphill side of the tree (FSVeg).

**DE (density)**: A FIREMON method used to assess changes in plant species density and height for a macroplot. This method uses quadrats and belt transects to sample within-stand variation and quantifies statistically valid changes in plant species density and height over time (FIREMON).

**Decay class**: A method used in FIREMON to determine the degree of decay present in coarse woody debris. Compare **snag class** (Ecology of coarse woody debris in temperate ecosystems).

**Deciduous**: Plants that shed their leaves annually; as opposed to evergreens (Webster's New World Dictionary).

**Declination**: The angle between the magnetic and geographical meridians, expressed in degrees and minutes east or west to indicate the direction of magnetic north from true north; also known as magnetic declination; variation (Dictionary of Scientific and Technical Terms).

**DEM (Digital Elevation Model)**: United States Geologic Survey geographic elevation data distributed in raster form. A digital representation of the shape of Earth's surface. Typically digital elevation data consists of arrays of values that represent topographic elevations measured at equal intervals on Earth's surface (FSVeg).

**Diameter tape**: A spring-wound tape measure that has one side in linear units that are converted to diameter and the other side in units for measuring feet and inches. Also called a D-tape or loggers tape (Measurements for Terrestrial Vegetation).

**Dot tally**: A method that uses a series of dots, lines, and boxes to tabulate numbers when sampling. This method is particularly useful when counting many small items.

**Duff**: The partially decomposed organic material of the forest floor that lies beneath the freshly fallen needles, twigs, and leaves, and above the mineral soil. This is the fermentation and humus layer where the vegetative material is broken down, and the individual pieces are no longer identifiable (FSVeg, Fire Effects Guide).

**Duff/litter profile**: The cross-sectional view of the litter and duff layers. It extends vertically from the top of the mineral soil to the top of the litter layer (FIREMON).

**DWD (dead and down woody debris)**: Dead woody pieces of trees and shrubs that have been broken, uprooted, or severed from their root system, not self-supporting, and are lying on the ground (no longer supporting growth). DWD are categorized in size classes of 1-hour, 10-hour, 100-hour, and 1,000-hour (and greater) (FSVeg).

**ECODATA**: Land planning inventory package including sampling methods, field forms, database software, and analysis software. Many FIREMON sampling methods are based on the ECODATA protocols.

**Ecotone**: The area influenced by the transition between plant communities or between successional stages or vegetative conditions within a plant community (Fire Effects Guide).

**Epicormic branching**: A branch that arises from latent, adventitious, or dormant buds within the bark of the tree. Epicormic branch development is often initiated by crown damage such as breakage or fire injury (The Practice of Silviculture).

**FARSITE (Fire Area Simulator)**: A fire growth model that uses fuels, topography, and weather to predict fire spread.

**Fern**: A nonflowering embryophyte having roots, stems, and fronds, and reproducing by spores instead of by seeds (Webster's New World Dictionary).

**Fire behavior**: The manner in which a fire burns in response to the variables of fuel, weather, and topography. A fire may be described as hot or cool, running or creeping, flaming or smoldering, or perhaps, torching or crowning. Compare to **fire effects** (Fire Effects Guide, Glossary of Wildland Fire Management Terms Used in the U.S.).

**Fire Behavior Fuel Model**: Mathematical descriptions of fuel properties (for example, fuel load and fuel depth) that are used in conjunction with environmental conditions to predict certain aspects of fire behavior (FSVeg, Introduction to Wildland Fire Behavior Calculations-Glossary).

**Fire effects**: Any consequence neutral, detrimental, or beneficial resulting from fire. Examples of first order fire effects are tree mortality, emissions, and fuel consumption. Examples of second order fire effects are trees damaged by fire that later succumb to insect infestations, sedimentation in streams from eroding soils, and plant succession. Compare to **fire behavior** (Glossary of Landscape and Vegetation Ecology for Alaska).

**Fire severity**: A qualitative indicator of the effect of fire on the ecosystem, whether it affects the forest floor, canopy, or some other part of the system. It is sometimes assessed based on postfire attributes such as char height or crown scorch. Compare to **burn severity** (Glossary of Landscape and Vegetation Ecology for Alaska, FIREMON).

**FL (fuel load)**: Methods used to sample dead and down woody debris, depth of the duff/litter profile, estimate the proportion of litter in the profile, and estimate total vegetative cover and dead vegetative cover. Down woody debris is sampled using the planar intercept technique (Handbook for Inventorying Downed Woody Material, FIREMON).

**Flame height**: The average height of flames as measured on a vertical axis. It may be less than flame length if flames are angled. Compare to **flame length** (Introduction To Wildland Fire Behavior Calculations-Glossary).

**Flame length**: The distance measured from the tip of the flame to the middle of the flaming zone at base of the fire. It is measured on a slant when the flames are tilted due to effects of wind and slope. Compare to **flame height** (Glossary, Wildland Fire Behavior Calculations).

**FOFEM (First Order Fire Effects Model)**: Model for predicting tree mortality, fuel consumption, emissions, and soil heating from preburn calculations (FOFEM ver. 4.0).

**Foliar cover**: The vertical projection of shoots (stems and leaves). The percentage of ground covered by the vertical projection of the aerial portion of the plants (foliage and supporting parts). Small openings in the canopy and intraspecific overlap are excluded. This is the cover assessment recommended in FIREMON (Glossary of Terms Used in Range Management, FIREMON).

**Forb**: A plant with a soft, rather than permanent woody stem, that is not a grass or grasslike plant. Compare to **graminoid** (Fire Effects Guide).

**FWD (fine woody debris)**: Dead woody debris less than 3 inches in diameter, including 1-hour, 10- hour, and 100-hour fuels (FIREMON).

**GIS (geographic information system):** Integrated software, hardware, and data to store and manipulate information that combines thematic and locational attributes about geographic features (FIREMON Landscape Assessment Sampling Methods).

**Go/no-go gauge:** A tool used to classify fuels into one of three classes: 1-hour, 10-hour, and 100-hour (Handbook for Inventorying Down Woody Debris).

**GPS (Geographic Positioning System):** A network of radio-emitting satellites from which your position can be triangulated in three dimensions (north, east, and elevation) to within 3 to 50 m of accuracy (FSVeg, FIREMON).

**Graminoid:** Grasslike plants, including grasses, sedges, rushes, reeds, and cattails (Fire Effects Guide).

**Grid frame:** Used with the PO method a grid frame is a frame of any shape or size where crosshairs formed by perpendicularly oriented strings are considered sampling points. Interceptions of crosshairs with plant parts are considered hits (FIREMON).

**Ground cover:** This includes cover of basal vegetation, moss/lichens, litter, rocks, gravel, and so forth, on a site (FIREMON).

**Herb:** A small, nonwoody, seed-bearing plant in which aerial parts die back at the end of each growing season. Compare to **shrub** (The Concise Oxford Dictionary of Ecology).

**Igneous:** Rocks that crystallized from molten magma, such as basalt, andesite, diorite to gabbro, latice, quartz monzonite, trachyte and syenite, rhyolite, granite, welded tuff (tufa), and scoria (porcellanite) (Roadside Geology of Montana, FIREMON).

**Increment borer:** An instrument used to bore into the pith of the tree and extract a core that can be used for determining age and growth rate by counting the rings in the extracted core (Forest Measurements).

**LA (land assessment):** A method that identifies and quantifies fire effects over large areas, at times involving many burns, by using satellite derived Normalized Burn Ratio (NBR) together with a ground-based indicator of fire severity, Composite Burn Index (CBI) (FIREMON).

**Ladder fuel:** Fuels that provide vertical continuity between strata, thereby allowing fire to carry from surface fuels into the crowns of trees and shrubs with relative ease (Glossary of Wildland Fire Management Terms Used in the U.S.).

**Landsat:** A satellite that carries a Thermatic Mapper sensor that records 30-m data in specific spectral bands, and 60-m data in one band. The bands measure reflected energy and heat (FIREMON Landscape Assessment Sampling Methods).

**Landscape:** All of the natural features, such as hills, forest, and water that distinguish one part of Earth's surface from another. A landscape can be any size and shape but it spatially defines stands. Compare to **stand** (Glossary of Landscape and Vegetation Ecology For Alaska, FIREMON).

**Layered canopy structure:** The vertical structural components of a community consisting of plants of approximately the same stature or height. For example, tree layer, shrub layer, herb layer, cryptogam layer. Compare **strata**, **vertical** (Glossary of Landscape and Vegetation Ecology for Alaska).

**LCP (live crown percent):** The percent of the total length of tree bole that is supporting live crown. This is assessed from the highest live foliage to the lowest live foliage or the base of the crown. Estimated by visually redistributing the live tree crown evenly around the tree so the branches are spaced at the same branch density as seen along the bole and form the typical conical crown (FIREMON).

**LI (line intercept):** The FIREMON line intercept method is used to assess changes in plant species cover for a macroplot. This method uses line transects to sample plot variation and quantify statistically valid changes in plant species cover and height over time. This method is especially useful for quantification of shrub cover greater than 3 ft tall. Canopy cover is recorded as the number of inches intercepted along a transect. Percent canopy cover is calculated by dividing the number of inches intercepted by each item by the total length of the transect. Compare to **planar intercept** (FIREMON).

**Lichen:** A nonvascular small plant composed of a particular fungus and a particular alga growing in an intimate symbiotic association and forming a dual plant, commonly adhering in colored patches to rock, wood, and soil (Webster's New World Dictionary).

**Litter**: The top layer of the forest floor composed of loose debris such as branches, twigs, and recently fallen needles and leaves; little altered by decomposition and still identifiable. This layer may also include debris from shrubs, grasses, and forbs that have recently died. The litter layer is directly above the duff layer (Fire Effects Guide, FIREMON).

**Live canopy base height**: In a stand, an estimate of the typical or common height of the lowest live branch having live foliage (FIREMON).

**Live crown base height**: For individual trees, the height of the lowest live branch whorl with live branches in two quadrants exclusive of epicormic branching and of whorls not continuous with the main crown (USFS Region One, Common Stand Exam Guide).

**Loam**: Soil mixture of sand, silt, clay, and humus (Dictionary of Scientific and Technical Terms).

**Loggers tape**: A spring-wound tape measure that has one side in linear units that are converted to diameter and the other side in units for measuring feet and inches. Also, called a diameter tape or D-tape (Measurements for Terrestrial Vegetation).

**Macroplot**: A term defining the greater sampling area in which all other sampling methods are nested. The size and shape of the macroplot is determined by sampling objective and available resources, but most macroplots are rectangular or circular encompassing about 0.04-0.1 ha (FIREMON).

**Mature tree**: In FIREMON a tree greater than breakpoint diameter. This class includes SAF (Society of American Foresters) standard pole trees, medium trees, large trees, and very large trees (FIREMON).

**Mean height**: In FIREMON, an estimate of the average height in meters for all individuals of a species or a species by size/age class; estimated by visualizing a plastic sheet draped over the vegetation in the class and recording the average height of the sheet above ground (FIREMON).

**Metadata**: Data about the data. In data processing, metadata are definitional data that provide information about documentation of the data managed within an application environment. Metadata may include descriptive information about the context, quality, condition, or characteristics of the data. In FIREMON this includes, among other things, specific details regarding the sampling design, approach, and particulars describing the application of methods (www.dictionary.com).

**Metamorphic**: Rock of any origin altered in mineralogical composition, chemical composition, or structure by heat and pressure. Nearly all such rocks are crystalline such as argillite, siltite, quartzite, slate, phyllite, schist, and gneiss (Glossary of Landscape and Vegetation Ecology for Alaska).

**Meter stick**: A metric measuring stick that is 1 m, equal to 39.37 inches.

**Microplot**: A sampling area that is smaller than the macroplot used for measuring small scale phenomenon, such as ground cover or individual plant or species attributes. Often square and about 1 $m^2$ in size. Microplots are usually located in a grid pattern nested within the macroplot. Compare **quadrat** and **subplot** (FIREMON).

**Mineral soil**: Soil composed principally of mineral matter, the characteristics of which are determined more by the mineral than the organic content. This soil is often gravelly or sandy and lighter than the duff layer (The Concise Oxford Dictionary of Ecology).

**Mode/modal**: The value that occurs most often in a frequency distribution (Measurements for Terrestrial Vegetation).

**Moss**: A nonvascular small, green bryophyte growing in velvety clusters on rocks, trees, and moist ground (Webster's New World Dictionary).

**NBR (normalized burn ratio)**: A methodology involving remote sensing; this uses Landsat 30-m data and a derived radiometric value. The normalized burn ratio is temporally differenced between pre- and postfire datasets to determine the extent and degree of detected change from burning (FIREMON Landscape Assessment Sampling Methods).

**NEXUS**: Algorithm package for predicting fire behavior for assessing crown fire hazard (Nexus).

**Nonvascular plants**: Plants without an internal vascular system (xylem and phloem) for the transport of nutrients, such as mosses and lichens (Webster's New World Dictionary).

**NRCS plant database**: The Natural Resources and Conservation Service supported plants database. It is used in FIREMON for consistent naming and coding of plant species (http://plants.usda.gov/).

FIREMON Glossary

**NRF (nested root frequency)**: Used when sampling plant frequency. NRF balances plant density and size by assigning frequency values based on the plant's presence in a nested set of plots corresponding to, usually, 1, 25, 50, and 100 percent of the quadrat.

**NRP (number of required plots)**: This is the number of plots required per stand per stratification category needed to meet different statistical objectives (FIREMON).

**PCS (percent crown scorched)**: A fire severity measurement that relates directly to tree mortality; an estimate of the amount of crown volume that was consumed or damaged by fire (FIREMON).

**PD (plot description)**: Methods used to describe general characteristics of the FIREMON macroplot to provide ecological context for data analyses. The PD data characterize the topographical setting, geographic reference point, general plant composition and cover, ground cover, fuels, and soils information. The PD method contains the only required fields in FIREMON (FIREMON).

**Peak canopy cover**: A method in FIREMON for estimating cover at peak phenological development of a plant. For example, if leaves have fallen off the plant and are on the ground, the projected cover has been mentally reconstructed with leaves on the plant (FIREMON).

**PF (project funds)**: The amount of money available to conduct the entire monitoring project (FIREMON).

**Phenological stage**: A specific phase within the cycle (usually annual) of a plant's leafing, flowering, fruiting, and so forth (The Concise Oxford Dictionary of Ecology).

**Pixel**: Literally, "picture element"; the smallest area for which data values are assigned. Pixels generally are all the same size and arranged in a contiguous rectangular grid of rows and columns. Spatial orientation of the grid can be registered to a map projection, so that individual pixels may be located on the ground (FIREMON Landscape Assessment Sampling Methods).

**Plain**: An extensive, level, and usually treeless area of land (www.dictionary.com).

**Planar-intercept**: For sampling down woody debris in FIREMON, it is a method in which the sampling area is an imaginary plane extending from the ground, vertically from horizontal (not perpendicular to the slope) to a height of 6 ft above the ground. Pieces of DWD (down woody debris) that intercept the sampling plane are measured and recorded. Frequently the term "line transect sampling" is used when discussing the planar intercept method. Compare to **line intercept** (FIREMON).

**Plot**: The basic sampling unit. This is an area of ground where FIREMON methods will be implemented. The plot is spatially defined by the macroplot (FIREMON).

**PO (point cover)**: This method is used to estimate vegetation and/or ground cover for a macroplot. Point estimates of cover are collected at fixed locations along randomly located line transects. Individual pins, pin frames, or point grids are placed at systematic intervals along a transect. Pins are lowered, and plant species and/or ground cover categories are recorded as the number of "hits" encountered along a transect. Cover is calculated by dividing the number of "hits" by the total number of points along a transect (FIREMON).

**Point frame**: Used with the PO method, a point frame is a wooden or metallic frame with two legs and two cross arms typically containing 10 pins. Steel rods or wire pins are lowered through the holes (FIREMON).

**Polygon**: Generally a discrete area defined by vectors or pixels electronically mapped in a Geographical Information System (GIS). In FIREMON polygons define areas with similar stratification attributes. A polygon can define a stand if the polygon boundaries are based on differences in vegetation characteristics (FIREMON).

**Potential lifeform**: A code that describes the community lifeform that would eventually inhabit the FIREMON plot in the absence of disturbance (FIREMON).

**PPR (plots per day)**: The number of plots that can be sampled by one crew; if unknown, estimate the rate of four plots per day (FIREMON).

**Prism (wedge prism)**: A tapered wedge of glass that bends or deflects light rays at a specific offset angle. When a tree stem is viewed through such a wedge, the bole appears to be displaced. All tree stems not completely offset when viewed through the wedge are counted. Trees that appear borderline should be measured and checked with the appropriate plot radius factor (Forest Measurements).

**Quadrat**: A small clearly demarcated plot or sample area of known size on which ecological observations are made. Quadrats may be square, rectangular, or circular, and are usually no more than 1 $m^2$. Compare **microplot** and **subplot** (Glossary of Landscape and Vegetation Ecology for Alaska).

**Raster**: A digital image stored in one of many grid cell formats, where the cells (that is, pixels) are represented as binary numeric values referenced by byte position within the file. Byte position can be translated into pixel row and column, such that the grid models some two-dimensional space (FIREMON Landscape Assessment Sampling Methods).

**Relevé approach**: A sampling method in which one plot is placed in a representative portion of a stand "without preconceived bias." The assumption in relevé sampling is that the plot is representative of a larger area (such as a stand or polygon), and therefore, conditions measured at the plot can be used to describe the stand or polygon as a whole (Measurements for Terrestrial Vegetation, FIREMON).

**Rhizomatous plants**: A plant that has a stem, generally modified, that grows below the surface of the ground and produces roots, scale leaves, and suckers irregularly along its length (Glossary of Landscape and Vegetation Ecology for Alaska).

**Riparian**: 1) Pertaining to streamside environment. 2) Vegetation growing in proximity to a watercourse, lake, or spring and often dependent on its roots reaching the water table (Glossary of Landscape and Vegetation Ecology for Alaska).

**RS (rare species)**: This FIREMON method is used specifically for monitoring rare plants such as threatened and endangered species. Plants are located using measurement along and perpendicular to the sampling baseline.

**SA (sample area)**: The established area used for sampling. This can be all of the stands on a selected landscape or targeted stands. For example, monitoring plots may only be needed on steep areas where rehabilitation efforts were prevalent, or it may be only forested areas that need to be sampled to monitor tree establishment after wildfire.

**Sampling plane**: Used in the FIREMON FL sampling. The imaginary plane is defined by a measuring tape laid on or near the ground and extends from the top of the litter layer, duff layer, or mineral soil—whichever is the highest layer—to a height of 6 ft (Handbook for Inventorying Downed Woody Material, FIREMON).

**Sapling**: A tree greater than 4.5 ft and less than the established breakpoint diameter (FIREMON).

**SC (species composition)**: This is a method used to provide ocular estimates of canopy cover and height for plant species on a macroplot. The SC sampling methods are suited for a wide variety of vegetation types and are especially useful in communities of tall shrubs or trees (FIREMON).

**Sedimentary**: Rocks made up of particles deposited from suspension in water. The main kinds of sedimentary rock are limestone, dolomite, sandstone, siltstone, shale, and conglomerate (Glossary of Landscape and Vegetation Ecology for Alaska).

**Seedling**: A tree less than 4.5 ft tall (FIREMON).

**Severity description**: A code used to quantify the degree of damage by biotic (insects, disease, browsing) and abiotic (wind, snow, fire) agents. Compare to **damage severity** (FIREMON).

**Shrub**: A woody plant that branches below or near ground level into several main stems, so has no clear trunk. It may be deciduous or evergreen. At the end of each growing season there is no die-back of the axes. Compare to **herb** (The Concise Oxford Dictionary of Ecology).

**Silt**: As a soil separate, individual mineral particles that range in diameter from the upper limit of clay (0.002 mm) to the lower limit of very fine sand (0.05 mm). As a soil textural class, soil that is 80 percent or more silt and less than 12 percent clay (Glossary of Landscape and Vegetation Ecology for Alaska).

**Slope**: Defined in FIREMON as the inclination of the land surface, measured in degrees, from the horizontal (Glossary of Landscape and Vegetation Ecology for Alaska, FIREMON).

**Slope shape**: Slopes may be characterized as uniform (linear or planar), concave, convex, undulating, flat, or patterned (FSVEG).

**Snag class**: A code used to describe the condition of a dead tree. Compare to **decay class** (FSVeg).

**Soil colloid**: Soil substance of very small particle size, mineral or organic (The Concise Oxford Dictionary of Ecology).

**SP (sampling potential)**: In general, sampling potential describes the number of standard plots that can be installed during the sampling effort. This statistic integrates most sampling resources into one index that describes the capacity to perform the monitoring project. The sampling potential (SP-number plots) is the project funds (PF-dollars) divided by crew costs (CC-dollars per day) multiplied by plot production rate (PPR-plots per day) (FIREMON).

**Spread Rate**: Relative activity of a fire in extending its horizontal dimensions, expressed as rate of increase of the perimeter, rate of increase in area, or rate of advance of its head, depending on the intended use of the information; generally in chains or acres per hour (Glossary of Wildland Fire Management Terms Used in the U.S.).

**SRF (stand replacing fire)**: Fire that kills all or most living overstory trees in a forest and initiates secondary succession or regrowth (Glossary of Wildland Fire Management Terms Used in the U.S.).

**Stand**: A spatially continuous group of trees and associated vegetation having similar vertical and spatial structure and species composition (examples are: pole, seedling, sapling, mature) usually growing under similar soil and climatic conditions. Compare to Strata, horizontal (FSVeg).

**Stand height**: The estimate of the height of the highest stratum that contains at least 10 percent crown cover measured across a stand (FIREMON).

**Statistical approach**: A method using random sampling, which is utilized in most natural resource inventories. The emphasis is on gaining a statistically sound estimate of variation and mean that can be used to make inference (FIREMON).

**Status**: In FIREMON this is a classification for the general health of a plant. There are four tree status codes: 1) Healthy, 2) Unhealthy, 3) Sick, and 4) Dead (FIREMON).

**Strata, horizontal**: In FIREMON these are areas with similar stratification attributes. Compare **polygon** (FIREMON).

**Strata, vertical**: Referring to one or more layers of a community, arranged vertically and having a continuous sequential order from below ground to ground level, and from ground level to the uppermost vegetative canopy. Strata typically are based on within-stratum similarities of physical organization, species composition, and/or microclimate (FIREMON Landscape Assessment Sampling Methods).

**Stratification factor**: Biotic or abiotic attribute such as fuel load, tree density, or treatment type used to stratify or divide a landscape into like polygons or strata (FIREMON).

**Stratified random sample**: Method of locating plots within a stratum used with the statistical approach to establish plots randomly across the landscape stands or based on some land type stratification factor; using any technique that distributes plots so that probability of all possible plot locations is equal (FIREMON).

**Stratified systematic sample**: Method of locating plots within a stratum used with the statistical approach. Establishes the first plot randomly where the probability of all possible locations are equal. Then all other macroplots are located with reference to the first, usually along a grid or network of, usually, equal spacing (FIREMON).

**Structural characteristics**: In FIREMON the five important tree characteristics are: DBH, height, live crown length, crown fuel base height, and crown class. These characteristics are used to compute properties such as crown biomass, vertical ladder fuels, and potential fire-caused mortality (FIREMON).

**Structural stage**: Describes a stand in terms of the primary elements of vegetation structure, which are growth form, vertical structure, and coverage (Glossary of Landscape and Vegetation for Alaska).

**Subplot**: A subplot is a microplot nested inside the macroplot for the purpose of measuring numerous individuals or other attributes that would be difficult to assess over the entire macroplot. In FIREMON, generally associated with the TD method. Compare **quadrat** and **microplot** (FIREMON).

**Surficial geology**: The description of the rock type on the surface of Earth.

**TD (tree data)**: Methods used to sample individual trees in a fixed-area plot to estimate tree density, size, and age class distributions before and after fire so that tree survival and mortality rates can be assessed. This method allows the measurement of diameter, height, age, growth rate, crown length, pathogen evidence, fire severity, and snag description for each tree above a user-specified diameter (FIREMON).

**Topography**: The configuration of Earth's surface, including its relief and the position of its natural and artificial features (Introduction to Wildland Fire Behavior Calculations-Glossary).

**Transect**: A theoretically nondimensional line that is located within the macroplot. Ecological attributes that intersect or cross the transect are tallied or measured. Compare to **belt transect** (FIREMON).

**Treatment**: Procedures applied on the landscape or stand level where the effects can be compared to other applied procedures. Examples include a fall burned prescribed fire, an unburned 'control,' or an area burned with a specific ignition method or pattern, or a harvested and burned area (Fire Effects Guide).

**UTM (Universal Transverse Mercator)**: A two-dimensional (flat-map) projection widely used in natural resource applications, suitable for maps of 1:100,000 and greater scale. Each hemisphere of the world is divided into 60, 6-degree, zones by longitude. Within each zone, the reference is an X, Y equidistant grid in meters, with origin at the lower left zone corner (western most point on the equator). Coordinate pairs are given in meters northing and easting (example: 5437689N, 278334E), increasing from the origin to the north and to the east, respectively (FIREMON Landscape Assessment Sampling Methods).

**Vascular plants**: A plant having specialized tissues (xylem and phloem) that conduct water and synthesized foods, such as a fern or seed plant (Webster's New World Dictionary).

**Vector**: Geographic data represented as numeric X, Y coordinates, and usually some attribute identifier. Vector data define features by point, line, or polygon topology, and are displayed as such on maps or graphics (FIREMON Landscape Assessment Sampling Methods).

**WD (work days)**: The number of 8-hour work days available to finish the monitoring project (FIREMON).

**Whorl**: An arrangement of three or more leaves, petals, or organs radiating from a single node (www.dictionary.com).

# FIREMON

## References

Allaby, M. 1994. The concise Oxford dictionary of ecology. Oxford, UK: Oxford University Press. 415 p.

Alt, D.; Hyndman, D. W. 1986. Roadside geology of Montana. Missoula, MT: Mountain Press Publishing. 427 p.

Anderson, H. E. 1983. Aid for determining fuel models for estimating fire behavior. Gen. Tech. Rep. GTR-INT-122. Ogden, UT: U.S. Department of Agriculture, Forest Service, Intermountain Forest and Range Experiment Station. 22 p.

Avery, T.E., Burkhart, H.E. 1983. Forest measurements. 3d ed. New York: McGraw-Hill, Inc. 331 p.

Bailey, A. W.; Poulton, C. E. 1968. Plant communities and environmental relationships in a portion of the Tillamook Burn, northwest Oregon. Ecology. 49: 1–14.

Barbour, M. G.; Burk, J. H.; Pitts, W. D. 1987. Terrestrial plant ecology. The Benjamin/Cummings Publishing Co. Inc.

Bartlett, M. S. 1948. Determination of plant densities. Nature. 162: 621.

Bauer, H. L. 1943. The statistical analysis of chaparral and other plant communities by means of transect samples. Ecology. 24: 45–60.

Blackman, G. E. 1935. A study of statistical methods of the distribution of species in grassland associations. Annals of Botany. 49: 749–777.

Bonham, C. D. 1989. Measurements for terrestrial vegetation. New York: John Wiley and Sons. 338 p.

Braun-Blanquet, J. 1965. Plant sociology: the study of plant communities. New York: Hafner Publishing. 437 p.

Brown, J. K. 1971. A planar intersect method for sampling fuel volume and surface area. Forest Science. 17: 96–102.

Brown, J. K. 1974. Handbook for inventorying downed woody material. Gen. Tech. Rep. GTR-INT-16. Ogden, UT: U.S. Department of Agriculture, Forest Service, Intermountain Forest and Range Experiment Station. 34 p.

Brown, J. K.; Roussopoulos, P. J. 1974. Eliminating the biases in the planar intersect method for estimating volumes of small fuels. Forest Science. 20: 350–356.

Busing, R.; Rimar, K.; Solte, K. W.; Stohlgren, T. J. 1999. Forest health monitoring: vegetation pilot field methods guide: vegetation diversity and structure down woody debris fuel loading. Research Triangle Park, NC: National Forest Health Monitoring Program.

Canfield, R. H. 1941. Application of the line interception method in sampling range vegetation. Journal of Forestry. 39: 388–394.

Coulloudon, B.; Podborny, P.; Eshelman, K.; Rasmussen, A.; Gianola, J.; Robles, B.; Habich, N.; Shaver, P.; Hughes, L.; Spehar, J.; Johnson, C.; Willoughby, J. Pellant, M. 1999. Sampling vegetation attributes. Tech. Ref. 1734-4. Denver, CO: Bureau of Land Management. 164 p.

Daubenmire, R. F. 1959. A canopy coverage method. Northwest Science. 33: 43–64.

Elzinga, C. L.; Salzer, D. W.; Willoughby, J. W. 1998. Measuring and monitoring plant populations. Tech. Ref. 1730-1, BLM/RS/ST-98/005+1730. Denver, CO: Bureau of Land Management, National Business Center.

Elzinga, C. L.; Salzer, D. W.; Willoughby, J. W.; Gibbs, J. P. 2001. Monitoring plant and animal populations. Malden, MA: Blackwell Sciences, Inc. 360 p.

Finney, M. A. 1998. FARSITE: Fire Area Simulator-model development and evaluation. Res. Pap. RMRS-RP-4. Fort Collins, CO: U.S. Department of Agriculture, Forest Service, Rocky Mountain Research Station. 47 p.

Fischer, William C.; Hardy, Charles E. 1976. Fire-weather observers' handbook. Agric. Handb. 494. Washington, DC: U.S. Department of Agriculture. 152 p.

Fisser, H. G.; Van Dyne, G. M. 1966. Influence of number and spacing of points on accuracy and precision of basal cover estimates. Journal of Range Management. 19: 205–211.

Gabriel, H. W.; Talbot, S. S. 1984. Glossary of landscape and vegetation ecology for Alaska. Tech. Rep. 10. Anchorage, AK: Bureau of Land Management. 137 p.

Goldsmith, F. B.; Harrison, C. M.; Morton, A. J. 1986. Description and analysis of vegetation. In: Moore, P. D.; Chapman, S. B., eds. Methods in plant ecology. Malden, MA: Blackwell Scientific Publications: 437–524.

Goodall, D. W. 1952. Some considerations in the use of point quadrats for the analysis of vegetation. Australian Journal of Science Research. Ser. B. 5: 1–41.

Grace, S.; Brownlie, D. 2002. Fuel and fire effects monitoring guide, [Online]. Available at: http://fire.fws.gov/ifcc/monitor/RefGuide/default.htm.

Gray, A. 2003. Monitoring stand structure in mature coastal Douglas-fir forests: effect of plot size. Forest Ecology and Management. 175: 1–16.

Greig-Smith, P. 1983. Quantitative plant ecology. 3d ed. Berkeley: University of California Press. 347 p.

Hall, F. C. 2002a. Photo point monitoring handbook: Part A—field procedures. Gen. Tech. Rep. PNW-GTR-526. Portland, OR: U.S. Department of Agriculture, Forest Service, Pacific Northwest Research Station. 48 p.

Hall, F. C. 2002b. Photo point monitoring handbook: Part B—concept and analysis. Gen. Tech. Rep. PNW-GTR-526. Portland, OR: U.S. Department of Agriculture, Forest Service, Pacific Northwest Research Station. 48 p.

Hann, W. E.; Jensen, M. E.; Keane, R. E. 1988. Ecosystem classification handbook. Chapter 4—ECODATA sampling methods. U.S. Department of Agriculture, Forest Service, Northern Region. 144 p.

Hardwick, P.; Lachowski, H.; Griffith, R.; Parsons, A. 1998. Burned area emergency rehabilitation project: an example of successful technology transfer. In: Greer, J. D., ed. Proceedings of the seventh Forest Service remote sensing applications conference; 1998 April 6–10; Nassau Bay, TX. Bethesda, MD: American Society for Photogrammetry and Remote Sensing.

Harmon, M. E.; Franklin, J. F.; Swanson, F. J.; Sollins, P.; Gregory, S. V.; Lattin, J. D.; Anderson, N. H.; Cline, S. P.; Aumen, N. G.; Sedell, J. R.; Lienkaemper, G. W.; Cromack, K.; Cummins, K. W. 1986. Ecology of coarse woody debris in temperate ecosystems. Advances in Ecological Research. 15: 133–302.

Hironaka, M. 1985. Frequency approaches to monitor rangeland vegetation. In: Kruger, William C., chairman. Symposium on use of frequency for rangeland monitoring. Proceedings, 38$^{th}$ annual meeting, Society for Range Management; 1985 February; Salt Lake City, UT. Society for Range Management: 84–86.

Holdorf, H.; Donahue, J. 1990. Landforms for soil surveys in the Northern Rockies. Misc. Publ. No. 51. Missoula: University of Montana, Montana Forest and Conservation Experiment Station, School of Forestry. 26 p.

Howard, J. O.; Ward, F. R. 1972. Measurement of logging residue—alternative application of the line intersect method. Res. Note RN-PNW-183. Portland, OR: U.S. Department of Agriculture, Forest Service, Pacific Northwest Forest and Range Experiment Station. 8 p.

Hyder, D. N.; Bement, R. E.; Remmenga, E. E.; Terwilliger, C., Jr. 1965. Frequency sampling of blue gramma range. Journal of Range Management. 18: 90–94.

Iqbal, Muhammad. 1983. An introduction to solar radiation. New York: Academic Press. 390 p.

Jensen, M. E.; Hann, W. H.; Keane, R. E.; Caratti, J.; Bourgeron, P. S. 1993. ECODATA—a multiresource database and analysis system for ecosystem description and analysis. In: Jensen, M. E.; Bourgeron, P. S., eds. Eastside forest ecosystem health assessment, volume II: Ecosystem management: principles and applications. Gen. Tech. Rep. GTR-PNW-318. Portland, OR: U.S. Department of Agriculture, Forest Service, Pacific Northwest Forest and Range Experiment Station: 203–216.

Keane, R. E.; Hann, W. J.; Jensen, M. E. 1990. ECODATA and ECOPAC: analytical tools for integrated resource management. The Compiler. 8(3): 24–37.

Kemp, C. D.; Kemp, A. W. 1956. The analysis of point quadrat data. Australian Journal of Botany. 4: 167–174.

Lesica, P.; Steele, B. M. 1997. Use of permanent plots in monitoring plant populations. Natural Areas Journal. 17: 331–337.

Lillesand, Tom. M.; Kiefer, Ralph W. 1994. Remote sensing and image interpretation. 3d ed. New York: John Wiley and Sons. 750 p.

Maser, C.; Anderson, R. G.; Cromack, K., Jr.; Williams, J. T.; Martin, R. E. 1979. Dead and down woody material. In: Wildlife habitats in managed forests: the Blue Mountains of Oregon and Washington. Agric. Handb. 553. Washington, DC: U.S. Department of Agriculture: 78–95.

Maser, C.; Tarrant, R. F.; Trappe, J. M.; Franklin, J. F. 1988. From the forest to the sea: a story of fallen trees. Gen. Tech. Rep. PNW-GTR-229. Portland, OR: U.S. Department of Agriculture, Forest Service, Pacific Northwest Forest and Range Experiment Station. 153 p.

McPherson, G. R.; Wade, D. D.; Phillips, C. D., comps. 1990. Glossary of wildland fire management terms used in the United States. SAF 90-05. Washington, DC: Society of American Foresters. 137 p.

Miller, M., ed. 1994. Fire effects guide. NFES 2394. Boise, ID: National Wildfire Coordinating Group. (Also available at: http://fire.r9.fws.gov/ifcc/monitor/FEG.pdf).

Morrison, R. G.; Yarranton, G. A. 1970. An instrument for rapid and precise sampling of vegetation. Canadian Journal of Botany. 48: 293–297.

Mueller-Dombois, D.; Ellenberg, H. 1974. Aims and methods of vegetation ecology. New York: John Wiley and Sons. 547 p.

National Resource Conservation Service. 1997. National range and pasture handbook. Washington, DC: U.S. Department of Agriculture, Natural Resource Conservation Service.

National Wildfire Coordinating Group. 1981. Introduction to wildland fire behavior calculations (S-390) student workbook. Boise, ID: National Interagency Fire Center.

National Wildfire Coordinating Group. 1994. Fire effects guide. NFES 2394. Boise, ID. National Interagency Fire Center.

Ottmar, R. D.; Vihnanek, R. E. 1999. Stereo photo series for quantifying natural fuels, volume V: red and white pine, northern tall grass prairie and mixed oak/hickory in the Midwest. PMS 834. Boise, ID: National Wildfire Coordinating Group. National Interagency Fire Center. 65 p.

Parker, K. W.; Savage, R. W. 1944. Repeatability of the line interception method in measuring vegetation on the Southern Great Plains. Journal of American Society of Agronomy. 36: 97–110.

Parker, S. B., ed. 1989. McGraw-Hill dictionary of scientific and technical terms. 4th ed. New York: McGraw-Hill. 2088 p.

Pickford, S. G.; Hazard, J. W. 1978. Simulation studies in the line intersect sampling of forest residue. Forest Science. 24: 469–483.

Pound, P.; Clements, F. E. 1898. A method of determining the abundance of secondary species. Minnesota Botanical Studies. 2: 19–24.

Reinhardt, E.; Keane, R. E.; Brown, J. K. 1997. First Order Fire Effects Model: FOFEM 4.0 user's guide. Gen. Tech. Rep. INT-GTR-344. Ogden, UT: U.S. Department of Agriculture, Forest Service, Intermountain Research Station. 65 p.

Reinhardt, Elizabeth; Crookston, Nicholas L., tech. eds. 2003. The Fire and Fuels Extension to the Forest Vegetation Simulator. Gen. Tech. Rep. RMRS-GTR-116. Ogden, UT: U.S. Department of Agriculture, Forest Service, Rocky Mountain Research Station. 209 p.

Reinhardt, Elizabeth D.; Keane, Robert E.; Brown, James K. 2001. Modeling fire effects. International Journal of Wildland Fire. 10: 373–380.

Schroeder, Mark J.; Buck, Charles C. 1970. Fire weather. Agric. Handb. 360. Washington, DC: U.S. Department of Agriculture, Forest Service. 229 p.

Schwarz, C. F.; Thor, E. C.; Elsner, G. H. 1976. Wildland planning glossary. Gen. Tech. Rep. GTR-PSW-13. Berkeley, CA: U.S. Department of Agriculture, Forest Service, Pacific Southwest Research Station. 252 p.

Scott, J. H. 1999. NEXUS: a system for assessing crown fire hazard. Fire Management Notes. 59: 20–24.

Smith D. M. The practice of silviculture. 8th ed. 1986. New York: John Wiley and Sons. 527 p.

Society for Range Management. 1989. Glossary of terms used in range managemet. Denver, CO: Society for Range Management.

Stanton, F. W. 1960. Ocular point frame. Journal of Range Management. 13: 153.

Stephenson, S. N.; Buell, M. F. 1965. The reproducibility of shrub cover sampling. Ecology. 46: 379–380.

Tansley, A. G.; Chipp, T. F., eds. 1926. Aims and methods in study of vegetation. London: British Empire Vegetation Committee. 383 p.

USDA Forest Service. 1996. ECADS: ecosystem characterization and description—user documentation. Missoula, MT: U.S. Department of Agriculture, Forest Service, Northern Region, Ecosystem Management Program. 78 p.

USDA Forest Service. 2001. FSVeg field sampled vegetation, data dictionary. U.S. Department of Agriculture, Forest Service, Natural Resource Information System. Unpublished report.

USDA Forest Service. 2001. Common stand exam userguide: version 1.5. [Online] Available: http://fsweb.ftcol.wo.fs.fed.us/fsveg/ [August 19, 2005] [NOTE: This site is available to USDA Forest Service employees only.]

USDA Forest Service. 2002. Timber cruising in the Pacific Northwest. R6-NR-TM-TP-21-97. Portland, OR: U.S. Department of Agriculture, Forest Service, Pacific Northwest Region.

USDA Forest Service. 2003. Common stand exam field guide for Region One, version R1 2.03.2003. Unpublished document on file at: U.S. Department of Agriculture, Forest Service Region One, Missoula, MT.

USDA Forest Service, Forest Management Service Center, (2002, October 11), [Online]. Available: http://www.fs.fed.us/fmsc/fvs/ [August 19, 2005].

USDA Natural Resources Conservation Service. 1997. National range and pasture handbook. Washington, DC: U.S. Department of Agriculture, Natural Resources Conservation Service.

USDA Natural Resources Conservation Service. 2002. National soil survey handbook, title 430-VI, [Online]. Available: http://soils.usda.gov/technical/handbook/ [August 19, 2005].

USDI Bureau of Land Management. 1996. Sampling vegetation attributes. Interagency technical reference. BLM/RS/ST-96/002+1730: 23–29

USDI Fish and Wildlife Service. Grace, S.; Brownlie, D., eds. [Draft]. Fuel and fire effects monitoring guide, [Online]. Available: http://fire.fws.gov/ifcc/monitor/RefGuide/default.htm. [August 19, 2005].

USDI National Park Service. 1992. Western Region fire monitoring handbook. San Francisco, CA: U.S. Department of the Interior, National Park Service, Western Region. 89 p.

USDI National Park Service. 2001. Fire monitoring handbook. Boise, ID: National Interagency Fire Center. 283 p.

Van Dyne, G. M.; Vogel, W. G.; Fisser, H. G. 1963. Influence of small plot size and shape on range herbage production estimates. Ecology. 44: 746–759.

Van Mantgem, P.; Schwartz, M.; Keifer, B. 2001. Monitoring fire effects for managed burns and wildfires: coming to terms with pseudoreplication. Natural Areas Journal. 21: 266–273.

Van Wagner, C. E. 1968. The line intersect method in forest fuel sampling. Forest Science. 14: 20–26.

Van Wagner, C. E.; Wilson, A. L. 1976. Diameter measurements in the line transect method. Forest Science. 22: 230–232.

Warren, W. G.; Olsen, P. F. 1964. A line intersect technique for assessing logging waste. Forest Science. 10: 267–276.

Webster's New World Dictionary, Third College Edition. 1988. New York: Simon and Schuster, Inc.

Winkworth, R. E. 1955. The use of point quadrats for the analysis of heathland. Australian Journal of Botany. 3(1): 68–81.

*The use of trade or firm names in this publication is for reader information and does not imply endorsement by the U.S. Department of Agriculture of any product or service*

You may order additional copies of this publication by sending your mailing information in label form through one of the following media. Please specify the publication title and series number.

**Fort Collins Service Center**

| | |
|---:|:---|
| **Telephone** | (970) 498-1392 |
| **FAX** | (970) 498-1122 |
| **E-mail** | rschneider@fs.fed.us |
| **Web site** | http://www.fs.fed.us/rm |
| **Mailing Address** | Publications Distribution |
| | Rocky Mountain Research Station |
| | 240 W. Prospect Road |
| | Fort Collins, CO 80526 |

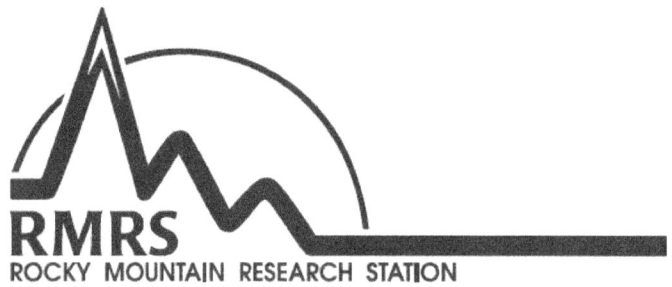

**ROCKY MOUNTAIN RESEARCH STATION**

The Rocky Mountain Research Station develops scientific information and technology to improve management, protection, and use of the forests and rangelands. Research is designed to meet the needs of National Forest managers, Federal and State agencies, public and private organizations, academic institutions, industry, and individuals.

Studies accelerate solutions to problems involving ecosystems, range, forests, water, recreation, fire, resource inventory, land reclamation, community sustainability, forest engineering technology, multiple use economics, wildlife and fish habitat, and forest insects and diseases. Studies are conducted cooperatively, and applications may be found worldwide.

### Research Locations

| | |
|---|---|
| Flagstaff, Arizona | Reno, Nevada |
| Fort Collins, Colorado* | Albuquerque, New Mexico |
| Boise, Idaho | Rapid City, South Dakota |
| Moscow, Idaho | Logan, Utah |
| Bozeman, Montana | Ogden, Utah |
| Missoula, Montana | Provo, Utah |

*Station Headquarters, Natural Resources Research Center, 2150 Centre Avenue, Building A, Fort Collins, CO 80526

The U.S. Department of Agriculture (USDA) prohibits discrimination in all its programs and activities on the basis of race, color, national origin, age, disability, and where applicable, sex, marital status, familial status, parental status, religion, sexual orientation, genetic information, political beliefs, reprisal, or because all or part of an individual's income is derived from any public assistance program. (Not all prohibited bases apply to all programs.) Persons with disabilities who require alternative means for communication of program information (Braille, large print, audiotape, etc.) should contact USDA's TARGET Center at (202) 720-2600 (voice and TDD).

To file a complaint of discrimination, write to USDA, Director, Office of Civil Rights, 1400 Independence Avenue, S.W., Washington, DC 20250-9410, or call (800) 795-3272 (voice) or (202) 720-6382 (TDD). USDA is an equal opportunity provider and employer.

Federal Recycling Program    Printed on Recycled Paper

www.ingramcontent.com/pod-product-compliance
Lightning Source LLC
Chambersburg PA
CBHW081233180526
45171CB00005B/410